The SAGE Handbook of
Television Studies

The SAGE Handbook of
Television Studies

Edited by
Manuel Alvarado†, Milly Buonanno,
Herman Gray and Toby Miller

SAGE reference

Los Angeles | London | New Delhi
Singapore | Washington DC

SAGE

Los Angeles | London | New Delhi
Singapore | Washington DC

SAGE Publications Ltd
1 Oliver's Yard
55 City Road
London EC1Y 1SP

SAGE Publications Inc.
2455 Teller Road
Thousand Oaks, California 91320

SAGE Publications India Pvt Ltd
B 1/I 1 Mohan Cooperative Industrial Area
Mathura Road
New Delhi 110 044

SAGE Publications Asia-Pacific Pte Ltd
3 Church Street
#10-04 Samsung Hub
Singapore 049483

Editor: Mila Steele
Assistant editor: Judi Burger
Production editor: Shikha Jain
Copyeditor: Rosemary Campbell
Proofreader: Dick Davis
Indexer: Avril Ehrlich
Marketing manager: Michael Ainsley
Cover design: Wendy Scott
Typeset by: Cenveo Publisher Services
Printed and bound by CPI Group (UK) Ltd,
Croydon, CR0 4YY [for Antony Rowe]

MIX
Paper from responsible sources
FSC www.fsc.org FSC® C013604

At SAGE we take sustainability seriously. Most of our products are printed in the UK using FSC papers and boards. When we print overseas we ensure sustainable papers are used as measured by the Egmont grading system. We undertake an annual audit to monitor our sustainability.

Library of Congress Control Number: 2014941608

British Library Cataloguing in Publication data

A catalogue record for this book is available from the British Library

ISBN 978-0-7619-4741-7

Contents

List of Figures and Tables

FIGURES

TABLES

Notes on the Editors and Contributors

THE EDITORS

Manuel Alvarado† was Secretary of the Society for Education in Film and Television; editor of *Screen Education*; Head of Education at the British Film Institute; an academic at the Institute of Education, the University of Luton, West Surrey College of Art and Design, City University of London, and the University of Sunderland; and a publisher with John Libbey Publishing, University of Luton Press, and Intellect Books. He was also a governor of Sir John Cass's Foundation. Manuel authored and edited many books, which include *Hazell: The Making of a TV Series* (BFI Publishing, 1978), *Doctor Who: The Unfolding Text* (St Martin's Press, 1984) and *The Media Reader* (The MIT Press, 2003). He died in 2010.

Milly Buonanno is Professor of Television Studies at La Sapienza University of Roma. She is the founder and director of the Observatory of Italian TV Drama and the head of GEMMA (GEnder and Media MAtter) research programme. Her scholarships include television theory and history, TV drama, feminist media studies, journalism. She is the author and the editor of more than fifty books and her work has been translated into English, French, Spanish, Portuguese. Recent publications are the monographs *Italian TV Drama and Beyond: Stories from the Soil, Stories from the Sea* (Intellect, 2012), and *The Age of Television. Experiences and Theories* (Intellect, 2008).

Herman Gray has held appointments on the faculties of Northeastern University and the University of Southern California's Annenberg School of Communication. He is currently Professor of Sociology at the University of California, Santa Cruz. He has published widely in academic journals and popular media on black cultural politics, representation, music and television. Gray's books include *Producing Jazz* (Temple University Press, 1988), *Watching Race* (University of Minnesota Press, 1995), *Cultural Moves* (University of California Press, 2005), and *Towards a Sociology of the Trace*, (co-edited with Macarena Gomez Barriś) (University of Minnesota Press, 2010). Gray's current research considers the limits of a cultural politics of media visibility and neoliberalism.

Toby Miller is Emeritus Distinguished Professor at the University of California, Riverside, the Sir Walter Murdoch Professor of Cultural Policy Studies at Murdoch University and Professor of Journalism, Media and Cultural Studies at Cardiff University/Prifysgol Caerdydd. The author and editor of over thirty books, his work has been translated into Spanish, Chinese, Portuguese, Japanese, Turkish, German, and Swedish. His two most recent volumes are *Greening the Media* (with Richard Maxwell) (Oxford University Press, 2012) and *Blow Up the Humanities* (Temple University Press, 2012). His adventures can be scrutinized at www.tobymiller.org.

THE CONTRIBUTORS

Mark Andrejevic is Associate Professor and ARC QEII Research Fellow at the Centre for Critical and Cultural Studies, University of Queensland (Australia). He is the author of *Reality TV: The Work of Being Watched* (Rowman & Littlefield Publishers, 2003), *iSpy: Surveillance and Power in the Interactive Era* and *Infoglut: How Too Much Information is Changing the Way We Think and Know* (Routledge, 2013), as well as numerous articles and book chapters on surveillance, popular culture and digital media.

Miranda J. Banks is Assistant Professor of Visual and Media Arts at Emerson College. Her research focuses on Hollywood's creative labour and American media industry history. She is the author of *The Writers: A History of American Screenwriters and Their Guild* (Rutgers, 2015), and co-editor of *Production Studies: Cultural Studies of Media Industries* (Routledge, 2009) as well as the forthcoming *Production Studies 2*. Her work has appeared in *Television & New Media, Popular Communication, The Journal of Popular Film and Television* and *Montage/AV*, as well as the anthologies *Teen Television: Genre, Consumption, and Identity* (BFI, 2008) and *How to Watch Television* (NYU, 2013).

Martín Becerra is a Professor at Universidad Nacional de Quilmes and Universidad de Buenos Aires on media policy and information and communication technologies, and researcher at the National Council for Scientific and Technological Research (CONICET). He has a PhD and a Masters in Communication Sciences from the Autonomous University of Barcelona, Spain. He is co-author of *Cajas Mágicas: el renacimiento de la televisión pública en América Latina* with Luis Arroyo, Ángel García Castillejo and Oscar Santamaría (2013) and three books on the politics, structure and concentration of the Latin American media system (*Los dueños de la palabra: Acceso, estructura y concentración de los medios* (2009); *Los monopolios de la verdad* (2009); and *Periodistas y magnates* (2006) with Guillermo Mastrini). He is author of the book *Sociedad de la Información: proyecto, convergencia y divergencia* (2003). Recently, he wrote, with Guillermo Mastrini and Santiago Marino, the paper 'Mapping digital media, Argentina', organized and edited by the Open Society Foundation (2012).

Oliver Boyd-Barrett acquired his PhD from the Open University (UK). His publications include 20 books and over 120 journal articles and book chapters, principally in the fields of international communication, news agencies, propaganda and educational communication. As author or co-author, editor or co-editor his books include *Hollywood and the CIA: News Agencies in the Turbulent Era of the Internet, Globalization, Communications Media and Empire, The Globalization of News, Educational Reform in Democratic Spain, Contra-Flow in Global News* and *The International News Agencies*. He has taught at the Open University (UK), California State Polytechnic University (Pomona, CA) and Bowling Green State University, Ohio.

Clare Cannon is a graduate student in the PhD programme, City, Culture + Community at Tulane University. Her areas of interest are political economy of the environment and media ecologies. She currently researches waste using quantitative and qualitative methods at the local, national and international levels. She received her MA from Union Theological Seminary and her BA from Scripps College.

Michael Curtin is the Duncan and Suzanne Mellichamp Professor of Global Studies in the Department of Film and Media Studies at the University of California, Santa Barbara. Curtin is also Director and co-founder of the Media Industries Project of the Carsey-Wolf Center. His books include *The American Television Industry*, *Playing to the World's Biggest Audience: The Globalization of Chinese Film and TV* and *Reorienting Global Communication: Indian and Chinese Media Beyond Borders*. He is currently working on a comparative analysis of media capitals around the world, including Hollywood, Miami, Lagos and Mumbai.

Matthew Delmont is an Associate Professor of History at Arizona State University. His research and teaching areas include popular culture and media studies, urban history, education and comparative ethnic studies. He has published articles in *American Quarterly, Journal of Popular Music Studies*, *Journal of Urban History*, *History of Education Quarterly* and the *Journal of Pan-African Studies*. His first book, *The Nicest Kids in Town: American Bandstand, Rock 'n' Roll, and the Struggle for Civil Rights in 1950s Philadelphia*, was published by University of California Press (American Crossroads series) in 2012. He is currently doing research on bussing for school desegregation in the 1970s, focusing on how anti-bussing activists successfully used television to turn public opinion against bussing.

André Dorcé is Professor at the Communication Sciences Department in Universidad Autonoma Metropolitana Cuajimalpa, in Mexico City. His research has addressed interdisciplinary questions of media and political culture, especially in relation to particular instances of media production and consumption in the context of historically grounded cultural practices of media convergence. His interests are focused on the role of both material and symbolic communication technologies as expressions – and constitutive elements – of contemporary subjectivities, social organization and power relations. He is a member of the executive board of the UCCS (Union of Socially Committed Scientists) where he coordinates the media and communication group. He also collaborates with Public Broadcasting Channel 22 as the Channel's Ombudsman. He is currently working in a book about telenovelas and politics in Latin America and another volume on critical perspectives on media convergence. He is collaborating with several local and regional initiatives for progressive change in media legislation.

Roderick Flynn lectures in the School of Communication, Dublin City University where he has also held posts as Chair of the MA in Film and Television Studies, and member of the School Executive. His research interests include: broadcasting, audiovisual and telecoms policy in Ireland, Europe and USA; social history of communications; and political economy of the media. His book-length publications include: *John Huston: Essays on a Restless Director* (McFarland, 2010); *Historical Dictionary of Irish Cinema*, co-edited with Pat Brereton (Scarecrow Press, 2007) and *Here's Looking at You Kid: Ireland goes to the pictures*, co-edited with Stephanie McBride (Wolfhound Press, 1996).

Des Freedman is Professor of Media and Communications at Goldsmiths, University of London. He is the author of *The Politics of Media Policy* (Polity, 2008) as well as co-author (with James Curran and Natalie Fenton) of *Misunderstanding the Internet* (Routledge, 2012) and co-editor (with Daya Thussu) of *Media and Terrorism: Global Perspectives* (Sage, 2012). His latest book is *The Contradictions of Media Power* (Bloomsbury, 2014) and he is the current chair of the Media Reform Coalition in the UK.

Laura Grindstaff is Professor of Sociology at the University of California, Davis. She is the author of *The Money Shot: Trash, Class, and the Making of TV Talk Shows* (2002) as well as numerous articles and essays on media and popular culture; she is also co-editor of the *Handbook of Cultural Sociology* (2010, 2012).

Felicia D. Henderson is currently a PhD candidate in Cinema and Media Studies at UCLA. She writes about cultural politics and creative economy in television writers' rooms and issues of race, gender and othering in this same space. She is also the creator of Showtime's award-winning, critically acclaimed, *Soul Food.* Other writing and producing credits include: *Fringe, Gossip Girl, Everybody Hates Chris, Sister-Sister, The Fresh Prince of Bel Air* and *Moesha.* Currently, she has one-hour drama pilots in development at BET and HBO, where she is partnered with Oprah Winfrey's Harpo Productions. Her critical essays have been published in the *Cinema Journal, Popular Communication, Media/Cultural Studies: Critical Approaches, Production Studies: Cultural Studies of Media Industries,* and *Emergences.* She has also endowed two scholarships at UCLA: The Four Sisters Fellowship (endowed along with her sister-colleagues, Sara Finney-Johnson, Mara Brock Akil and Gina Prince-Bythewood); and the Felicia D. Henderson Screenwriting Scholarship.

Darnell Hunt is Director of the Ralph J. Bunche Center for African American Studies and Professor of Sociology at UCLA. He has written extensively on race and media, including four books and numerous articles and book chapters. He is also author of a series of reports examining the state of diversity in the Hollywood entertainment industry. These include *The Hollywood Writers Report*, issued by the Writers Guild of America in 2005, 2007, 2009 and 2012, and *The African American Television Report*, released by the Screen Actors Guild in June of 2000. Prior to his academic positions, he worked in the media and as a media researcher for the US Commission on Civil Rights' 1993 hearings on diversity in Hollywood.

Jason Jacobs is Associate Professor and Head of the School of English, Media Studies and Art History, the University of Queensland. He has an international reputation as a historian of television drama, its institutions, technology and aesthetics. He is author of *The Intimate Screen* (Oxford University Press, 2000), *Body Trauma TV* (BFI, 2003) and *Deadwood* (BFI, 2012). He is currently working on an Australian Research Council funded project called, 'The persistence of television: how the medium adapts to survive in a digital world'. He is also completing a book for Manchester University Press on the writer David Milch.

Jennifer Lynn Jones is a doctoral candidate in Film and Media Studies in Indiana University's Department of Communication and Culture. She is a feminist media scholar working in comparative media and critical cultural studies. Her research centres on identity, embodiment and representation, and her dissertation is on celebrity, convergence and corpulence in contemporary American media. She has published in *Antenna* and *In Media Res*, and has an article forthcoming at *Camera Obscura.*

L.S. Kim is Associate Professor of Film and Digital Media at UC Santa Cruz. Her areas of teaching, writing, and public scholarship are in television, racial discourse, feminist film/media criticism, media and social change, and Asian American cultural production; her related interests include music and dance, and genres. She has taught at Northwestern, Berkeley, UC Santa Barbara, and UCLA. L.S. Kim appears in the new film, *Wonder Women! The Untold Story of American Superheroines* (2012), directed by Kristy Guevara-Flanagan and *Uploaded: The Asian American Movement* (2012), a feature-length documentary about Asian American artists utilizing new media and transmedia modalities.

Guillermo Mastrini is a Professor at Universidad Nacional de Quilmes, where he is also the Head of the Masters in Cultural Industries, and at Universidad de Buenos Aires. Together with Martín Becerra, he published *Los dueños de palabra, Los monopolios de la Verdad* (both in

2009) and *Periodistas y magnates. Estructura y concentración de las industrias culturales en América Latina* (2006). He also published *Mucho ruido, pocas leyes. Economía y política en la comunicación en la Argentina (1920–2004)* (2005), among other books. He has been the President of the Argentine Federation of Social Communication Careers and Head of the Communication Science School at Universidad de Buenos Aires.

Vicki Mayer is Professor of Communication at Tulane University, New Orleans, Louisiana. She is author of two books: *Producing Dreams, Consuming Youth: Mexican Americans and Mass Media* (Rutgers University Press, 2003) and *Below the Line: Producers and Production Studies in the New Television Economy* (Duke University Press, 2011). She has also edited two volumes: *Media Production* (Wiley Blackwell 2013) and *Production Studies: Cultural Studies of Media Industries* (Routledge 2010), and has authored over 30 publications on media industries, media production, media consumption, and the identities of producers and consumers. She edits the journal *Television & New Media* (SAGE) and directs the digital humanities project MediaNOLA (medianola.org).

Lisa McLaughlin is Associate Professor with appointments in the Department of Media, Journalism and Film and the Women's, Gender and Sexuality Studies programme at Miami University-Ohio, USA. She has published on transnational feminism, the public sphere, political economy, and women, work, and information technologies. She is founding co-editor, with Cynthia Carter, of *Feminist Media Studies*, an international, peer-reviewed journal which she co-edited for 16 years. She has published articles in journals including *Media, Culture & Society*, *Sociological Review* and *Television & New Media*, along with a number of chapters in edited collections. Recent publications include *Current Perspectives in Feminist Media Studies* (co-edited with Cynthia Carter, 2013) and *The Routledge Companion to Media and Gender* (co-edited with Cynthia Carter and Linda Steiner, 2014).

Albert Moran is Professor in Screen Studies at Griffith University in Brisbane. His earlier research concerned the history of Australian film and television, while more recently his attention has focused on programme formats and transnational television. *TV Format Mogul: Reg Grundy's Transnational Career* (Intellect, 2013) is one of his latest publications. Currently he is writing a history of transnational television development as well as a study of Australian television from a transnational perspective.

Michael Morgan is Professor in the Department of Communication at the University of Massachusetts Amherst. His research interests include television, socialization and enculturation; international and intercultural effects of mass media; new media technology and policy; and social and familial contexts and consequences of media exposure. He has authored or co-authored over 100 journal articles and book chapters on cultivation analysis and media effects. He has directed or collaborated on numerous international comparative research projects, and he was a Fulbright Research Scholar in Argentina. His most recent books are *George Gerbner: A Critical Introduction to Media and Communication Theory* (Peter Lang, 2012) and *Living with Television Now: Advances in Cultivation Theory and Research* (edited with James Shanahan and Nancy Signorielli, Peter Lang, 2012).

Guillermo Orozco is a Senior Researcher of TV Reception Analysis and Media Education at the Social Communication Studies Department of the University of Guadalajara, México. He is UNESCO Professor of Communication, International Co- Coordinator of OBITEL: Iberoamerican Observatory of TV Fiction and Academic Coordinator of TVMORFOSIS, a major annual university forum focused on research and broadcast seminars about transformations of TV and its audiences. He is the editor of *Historias de la TV en América Latina* (Gedisa 2002), co-author of *Televisiones en México, un recuento histórico* (Universidad de Guadalajara 2007), and coeditor of *Social Memory and TV Fiction in Ibero America, OBITEL Yearbook (2013)* (Globo Universidade).

Stylianos Papathanassopoulos is Professor at the Faculty of Communication and Media Studies at the National and Kapodistrian University of Athens. He has written extensively on media developments in Europe and especially on television issues in various journals. He also edits the Greek communication journal *Zitimata Epikoinonias/Communication Issues*. He is the author, co-author or editor of 19 books about the media. Among his most recent books are (with Ralph Negrine) *European Media, Structures, Policy & Identity* (Polity Press, 2011); *Media Perspectives* (editor, Routledge, 2011), *Television in the 21st Century* (Kastaniotis, 2005); and *European Television in the Digital Age: Issues, Dynamics and Realities* (Polity Press, 2002).

Allison Perlman is an Assistant Professor in the Departments of Film and Media Studies and History at the University of California, Irvine. She is currently completing a book on the intersections between media activism, broadcasting policy and social movements in the US. Her publications include articles in *Critical Studies in Media Communication, Television & New Media, Cinema Journal, Feminist Media Studies, The Journal of Popular Film and Television* and *Communication, Culture and Critique*.

Paschal Preston holds a research professorship at the School of Communications in Dublin City University. He is the founding director of the Communication, Technology, Culture (COMTEC) research unit which has a distinguished 25-year record of collaborative research, including many multi-country or comparative research projects. He is also a member of the Management Board of the Irish Social Sciences Platform (ISSP), and co-ordinator of the Knowledge Society research pillar. His books include: *Making the News* (Routledge, 2009) and *Reshaping Communications* (Sage, 2001).

Arvind Rajagopal is a Professor in the Department of Media, Culture and Communication, and an affiliate faculty in the Department of Sociology, and the Department of Social and Cultural Analysis, at New York University. His book *Politics After Television: Hindu Nationalism and the Reshaping of the Public in India* (Cambridge, 2001) won the Ananda Kentish Coomaraswamy Prize from the Association of Asian Studies in 2003, and his edited volume *The Indian Public Sphere* appeared in 2009. His recent articles include 'The Emergency and the New Indian Middle Class' in *Modern Asian Studies,* 2011, and 'Special Political Zone' on the anti-Muslim violence in Ahmedabad, Gujarat, in *South Asian Multidisciplinary Academic Journal* from the Ecole des Hautes Etudes en Sciences Sociale, Paris, in their special issue on Rethinking Urban Democracy in South Asia, 2011. His latest book, under contract with Duke University Press, is on the political culture of post-independence India. He has held fellowships at the Center for Advanced Study in the Behavioral Sciences at Stanford University, at the Institute for Advanced Studies in Princeton, and at the Woodrow Wilson International Center for Scholars, and visiting professorships at the University of Goettingen, Germany, the Delhi School of Economics at the University of Delhi, and the Central University of Hyderabad.

David Rowe is Professor of Cultural Research, Institute for Culture and Society, University of Western Sydney, Australia. His principal research interests are in media and popular culture, especially sport, journalism and urban leisure. Among Professor Rowe's current research projects is a study of the relationships between sport, nation and cultural citizenship. His most recent books are *Global Media Sport: Flows, Forms and Futures* (Bloomsbury Academic, 2011); *Sport Beyond Television: The Internet, Digital Media and the Rise of Networked Media Sport* (authored with Brett Hutchins, Routledge, 2012); *Digital Media Sport: Technology, Power and Culture in the Network Society* (edited with Brett Hutchins, Routledge, 2013) and *Sport, Public Broadcasting, and Cultural Citizenship: Signal Lost?* (edited with Jay Scherer, Routledge, 2014). His work has been translated into several languages, including Chinese, French, Turkish, Spanish, Italian and Arabic.

Kathleen M. Ryan (PhD University of Oregon; MA University of Southern California; BA University of California Santa Barbara) is an Associate Professor of journalism and mass communication in the College of Media, Communication, and Information at the University of Colorado Boulder. Her areas of research include visual communication, freelance news labour, oral history and media production practices. She has spent nearly 20 years in television news production before moving to academia full time. She is also an active documentary filmmaker, with films including *Homefront Heroines: The WAVES of World War II* (2013) and *Pin Up! The Movie* (in production). She is the co-editor of *Television and the Self: Knowledge, Identity and Media Representation* (Lexington, 2013) and *How Television Shapes our Worldview: Media Representations of Social Trends and Change* (2014). Journal articles have appeared in *Electronic News*, *Visual Studies* and other journals.

Kevin Sanson is the Research Director of the Carsey-Wolf Center's Media Industries Project at the University of California, Santa Barbara, where he also teaches in the Department of Film and Media Studies. His current book project focuses on the shifting relationships between locations and labour in global film and television production. He is co-editor of *Connected Viewing: Selling, Streaming, & Sharing Media in the Digital Era* (Routledge, 2014) and *Distribution Revolution: Conversations about the Digital Future of Film and Television* (University of California Press, 2014). He also serves as part of the founding editorial collective of *Media Industries*, the first peer-reviewed, open-access journal for media industries research.

James Shanahan is a mass media effects researcher. He is the Associate Dean of the College of Communication at Boston University. He holds a PhD in Communication from the University of Massachusetts at Amherst. His research interests focus on cultural indicators, cultivation theory, media effects and public opinion. Special areas of focus are communication in relation to science and the environment. His studies in this area have dealt with the impact of media (especially television) on environmental attitudes, the environmental content of media, and the narrative structure of environmental news. His books include *Democracy Tango* (1995) and *Television and its Viewers* (both with Michael Morgan, 1999) and *Nature Stories* (with Katherine McComas, 1999).

David Sholle is an Associate Professor in the Department of Media, Journalism and Film at Miami University-Ohio and teaches television and film production, documentary and media theory. He is the co-author of *Media Education and the Reproduction of Culture* (1994) and *Rethinking Media Literacy* (1995). He has written articles on alternative media, Negt and Kluge's ideas on the public sphere, critical media pedagogy, and new media technology. His articles have appeared in *Javnost: the Public*, *Critical Studies in Media Communication*, *Cultural Studies* and *The Journal of Education*. Sholle also has directed several documentaries, including *A Cold Day in D.C.* (on the 2nd Bush inauguration), *The Hour Has Come* (on the 97th Bomb Squad from World War II) and *What Good are the Humanities*?

Nancy Signorielli (PhD, University of Pennsylvania, 1975) is Professor of Communication and Director of the MA programme in Communication at the University of Delaware (USA). Her research focuses on images in the media, specifically on gender roles, television violence, ageing, minorities and health and how these images are related to people's conceptions of social reality. She has authored or co-authored over 100 articles or book chapters in this research area as well as written or edited eight books. She was named a distinguished scholar of the Broadcast Education Association in 2010 and a centennial scholar of the Eastern Communication Association in 2009.

Ruth Teer-Tomaselli is the UNESCO Chair of Communication in the Centre for Communications and Media in South Africa, University of KwaZulu-Natal, South Africa. Her research interests include broadcast media, specially public service broadcasting, broadcasting history and news and drama on radio and television. She is an avid reader of crime fiction and grows bonsai trees for pleasure.

Deborah Tudor is the Associate Dean of the College of Mass Communication and Media Arts at Southern Illinois University. She received her doctorate from Northwestern and taught at DePaul University prior to coming to SIU in 2006. She served as the Chair of the Department of Cinema and Photography before moving to the MCMA Associate Dean position in 2009. Dr Tudor's research interests lie in two areas: culture and technology, and globalization and cinema. Her most recent publications include a chapter on *Star Trek* in *Neoliberalism and Global Cinema*, (2011), and articles on digital cinema production tools in *Cinema Journal* (2008) and *Jump Cut* (2011). Additional articles include publications on digital audience monitoring tools in *Media Culture & Society*, and neoliberalism and media in *Society* (2012) and *Cinema Journal* (2013).

Silvio Waisbord is Professor in the School of Media and Public Affairs at George Washington University. He earned a PhD in sociology from the University of California, San Diego. Dr Waisbord's books include *Reinventing Professionalism: Journalism and News in Global Perspective* (Polity, 2013), *Media Sociology: A Reappraisal* (editor, Polity, 2014) and *Vox Populista: Medios, periodismo, democracia* (Gedisa, 2013). He was editor-in-chief of the *International Journal of Press/Politics* between 2008 and 2014, and currently serves as editor-in-chief of the *Journal of Communication.*

Mike Wayne is a Professor in Screen Media at Brunel University. He has written widely on Marxist theories of the media and cultural theory. He is the author of *Marxism and Media Studies: Key Concepts and Contemporary Trends* (Pluto, 2003) and *Marx's Das Kapital For Beginners* (Steerforth Press, 2012). He is the co-director (with Deirdre O'Neill) of a 2012 feature-length documentary film called *The Condition of the Working Class* (see www.conditionoftheworkingclass.info). His latest book is *Red Kant: Aesthetics, Marxism and the Third Critique* (Bloomsbury, 2014).

Brenda R. Weber is an Associate Professor and Director of Graduate Studies in Gender Studies at Indiana University, with adjunct appointments in American Studies, Cultural Studies, Communication and Culture, English, and the Kinsey Institute for Research in Sex, Gender, and Reproduction. Her books include *Makeover TV: Selfhood, Citizenship, and Celebrity* (Duke, 2009), *Women and Literary Celebrity in the Nineteenth Century: The Transatlantic Production of Fame and Gender* (Ashgate, 2012) and *Reality Gendervision:*

Sexuality and Gender on Transatlantic Reality TV (Duke, 2014). She is presently working on a monograph entitled *Mormons on Our Minds: Gender, Modernity, and Mediated Mormonism.*

Helen Wood is Professor of Media and Communication at the University of Leicester, UK. She is author of two audience research books, *Talking with Television* (2009) and, with Beverley Skeggs, *Reacting to Reality Television* (2012) and with Beverley Skeggs edited the collection *Reality Television and Class* (2011). She has published widely on television and audiences and is associate editor of the journals *Ethnography* and the *European Journal of Cultural Studies*. With Helen Wheatley and Rachel Moseley she has been working on the project 'A history of television for women'.

Introduction

Toby Miller[1]

> [T]elevision ... is both an industry and a set of state institutions ... whose purpose is to present itself, to expose itself continuously and conspicuously as no other set of institutions does, and yet which constantly effaces its own practices and methods. (Manuel Alvarado, 1983)

This *Handbook* engages perennial questions that have been posed by and of television studies: Who runs TV? Who makes TV? What is on TV? And who is the audience to TV? The specific headings that divide the volume cover ownership and regulation, makers and making, cultural forms, audiences, reception and consumption. Our goal has been to acknowledge the equal importance and achievements of work done in political economy, textual analysis and reception, the principal fields that address such matters.

Each section of the volume:

- presents the main topics for debate and offers original and challenging contributions that pose new questions and open new horizons
- contextualizes theories, methods and traditions of intellectual reflection and empirical research, giving recognition both to mainstream and marginal, established and emergent voices and norms; and
- explains the historical, social, economic and political conditions that have contributed to the inception, support and decline of approaches, where appropriate

In accordance with the publisher's *Handbook* norms, we wanted authors to follow some basic guidelines, but to do so in ways that gave them freedom to decide where to head next, both for themselves and the field more generally.

Inevitably, there are gaps in terms of places, politics, people, epistemologies and empirics. Some of this is due to our limitations as editors, and some to the exigencies that accompany any project of this scope. We are especially aware of certain absences in our coverage of gender issues that derive from these difficulties. In addition, we lack a truly global set of contributors. That said, we are glad to say that their addresses and origins include not only the usual suspects (the US and UK) but also Argentina, South Africa, Greece, Ireland, India, Australia, Italy and Mexico, and there is a reasonable gender balance.

This, *The SAGE Handbook of Television Studies* has been a long time in the making, due to the contingencies of living, teaching, other writing projects, and, sadly, loss. Over the course of the long germination and fruition of the project, we gained and lost momentum, felt frustration when we failed to fill a hole in the project, and were exuberant when a new author came aboard or existing authors confirmed their commitment.

Through it all, we continued to collaborate, support each other, and reaffirm our collective commitment to the project and those who signed on to work with us to realize it. So in every sense – from the conception to the solicitation of chapters, from editorial tasks to the final stages of production – the *Handbook* is the product of our collective and collaborative work. Toby Miller wrote the introduction and served as our primary representative with SAGE, speaking and acting on our behalf. And while we lost the central voice of our dear friend and colleague

Manuel Alvarado much too soon in the project, his influence, voice and contribution to our collaboration is irreplaceable and unmistakable.

Clearly, drawing an enterprise such as this one to a close is a source of relief as well as pleasure for all concerned. In the case of the *Handbook*, however, there is an even stronger emotion: grief. For our co-editor Manuel died in 2010 (Buscombe, 2010) when the volume was already well underway; as noted above, he was a formative influence on its ultimate shape. I want to acknowledge Manuel's intellectual legacy to TV studies. I hope that a brief exegesis of his ideas will illustrate their ongoing value and encourage readers to allow Manuel's work to reverberate as they make their way through this *Handbook*.

Following that tribute, I offer an account of how television has been theorized, then conclude by engaging with assertions, which we have heard and read for almost twenty years, that TV is on the way out. It may seem odd to mount a defence of the relevance of a *Handbook* in terms of its object of study, but in this case, it feels warranted, so widespread and inaccurate are the claims made about television's demise. To make the point, this Introduction ends with an attempt to apply some of Manuel's favoured methods to a globally significant television programme that has encouraged public debate over both torture and the environment.

MANUEL ALVARADO†

Manuel had a passionately held and powerfully expressed intellectual agenda. It was thematically internationalist, politically socialist, methodologically promiscuous, and dedicated to teaching school pupils as much as university students. Manuel's blend of market and non-market principles derived from the French Revolutionary cry *liberté, égalité, fraternité* [liberty, equality, solidarity] and the Argentine left's contemporary version: *ser ciudadano, tener trabajo, y ser alfabetizado* [citizenship, employment, and literacy]. The first category concerned political rights; the second, material interests; and the third, cultural representation. This agenda was equally influenced by Marx, Gramsci, Althusser, Foucault and the Global South. Manuel had no time for infantile anti-statism, Panglossian market populism, romantic investments in the 'other' – or their opposites, such as a blind faith in public broadcasting or the narcissistic reproduction of cultural sameness.

And Manuel was far from a banal careerist. His legacy is not the outcome of climbing a ladder towards formal recognition in scholarly journals, granting bodies, professional associations, or university departments – a trajectory that generally produces 'normal science'.[2] His cohort mostly regarded universities as 'repositories of reactionary thought' (Alvarado and Buscombe, 2008: 137).

Manuel was fond of the German Socialist rallying cry, 'The Long March of the Institutions', which invoked a devolved but carefully calculated takeover of cultural bodies. For, despite their love of anarchic spectacle, smart *soixante-huitards* like him rejected the civil-society mythology of volunteerism that is so improbably shared by gullible progressives who imagine utopia around the corner from immiseration, anti-Marxist intellectuals who valorize private enterprise, and coin-operated US client states and organizations.

Manuel was part of a formation that sought and gained an often contingent, sometimes shaky, but always lively control of cultural apparatuses as extra-parliamentary political projects: schooling, via the Society for Education in Film and Television (SEFT), the British Film Institute (BFI), and the United Nations Educational, Scientific and Cultural Organization (UNESCO); and scholarly and general-interest publishing, through those organizations and other private, not-for-profit, or semi-public ventures, such as John Libbey, Intellect and the University of Luton Press.

Believing that people made their way through education and read books and periodicals in order to learn things they didn't already know, Manuel doubted whether pupils' and students' existing inner processes and preferences were any better suited to understanding social relations or television genres than they would be to building bridges or piloting planes. The formation of taste and knowledge about television had to be theorized, engaged and changed by comprehending and explicating the 'interrelationship of production, distribution, broadcasting, [and] advertising' that generated pleasures and beliefs. It was vital to appreciate the constraints and opportunities presented by young people's 'class/cultural background … the containing society … [and] their mode of insertion into the school system' (Alvarado, 1977). As Foucault put it: 'There are times in life when the question of knowing if one can think differently than one thinks, and perceive differently than one sees, is absolutely necessary if one is to go on looking and reflecting at all' (1985: 8).

Thinking differently might transcend the graceless antinomies of TV policy identified by Bourdieu: either a 'paternalistic-pedagogic' address or 'populist spontaneism and demagogic capitulation to popular tastes' (1998). Manuel liked to complicate this binarism: he enjoyed the most basic situation comedies as well as the *avant-garde* and abjured what he called 'patrician, class-ridden, historical, costume drama' (2000: 315).

For Manuel and other activist intellectuals, 'theory was never justified for its own sake' but because it could produce new 'knowledge, since knowledge, unlike taste, was verifiable and transferable' (Alvarado et al., 1993). His use of 'Critical Media Theory' opposed a 'yea-saying attitude … [c]elebratory' orientation towards new forms of communication. The revelation that 'popular culture [is] wonderful! It's so *complicated*' (Alvarado and Thompson, 1990a, 1990b) didn't impress him; nor did the vaguely formulated assumption of a post-state, post-market return to Eden.

Rather than centralized state control necessarily constraining choice by people, Manuel made the point that choice is generally constrained by centralized commercial control. The marginal propensity to consume is very marginal indeed for the vast majority! In some astute comments on what used to be known in scholarly circles as cultural imperialism (and is still called that outside Anglo academia) he insisted that 'the international exchange of TV programmes is not based on the conventional principles of commodity valuation (i.e. assessing the marginal cost of production).' Instead, it must be understood 'in terms of the political and economic position of the buying country' (Alvarado, 1996).

This resonates with an earlier essay he wrote about television's duality: its Janus-faced capacity to witness and embody capitalism's paradoxical, simultaneous, sinuous and coeval desires for publicity and secrecy, marketing and privacy. Manuel recognized that TV owners and programme makers want to appear open to viewers in terms of texts, genres and channels – but insist on being closed to them as a set of political-economic interests, methods and commitments (Alvarado, 1983). Since he first created that striking homology, television has opened up to the point where it now appears to welcome academic researchers, for example, provided that they buy into its *faux* responsiveness to commodified audience reactions (Alvarado, 2009b).

The scepticism that Manuel expressed about accepting viewer interpretations, industry shibboleths or high-blown theory and science have informed our final, faltering steps towards the conclusion of this *Handbook*. It is dedicated to his memory, with thanks for enriching our world.

TELEVISION STUDIES

What *is* television? I thought we'd look at some newspapers to see what they said. *The Guardian* (of London, New York, Sydney, and counting) thinks writing about television involves promoting and reviewing programmes. That means telling readers what shows are about and whether they are good, either via the eager ears of star-struck interviewers or the acerbic eyes of lounge-lizard critics (www.theguardian.com/tv-and-radio). The approach is largely nationalistic—if you're not in English and on 'our' screens, you don't matter. Spain's *El País*, which also publishes a global edition across the Americas and Europe, is much more catholic, if you'll pardon the phrase, in that news can actually be made in different languages and states; but *El País*, too, thinks TV is largely about programming (www.elpais.com/tag/television/a/). *Le Monde* subsumes TV under 'culture', quite rightly, though its themes are akin to the rest (www.lemonde.fr/culture/). The *New York Times* is sufficiently anachronistic to include 'local listings'. How quaint. But again, its stock in trade is promotion and critique (www.nytimes.com/pages/arts/television/index.html?8qa). Mexico's *La Jornada*, by contrast, generally addresses television as an object of news (www.jornada.unam.mx/ultimas/@@search?SearchableText=television). It stands out in this group.

My far-from-random sample suggests that television critics write profiles of celebrities we'll never meet and watch programmes before we do. TV may also appear in newspapers' business and policy columns, but under different headings, more akin to *Jornada*.[3] That sort of division is acknowledged but disobeyed in the collection you have before you.

Television has given rise to three related topics of scholarly inquiry that stretch across such concerns:

- technology, ownership, and control – its political economy
- textuality – its content; and
- audiences – its public

Within these categories lie three further divisions:

- approaches to technology, ownership, and control vary between neoliberal endorsements of limited regulation by the state in the interests of protecting property and guaranteeing market entry for new competitors, Marxist critiques of the *bourgeois* media for controlling the socio-political agenda, and environmental investigations of the impact of TV gadgetry on energy use and electronic waste
- approaches to textuality vary between hermeneutics, which unearths the meaning of individual programmes and links them to broader social formations and problems, such as the way that social identities are represented, and content analysis, which establishes patterns across significant numbers of similar texts, rather than close readings of individual ones; and
- approaches to audiences vary between social-psychological attempts to validate correlations between television and social conduct, political-economic critiques of imported texts threatening national culture, and celebrations of spectators' interpretations

These tasks in turn articulate to particular academic disciplines, which are tied to particular interests of state and capital:

- engineering, computing, public policy, journalism, and 'film' schools create and run TV production and reception via business, the military, the community and the public service
- communication studies focuses on such social projects as propaganda, marketing and citizenship
- economics theorizes and polices doctrines of scarcity, as well as managing over-production through overseas expansion
- Marxism points to the impact of ownership and control and cultural imperialism on television and consciousness; and
- cultural criticism evaluates representation, justifies protectionism, and calls for content provision

Today, major engagements with TV come from the psy-function (psychology, psychiatry and psychoanalysis), other social sciences (sociology, economics, communication studies,

Table I.1. Objects, methods and disciplines

Topics	*Objects*	*Methods*	*Disciplines*
Regulation, Industry Development, New Technology	State, Capital, Labour	Political Economy, Neoliberalism	Engineering, Computer Science, Economics, Political Science, Law, Communication Studies
Genre	Text	Content Analysis	Communication Studies, Sociology
Genre	Text	Textual Analysis	Literary/Cultural/Media Studies
Uses	Audience	Uses and Gratifications	Communication Studies, Psychology, Marketing
Uses	Audience	Ethnography	Anthropology, Cultural/Media Studies, Communication Studies
Effects	Audience	Experimentation, Questionnaire	Psychology, Marketing, Communication Studies

anthropology, political science, and law) and the humanities (literature, cinema studies, media studies, languages, and cultural studies). There are seven principal forms of inquiry, which:

- borrow ethnography from sociology and anthropology to investigate the experiences of audiences
- use experimentation and testing methods from psychology to establish cause-and-effect relations between television viewing and subsequent conduct
- adapt content analysis from sociology and communication studies to evaluate programming in terms of generic patterns
- adopt textual analysis from literary theory and linguistics to identify the ideological tenor of content
- apply textual and audience interpretation from psychoanalysis to speculate on psychic processes
- deploy political economy to examine ownership, control, regulation and international exchange; and
- utilize archival, curatorial, and historiographic methods to give TV a record of its past

Relevant professional associations are shown in Table I.2.

So where did television studies, in the broadest sense, come from? It emerged from a complex blend of the social sciences and the humanities, from qualitative and quantitative sociology and communications, from language and literature departments, and various formations that arose across these and other domains.

A key source has been speech communication, which was founded in the US a century ago to assist with the assimilation of white, non-English-speaking migrants into the workforce. It

Table I.2. Relevant professional associations

• International Association for Media & Communication Research	• Union for Democratic Communications
• Broadcast Education Association	• Association for Education in Journalism and Mass Communication
• National Communication Association	• International Communication Association
• Society for Cinema and Media Studies	• American Association for Public Opinion Research
• American Journalism Historians Association	• Asociación Latinoamericana de Investigadores de la Comunicacion
• Association for Chinese Communication Studies	• Association of Internet Researchers International Association for Media and History
• Chinese Communication Association	• European Consortium for Communications Research
• European Society for Opinion and Marketing Research	• Global Communication Research Association
• International Association for Media History	• Southern African Communication Association
• Media, Communications & Cultural Studies Association	• Canadian Communication Association

became the first home of media education, because the engineering professors who founded radio stations in colleges during the 1920s needed programme content, and drew volunteers from that area. These stations became laboratories, with research undertaken into technology, content and reception (Kittross, 1999). This was also a period of massively complex urbanization and the spread of adult literacy, democratic rights, labour organization and socialist ideas, which gave rose to a social-science equivalent to the study of speech: mass communication. First radio then TV were simultaneously prized and damned for their demagogic qualities, which it was hoped and feared could turn people into consumers or communists alike.

With the standardization of social-science method and its uptake and export by US military, commercial, scholarly and governmental interests across the 20th century, television audiences came to be conceived as empirical entities that could be known via research instruments derived from communication, sociology, demography, the psy-function and marketing. Such concerns were coupled with a secondary concentration on content. Texts, too, were conceived as empirical entities that could be known, via research instruments derived from sociology, communication and literary criticism. Universities across the US introduced schools that would prepare students for media work. In Britain, Granada TV endowed a research position into television at Leeds University in 1959. Then SEFT and the BFI began sometimes separate, sometimes overlapping forms of stimulus in the 1960s and 70s, from seeding teaching posts to publishing critical dossiers. This ultimately fed into major formations of UK media studies influenced by continental Marxism and feminism, then by similar social movements as the US[4] (Bignell et al., 2000: 81; Bolas, 2009).

Some of the bodies listed in Table I.2 see themselves as feeder groups and even advocates for the industry; some identify as purely scholarly entities; and others call for progressive change. Let's consider key instances, mostly from the Anglo world.

The Association for Education in Journalism and Mass Communication (AEJMC) describes itself as 'a multicultural network of practitioners'. Founded in 1912, AEJMC seeks to 'advance education in journalism, cultivate better professional practice and promote the free flow of information, without boundaries' (www.aejmc.org/), while the National Communication Association, formerly the Speech Communication Association (SCA, which began in 1914) says it is '[d]edicated to fostering and promoting free and ethical communication' (www.natcom.org/about/).

The Broadcast Education Association (BEA) commenced in 1948 as an educational arm of the US radio then TV industry through the National Association of Broadcasters, which has provided it with resources to fund events and publications (Kittross, 1999). The BEA bears the lineaments of a heritage 'preparing college students to enter the radio & TV business' (www.beaweb.org/). The Society for Cinema and Media Studies (SCMS) says it is 'devoted to the study of the moving image'. Founded in 1959, the Society aims to 'promote all areas of media studies' and 'advance multi-cultural awareness and interaction' (www.cmstudies.org/). Attempts to bring television studies into the former Society for Cinema Studies were roundly rejected in the 1990s by *cinéphiles*. It may be that the eventual expansion of their rubric to incorporate 'media studies' derived not from an appreciation of the importance of TV, but because of 'the propinquity of television studies to the higher-prestige if loosely defined field of new media studies' (Boddy, 2005: 81): US universities 'tend to value anything called new media' thanks to its applications to militarism, and its ability to draw hefty research money through governmental and commercial fetishes for new technology. The upshot is that 'studying anything that comes over the Internet … has somehow become more legitimate than studying television itself' (Spigel, 2005: 84).

The International Communication Association, which started in 1950 as a breakaway from SCA focused on mass communication, avows that it exists:

1. to provide an international forum to enable the development, conduct, and critical evaluation of communication research;
2. to sustain a programme of high quality scholarly publication and knowledge exchange;
3. to facilitate inclusiveness and debate among scholars from diverse national and cultural backgrounds and from multi-disciplinary perspectives on communication-related issues; and
4. to promote a wider public interest in, and visibility of, the theories, methods, findings and applications generated by research in communication and allied fields. (www.icahdq.org/about_ica/mission.asp)

Britain's Media, Communications and Cultural Studies Association (MeCCSA) is the newest of these bodies (it was established in 2007). MeCCSA presents itself as a service to students, rather alarmingly suggesting that '[m]any of the jobs you will go into once you have finished your degree have not yet been invented'. The Association suggests that obtaining employment in television may flow from 'the ability to produce high[-]quality research, to analyze sociological trends, to work effectively with people, to organize events, to think creatively and to write well' (www.meccsa.org.uk/).

The International Association for Media Communication and Research (started by UNESCO in 1957 and the only body discussed here that is not Anglo-dominated) says that it 'aims to support and develop media and communication research throughout the world. It particularly encourages the participation of emerging scholars, women and those from economically disadvantaged regions' (www.iamcr.org/welcome-to-iamcr-aboutiamcr-375)

By contrast with the jobs, legitimacy and inclusivity emphases of the other Associations and the aesthetico-historical goals of SCMS, the Union for Democratic Communications (UDC, founded in 1981) sees itself as an organization of communication researchers, journalists, media producers, policy analysts, academics and activists dedicated to:

- critical study of the communications establishment;
- production and distribution of democratically controlled and produced media;
- fostering alternative, oppositional, independent and experimental production;
- development of democratic communications systems locally, regionally and internationally. (www.democraticcommunications.net/about)

This is a transformative rather than a parthenogenetic project: UDC works towards a different world, as opposed to generating new cohorts of producers, scientists, rhetoricians, journalists and aesthetes.

Clearly, there is no single professional association to go to in order to see how academia makes sense of TV – I could just as easily have listed relevant segments of professional bodies within political science (www.politicalcommunication.org/), anthropology (www.societyforvisualanthropology.org/), sociology (www.asanet.org/sections/CIT.cfm), education (www.aera.net/DivisionC/CommunicationandSocialMedia/tabid/15413/Default.aspx), law (www.lawandsociety.org/crn.html), pediatrics (www.aap.org/en-us/about-the-aap/Committees-Councils-Sections/Council-on-Communications-Media/Pages/default.aspx) or psychology (www.apa.org/about/division/div46.aspx). There would be some overlap and many epistemological, methodological, focal and political differences.

The same applies to journals. Table I.3 lists a huge, diverse, yet only partial, list! But consulting these and other titles will keep readers abreast of debates in television studies, as it is undertaken across and within disciplines.

Some of these journals are the organs of professional associations: authors may be obliged to join in order to publish there and will be expected to cite the work of powerful fellow-members. Others are journals of tendency, which seek new and transformative work rather than the reiteration of normal science.[5]

Table I.3. List of Journals

International Journal of Cultural Policy, Entertainment Law Review, Transnational Television Studies, Global Media Journal, Television & New Media, Global Media and Communication, Poetics, Journal of Media Economics, Media International Australia, European Journal of Communication, Media Culture & Society, International Communication Gazette, Media Law and Practice, Feminist Media Studies, Comunicaço & Politica, International Journal of Communication, International Journal of Communications Law and Policy, Asian Journal of Communication, Games & Culture, Journal of Broadcasting & Electronic Media, Revista Electrónica Internacional de Economía Política de las Tecnologías de la Información y de la Comunicación, Entertainment and Sports Law Journal, Asian Media, Comunicaçao e Sociedade, Convergence, Loyola Entertainment Law Journal, Columbia VLA Journal of Law and the Arts, Loyola Entertainment Law Journal, Cultural Studies Review, Mediascape, Communication Review, Cultural Politics, Critical Studies in Media Communication, Quarterly Review of Film and Video, Cinema Journal, Journal of Media Sociology, Democratic Communiqué, Television Quarterly, Cultural Sociology, Journal of Arab and Muslim Media Research, Journal of Creative Communications, Media Development, Canadian Journal of Communication, Visual Anthropology, Visual Anthropology Review, NORDICOM Review of Nordic Research on Media and Communication, Journal of International Communication, Asian Journal of Communication, New Media & Society, Journalism & Mass Communication Quarterly, Journal of Communication Inquiry, Historical Journal of Radio, Film & Television, Journal of Communication, European Journal of Cultural Studies, Journalism History, Journalism: Theory, Practice and Criticism, Media History, Women's Studies in Communication, Public Opinion Quarterly, Political Communication, Gamasutra, Federal Communications Law Journal, Fordham Intellectual Property, International Journal of Press/Politics, Popular Communication, Media & Entertainment Law Journal, Topia, Cultural Studies, Communications, International Journal of Cultural Studies, Journal of British Cinema & Television, Social Semiotics, Journal of E-Media Studies, Critical Studies in Television, Chinese Journal of Communication, Jump Cut, Screen Education, Screen, Velvet Light Trap, Flow, Journal of Film & Video, New Review of Film and Television Studies, Journal of Popular Film & Television, Middle East Journal of Culture and Communication, Political Communication, Media Psychology, Popular Communication

It should be clear from this schematic account of scholarly bodies and publications that the study of television is fractured by objects of study, politics, nations, disciplines, theories and methods. Why? A dispersed field of knowledge, it derives from the spread of new media technologies over the past two centuries into the lives of urbanizing populations, and the policing questions that has posed to both state and capital: what would be the effects of these developments, and how would they vary between those with a stake in maintaining society versus transforming it?

By the early 20th century, academic experts had decreed media audiences to be passive consumers, thanks to the missions of literary criticism (distinguishing the aesthetically cultivated from others) and psychology (distinguishing the socially competent from others). Decades of social science have emphasized audience reactions to audiovisual entertainment: where they came from, how many there were, and what they did as a consequence of being present.

More than half a century ago, Dallas Smythe explained that audience attention – presumed or measured – was the commodity that TV stations sold to advertisers (2004: 319–20). Programmes are therefore not so much consumer items as 'symbols for time' (Hartley, 1987: 133). Television audiences participate in the most global (but local) communal (yet individual) and time-consuming practice of making meaning in the history of the world. The concept and the occasion of being an audience are links between society and person, at the same time as viewing and listening involve solitary interpretation as well as collective behaviour.

Production executives invoke the audience to measure success and claim knowledge of what people want, regulators to organize administration, psychologists to produce proofs, and lobby groups to change content. Hence the link to panics about education, violence and apathy supposedly engendered by television and routinely investigated by the state, psychology, Marxism, feminism, conservatism, religion and others. The audience as consumer, student, felon, voter, sexist, heathen, progressive and fool engages such groups. Effects and ratings research traverses the industry, the state and criticism. Academic, commercial, and regulatory approaches focus most expansively on audiences to television as citizens and consumers, far more than the medium's technology, law or even content.

For example, female viewers were central to corporate calculations about US television from the very first, because they were expected to spend more time in the home than other potential spectators. In the early days of tuning sets, it was thought that women would be unable to cope with the technical challenges of reception. Then there was the question of their unpaid domestic labour – how could this crucial economic and social service be safeguarded while they fulfilled their other useful role as captives of commercials? The US strategy, which became orthodox elsewhere, was to drop plans for reconstructing cinema in the home via this new, exciting apparatus. Earlier assumptions about seeking to repeat the immersive experience of movies—lights off, full attention, and immobile – were rejected in favour of a distracted experience. Like radio, TV would be just one aspect of home life, alongside demanding children, husbands and tasks. Its visuals would reinforce a message that could be understood in another room or while doing chores – the volume would go up when the commercials came on (Morley, 2007: 277).

This focus on the TV public and how to corral and control it is in keeping with a long history that predates television's advent. The pattern has been that when new media technologies emerge, young people in particular are identified as both pioneers and victims, simultaneously endowed by manufacturers and critics alike with power and vulnerability – the first to know yet the last to understand cheap novels during the 1900s, silent then sound film during the teens and 1920s, radio in the 1930s, comic books of the 1940s and 50s, pop music and television from the 1950s and 60s, satanic rock as per the 1970s and 80s, video cassette recorders in the 1980s, and rap music, video games and the internet since the 1990s.

Here are some concerned Australian pundits pondering what the new arrival, television, might do to 1950s' adolescents:

> In September 1956 many Sydney residents had their first opportunity to experience first-hand contact with the new mechanical 'monster' – TV – that had for the last seven or eight years been dominating the lounges of English and American homes. Speculation on its effects had run high. On the one hand it was claimed: it would eventually destroy the human race since young couples would prefer viewing to good honest courting; children would arrive at school and either go to sleep or disgorge half-baked concepts about the Wild West and the 'gals' who inspired or confused the upholders of law; it would breed a generation of youngsters with curved spines, defective eyesight, American vocabulary but no initiative; it would result in a fragmentation of life whereby contact among, and even within, families would be reduced to the barest minimum. (Campbell assisted by Keogh, 1962: 9)

Television and the Australian Adolescent, the source of this quotation, finds its authors worried about the 'habits of passivity' that TV might induce, and the power of particular genres to 'instil certain emotions, attitudes and values'. The upshot of all this, they feared, might be 'a generation of people who are content to be fed by others' (Campbell assisted by Keogh, 1962: 23).

Each new media innovation since the advent of print has brought an expanded horizon of texts to audiences. In keeping with this history, television programmes and viewers come to be defined in both market terms and through a regulatory morality of conscience and taste, because 'a new practice of piety' accompanies each 'new communications technology' (Hunter, 1988: 220). As a consequence, moral panics are common amongst the denizens of communication studies, paediatrics, psychology, and education, who largely abjure cultural and political matters to do with television in favour of experimenting on its viewers. This is the psy-function at work.

Television studies also covers political economy, which focuses on ownership and control rather than audience responses. Because the demand for television is dispersed but its supply is centralized, political economy argues that TV is one more industrial process subordinated to dominant economic forces within society that seek standardization of production. Far from reflecting already-established and -revealed preferences of consumers in reaction to tastes and desires, television manipulates audiences from the economic apex of production, with coercion mistaken for free will. The only element that might stand against this levelling sameness is said to be individual consciousness. But that consciousness has itself been customized to the requirements of the economy and television programmes: maximization of sameness through repetition and minimization of innovation and newness in order to diminish risk and cost (Adorno and Horkheimer, 1977).

There are significant ties between critical theory, which calls for a resistive consciousness through artisanal rather than industrial created texts, and political economy, which calls for diverse ownership and control of the industry. The first trend is philosophical and aesthetic in its desire to develop modernism and the *avant-garde*, the second policy-oriented and political in its focus on institutional power. But they began as one with lamentations for the loss of a self-critical philosophical address and the triumph of industrialized cultural production. The two approaches continue to be linked via a shared distaste for what is still often regarded as mass culture (Garnham, 1987). TV is said to force people to turn away from precious artistic and social traces of authentic intersubjectivity as it takes control of individual consciousness. Like the psy-function, this part of television studies is frequently functionalist, neglecting struggle, dissonance, and conflict in favour of a totalizing narrative in which television dominates everyday life and is all-powerful.

Something happened in the mid-1960s to counter these two forms of knowledge via a more conflictual version of television studies. The medievalist, semiotician, columnist, and novelist Umberto Eco developed notions of encoding-decoding, open texts, and aberrant readings by audiences (Eco, 1972). He looked at differences between the way meanings were put into Italian TV programmes by producers and how they were deciphered by viewers. His insights were picked up by the political sociologist Frank Parkin (1971) and the cultural studies theorist Stuart Hall (1980).

There have been two principal methodological iterations of the encoding-decoding approach: uses and gratifications (U&G) and ethnography/cultural studies. Uses and gratifications operates from a psychological model of needs and pleasures; cultural studies from a political one of needs and pleasures. U&G focuses on what are regarded as fundamental psychological drives that define how people use the media to gratify themselves. Conversely, cultural studies' ethnographic work has shown some of the limitations to claims that viewers are stitched into certain perspectives by the interplay of narrative, dialogue and image. Together, they have brought into question the notion that audiences are blank slates ready to be written on by media messages.

Drawing upon these insights, some denizens of television studies argue that TV represents the apex of modernity, the first moment in history when central political and commercial organs and agendas became receptive to the popular classes. This perspective differs from the idea that the apparatus is all-powerful. It maintains instead that the all-powerful agent is the audience: the public is so clever and able that it makes its own meanings, outwitting institutions of the state, academia and capitalism that seek to measure and control it. In the case of children and the media, anxieties about turning Edenic innocents into rabid monsters or capitalist dupes are dismissed.[6]

Television supposedly obliterates geography, sovereignty and hierarchy in an alchemy of truth and beauty. Today's deregulated, individuated TV world allegedly makes consumers into producers, frees the disabled from confinement, encourages new subjectivities, rewards intellect and competitiveness, links people across cultures, and allows billions of flowers to bloom in a post-political cornucopia. It's a kind of Marxist/Godardian wet dream, where people fish, film, fornicate, frolic and finance from morning to midnight.

Sometimes, faith in the active audience reaches cosmic proportions, such that TV is not responsible for – well, anything. Consumption is the key – with production discounted, work neglected, consumers sovereign and research undertaken by observing one's own practices of viewing and one's friends and children. This is narcissography at work, with the critic's *persona* a guarantor of assumed audience revelry and Dionysian joy (Morris, 1990). Welcome to 'Readers Liberation Movement' media studies (Eagleton, 1982): everyone is creative and no one is a spectator. Internally divided – but happily so – each person is 'a consumer on the one hand, but … also a producer' (Foucault, 2008: 226).

In one sense, then, TV studies buys into corporate fantasies of control – the political economist's arid nightmare of music, movies, television and everything else converging under the sign of empowered firms. In another, it incarnates individualist fantasies of reader, audience, consumer, or player autonomy – the libertarian intellectual's wet dream of music, movies, television, and everything else converging under the sign of empowered fans. Those antinomies shadow the fetish of innovation that informs much talk of media technology and consumerism while ignoring the environmental destruction and centralized power that underpin them (Maxwell and Miller, 2012).

In its uneasy blend of these perspectives, television studies has been partially colonized by a progressive agenda that is sceptical without being cynical, rigorous without losing optimism, and committed to popular democracy. The result has seen analyses devoted to some key issues that go beyond the psy-function, political economy and active audiences, while drawing on their insights. A brief list might include: feminist concerns over the representation of women, both on- and off-screen; critics' desires to reach beyond *bourgeois*-individualistic accounts of creativity in favour of generic analysis; studies of post-imperial social control in the Global South via domestic and global media dominance; Marxist aesthetics reading story against ideology; and voices from below, heard through the participant observation of workers and audiences. Such staples as cultural imperialism critique and national television history have been supplemented by work on national, regional, global, diasporic, First Peoples, women's, and activist television. Ideology critique has been enriched by Gramscianism, racialization analysis, queer theory and policy studies. This is in keeping with intellectual developments and political trends, such as social movements, the globalization and privatization of television in the wake of the Cold War, and the rise of neoliberalism. Foundational debates since 1990 have put leftist, queer, disabled, feminist, multicultural and post-colonial formations in play.

As higher education has grown and opened up to these critical tendencies within the human sciences, social movements, and more instrumental, conservatory-style training, TV studies has often been deemed simultaneously too progressive and too applied by many traditionalists. Unlike, for instance, neoclassical economics or the psy-function, media studies has attracted some intense opprobrium.

Within the *bourgeois* media, the *Village Voice* dubs TV studies 'the ultimate capitulation to the MTV mind … couchpotatodom writ large … just as Milton doesn't belong in the rave scene, sitcoms don't belong in the canon or the classroom' (Vincent, 2000). For the *Times Literary Supplement*, media and cultural studies form the 'politico-intellectual junkyard of the Western world' (Minogue, 1994: 27). The *Wall Street Journal* describes our work as 'deeply threatening to traditional leftist views of commerce,' because its notions of active consumption are close to those of the right: 'cultural-studies mavens are betraying the leftist cause, lending support to the corporate enemy and even training graduate students who wind up doing market research' (Postrel, 1999). The *Observer* scornfully mocks via a parental parody: 'what better way to have our little work-shy scholars rushing off to read an improving book than to enthuse loudly in their presence about how the omnibus edition of *EastEnders* [1985–] is the new double physics?' (Hogan, 2004). The *Telegraph* thunders that media studies is 'quasi-academic' and delights whenever enrolment diminishes (Lightfoot, 2005; Cairns, 2013), while the *Daily Express* deems it 'worthless' (Douglas, 2011).

Alan Sugar, UK inquisitor for *The Apprentice* (2005–) worries that TV studies 'may be putting future scientific and medical innovation under threat' and 'undermining the economy' (Paton, 2007). Britain's former Inspector of Schools denounces it as 'a subject with little intellectual coherence and meager relevance to the world of work' (Woodhead, 2009). Chris Patten, the last Governor of Hong Kong and a former Chair of the BBC Trust, called us 'Disneyland for the weaker minded' (quoted in Morley, 2007: 17) and the late *Guardian* newspaper columnist Simon Hoggart could be seen on British television in 2000 chiding local universities for wasting time on this nonsense when they should be in step with Harvard and MIT.[7] Similar criticisms come from Adam Gopnik in the *New Yorker*, who expresses alarm that so much 'energy on the American left is in cultural studies, not health care' (1994: 96). National Public Radio places us among the 'softer majors' whose *alumni* are fated to 'pretty low salaries, and that's where the trouble really is' for 2014's 'nearly two million [US] college graduates' (Editor, 2014).

In 2014, the British government's Office of Qualifications and Examinations deemed media studies to be one of the country's high school subjects that must be made more 'demanding' and subject to 'reform' or be discontinued within four years, despite the fact that over 55,000 pupils took it (high numbers in comparison with most other areas, but heading downwards as attacks on it gained traction) (Ofqal, 2014). The squeals of delight from the *bourgeois* press were immediate: the *Daily Mail* singled out media studies in welcoming a 'crackdown on easy subjects' via a 'bonfire' (Harris, 2014), the *Guardian* highlighted the drive to make our field 'new, tougher' (Adams, 2014), and the *Telegraph* headlined the notion of being 'dramatically toughened up' (Paton, 2014). The *Independent* also used this hyper-masculinist language, describing the reforms as directed at 'soft' subjects (Dugan, 2014).

Within academia, MeCCSA argues that media studies gets 'negative publicity' because it 'involves studying things which are generally seen as entertaining but trivial' or 'by using complicated theoretical language' (www.meccsa.org.uk/). Robert W. McChesney laments that we are 'regarded by the pooh-bahs in history, political science, and sociology as having roughly the same intellectual merit as, say, driver's education' (2007: 16). Similar attitudes abound within the humanities (Hilmes, 2005: 113). *The Simpsons and Philosophy* (Irwin et al., 2001), probably the biggest-selling academic work ever on TV, sold a quarter of a million copies within six years without any relationship to work done in television studies (Asma, 2007).

The news is not all bad! It is worth recalling that new subject areas tend to cause controversy when they enter universities, as per the British experience with the introduction of the natural sciences in the 19th century, and politics, philosophy, English and sociology in the 20th. These innovations were practical responses to major socio-economic transformations – industrialization, state schooling, class mobility and public welfare (Fox, 2003; Whittam Smith, 2008). They were far from welcome at the time of their advent, yet have become key elements of a liberal education since. As the *Scotsman* acknowledged in its defence of studying the media, 'Mickey Mouse is no cow'rin, tim'rous beastie' (Instrell, 2014).

Moreover, within the industry, Greg Dyke, a leading British media and sports executive, says we are good for both citizenship and professional awareness (Burrell, 2008) and Britain's communications regulator Ofcom has a *Media Literacy E-Bulletin*, amongst other initiatives, while UK employment data show very positive signs for media studies graduates, albeit at low rates of pay (Anderson, 2013; Office for National Statistics 2013: 19). Mexican *telenovelas*, now seen in more than a hundred countries, are researched and revised by TV Azteca via a blend of genre study and *análisis semántico basado en imagines* (semantic analysis based on images) which uses viewer interviews to uncover cultural responses to stories as they unfold. Data and analysis from television studies help to determine plot lines (Clifford, 2005; Slade and Beckenham, 2005: 341 n. 1).

Away from debates about the value of studying the media, it is abundantly clear that TV causes anxiety, whether about its own audiences or audiences to classes about it. Why? Because television is powerful. As Smythe said sixty years ago, it channels an immense 'flow of representations of the human condition' (1954: 143). John Hartley suggests that 'the energy with which audiences are pursued in academic and industry research' is 'larger and more powerful than the quest for mere data', because it seeks 'knowledge of the *species*' (1992: 84; also see Ang, 1991).

TV has probably received the greatest attention (and frequently demonization) of any cultural medium. The opulence of television as a technology is supposedly matched by a barrenness of civilization. Critics find it responsible for deficits in knowledge, concentration and responsibility among populations. Aesthetically, it is said to appeal to base instincts and lowest-common denominators. Politically, it is seen as instilling either quietude or hysteria. Criticisms come from both left and right that a surfeit of signage and a deficit of understanding cheapen public culture, as *Kitsch* overruns quality (Martín-Barbero and Rey, 1999: 15–16, 22, 24).

Most television programming has been dedicated to entertainment, and that focus, along with the ease of use and the double pull of vision and sound, have long produced embarrassment and even shame on the part of some viewers. Consider this Ivy League product recalling his New Haven follies of 1953:

> In those days a Yale faculty member who owned a television set lived dangerously. In the midst of an academic community, he lived in sin. Nevertheless, in an act of defiance, we put our television set in the living room instead of the basement or the garage where most of the faculty kept theirs, and we weathered the disapprobation of colleagues who did not own or would not admit to owning this fascinating but forbidden instrument. (Silber, 1968: 113)

Communitarian US sociologists contrast an allegedly active public with a putatively passive audience:

> we are not happy when we are watching television, even though most of us spend many hours a week doing so, because we feel we are 'on hold' rather than really living during that time. We are happiest when we are successfully meeting challenges at work, in our private lives, and in our communities. (Bellah et al., 1992: 49)

US producer David Susskind confided to 1950s readers of *Life* magazine that he was 'mad at TV because I really love it and it's lousy. It's a very beautiful woman who looks abominable' (quoted in Schramm et al., 1961: 3).

This sexist metaphor exemplifies the seemingly ineradicable fear of televisual degradation. Such nightmares about TV have never receded. Fifty years later, then-Fox Entertainment president and former NBC executive Kevin Reilly said 'NBC is like the crazy ex-wife I can't get away from' (quoted in Friedman, 2009).[8]

THE END OF TELEVISION? OR A NEW BEGINNING?

But perhaps he *can* get away from television. For, despite the level of policy and scholarly activity swirling around the medium, many people today say that TV is finished, that it no longer matters. The rhetoric of the newer audiovisual media, notably mobile telephones and the internet in general, is inflected with the phenomenological awe of a precocious child set to heal the wounds of modern life, magically reconciling public and private, labour and leisure, commerce and culture, citizenship and consumption. The alleged upshot? *La televisión ha muerto* [*Television is dead*] (De Silva, 2000). A synoptic survey in the *Annual Review of Sociology* proposes that 'changes in the medium threaten to make past research on TV appear quaint and anachronistic' (Grindstaff and Turow, 2006: 103). The grand organizer of daily life over half a century has lost its proud place in the layout of homes and the daily order of drama and data.

Dual monopolies have been broken – the physical object no longer dominates, and nor does its model of unidirectional production. TV has lost its identity. We must all say 'Bienvenidos al mundo de la postelevisión' [Welcome to the post-television world] (Verón, 2008).

Jinna Tay and Graeme Turner have coined the terms 'broadcast pessimism' and 'digital optimism' (2010: 32) to encapsulate two differing positions on the future of television. Proponents of broadcast pessimism argue that we are witnessing the inexorable obsolescence of traditional TV – the television of sharedness, of family togetherness – under the disrupting, disuniting impact of media digitization. Digital optimists, by contrast, welcome a post-broadcast era because it offers an unprecedented range of content and allows audiences unrestrained time, space and modes of access to an array of platforms, screens and outputs, which is deemed to satisfy individual needs for choice and control.

Time magazine exemplified this fetish when it chose 'You' as 2006's 'Person of the Year,' because 'You control the Information Age. Welcome to your world' (Grossman, 2006). The *Guardian* is prey to the same touching magic: someone called 'You' headed its 2013 list of the hundred most important folks in the media (www.theguardian.com/media/2013/sep/01/you-them-mediaguardian-100-2013).

This discourse buys into fantasies of reader, audience, consumer, player and activist autonomy. True believers invest with unparalleled gusto in Schumpeterian entrepreneurs, evolutionary economics, creative industries and internet politics. They've never seen an 'app' they didn't like, or a socialist party they did. Faith in devolved media-making amounts to a secular religion, offering transcendence in the here and now via a 'literature of the eighth day, the day after Genesis' (Carey, 2005).

The new era seems to have defanged media and political gatekeepers and restored popular culture to its original, organic quality as the property of ordinary people. Customs override consumption, capitalism is no longer corporate, and citizens govern suzerains. Netflix proclaims that '[i]nternet TV is replacing linear TV. Apps are replacing channels, remote controls are disappearing, and screens are proliferating' (2013). IBM disparages 'Massive Passives … in the living room … a "lean back" mode in which consumers do little more than flip on the remote and scan programming'. By contrast, it valorizes and desires 'Gadgetiers and Kool Kids' who 'force radical change' because they demand 'anywhere, anytime content' (2006: 1, 10).

Yet the evidence for television's demise is sparse and thin. Historically, most new media have supplanted earlier ones as central organs of authority or pleasure: books versus speeches, films versus plays, singles versus sheet music. But TV blended all of them. A warehouse of contemporary culture, it merged what had come before, and now it is merging with personal computers (which were modelled on it) to do the same (Newcomb, 2005: 110; Standage, 2006). The *New York Times* presciently announced this tendency over thirty years ago with the headline 'Television Marries Computer' (Gardner, 1983).

New technologies *are* changing contemporary culture and communication. For example, the corporate music industry is shrinking due to the internet, with massive decreases in compact disc sales, and the internet is displacing physical newspapers in the Global North as a key source of political and commercial information. But corporate and public TV continues to grow by every indicator imaginable (Friedman, 2008; Pew Research Center, 2008; Spangler, 2009). *La Tempestad* magazine's 2008 dossier on the subject asked, for instance, 'Está en la televisión el futuro del cine?' [does the future of cinema lie in TV?].

It remains the case that establishing, maintaining, and upgrading a national television service is a fundamental part of successful government. Television still occupies vast amounts of people's time and money, because it delivers information and entertainment with astonishing speed and ease. There are tens of thousands of broadcast, cable and satellite TV stations: over 7,000 in Russia; 3,000 in China; 2,700 in the European Union; and 2,200 in the US. Worldwide subscriptions to television via satellite and cable surged by 8% to 800 million in 2012, when global TV revenue grew by 4.1%, to £252 billion, thanks to these increases and advertising. The Global Internet TV portal lists over 8,000 TV stations available on the web in 2014 (Central Intelligence Agency, 2007; Euroconsult, 2008; Bilbao-Osorio et al., 2013; Friedman, 2013; Ofcom, 2013; www.global-itv.com/).

The Organisation for Economic Co-operation and Development (OECD), the top body of advanced capitalist democracies from the European Union to Australasia, says its members have witnessed an explosion of television stations across the first years of the 21st century, mostly via satellite and cable: from 816 in 2004 to 1,165 in 2006 – 43% growth (OECD, 2007: 175).

The number of Indian TV households grew by 11 million in 2012 (FICCI-KPMG, 2013). Residents are likelier to own television sets than have access to indoor plumbing, and politicians devote advertising money to TV ahead of all other options, drawing on its confessional qualities via close-ups and generic religiosity to appeal to voters in a non-secular, highly personalized, way (Rajagopal, forthcoming). In Mexico, as digital media proliferate, so does TV – always, and if anything increasingly, the dominant medium (Gomez et al., 2011).

Consider the United States, which is often a harbinger of media futures (for better or worse); as the editors of *A European Television History* put it, 'American television becomes a horizon towards which all television seems to progress' (Bignell and Fickers 2008: 4). In 2006, more than 98% of US homes had at least one television set, while 64% had cable, up twenty points in twenty years. Consumers spent US$20 billion buying new TVs that year. By 2007, 51% of people had three or more sets, a figure that has remained constant since (Motion Picture Association of America, 2007: 35, 37; Borland and Hansen, 2007; Ellis, 2007: 40; De Roo et al., 2013). In 2008, the number of US households owning televisions increased by 1.5%, with particularly significant growth among migrants and their recent descendants. Nielsen suggests 115.6 million US homes had sets in 2013–14, an increase on 2012–13's 114.2 million ('Nielsen Estimates', 2013). In the last twenty years, the number of black households with TVs grew nearly 40%, while the figure for Latin@ homes doubled (Nielsen, 2014).

The US population watched more TV in 2005 than a decade earlier – *an hour more* than in that basically pre-web era. Children between the ages of six and 14 were tuned to television at rates unprecedented for twenty years; 69% of them had sets in their bedrooms, versus 18% with internet access and 49% owning or subscribing to videogames (Pew Research Center, 2005). Children between the ages of two and 11 watched 17.34 hours of TV a week in 2006, an increase on the previous year; in Britain, the number was in excess of 15 hours. The keenest viewers in the US are young girls. They quite like new technology, and adopt it at a frenetic pace—but 'TV is king', in the words of the old song by *The Tubes*. Teens have always watched less television than other age groups, but the amount they view has barely changed from 1953 to 2013. Meanwhile, 93% of adults watch at least an hour of TV a day, but just 4% watch an hour of video on line each day (Downey, 2007; 'Majority of Americans', 2007; Ofcom, 2009: 109; Thompson, 2013).

AOL Television and the Associated Press polled US residents on their viewing habits in 2007. Over a quarter of the population said they watched more than three hours a day, while 13% watched more than 30 hours a week, up five points on 2005. And in 2008, the most statistically significant change in how people passed their leisure time was the increase in time watching TV. Three-quarters of people had viewed television at one time or another on line – but they spent seventy times more hours a month doing so via a conventional set ('Precious Little', 2008; 'What Impact', 2008). Ratings disclose that the average household watched eight and a quarter hours of television and individuals four and a half hours daily in 2006 – record numbers. Even the venerable if unvenerated Academy Awards saw a ratings increase of 13% in 2009 (Grindstaff and Turow, 2006: 119; 'Nielsen Media', 2006; Cabletelevision Advertising Bureau, 2007; Rash, 2009).

In 2004, people in OECD countries spent a third of their time using the media watching television and a fifth on line. Although other OECD members exhibit only half the loyalty to television of US viewers, it remains the most popular medium, and viewing time has remained consistent over the decade since (OECD, 2007: 177, 2013: 184). In Britain:

> [i]n 2013, the average … viewer watched three hours and 52 minutes of linear television a day on a TV set, and just three minutes and 30 seconds on other devices such as tablets, smartphones and laptops. Viewing on devices other than TVs of video on demand services like ITV Player, Sky Go and the BBC iPlayer accounted for just 1.5% of overall TV consumption. That was up from 1.2% in 2012, but it is still a minority pursuit. (Cellan-Jones, 2014)

People with digital video recorders barely use them. Audiences turned more and more to television as the global recession deepened, with distinct increases in ratings across 2008 (Ellis, 2007: 23; Graf, 2008; Fitzsimmons, 2009; Gray, 2009; Cellan-Jones, 2014). In Australia, even a study that seems intent on demanding that multiple screens are the reality (almost bringing them into being through the will of desire) lies down in a post-orgasmic froth of exhaustion and ruefully admits that 'all age groups continue to spend the majority of their screen time with the in-home TV set' ('Nielsen: Ten', 2013).

Of the top hundred brands recognized by US residents in 2009, 16 were TV or film-related, with CNN and MTV in the top 10 (Reynolds, 2009). Half the internet sites that US children aged between six and 11 visit first attracted their attention through advertising on TV or in print. Right across the age spectrum, television is the most influential advertising medium, more than during the pre-web period. Hundreds of case studies undertaken over the past two decades confirm that it is the principal means of raising brand awareness through advertising (Gonsalves, 2008; 'Nielsen Reports TV', 2008; 'Nielsen Reports Growth', 2008; 'Kids Motivated,' 2009; Neff, 2009; Shields, 2009; Thomasch, 2009).

Thanksgiving 2013 saw US residents in the hundreds of millions watching an average of 15 hours of television per person. That year's Macy's Day Parade, which commenced at 9 am eastern, has seen audiences increasing ever since 2001. Nineteen million hung in there in 2013 to watch the dog show that followed NBC's official coverage (Lynch, 2013).

Just weeks after the national holiday, the Consumer Electronics Show ushered in 2014. For all its cult of newness, the Las Vegas convention acknowledged, yet again, that as far as advertisers are concerned, television remains the holy grail. It moves people, which means it moves products. Even digital specialists amongst marketers faced the truth – people keep watching television, on a set, at home, with other people, based on the schedule constructed and constricted by networks and cable companies. Advertising agencies recognize that college students are promiscuous viewers, but there's nothing new in that. As always, as thirty years ago, once they graduate and get jobs, they subscribe to satellite or cable. They don't *cut* cords; they *order* them (Poggi, 2014). This isn't just about finding TV 'OK'. People even like it more than they used to ('Deloitte', 2009).

Ninety-four per cent of the population watch TV news, which for years has remained their principal resource for understanding both global events and council politics. Political operatives pay heed to the reality (Saad, 2013). During the 2004 US Presidential election, 78% of the population followed the campaign on television, up from 70% in 2000 (Pew Research Center for the People and the Press, 2005; Project for Excellence in Journalism, 2005). Between the 2002 and 2006 mid-term elections, and across that 2004 campaign, TV expenditure on political advertising grew from US$995.5 million to US$1.7 billion – at a time of minimal inflation. That amounted to 80% of the growth in broadcasters' revenue in 2003–4 (vanden Heuvel, 2008: 34). The 2002 election saw US$947 million spent on TV election advertising; 2004, US$1.55 billion; and 2006, US$1.72 billion. The correlative numbers for the internet were US$5 million in 2002;

US$29 million in 2004; and US$40 million in 2006 (Gueorguieva, 2007). The vast majority of electronic electoral campaigning takes place on local TV – 95% in 2007 (Bachman, 2007; TNS Media Intelligence, 2008).

Consider the famous Barack Obama campaign of 2008 and its much-vaunted use of the internet. Obama's organization spent the vast majority of its energy and money on television. The internet was there to raise funds and communicate with supporters. The US Presidency cycles with the summer Olympics, broadcast by NBC, but few candidates commit funds to commercials in prime time during this epic of capitalist excess, where the classic homologues of competition vie for screen time – athletic contests versus corporate hype. Obama, however, took a multi-million-dollar package across the stations then owned by General Electric: NBC (Anglo broadcast), CNBC (business-leech cable), MSNBC (news cable), USA (entertainment cable), Oxygen (women's cable), and Telemundo (Spanish broadcast). TV was on the march, not in retreat: on election night 2008, CNN gained 109% more viewers than the equivalent evening four years earlier (Gough, 2008; Teinowitz, 2008).

The 2012 US Presidential election was again a televisual one. How many US residents who watched the debates between Mitt Romney and Obama preferred the internet to TV as their source? Answer: 3%. How many watched on both TV and the internet? Answer: 11%. How many people shared their reactions online? Answer: 8% (Pew Research Center for the People and the Press, 2012).

In Europe as well as the US, TV rules the roost by a long way when viewers seek news (Eggerton, 2014; Newman and Levy, 2013). Worldwide, owners of tablets such as iPads are the keenest consumers of television news worldwide. These tablets are adjuncts, gadget partners, to the main source. If anything, they stimulate people to watch more television ('BBC World', 2013).

This is not to deny that change is afoot, as Lynn Spigel explained a decade ago:

> [I]ncreasing commercialization of public-service/state-run systems, the rise of multichannel cable and global satellite delivery, multinational conglomerates, Internet convergence, changes in regulatory policies and ownership rules, the advent of high-definition TV, technological changes in screen design, digital video recorders, and new forms of media competition – as well as new forms of programming … and scheduling practices … have all transformed the practice we call watching TV. This does not mean all of television is suddenly unrecognizable – indeed, familiarity and habit continue to be central to the TV experience—but it does mean that television's past is recognizably distinct from its present. (2005: 83)

As intimated earlier, the signs are that the internet will merge with television, and the two technologies will transform one another. In the words of Steve Ballmer, Microsoft's former chief executive, 'we will see TVs become more sophisticated and more connected. The boundary between the PC and the TV will dissolve' (quoted in Moses, 2009).

Without denying the substantial change brought about by the spread of digital technologies and deregulation, I challenge the universalizing claim that broadcast, cable, and satellite television are over, given multiple signs that TV is alive and well in most countries around the world, and holds a central, even dominant cultural position.

Television 'seems designed, no matter what its platform of delivery, to generate new ways of being-together-while-apart' (Pertierra and Turner 2013: 66). In Horace Newcomb's view, 'the future of television will be essentially the same as its past' via 'strategies of adjustment' to change (2014). TV remains the dominant source of our truth and object of our consumption, in dual senses – the sets cost more, and we spend more time with them than other devices. All in all, there simply isn't evidence that newer technologies have displaced or are displacing the traditional cultural *bodega* of the last half-century.

CONCLUSION

I'll close this introduction with further inspiration from our late co-editor. Manuel Alvarado was animated by a scholarly and political impulse to unearth 'the actual and potential cracks, fissures, dislocations, and absences that exist within television and television programmes' and ask '*how* things could be done differently' in 'oppositional space' (Alvarado, 1983). And he called for an increasingly comprehensive interdisciplinarity to cross boundaries in just the way that people, programmes, potentates and proprietors did (Alvarado, 2009a).

TV studies can follow his example by teasing out the manifold complexities of media production, meaning and reception in order to indicate their socio-cultural significance. To do so requires a sometimes bewildering array of skills, which I endeavour to draw on here by borrowing from numerous analysts in order to offer a tentative example. My case study is *24*, one of the longest-running and most internationally successful US spy shows (2001–10, revived in 2014 as *24: Live Another Day*).

24 began in the fateful northern-hemisphere fall of 2001, right after airplane missiles had struck the northeast of the US. By 2009, a hundred million people were watching it across the world on 236 channels. The programme binds together two senses of realism in a classic dual verisimilitude that draws both on faithfulness to a genre (espionage) and on narrative cues, images, sounds and editing that are frequently associated with documentaries or news programmes. This is in keeping with its central conceit of a season's action taking place over the 24 hours it takes to watch each season of episodes.

24 has been welcomed as a return of high-quality drama that runs counter to the hegemony of reality television, and even celebrated as a grand piece of existential philosophy – a solitary figure standing against an array of untrustworthy institutions. Yet it clearly borrows devices and storylines from more critically derided genres, such as soap opera, reality and vigilante action adventure, thanks to its cliff-hanger episodic stories and *macho* violence – in addition to pirating the *avant-garde*, courtesy of fractured storylines and points of view. Of course, this is all underwritten by corporate messages; the first episode of 2003 began and ended with a six-minute film promoting a Ford car. And *24*'s uniqueness became a formula when CBS announced *Harper's Island* (2008–9), an overtly self-destructing series in which viewers were guaranteed that a central character would die each week (Attallah, 2006; Lotz, 2008: 173; McMahon, 2008; McPherson, 2008; Miklos, 2008; Aitkenhead, 2009; Steinberg, 2009; Shimpach, 2010).

Then there is the question of *24*'s politics. Produced by Republicans (the show's creator, Joel Surnow, boasts of being a 'rightwing nut job' [quoted in Aitkenhead, 2009]), it has featured cameos by their ideological *confrères* in Congress (John McCain) and the news media (Laura Ingraham and Larry Elder) and was endorsed by such intellectual lackeys of the George W. Bush regime as the *ur*-disgraced-academic John Woo, who wrote legal justifications for inhumane brutality. The Heritage Foundation, a reactionary, coin-operated think tank, held a press conference in 2006 in celebration of the series that featured Michael Chertoff, the Secretary of Homeland Security, and talk-radio host Rush Limbaugh, who announced that then-Vice-President Dick Cheney and Defense Secretary Donald Rumsfeld were fans of the programme (Lithwick, 2008).

24 clearly endorses torture as a means of extracting information from terrorists. For some critics, it represents 'la suma de los miedos americanos' [the sum of American fears] (Miklos, 2008: 79). John Downing has termed the programme 'the most extended televisual reflection to date on the implications of 9/11' and an egregious argument in favour of the 'need' for immediate and illegal action in the 'public interest' (2007: 62). It's fine for the hero, Jack Bauer, 'a man never at a loss for something to do with an electrode', to deny medical assistance to a terrorist whom he has wounded, shoot another's wife in the leg, then threaten a second shot to the knee unless her husband confides in him; and fine for the US President to subject a Cabinet member to electric shocks to interrogate him (Downing, 2007: 72, 77; Lithwick, 2008) as Bauer endlessly intones 'Whatever it takes.'

A delegation from the major US officer-training site, West Point, visited *24*'s producers in 2007 to express anxiety that many military recruits had adopted illegal, immoral attitudes to torture based on their interpellation by the series, while interrogators reported a direct mimesis between the show and actual practices in Iraqi prisons by US forces inspired by it. Human Rights Watch also weighed in. And when the programme returned in 2014, Amnesty International noted its popularity in African nations that have not outlawed torture and claimed that its glamorization of the activity desensitized viewers internationally. But executive producer Kiefer Sutherland, the highest-paid TV actor in the world, who is liberal in his politics, disavows the notion that the programme works ideologically at all – 'it's good drama. And I love this drama!' Further, *24* became the first carbon-neutral US TV fiction show in 2009, with offsets calculated against the impact of car chases, air travel, and use of coal-generated electricity, in addition to favouring wind and solar power (Miller, 2007; Kaufman, 2009; Sutherland quoted in Aitkenhead, 2009; Glez, 2014; Toomer, 2014; 'Torture Spreading', 2014). Finally, it is worth noting that in the US, *24*'s staple audience was highly educated and affluent (Sconce, 2004: 99) – not people usually associated with the violence and anti-intellectualism of the programme's far-right producers and public supporters.

24 is an immensely rich, vulgar, polysemic, contradictory, educational, misguided, misguiding text. It is laden with imperialistic, messianic, vigilante messages – and underwritten by progressive ecology and a liberal star. Thank heavens for Stella Artois' Godardian spoof (www.theguardian.com/media/video/2009/mar/23/stella-artois-viral-ad). Manuel would have appreciated the skit – and the need to blend political-economic, textual and audience knowledge to understand what gave rise to *24* in the first place and animated its successes.

Television's patron saint, Clare of Assisi, a teen runaway from the 13th century, was the first Franciscan nun. She was canonized in 1957 for her prescient bed-ridden vision of images from a midnight mass cast upon a wall, which Pius XII decreed to have been the first TV broadcast (www.catholic.org/saints/saint.php?saint_id=215#wiki; Pius XII, 1958). Perhaps Clare would smile benignly at today's television studies, given her commitment to truth and love. We hope this book does justice to Manuel's vision as well as hers – and to television itself.

NOTES

1 While I am the author of this piece, it has been read by Herman and Milly, who have both *im*proved and *ap*proved it.
2 His published work was nonetheless extensive (Miller, 2010).
3 A more scientific probe suggests that the fetishization of art from industry is an incomplete project in the US *bourgeois* media's coverage of TV (Bielby, 2014). I hope this remains the case.
4 As we have seen, Manuel Alvarado was centrally involved. In the US, the challenge from the left was more muted, with disciplines dedicated to the industry and normal science.
5 The latter characterized publishing projects that Manuel worked on, notably *Screen Education*.
6 Perhaps there should be more of a focus on the material dangers posed to children by TVs: well over 385,000 young people have been admitted to US emergency rooms in the last two decades due to physical injuries caused by sets (De Roo et al., 2013).
7 As Harvard long hosted a journal of media studies (the ungainly-titled *Harvard International Journal of Press/Politics*, now thankfully free of its oxymoronic Yanqui moniker) and a New Approaches to International Law

colloquium that engaged with cultural studies, and MIT has held major conferences called 'Media In Transition' to trope its acronym, Hoggart's dismissiveness was ill-informed – but quite representative.

8 Calm down, Kevin. Women continue to be excluded from key positions in TV newsrooms around the globe in favour of people like you (Byerly, 2013). In any event, since you lost your job in 2014 as Fox ratings descended into the mire, perhaps you'll find more time to appreciate NBC (James, 2014).

REFERENCES

Adams, Richard (2014, June 4). 'Home Economics GCE Set to be Scrapped'. *Guardian* www.theguardian.com/education/2014/jun/04/home-economics-gcse-scrapped-a-levels

Adorno, Theodor W. and Max Horkheimer (1977). 'The Culture Industry: Enlightenment as Mass Deception'. *Mass Communication and Society*. Ed. James Curran, Michael Gurevitch and Janet Woollacott. London: Edward Arnold. 349–83.

Aitkenhead, Decca (2009, February 2). 'One Hour with Kiefer Sutherland'. *Guardian* www.theguardian.com/lifeandstyle/2009/feb/02/kiefer-sutherland-interview-tv-film

Alvarado, Manuel (1977). 'Class, Culture and the Education System'. *Screen Education* 22: 49–55.

Alvarado, Manuel (1983). 'The Question of Media Studies'. *Meridian* 2(1): 182–4.

Alvarado, Manuel (1996). 'Selling Television'. *Film Policy*. Ed. Albert Moran. London: Routledge. 62–71.

Alvarado, Manuel (2000). 'The "Value" of TV Drama: Why Bother to Produce it?' *Television & New Media* 1(3): 307–19.

Alvarado, Manuel (2009a). 'Rethinking the Media Curriculum'. *Television & New Media* 10 (1): 10–12.

Alvarado, Manuel (2009b). 'Afterword'. *TV Formats Worldwide: Localizing Global Programs*. Ed. Albert Moran. Bristol: Intellect. 309–11.

Alvarado, Manuel and Edward Buscombe (2008). 'Cultural Strategies: Publishing at the British Film Institute'. *Cinema Journal* 47(4): 135–40.

Alvarado, Manuel and John O. Thompson (1990a). 'General Introduction'. *The Media Reader*. Ed. Manuel Alvarado and John O. Thompson. London: British Film Institute. 1–4.

Alvarado, Manuel and John O. Thompson. (1990b). 'Untitled'. *The Media Reader*. Ed. Manuel Alvarado and John O. Thompson. London: British Film Institute. 105.

Alvarado, Manuel, Edward Buscombe and Richard Collins (1993). 'Introduction'. *The Screen Education Reader: Cinema, Television, Culture*. Ed. Manuel Alvarado, Edward Buscombe and Richard Collins. New York: Columbia University Press. 1–22.

Anderson, Elisabeth (2013, November 19). 'Media Studies "The UK's Second-Most Employable Degree"'. *Management Today* www.managementtoday.co.uk/news/1221489/

Ang, Ien (1991). *Desperately Seeking the Audience*. London: Routledge.

Asma, Stephen T. (2007, October 12). 'Looking Up From the Gutter: Philosophy and Popular Culture'. *Chronicle of Higher Education* www.chronicle.com/article/Looking-Up-From-the-Gutter-/13840

Attallah, Paul (2006). Review. *Canadian Journal of Communication* 31(4) www.cjc-online.ca/index.php/journal/article/view/1706/1826

Bachman, Katy (2007, October 15). 'Candidates Still Favoring Local TV Ads'. *MediaWeek* www.adweek.com/?vnu_content_id=1003658398&utm_source=feedburner&utm_medium=feed&utm_campaign=Feed%253A+Mediaweek-Tv-Radio-Stations-And-Outdoor+%2528Mediaweek+News+-+TV%252C+Radio+Stations+and+Outdoor%2529

Bellah, Robert N., Richard Madsen, William M. Sullivan, Ann Swidler and Steven M. Tipton (1992). *The Good Society*. New York: Alfred A Knopf.

Bielby, Denise (2014). 'Who's to Say? Why Television Criticism is Complicated'. *Flow* 19(11) www.flowtv.org/2014/05/who%E2%80%99s-to-say-why-television-criticism-is-complicated-denise-bielby-university-of-california-santa-barbara/

Bignell, Jonathan and Andreas Fickers (2008). 'Introduction: Comparative European Perspectives on Television History'. *A European Television History*. Ed. Jonathan Bignell and Andreas Fickers. Malden: Wiley-Blackwell. 1–54.

Bignell, Jonathan, Stephen Lacey and Madeleine Macmurraugh-Kavanagh, eds (2000). *British Television Drama: Past, Present and Future*. Houndmills: Palgrave.

Bilbao-Osorio, Beñat, Soumitra Dutta and Bruno Lanvin, eds (2013). *The Global Information Technology Report: Growth and Jobs in a Hyperconnected World*. Geneva: World Economic Forum/INSEAD.

Boddy, William (2005). 'In Focus: The Place of Television Studies'. *Cinema Journal* 45(1): 79–82.

Bolas, Terry (2009). Screen Education: From Film Appreciation to Media Studies. Bristol: Intellect.

Borland, John and Evan Hansen (2007, April 6). 'The TV is Dead. Long Live the TV'. *Wired* http://wired.com/entertainment/hollywood/ news/2007/04/tvhistory_0406

Bourdieu, Pierre (1998). *On Television*. Trans. Priscilla Parkhurst Ferguson. New York: New Press.

Burrell, Ian (2008, March 17). 'Greg Dyke's Back: The Former BBC Director General on His New Job – And What's Wrong with Britain's Media'. *The Independent* www.independent.co.uk/news/media/greg-dykersquos-back-the-former-bbc-director-general-on-his-new-job-ndash-and-whatr-squos-wrong-with-britainrsquos-media-796749.html

Buscombe, Edward (2010, June 1). 'Manuel Alvarado Obituary'. *Guardian* www.theguardian.com/media/2010/jun/01/manuel-alvarado-obituary

Byerly, Carolyn L. (2013). *Global Report on the Status of Women in News Media*. Washington: International Women's Media Foundation.

Cabletelevision Advertising Bureau (2007). *Power of the Young*. New York: CAB.

Cairns, Richard (2013, August 16). 'Gloomy Times are Pushing Today's Pupils in Right Direction'. *The Telegraph* www.telegraph.co.uk/education/universityeducation/10247291/Gloomy-times-are-pushing-todays-pupils-in-right-direction.html

Campbell, W.J. assisted by Rosemary Keogh (1962). *Television and the Australian Adolescent: A Sydney Survey*. Sydney: Angus and Robertson.

Carey, James W. (2005). 'Historical Pragmatism and the Internet'. *New Media & Society* 7(4): 443–55.

Cellan-Jones, Rory (2014, February 17). 'TV's Changing? Not So Fast'. *BBC News* www.bbc.co.uk/news/technology-26224625

Central Intelligence Agency (2007). 'The World Factbook Page on Field Listing, Television Broadcast Stations'. www.cia.gov/library/publications/the-world-factbook/fields/2213.html

Clifford, Reginald (2005). 'Engaging the Audience: The Social Imaginary of the Novela'. *Television & New Media* 6(4): 360–69.

De Roo, Ana C., Thiphalak Chounthirath and Gary A. Smith (2013). 'Television-Related Injuries to Children in the United States, 1990–2011'. *Pediatrics* 132(2): 267–74.

De Silva, J.P. (2000). La televisión ha muerto: La nueva producción audiovisual en la era de Internet: La tercera revolución industrial. Barcelona: Editorial Gedisa.

Douglas, Hilary (2011, March 13). '3000% Increase in "Worthless" Subjects'. *Daily Express* www.express.co.uk/news/uk/234186/3000-increase-in-worthless-subjects

Downey, Kevin (2007, February 8). 'TV Viewing is Up, Despite Online Video'. *Media Life* www.medialifemagazine.com/tv-viewing-is-up-despite-online-video/

Downing, John (2007). 'Terrorism, Torture, and Television: 24 in its Context'. *Democratic Communiqué* 21(2): 62–82.

Dugan, Emily (2014, May 25). '"Soft GCSEs" Could Face Axe from Exam Watchdog Ofqal'. *Independent* www.independent.co.uk/news/education/education-news/soft-gcses-could-face-the-axe-from-exam-watchdog-ofqual-9433736.html

Eagleton, Terry (1982). 'The Revolt of the Reader'. *New Literary History* 13(3): 449–52.

Eco, Umberto (1972). 'Towards a Semiotic Inquiry into the Television Message'. Trans. Paolo Splendore. *Working Papers in Cultural Studies* 3: 103–21.

Editor (2014, June 4). 'What's Keeping Some Graduates from Getting Hired?' *KUOW* www.kuow.org/post/whats-keeping-some-graduates-getting-hired

Eggerton, John (2014, March 17). 'Survey: TV Remains Top News Device'. *Broadcasting & Cable* www.broadcastingcable.com/news/washington/survey-tv-remains-top-news-access-device/129847

Ellis, John (2007). *TV FAQ*. London: IB Tauris.

Euroconsult (2008). *Satellite TV Platforms World Survey & Prospects to 2017* file:///Users/tobymiller1/Desktop/8-sat-tv-platforms-2008-pr.pdf

FICCI-KPMG (2013). *The Power of a Billion: Realising the Indian Dream* www.ficci.com/spdocument/20221/FICCI-KPMG-Report.pdf

Fitzsimmons, Caitlin (2009, January 20). 'Recession Could be "Good News" for TV, Suggests Deloitte Report'. *Guardian* www.theguardian.com/media/2009/jan/20/recession-good-news-for-tv-deloitte

Foucault, Michel (1985). *The Uses of Pleasure: The History of Sexuality Volume Two*. Trans. Robert Hurley. Harmondsworth: Penguin.

Foucault, Michel (2008). *The Birth of Biopolitics: Lectures at the Collège de France, 1978–79*. Trans. Graham Burchell. Ed. Michel Senellart. Houndmills: Palgrave Macmillan.

Fox, Adam (2003, May 23). 'Talking About My Generation'. *The Guardian.*

Friedman, Wayne (2008, November 26). 'Digital Lessons TV Should Learn from Music Companies'. *MediaPost's TV Watch* www.mediapost.com/publications/article/95528/digital-lessons-tv-should-learn-from-music-compani.html

Friedman, Wayne (2009, January 13). 'Critics Tour: Fox Favors Limited TV Ad Inventory'. *MediaDailyNews* www.mediapost.com/publications/article/98286/critics-tour-fox-favors-limited-tv-ad-inventory.html

Friedman, Wayne (2013, May 23). 'Worldwide Pay TV On the Rise, Big Growth in Asia'. *MediaPost* www.mediapost.com/publications/article/201062/worldwide-pay-tv-on-the-rise-big-growth-in-asia.html

Gardner, Howard (1983, March 27). 'When Television Marries Computer'. *New York Times* www.nytimes.com/1983/03/27/books/when-television-marries-computer-by-howard-gardner.html

Garnham, Nicholas (1987). 'Concepts of Culture: Public Policy and the Cultural Industries'. *Cultural Studies* 1(1): 23–37.

Gomez, Rodrigo, Gabriel Sosa-Plata, Primavera Téllez Girón and Jorge Bravo (2011). *Mapping Digital Media: Mexico* www.opensocietyfoundations.org/reports/mapping-digital-media-mexico

Gonsalves, Antone (2008, December 31). 'The Internet's Cool, but TV Remains Ad King'. *InformationWeek.com* www.informationweek.com/the-internets-cool-but-tv-remains-ad-king/d/d-id/1075225

Glez, Damien (2014, May 14). 'L'oeil de Glez: Jack Bauer vs Amnesty'. *Jeune Afrique* www.jeuneafrique.com/Article/ARTJAWEB20140514115527/torture-droits-de-l-homme-amnesty-international-television-l-oeil-de-glez-jack-bauer-vs-amnesty.html

Gopnik, Adam (1994). 'Read All About It'. *New Yorker* 70(41): 84–102.

Gough, Paul J. (2008, December 30). 'In '08, Big Headlines for Everybody'. *Hollywood Reporter* www.hollywoodreporter.com/news/08-big-headlines-everybody-124988

Graf, Philip (2008). 'Foreword'. *Lifeblood of Democracy? Learning About Broadcast News*. Cary Bazalgette, John Harland and Christine James. BFI Research Report for Ofcom 1 http://stakeholders.ofcom.org.uk/binaries/research/media-literacy/lifeblood.pdf

Gray, Laura Craig (2009, February 18). 'Media Revolution: Tomorrow's TV'. *BBC News* http://news.bbc.co.uk/1/hi/business/7897212.stm

Grindstaff, Laura and Joseph Turow (2006). 'Video Cultures: Television Sociology in the 'New TV' Age'. *Annual Review of Sociology* 32: 103–25.

Grossman, Lev (2006, December 13). 'Time's Person of the Year: You'. *Time* www.time.com/time/magazine/article/0,9171,1570810,00.html

Gueorguieva, Vassia (2007). 'Voters, MySpace, and YouTube: The Impact of Alternative Communication Channels on the 2006 Election Cycle and Beyond'. *Social Science Computer Review* 26(3): 288–300.

Hall, Stuart (1980). 'Encoding/Decoding'. *Culture, Media, Language*. Ed. Stuart Hall, Dorothy Hobson, Andrew Lowe and Paul Willis. London: Hutchinson. 128–39.

Harris, Daisy (2014, June 1). 'Bonfire of "Soft" GCSEs: Media Studies, Astronomy and Tourism Could be Axed in a Bid to Make Qualification More Rigorous'. *Daily Mail* www.dailymail.co.uk/news/article-2645486/Michael-Gove-set-axe-soft-GCSEs-including-media-studies-astronomy-tourism-bid-make-qualification-rigorous.html?ITO=1490&ns_mchannel=rss&ns_campaign=1490

Hartley, John (1987). 'Invisible Fictions: Television Audiences, Paedocracy, Pleasure'. *Textual Practice* 1(2): 121–38.

Hartley, John (1992). The Politics of Pictures: The Creation of the Public in the Age of Popular Media. London: Routledge.

Hilmes, Michele (2005). 'The Bad Object: Television in the American Academy'. *Cinema Journal* 45(1): 111–17.

Hogan, Phil (2004, February 1). 'Television Studies'. *Observer* www.theguardian.com/lifeandstyle/2004/feb/01/features.familyandrelationships

Hunter, Ian (1988). 'Providence and Profit: Speculations in the Genre Market'. *Southern Review* 22(3): 211–23.

IBM Business Consulting Services (2006). *The End of Television as We Know It: A Future Industry Perspective* www-935.ibm.com/services/us/imc/pdf/ge510-6248-end-of-tv-full.pdf

Instrell, Rick (2014, May 13). 'Media Studies Course is Gaining in Value'. *Scotsman* www.scotsman.com/news/media-studies-course-is-gaining-in-value-1-3408501

Irwin, William, Mark T. Conard and Aeon J. Skoble, eds (2001). *The Simpsons and Philosophy*. Chicago: Open Court.

James, Meg (2014, May 29). 'Kevin Reilly to Step Down as Fox Entertainment Chairman'. *Los Angeles Times* www.latimes.com/entertainment/envelope/cotown/la-et-ct-fox-entertainment-kevin-reilly-steps-down-20140529-story.html

Kaufman, Leslie (2009, March 2). 'Car Crashes to Please Mother Nature'. *New York Times*: C3.

Kittross, John Michael (1999) 'A History of the BEA'. *Feedback* 40(2) www.beaweb.org/pdfs/beahistory.pdf

Lightfoot, Liz (2005, September 8). 'Students Mark Down Media and Tourism Degrees'. *Telegraph* www.telegraph.co.uk/news/uknews/1497885/Students-mark-down-media-and-tourism-degrees.html

Lithwick, Dahlia (2008, July 26). 'The Bauer of Suggestion'. *Slate* www.slate.com/articles/news_and_politics/jurisprudence/2008/07/the_bauer_of_suggestion.html

Lotz, Amanda D. (2008). *The Television Will Be Revolutionized*. New York: New York University Press.

Lynch, Jason (2013, November 25). 'Americans' New Thanksgiving Tradition: Watching TV All Day Long'. Quartz www.qz.com/150790/americans-new-thanksgiving-tradition-watching-tv-all-day-long/

Martín-Barbero, Jésus and Germán Rey (1999). *Los ejercicios del ver: Hegemonía audiovisual y ficcíon televisiva*. Barcelona: Gedisa Editorial.

Maxwell, Richard and Toby Miller (2012). *Greening the Media*. New York: Oxford University Press.

McChesney, Robert W. (2007). Communication Revolution: Critical Junctures and the Future of Media. New York: New Press.

McMahon, Jennifer L. (2008). '24 and the Existential Man of Revolt'. *The Philosophy of TV Noir*. Ed. Steven M. Sanders and Aeon J. Skoble. Lexington: University Press of Kentucky. 115–29.

McPherson, Tara (2008). '"The End of TV As We Know It": Convergence Anxiety, Generic Innovation, and the Case of 24'. *The Oxford Handbook of Film and Media Studies*. Ed. Robert Kolker. New York: Oxford University Press. 306–26.

Miklos, David (2008). 'El império en peligro'. *La Tempestad* 59: 78–9.

Miller, Martin (2007, February 14). '"24" and "Lost" Get Symposium on Torture'. *Seattle Times* http://seattletimes.com/html/entertainment/2003570697_tvtorture14.html

Miller, Toby (2010). 'Manuel Alvarado's Publications'. *Television & New Media* 11(6): 497–99.

Minogue, Kenneth (1994, November 25). 'Philosophy'. *Times Literary Supplement*: 27–8.

Morley, David (2007). Media, Modernity and Technology: The Geography of the New. London: Routledge.

Morris, Meaghan (1990). 'The Banality of Cultural Studies'. *Logics of Television: Essays in Cultural Criticism*. Ed. Patricia Mellencamp. Bloomington: Indiana University Press. 14–43.

Moses, Asher (2009, January 9). 'TV or not TV?' *Sydney Morning Herald* www.smh.com.au/news/technology/biztech/tv-or-not-tv/2009/01/09/1231004268540.html

Motion Picture Association of America (2007). *U.S. Entertainment Industry: 2006 Market Statistics* www.robertoigarza.files.wordpress.com/2008/11/rep-entertainment-industry-market-statistics-mp-2007.pdf

Neff, Jack (2009, February 23). 'Guess Which Medium is as Effective as Ever: TV'. *AdvertisingAge* www.adage.com/article/media/guess-medium-effective-tv/134790/

Netflix (2013). 'Netflix Long Term View' http://ir.netflix.com/long-term-view.cfm

Newcomb, Horace (2005). 'Studying Television: Same Questions, Different Contexts'. *Cinema Journal* 45(1): 107–11.

Newcomb, Horace (2014). 'The More Things Change ...'. *Flow* 19(5) www.flowtv.org/2014/01/the-more-things-change/

Newman, Nic and David A.L. Levy, eds (2013). *Digital News Report 2013*. Oxford: Reuters Institute for the Study of Journalism.

Nielsen (2014). *Advertising and Audiences: State of the Media*. www.nielsen.com/us/en/reports/2014/advertising-and-audiences-state-of-the-media.html

Ofcom (2009). *Ofcom's Second Public Service Broadcasting Review: Putting Viewers First* http://stakeholders.ofcom.org.uk/consultations/psb2_phase2/statement/

Ofcom (2013). *International Communications Market Report* http://stakeholders.ofcom.org.uk/market-data-research/market-data/communications-market-reports/cmr13/international/

Office for National Statistics (2013). *Full Report – Graduates in the UK Labour Market 2013* www.ons.gov.uk/ons/dcp171776_337841.pdf

Ofqal (2014). *Completing GCSE, AS and A Level Reform* www.ofqual.gov.uk/news/gcse-level-reform-consultation/

Organisation for Economic Co-operation and Development (2007). *OECD Communications Outlook 2007*. Paris: OECD.

Organisation for Economic Co-operation and Development (2013). *OECD Communications Outlook 2013*. Paris: OECD.

Parkin, Frank (1971). *Class Inequality and Political Order*. London: MacGibbon & Kee.

Paton, Graeme (2007, May 7). 'Media Studies Wastes Good Brains, Says Sugar'. *Telegraph* www.telegraph.co.uk/news/uknews/1550580/Media-studies-wastes-good-brains-says-Sugar.html

Paton, Graeme (2014, June 4). 'Ofqal: Dozens of "Soft' GCSEs and A-levels to be Axed'. *Telegraph* www.telegraph.co.uk/education/educationnews/10873368/Ofqual-dozens-of-soft-GCSEs-and-A-levels-to-be-axed.html

Pertierra, Anna Cristina and Graeme Turner (2013). *Locating Television: Zones of Consumption*. London: Routledge.

Pew Research Center (2005). *Trends 2005* www.pewresearch.org/2005/01/20/trends-2005/

Pew Research Center for the People and the Press (2005). *The Internet and Campaign 2004* www.pewinternet.org/2005/03/06/the-internet-and-campaign-2004/

Pew Research Center for the People and the Press (2008). *Internet Overtakes Newspapers as News Source* www.people-press.org/2008/12/23/internet-overtakes-newspapers-as-news-outlet/

Pew Research Center for the People and the Press (2012). *One-in-Ten 'Dual-Screened' the Presidential Debate* www.people-press.org/2012/10/11/one-in-ten-dual-screened-the-presidential-debate/

Pius XII (1958, August 21). 'La lettre Apostolique'. *Acta Apostolica Sedis* 50: 512–13.

Poggi, Jeanine (2014, January 8). 'Amid CES Gadgets, Television Still Works for Marketers'. *Advertising Age* www.adage.com/article/consumer-electronics-show/amid-ces-gadgets-tv-works-marketers/290965/

Postrel, Virginia (1999, August 2). 'The Pleasures of Persuasion'. *Wall Street Journal*: A18.

Project for Excellence in Journalism (2005). *The State of the News Media: An Annual Report on American Journalism* file:///Users/tobymiller1/Desktop/Journo2005%20execsum.pdf

Rajagopal, Arvind (forthcoming). 'India's Televisual Populism'. *Journal of Asian Studies*.

Rash, John (2009, February 23). 'Oscar Night Ratings Improved, But Far from Golden'. *AdvertisingAge* www.adage.com/article/media/oscar-night-ratings-improved-golden/134822/

Reynolds, Mike (2009, February 11). 'CNN, MTV Among Top 10 Most Social Brands: Vitrue Study'. *Multichannel News* www.multichannel.com/news/social-media/cnn-mtv-rank-among-top-10-most-social-brands-vitrue-study/305731

Saad, Lydia (2013, July 8). 'TV is Americans' Main Source of News'. *Gallup* www.gallup.com/poll/163412/americans-main-source-news.aspx

Schramm, Wilbur, Jack Lyle and Edwin B. Parker (1961). *Television in the Lives of Our Children*. Stanford: Stanford University Press.

Sconce, Jeffrey (2004). 'What If? Charting Television's New Textual Boundaries'. *Television After TV: Essays on a Medium in Transition*. Ed. Lynn Spigel and Jan Olsson. Durham: Duke University Press. 93–112.

Shields, Mike (2009, January 8). 'Nielsen: Mobile Video Usage Small, But Growing.' *MediaWeek.com* www.creativeplanetnetwork.com/2-pop/news/nielsen-mobile-video-usage-small-growing/33914

Shimpach, Shawn (2010). Television in Transition: The Life and Afterlife of the Narrative Action Hero. Malden: Blackwell.

Silber, John R. (1968). 'Television: A Personal View'. *The Meaning of Commercial Television: The Texas-Stanford Seminar*. Ed. Stanley T. Donner. Austin: University of Texas Press. 113–39.

Slade, Christina and Annabel Beckenham (2005). 'Introduction: Telenovelas and Soap Operas: Negotiating Reality'. *Television & New Media* 6(4): 337–41.

Smythe, Dallas (1954). 'Reality as Presented by Television'. *Public Opinion Quarterly* 18(2): 143–56.

Smythe, Dallas (2004). 'The Consumer's Stake in Radio and Television'. *Mass Communication and American Social Thought: Key Texts, 1919–1968*. Ed. John Durham Peters and Peter Simonson. Lanham: Rowman & Littlefield. 318–28.

Spangler, Todd (2009, February 19). 'ESPN, Discovery are Top "Must-Keep" Cable Nets: Survey'. *Multichannel News* http://multichannel.com/news/content/espn-discovery-are-top-must-keep-cable-nets-survey/364701

Spigel, Lynn (2005). 'TV's Next Season?' *Cinema Journal* 45(1): 83–90.

Standage, Tom (2006, October 12). 'Your Television is Ringing'. *Economist* http://www.economist.com/node/7995312

Steinberg, Brian (2009, February 20). 'Is CBS's "Harper's Island" a New Broadcast Model?' *AdvertisingAge* www.adage.com/article/media/cbs-s-harper-s-island-a-broadcast-model/134739/

Tay, Jinna and Graeme Turner (2010). 'Not the Apocalypse: Television Future in the Digital Age'. *International Journal of Digital Television* 1(1): 31–50.

Teinowitz, Ira (2008, July 23). 'Olympic Deal Sealed: Obama Makes $5 Million Buy'. *AdvertisingAge* www.adage.com/article/news/olympic-deal-sealed-obama-makes-5-million-buy/129853/

Thomasch, Paul (2009, February 20). 'As Advertisers Scrap Deals, TV Faces Cold Spring'. *Reuters* www.uk.reuters.com/article/2009/02/20/businesspro-us-advertising-idUKTRE51J5PW20090220

Thompson, Derek (2013, December 5). 'The Myth of Teens Rejecting Television'. *The Atlantic* www.theatlantic.com/business/archive/2013/12/the-myth-of-teens-rejecting-television/282084/

TNS Media Intelligence (2008). *An Analysis of 2007 and 2008 Political, Issue and Advocacy Advertising* www.bransonagentnewsline.blogspot.co.uk/2007/10/analysis-of-2007-and-2008-political.html

Toomer, Jessica (2014, May 16). 'Human Rights Group Skewers "24" for Graphic Torture Scenes'. *Huffington Post* www.huffingtonpost.com/2014/05/16/amnesty-international-24-torture_n_5337058.html

vanden Heuvel, Katrina (2008, July 21–28). 'Just Democracy'. *The Nation*: 31–40.

Verón, Eliseo (2008, September 28). 'Réquiem para una televisión difunta'. *Perfil* www.gacemail.com.ar/notas.php?idnota=10699

Vincent, Norah (2000, February 2–8). 'Hop on Pop: Lear, Seinfeld, and the Dumbing Down of the Academy'. *Village Voice* www.villagevoice.com/2000-02-01/nyc-life/hop-on-pop/full/

Whittam Smith, Andreas (2008, February 25). 'Media Studies is No Preparation for Journalism'. *Independent* http://independent.co.uk/voices/commentators/andreas-whittam-smith/andreas-whittam-smith-media-studies-is-no-preparation-for-journalism-786785.html

Woodhead, Chris (2009, March 8). 'Dive for Cover if You See "Studies": It Means Bogus'. *Times*: 9.

Other Resources

'BBC World News and bbc.com Release World's Largest Global Study of News Consumption Habits Across Mobile Devices' (2013, March 26). *BBC* http://www.bbc.co.uk/mediacentre/worldnews/news-consumption.html

'Deloitte: State of the Media Democracy' (2009, December 15). *Deloitte* www.deloitte.com/view/en_US/us/industries/Media-Entertainment/dce196d6a8295210VgnVCM100000ba42f00aRCRD.htm.

'Kids Motivated by TV and Print to Visit Web' (2009, January 1). *Center for Media Research* www.mediapost.com/publications/article/97459/kids-motivated-by-tv-and-print-to-visit-web.html.

'Majority of Americans Believe Television Programming is Getting Worse, Yet They're Watching More Than Ever' (2007, September 13). *AOL Television and Associated Press* http://ir.aol.com/phoenix.zhtml?c=147895&p=irol-newsArticle_print&ID=1353995&highlight=

'Nielsen Estimates 115.6 Million TV Homes in the U.S., Up 1.2%' (2013, May 7). *Nielsen* www.nielsen.com/us/en/newswire/2013/nielsen-estimates-115-6-million-tv-homes-in-the-u-s—up-1-2-.html

'Nielsen Media Research Reports Television's Popularity is Still Growing' (2006, September 21). *Nielsen Media Research* www.prnewswire.com/news-releases/nielsen-media-research-reports-televisions-popularity-is-still-growing-57098217.html

'Nielsen Reports Growth of 4.4% in Asian and 4.3% in Hispanic U.S. Households for 2008–2009 Television Season' (2008, August 28). *Nielsen Media Research* www.yumpu.com/en/document/view/13968809/press-release-nielsen

'Nielsen Reports TV, Internet and Mobile Usage Among Americans' (2008, July 8). *Nielsen Media Research* www.bloomberg.com/apps/news?pid=conewsstory&refer=conews&tkr=NTRT:US&sid=a_MAem625YOI

'Nielsen: Ten Year Analysis Reveals Australians' Thirst for Connected Knowledge' (2013, November 18). *Nielsen* www.nielsen.com/au/en/press-room/2013/ten-year-analysis-reveals-australias-thirst-for-connected-devices.html

'Precious Little Time' (2008, December 24). *Center for Media Research* www.mediapost.com/publications/article/97118/precious-little-time.html

'La televisión y el futuro del cine' (2008). *La Tempestad* 59: 67.

'Torture Spreading as "Glorified" by TV Series Like "24": Amnesty' (2014, May 13). *Jakarta Globe* www.thejakartaglobe.com/international/torture-spreading-glorified-television-series-like-24-amnesty-intl/

'What Impact Does Broadband Have on TV Viewing?' (2008, December 23). *VideoNuze* www.iq.videonuze.com/article/what-impact-does-broadband-have-on-tv-viewing-

PART I

Ownership and Regulation

1

How to Study Ownership and Regulation

Des Freedman

WHY BOTHER?

Outbreaks of interest in questions of media ownership or broadcast regulation have long been confined to controversies about 'inappropriate' media content and the use of 'indecent' language. George Carlin's broadcast profanities in the 1970s and Janet Jackson's exposed breast at the 2004 Superbowl are far more likely triggers of public concern about the structure and behaviour of television industries than a sober commitment to viewpoint or ownership diversity. But perhaps this is changing. The explosion of public anger that greeted the Federal Communications Commission's attempt in 2003 to liberalize broadcast ownership rules (explored in McChesney, 2004) has been followed by the unprecedented spectacle of a Presidential candidate actually going to the stump armed with speeches on media ownership.

This is precisely what Barack Obama did in his 2008 presidential campaign where his election pamphlets promised that a President Obama would 'encourage diversity in the ownership of broadcast media, promote the development of new media outlets for expression of diverse viewpoints, and clarify the public interest obligations of broadcasters who occupy the nation's spectrum' (Obama, 2008). In an interview with *Broadcasting & Cable* during the campaign, Obama argued that the consolidation of media markets was leading to 'less diversity of opinion, less local news coverage, replication of the same stories across multiple outlets' and promised that '[w]e can do better' (quoted in Eggerton, 2008).

Obama would most likely have wanted to wait a little longer before this commitment was tested. In December 2009, the nation's largest cable and broadband provider, Comcast, agreed to merge with the Hollywood giant NBC Universal in a $30 billion deal. The new company would control 20 per cent of all viewing hours in the US and one of every seven television channels and would have a dominant position in existing and emerging content markets. Political battle lines were soon drawn: *The New York Times*, on one side, argued that the merger 'could limit choices and raise prices for viewers and advertisers' (editorial, 7 December 2009) while the *Wall Street Journal*, on the other, insisted that any evaluation of the merger was 'a determination best made by the marketplace, not by anticorporate activists' (editorial, 6 December 2009). Opponents of media consolidation lined up against supporters of greater deregulation, with the implications for democracy and diversity placed at the heart of the debate.

The problem for Obama was that he was immediately implicated in the networks of influence and power that surrounded the deal. The Comcast chief executive had made significant donations to the Democrats since 2006 and had helped raise approximately $6 million dollars for Obama's 2008 election campaign; Comcast's VP had contributed some $180,000 in the same period while the chief executive of General Electric, owner of NBC, was a regular visitor to the Obama White House and a member of his Economic Recovery Advisory Board (Hart, 2009). The merger was eventually waived through in 2011 and was followed by the hiring by Comcast of Meredith Attwell Baker, one of the FCC commissioners who approved the deal, as a top lobbyist. This is just one small anecdote that illustrates the extent to which the ownership and regulation of broadcast outlets are deeply political phenomena that are connected to wider concerns about how we sustain and protect democratic institutions that are able to act in something called 'the public interest'. Put simply, there is a widely-held belief that democracy is throttled if the number of media outlets and distinctive voices is restricted – if, in other words, media power is concentrated in the hands of the few and not the many and if it is able to evade public oversight and regulatory control.

So let us suppose for a moment, as some people do, that Rupert Murdoch, the chairman of both 21st Century Fox and News Corporation, the transnational conglomerates that include amongst their portfolios, Fox Television, the *Wall Street Journal*, BSkyB, *The Times*, *The Sun*, Harper Collins publishing and Star Television, is a useful metaphor for the dangers of media concentration. Mr Murdoch is known not simply for his business acumen but for his advocacy of a series of neo-liberal economic and conservative social policies including his unwavering support for the US and UK 'war on terror' and his staunch opposition to the European Union (Cassidy, 2006). In response to Murdoch's ability to dominate the means through which public conversations take place, the Virgin boss Richard Branson once described him as a 'threat to democracy' (quoted in Puttnam, 2006). Such a description rests on the conviction that media ownership *matters* and that, where it is concentrated, it represents an unacceptable challenge to pluralist principles of a diverse and competitive media system and, by extension, to the very legitimacy of a democratic system.

In this argument, media concentration is anti-democratic not simply because it undermines the ability of citizens to acquire and exchange the information and ideas necessary to take informed decisions about public life, but because it distorts the logic of the media industries themselves, transforming them from vehicles of symbolic interaction to engines of capital accumulation. Consider Murdoch's response in 2009 to calls for a bail-out of news organizations as they struggle to cope with the combined impact of a huge drop in advertising revenue and the structural challenge posed by the internet. Welcoming the collapse of companies that fail to adapt to the new digital age, Murdoch argued that 'they should fail, just as a restaurant that offers meals no one wants to eat or a car-maker who makes cars no one wants to buy should fail' (Murdoch, 2009). This is the same vision of media – as commodities that measure their success simply using market criteria – as that espoused by FCC chairman Mark Fowler in the early 1980s when he described television as 'just another appliance. It's a toaster with pictures' (quoted in Horwitz, 1989: 245).

But, as stated above, there is another approach to the issue of media ownership that ties questions of structure to those of performance and accountability in relation to democracy (as distinct from profitability or efficiency). Some of the founding scholars of media, communications and culture identified issues of ownership as vital to the ability of media to pursue an independent, imaginative and critical role in public life. Back in 1948, Lazarsfeld and Merton noted the importance of locating mass media within the specific social and economic structures in which they operated and argued that 'the social effects of the media will vary as the system of ownership and control varies' (Lazarsfeld and Merton, 2004 [1948]: 236). More significantly, they insisted that '[s]ince the mass media are supported by great business concerns geared into the current social and economic system, the media contribute to the maintenance of this system' (2004 [1948]: 236). In his important 1962 book *Communications*, Raymond Williams, the great British theorist of culture and media, highlighted the emergence of new forms of media ownership that were contributing to a growing commodification of audiences and content. He argued that the 'methods and attitudes of capitalist business have established themselves near the centre of communications' (Williams, 1968 [1962]: 31), even in a media economy that contained a significant not-for-profit, public service core. 'From this it becomes one of the purposes of communication to sell a particular paper or programme. All the basic purposes of communications – the sharing of human experience – can become subordinated this drive to sell (1968 [1962]: 32).

This chapter pursues Williams's interest in the relationship between questions of ownership and the media's ability to facilitate democratic conversations by considering not just why but *how* we should study media ownership. It supplements this by reflecting on how we should make sense of the key mechanisms for shaping and monitoring this relationship: through the regulatory processes that continue to maintain oversight of the contemporary television environment. It considers traditional perspectives on ownership and regulation by providing a typology of different methods of analysis and goes on to argue for an approach to ownership and regulation that interrogates them as systems of thought and action that privilege particular ways of thinking about and ordering the world. The chapter attempts, in other words, to make the analysis of ownership and regulation not simply necessary but actually provocative and empowering.

STUDYING OWNERSHIP AND REGULATION AS DATA

The first approach to the study of media ownership and regulation is to treat them as sources of quantitative data that, when assessed empirically and presented methodically, produce the evidential basis for further types of action (i.e. ownership rules or regulatory interventions). Data in relation to broadcast ownership may include a breakdown of different types of ownership structure, figures concerning revenue and profit, the share of the market in terms of turnover or audience, levels of concentration, details of income streams and market predictions. Data on regulation often includes listing the various mechanisms through which regulation is able to take place (for example, quotas, content rules, licensing, adherence to codes and protocols, antitrust procedures and, in the case of negative regulation, an absence of intervention) and describing their specific operations. Many studies have provided this information in an attempt to map out key features of specific broadcast environments, for example Compaine and Gomery (2000 [1979]) and Noam (2009) with reference to US media ownership, Kelly, Mazzoleni and McQuail (2004), Open Society Institute (2005) and Sanchez-Taberno et al. (1993) in relation to European ownership while Creech (2007) and Napoli (2001) have produced encyclopaedic volumes which detail all the various interventions into broadcast speech and ownership.

In the midst of the often polarized debates on media ownership described earlier, quantitative data may be seen to provide a reassuring source of certainty and retains an ability to be mathematically precise in the midst of what some argue are instinctual claims and unsupported assertions. Noam, for example, insists that 'what the world needs on the subject of media concentration is more facts, not more opinions' (2009: 34) while Compaine talks of his own 'data-mongering, stick-to-the-facts approach' (Compaine and Gomery, 2000: 21). Concentration can be pinpointed not through a vague sense that one company seems to be dominant but through the systematic use of economic analysis, most often using the Herfindahl-Hirschmann Index (HHI) which assesses market competition in a particular sector and quantifies the results. An HHI score of 1,800 or above describes a sector with high levels of concentration while a score of 1,000 or below suggests a relatively unconcentrated sector. The FCC, in its desire to relax ownership rules in 2003, employed this approach when establishing a 'Diversity Index' that measured the degree of concentration in a typical market, 'Anytown USA', which was found, most conveniently, to have a score of exactly 738 – well below warning levels.

Noam, in his 500-page survey of the ownership of contemporary US communications industries, presents data in order to negotiate between what he sees as two opposing, and flawed, hypotheses: first, the 'destined to diversity' scenario in which digital technologies are ushering in a communications cornucopia, and second, the 'doomed to concentration' scenario (2009: 34) in which the media environment will inevitably be reduced to a small handful of all-powerful fiefdoms. Instead, he argues that what we find is a media structure governed by three trends – growth in economies of scale, the lowering of entry barriers and digital convergence – that will see it split into large 'integrator' firms who dominate conception and distribution and a whole slew of specialist companies which cluster around them (2009: 39). Noam concludes, after presenting a truly enormous amount of data, that while broadcast concentration – and more specifically multichannel concentration – is increasing, it is still well below levels 'that would normally raise antitrust action if encountered in other industries' (2009: 423). Compaine and Gomery, whose book is of a similar size and scope, are even more optimistic. Their analysis of the data leads them to conclude that the emerging media marketplace 'may be noted more for information overload and fragmentation than for concentration and scarcity' (2000: 578).

It is not just pro-market scholars who turn to data to make their arguments. Figures like Bagdikian (1990[1983]), Herman and Chomsky (1988) and Herman and McChesney (1997) all provide statistical evidence to support claims of increased media concentration and commercialism. Indeed, Robert McChesney, one of the founders of the Free Press media reform group, has repeatedly called for economists and legal scholars to furnish the reform movement with evidence that can be presented to policymakers to substantiate the allegations of a link, for example, between increased concentration and reduced diversity. In his book, *Communication Revolution*, he writes that 'the movement grasped how important it was to have first-rate credible media research' and noted that there was a pressing need 'for traditional quantitative communication scholars, for economists, and for legal scholars. And these scholars needed to work with us so that they didn't get swallowed up by baseless presuppositions' (2007: 172). Economists like Chang (2011), Gentzkow and Shapiro (2006) and Waldfogel (2007) have certainly taken up this challenge.

Quantitative data, therefore, is very much in fashion in the analysis of media ownership and in the discussions that guide policymakers and regulators in responding to any problems that may arise from concentrated ownership or market failure. It is the 'gold standard by which policy judgments are made' (Napoli and Karaganis, 2007: 56) and is at the heart of the 'evidence-based approach' of the FCC and the British communications regulator Ofcom, both of which rely heavily on consumer research, market data, investment reports and econometric analyses. Indeed, Ofcom's annual *Communications Market* reports are probably the most valuable source of data for any analyst of the British media.

There are two problems, however, with this fetishization of what has come to be known as 'media metrics', described by pro-liberalization campaigners as 'objective measures in contrast to subjective barometers' (Thierer and Eskelsen, 2008: 9). First there is the fact that such data is far more likely to be demanded of or supplied by commercial organizations. According to Napoli and Karaganis (2007) in their persuasive call for a 'federal data agenda for communications policymaking', policymakers tend to draw their data from a narrow range of elite sources. 'Today, communication policymakers rely heavily on the datasets developed by commercial data providers for their clients and the investment community, and, therefore, neglect their own substantial data collection capabilities and responsibilities' (2007: 56). Second, while a data-driven approach is designed theoretically to insulate media policymaking and regulatory domains from partisanship and bias, there is little to suggest that a call to 'objectivity' will necessarily undermine the use of selective facts and subjective judgments. Countervailing ideas can just as easily be marginalized, ignored or buried, as happened when FCC officials refused to distribute a piece of research that demonstrated a link between increased consolidation and decreased amount of local news (Associated Press, 2006). Indeed, the Diversity Index employed by the FCC in 2003 to measure degrees of concentration in local markets was eventually discredited by the Appeals Court for relying too heavily on quantitative methods and for not paying enough attention to issues of content and the impact of different types of speech. It seems that not everything in the media environment can be easily quantified.

STUDYING OWNERSHIP AND REGULATION AS NORMATIVE MODELS

A contrasting approach to tackling questions of ownership and regulation is to be found in the pursuit not of empirical data but of normative models, 'ideal types' of ownership structure or regulatory behaviour. These are most often expressed in terms of arrangements that best facilitate democratic practice, for example with the liberal market conception that the media collectively should act as a 'watchdog' where its prime role is to 'monitor the full range of state activity, and fearlessly expose abuses of official authority' (Curran, 2002a: 217). In this case, forms of media ownership and regulation are required that guarantee both autonomy from the state and free and fair competition in the allocation and distribution of resources. Similarly, a pluralist perspective on ownership holds that media should articulate the widest possible range of views in order to allow citizens to seek out and act on all viable information as they participate in public life. In both cases, rules may be necessary to break up unaccountable and unacceptable concentrations of power that distort or undermine the ability of media to achieve these noble objectives.

The legal scholar Ed Baker hints at such an ownership model when he provides three main reasons for opposing concentration. First, he insists that we will arrive at a more 'democratic distribution of communicative power (2007: 6) with a *dispersed* distribution structure that allows for a breadth of voices and representations in order that no one voice is excluded. Second, he argues that the 'widest possible dispersion of media power reduces the risk of the abuse of communicative power' (2007: 16), and notes how the former Italian premier, Silvio Berlusconi, has amassed significant economic and political power through his domination of the Italian media. Finally, he claims that a more egalitarian media structure is likely to improve quality and reduce the risk of market failure as it will check the urge of concentrated organizations to put profit and shareholder value above all else. There are, he concludes, 'fundamental democratic and significant structural economic reasons' (2007: 53) for introducing rules that will disperse media ownership and consolidate the democratic benefits of a pluralistic media system.

Indeed some of most lucid accounts of the relationship between a competitive media system and a robust democracy are to be found in official justifications for precisely these ownership rules. Consider this statement by the then Conservative government when reviewing British media ownership rules back in 1995:

> A free and diverse media are an indispensable part of the democratic process. They provide the multiplicity of voices and opinions that informs the public, influences opinion, and engenders political debate. They promote the culture of dissent which any healthy democracy must have. In so doing, they contribute to the cultural fabric of the nation and help define our sense of identity and purpose. If one voice becomes too powerful, this process is placed in jeopardy and democracy is damaged. Special media ownership rules, which exist in all major media markets, are needed therefore to provide the safeguards necessary to main diversity and plurality. (Department for National Heritage, 1995: 3)

Even the FCC in its attempt in 2003 to *relax* broadcast ownership restrictions nevertheless insisted on the normative importance of retaining measures to promote viewpoint diversity in the interests of democracy. Despite the huge implications of the advent of the internet and new sources of plurality, '[w]e therefore continue to believe that broadcast ownership limits are necessary to preserve and promote viewpoint diversity. A large number of independent owners will tend to generate a wider array of viewpoints in the media than would a comparatively smaller number of owners' (2003: 11). These views are shared by many scholars writing on the significance of media ownership rules, for example Barron (2000), Doyle (2002) and Overbeck (2007).

A related way in which scholars have approached ownership and regulation has been to identify the normative bases of traditional public policy approaches to these questions. Philip Napoli, for example, talks of the 'enduring normative principles' (2001: 21) that underlie communications policymaking: concepts such as localism, diversity, universal service, competition and the public interest. McQuail (1992, 68–77) discusses freedom, equality and social order as mobilizing principles for attempts to regulate media systems, while Tom Gibbons highlights values such as free speech, independence and accountability (1998: 35–54). We could add to these lists a whole host of other normative commitments – to pluralism, citizenship, welfare and social justice as well as to efficiency, productivity and profitability – and assess the ways in which different ownership structures and arrangements impact on these commitments.

Other scholars are keen to enshrine these kinds of normative principles in actual designs for democratic media practices. Georgina Born, for example, develops a 'normative architecture' (2006: 116) in which a range of communicative conversations may be fostered by the peculiar possibilities of a public service, non-profit broadcaster. The ability to stimulate these different dialogues is related not just to questions of ownership but to the characteristics of particular platforms (whether terrestrial television or the internet) in establishing 'dialogical flows between the sociocultural majority and minorities, or dominant and subordinate groups (2006: 116). James Curran has devised a 'working model' (2002a: 240) of a democratic media system that is stratified by different forms of ownership. At the core of his system, is a public service broadcaster responsible, above all, to its viewers and listeners (like, Curran argues, public service channels in Germany and Scandinavia with their broadcasting councils and democratic representation). This core is surrounded by four 'peripheral' media sectors – civic, professional, social market and private spaces – which, taken together, will guarantee not only the communicative rights of minorities but 'act as a restraint on the over-entrenchment of minority concerns to the exclusion of majority pleasures' (2002a: 241). While this is a distinctively European-looking project, William Hoynes has sketched out a vision for public broadcasting in the US which requires re-thinking the 'core mission of public broadcasting in the digital age' (2007: 375; see also Center for Social Media, 2009). Hoynes suggests that public broadcasting should adopt a remit that clearly distinguishes itself from commercial broadcasting and instead focuses on local programming as well as independent news and current affairs.

A focus on the normative and conceptual is valuable in allowing us to imagine democratic ownership structures and regulatory practices (or to criticize structures and practices that are not) from the point of view of firmly-held values and perspectives. There is little doubt, for example, that the media reform legislation passed in Argentina in 2009 and finally validated by the courts in 2013 – which, in its commitment to diversify the airwaves by assigning one-third of licenses each to community organizations, state broadcasters and private companies' (Reporters without Borders, 2013), is strikingly similar to Curran's model outlined above – was partly informed by a vision of what a democratic media system should look like. There is, however, always a danger that by choosing not to focus on empirical data, some normative accounts lack verifiability, or more precisely, relevance to particular political and economic contexts. Of course, there is no reason why normative accounts should be counterposed to empirically-based studies, and, indeed, many of the texts quoted above combine normative reflections with at least some detail about specific media environments. Georgina Born best expresses this approach in her anthropological analysis of a changing BBC (Born, 2004), and later concludes that scholars:

> need to break down the boundaries between normative theories and the design of democratic institutions, including media systems suited to democratic pluralism. From a policy perspective, we need to take political philosophies seriously – to realize that they offer tangible bases on which to construct institutional arrangements; but also to acknowledge that our existing institutions embody political philosophies that themselves deserve scrutiny and updating. (2006: 119)

STUDYING OWNERSHIP AND REGULATION AS IDEOLOGICAL PROCESSES

This leads us to a third way of studying ownership and regulation: not as the mere embodiment of quantitative data nor as the receptacle of theoretically sophisticated normative positions, but as systems of thought and action that are related to specific ways of ordering the world. From this perspective, attitudes towards conglomeration or competition, for example, are revealing for the wider ideological positions that they point to (for example, about the market as an enabler of productive symbolic activity); the advocacy of particular regulatory mechanisms and objectives is far from a simple technical or administrative matter, but is connected to values and priorities that are by no means natural or inevitable. This is an approach adopted by a whole number of critical communication scholars who identify forms of media ownership and regulation as central devices for securing consent to 'market-driven politics' (Leys, 2001). For Ben Bagdikian, whose book *The Media Monopoly* (1990 [1983]) was one of the first to combine data on media concentration with a sociological analysis of media influence, ownership is a mechanism for managing the exclusion of certain voices from political power. Media ownership, he argues, is connected to the ability 'to treat some subjects briefly and obscurely but others repetitively and in depth', a seemingly 'normal and necessary' (1990 [1983]: 16) journalistic practice that skews public opinion in the interest of powerful corporations.

Many other figures have followed in Bagdikian's footsteps, updating and refining his initial arguments in the face of sustained pressure from critics (such as Compaine and

Gomery, 2000 and Thierer, 2005). In the field of media ownership, notable contributions include Herman and Chomsky (1988) and their development of a 'propaganda model' in which concentrated ownership plays a significant role, as well as the work of Robert McChesney (1999, 2004) who argues that struggles over media ownership have long been battles over the soul of American democracy. With reference to media policymaking and regulation, there is a rich seam of literature that explores the ideological basis of policy and regulatory decisions that serve to naturalize capitalist property relations and systems of thought, including Aufderheide (1999), Braman (2007), Freedman (2008), Horwitz (1989), Schiller (1999) and, perhaps most persuasively, Streeter (1996), who argues that policymaking is an ideological means of resolving differences and enforcing order in the communications environment and beyond: 'underlying all the (very real) disagreements and debates is a relatively constant structure of expectations that limit discussion, not by coercion, but by way of the subtle but profound power of interpretation' (1996: 117). The rest of this chapter seeks to highlight this interpretive activity by calling for an analytical emphasis on *agency* and a recognition of the significance of *interests* in three particular ways: by treating media ownership and regulation as subjects for debate (rather than as a given); by examining dominant arguments in ownership and regulatory debates; and, finally, by focusing conceptually on processes of inaction and silence at the heart of these debates.

HIGHLIGHT OWNERSHIP AND REGULATION AS TOPICS FOR DEBATE

The classroom debate has long been one of the most popular ways of clarifying complex issues: organize students into different groups, ask them to propose contrasting arguments and to marshal the appropriate evidence, and then assess the strengths and weaknesses of all sides. Applied to media ownership and regulation, this approach can be extremely useful in that it refuses to take established starting positions – for example about the desirability or undesirability of regulation – for granted. When Rupert Murdoch announced to the Federal Trade Commission that 'the government needs to clear the path for companies to invest and innovate – by reducing unnecessary regulation and eliminating obstacles to growth and investment' (Murdoch, 2009), we should treat this both as a claim that needs substantiating as well as an expression of a particular, neo-liberal world view. Similarly, when Robert McChesney challenges the idea that we are witnessing a period of deregulation and that instead we are seeing 'government regulation that advances the interests of the dominant corporate players' (2004: 19–20), we should check the evidence and consider this statement in relation to his commitment to a form of radical democracy. Indeed, we should investigate both claims side by side in order to achieve a grounded understanding of contemporary arguments concerning regulation.

A stimulating debate along these lines was hosted by the openDemocracy website (www.opendemocracy.net/) where, broadly speaking, three positions were set out in relation to the question of media ownership: that we are seeing decreasing levels of concentration that pose no threat to media pluralism and diversity; that we are seeing *increasing* levels of concentration that represent a disturbing trend; and that a focus on levels of concentration is unwarranted and insufficient to explain the dynamics of the contemporary media.

The first position was most forcefully put by Ben Compaine (2001) and David Elstein (2002) who argued that, given the impact of new technologies and the increased number of media sources and outlets, ownership is becoming more dispersed, control of media markets more fragile and the dangers of oligopoly more distant. Compaine presents evidence (2001: 2) that the influence of the fifty largest media corporations in the US increased only fractionally from 1986 to 1997 and that while some sectors are more concentrated, particularly those like daily newspapers that are in decline, the general picture is of more choice and more competition. Compaine also argues that media ownership is generally 'extremely democratic' (2001: 4) because of the ability of shareholders to apply pressure on management and that, while not perfect, existing patterns of ownership provide audiences with more than a reasonable amount of diversity and quality. Elstein, a former head of programming at BSkyB, concurs with this view that the market has facilitated choice and quality and that the idea of omnipotent media moguls seriously underestimates the extent to which these entrepreneurs face uncertain technological futures, regulatory impediments and high-risk investments. 'The global media game' he concludes 'is far more complex than a simple David and Goliath story' (2002: 7).

These arguments were rebutted by commentators who argued that the expanding market power of a handful of transnational media corporations (for example Time Warner, Disney, Sony, Viacom and News Corporation) distorts competition, undermines independent journalism and restricts diversity. According to Robert McChesney, the growth of these media conglomerates that have cross-media interests in the world's largest media markets has led to a situation in which these 'giants do compete ferociously, but they do so under the rules of oligopolistic markets, meaning they have far greater control over their fate than those in truly competitive markets' (2001: 1). This concentration of economic and symbolic power is disastrous for democracy, not simply because it restricts the number and variety of information sources but also because it provides unaccountable media owners with extensive political influence. Witness (once again) the rise of former Italian prime minister Silvio Berlusconi who would not have reached that position 'if he did not dominate a massive media empire that enabled him to manufacture a political party' (Curran, 2002b: 3). Curran also argues that, *contra* Compaine's claim (2001, 4) that media businesses tend 'to be run not to promote an ideology but to seek profit', media concentration is likely to promote the dissemination of 'conservative or market-friendly positions, but more rarely their antithesis' (Curran, 2002b: 3). McChesney (2002) illustrates this point with reference to the generally compliant press coverage in the US of the decision to go war in Iraq, the Enron scandal and the 2000 presidential election, all of which privileged official sources and lacked critical perspectives and non-routine voices. The logic of McChesney and Curran's arguments is that restrictions on media ownership are therefore necessary to curb concentrated media power and to foster a more open and *genuinely* competitive environment.

David Hesmondhalgh (2001) provides a critique of both sets of arguments and an alternative focus to one on ownership. Whilst accusing Compaine of neglecting the negative consequences of media conglomeration and marketization, he also criticizes McChesney for painting an 'inaccurate and unrecognizable picture of complete corporate control and uniform output' (McChesney, 2001). Instead, Hesmondhalgh argues that media producers enjoy sufficient autonomy allowing them not to follow meekly or consistently the dictates of owners and executives and that even following a ratings-led perspective is likely to enable the production of challenging and alternative forms of content just as it is also likely to lead to the emergence of trivial and sensationalized material. An emphasis on ownership alone fosters an instrumentalist approach (on both sides of the argument) that marginalizes more difficult questions concerning the management of 'risk and uncertainty within the media business' (Hesmondhalgh, 2001). The unpredictability of audiences and the high degree of failure means that a certain amount of innovation and creativity is built into media production in such a way that media content cannot be reduced to questions of ownership alone. Furthermore, the emphasis on news and current affairs in the debates on ownership misses out on the role of other formats and genres and diminishes the importance of popular culture. Finally, the fact that there is 'no necessary link between oligopoly and reduced diversity' (Hesmondhalgh, 2001) – witness the explosion of niche musics presided over by both large and small corporations – suggests to Hesmondhalgh that 'there are other factors, besides ownership, at work in explaining media output' (Hesmondhalgh, 2001).

Hesmondhalgh makes some valuable points about the contradictions of media power but his insistence that we need to move on from a focus on ownership is more problematic. Firstly, as we shall see later, this is a view shared by proponents of ownership liberalization who wish to de-prioritize concentration as a controversial issue. Indeed Compaine is quite happy to use Silvio Waisbord's nuanced argument, that *despite* unwelcome concentration there remain political opportunities and 'spaces for conflict and change' (Waisbord, 2002: 1), as evidence of the innate dynamism and diversity of media markets. Secondly, it is not the case that those who harp on about ownership claim that it explains everything about media performance. McChesney (2004: 208), for example, repeatedly points out that there are many factors that shape media output apart from ownership but he argues nevertheless that a public policy emphasis on stimulating competition within media markets by curbing excessive concentration is needed precisely in order to facilitate diverse, creative, unexpected and antagonistic content. Thirdly, for all those interested in an independent, critical and relevant media system, is it the right time to move on from an interest in ownership, particularly when the relaxation of media ownership rules continue to be debated by policymakers and regulators in both Britain and the US?

The lively debate hosted by open Democracy is a productive way of engaging with foundational propositions concerning media ownership. It reminds us that ownership matters not only because of the largely unaccountable political and economic power that accrues to those individuals and corporations with extensive media interests, and not only because they are able to deploy their market power to act as influential cultural gatekeepers, but also because administrations and regulators in the UK and US have been particularly absorbed with trying to relax media ownership regulations over the last ten years. If media ownership is only a peripheral influence on media content, why are politicians, policymakers and media moguls so preoccupied with it? Debating what are often presented as fixed positions is both a stimulating and an analytically rewarding way of reflecting on approaches to ownership and regulation.

EXAMINE MOBILIZING ARGUMENTS

As well as treating ownership and regulation as subjects for debate, we should interrogate very carefully the arguments of elite actors in the contexts *in which they are proposed.* This involves focusing on those documents and processes where arguments are most forcefully mobilized: in corporations' annual reports, parliamentary transcripts, submissions to consultations and reviews, contributions to regulator or industry panels, evidence to inquiries and so on. Of course, not all information is made public and the often subterranean connections between lobbyists and government, in particular, make it difficult to identify all relevant and influential interventions (see Freedman, 2008: 93). It is, nevertheless, worth highlighting the claims of key participants involved in specific ownership-related issues and this next section focuses on four of the most significant justifications articulated by advocates of liberalization in their campaigns to scale back broadcast ownership regulations.

The most common argument is technological: that in a brave new world of digital developments and consumer choice, there is little need to worry about oligopoly or a lack of diversity. The dizzying speed of technological innovation and market adaptation make redundant most attempts to control artificially the structure of markets or the preferences of consumers. This is especially true for television where spectrum scarcity, the phenomenon that provided the historic justification for regulating broadcasting, has been largely abolished in the digital age of virtually limitless bandwidth. 'Given the overwhelming wealth of both broadcast and non-broadcast media options available to consumers today', argued some of the largest entertainment companies in the US back in 2003, 'the factual underpinnings of the spectrum scarcity rationale of broadcast regulation ... no longer are valid (if they ever were)' (Joint Commenters, 2003: v).

Referring to the communication 'abundance' specifically facilitated by the internet, former FCC chair Michael Powell (2003) argued that:

> the most striking difference between the world today and the world pre-remote [control] is that Americans now have access to a bottomless well of information called the Internet ... The time has come to honestly and fairly examine the facts of the modern marketplace and build rules that reflect the digital world we live in today, not the bygone era of black-and-white television.

Adam Thierer, an influential Washington think tank analyst and firm supporter of ownership liberalization, echoes this approach when arguing that we are currently living in a 'golden age' of media.

> There has never been a time in our nation's history when the citizens had access to more media outlets, more news and information, or more entertainment. Abundance, not scarcity, is the defining fact of our current media age... the question of who owns what, or how much they own is utterly irrelevant (Thierer, 2005: 161).

In recent FCC proceedings on media ownership, the general counsel for the National Association of Broadcasters (NAB) attacked the 'asymmetric regulation of broadcast outlets in comparison to their [cable and online] competitors' and called for the 'modernization of out-of-date restrictions that do not reflect current competitive realities in the Internet age' (Mago, 2009: 8).

Essentially, the argument is that the internet has stolen audiences, attention and advertising away from television in particular and that, therefore, the 'rules of the game' have changed forever. BSkyB's senior legal adviser, for example, argues that British regulators have 'played down the significant increase in the use of the internet as a main source of news' (Austin, 2009) in their decision *not* to scrap national media ownership rules. Ken Ferree, formerly head of the FCC's Media Bureau and then a fellow at the now-defunct Progress and Freedom Foundation, declared to a 2009 ownership hearing that 'there can be no denying that radio and television broadcast is a much less significant part of the media universe than they once were' (Ferree, 2009: 1), leading him to ask one simple question: 'are these ownership rules even relevant in today's media marketplace?' (Ferree, 2009: 1).

Liberalization is also justified on economic grounds: that the profitability of media businesses would be enhanced through the relaxation of ownership rules. This was most crudely put by Viacom president Mel Karmazin who noted that a wave of deregulation followed the Gulf War in 1991 and therefore argued for further ownership rule changes to compensate broadcasters for losses suffered after 9/11. 'We believe acquisition opportunities could present themselves', he said. 'There are a lot of very leveraged companies out there. If there is a silver lining in all of this, it could be that some of our competition is not as strong' (quoted in Freedman, 2003: 297). According to Bob Okun, a lobbyist for NBC, the economic climate for broadcasters 'only got worse after Sept, 11. As we look to 2002, it certainly makes sense to re-examine any and all regulations that continue to hinder broadcasters' ability to grow and survive an increasingly difficult marketplace' (quoted in McClintock, 2001).

Reviews of media ownership rules are frequently dominated by claims by 'old media' companies that by continuing with existing prohibitions on ownership, traditional media businesses will be at a disadvantage in the new digital environment This argument was made with particular vigour by CBS in the 2006 ownership review, warning that without the ability to manoeuvre freely in the marketplace, 'we put at risk the rich American tradition of free, over-the-air broadcasting, including its universal availability, commitment to public service and localism, and its high quality entertainment, sports and news programming' (CBS, 2006: 3). Many other US broadcast interests, particularly those representing smaller station owners, have made the same point: that liberalization is increasingly necessary to maintain 'economic viability in a market dominated by consolidated multichannel providers and other competitors' (NAB, 2006: iv). The director of Heart Television put it this way: 'By consolidating ownership, operational and technological resources, local television stations can achieve new economic efficiencies and will be able to compete more effectively in the digital era' (Barrett, 2009: 4). The British communications regulator also agrees that consolidation will allow companies to realize economies of scale and scope, to recruit the best managers and provide them with better access to overseas capital (Ofcom, 2006: 7).

A third source of pressure to change existing ownership rules is the increasingly fierce interrogation by media businesses and regulators of the link between plurality of ownership and diversity of viewpoint, a connection that is at the heart of democratic accounts of media policy but dismissed by the NAB as 'unproven' (2006: 42). Ofcom too notes that '[o]wnership plurality *does not necessarily ensure editorial or viewpoint diversity*' (2006: 6, emphasis in original) and suggests that 'it may also be the case that different sources of news offer the same perspective' (2006: 6). According to this logic, the best guarantee of diversity is not statutory restrictions on ownership but an environment in which the market is allowed to operate as freely as possible and in which competing viewpoints will be publicized as a result of consumer sovereignty and not regulatory intervention.

This links to a fourth justification for liberalization: a belief that regulation is likely only to constrain the innate creativity unleashed by market forces. This reflects a neo-liberal conception that the state should have only a limited role in directing productive activity, that innovation is best realized when left to its own devices and that restrictions on media ownership are, ultimately, a violation of speech rights. 'My notion of the First Amendment', wrote Shaun Sheehan, chief lobbyist for the giant Tribune media corporation, in a letter to William Safire, the *New York Times* columnist and supporter of ownership rules,

> is that the framers feared excesses of government and that the press would be the countervailing force, the unfettered watchdog. Without scale and with the White House gunning for its broadcast licences, could the Washington Post have pursued Watergate? The New York Times, the Pentagon Papers? The more we get back to 'Congress shall make no laws' the better we will be. (Sheehan, 2003)

Ownership rules, according to this rhetoric, conflict with basic First Amendment rights to freedom of expression and should be minimized in order to protect the speech rights of corporations. Market forces, according to this perspective, are not only *not* inimical to pluralist goals of source and viewpoint diversity but are increasingly the main guarantors of such aspirations.

Identifying the key arguments posed by pro-liberalization voices is a fantastic way not simply of bringing to life what can be rather dry debates on ownership but also of helping to know where to start in challenging these perspectives. Opponents of concentration can point to the strong strand of technological determinism that exaggerates the impact of new technologies in a particularly one-sided way, as *requiring* liberalization. Arguments that the internet has done away with the need for ownership rules aimed at ensuring media pluralism and diversity are as much ideologically driven as they are the result of technological necessity or empirical accuracy (Curran et al., 2012). 'The sceptical observer', according to Jean Seaton (1998: 123), may 'conclude that the frequently heard claim that technology "requires" this or that policy, is just a way of masking a vested interest in the handy disguise of modernity'. Those interested in ownership questions can point to conceptual debates about the democratic significance of the media and use empirical data to reject the argument that the market, as a matter of course, gives voice to divergent groups and perspectives (see Di Cola, 2006 for research that points to a link between ownership consolidation and 'homogenized programming').

A further response to pro-liberalization arguments concerns their underlying assumptions about ownership rules themselves: that they constitute a 'barrier' to growth, a 'shackle' on investment, a 'violation' of free speech, and a 'restraint' of free trade. The equation of ownership rules with overwhelmingly negative and restrictive characteristics is designed to shift the focus away from their original intention: to promote a wider range of voices and perspectives than would be provided by market forces alone. In fact, it is more than likely that in an internet-dominated future there will still be the need for special controls to promote diversity. Given the amount of internet traffic dominated by Google, YouTube and Facebook as well as by established media companies, a broadband internet future is certain to produce new types of monopoly and new forms of exclusion that can only be tackled with purposeful and positive intervention into media markets. It seems rather obvious that if we are still committed to pluralism and diversity, then there is little point in junking the traditional mechanisms for achieving these outcomes simply because we are faced with *different* technologies. However, in order to be able successfully to rebut pro-liberalization arguments, those opposed to further liberalization need to be familiar with the detail of their arguments.

FOCUS ON INACTION AND SILENCE

At first glance, a call to focus on *inaction* in relation to the ownership and regulation of the television sector makes very little sense when you consider the hyperactivity of governments, civil servants, regulators and broadcasters around the world as they prepare to shift from analogue to digital television or to update their media ownership rules for a digital future. In the UK, for example, the previous Labour government published a report, *Digital Britain* [*DB*] (BIS/DCMS, 2009) that provided the legislative framework for communications at a critical moment in the country's digital development. *DB* was very precise in some of its recommendations: it sanctioned a small monthly levy on every fixed copper telephone line to make sure that the whole of the UK population had access to a broadband service by 2012; it further proposed to set up three pilot Independently Funded News Consortia (IFNC) schemes by 2012, one in Scotland, one in Wales and one in an English region, to deliver television news as a result of its acceptance that the main commercial channel, ITV1, will no longer be able to afford to produce regional bulletins. *DB*'s plans for radio involved an equally activist approach based on a 'clear direction from Government' (2009: 75) to move national FM stations over to digital broadcasting (DAB) and to shift medium wave services to FM by 2015. Many of these schemes have since either been dropped or deemed to be 'unrealistic' but, by any account, this is not a regulatory environment marked by sloth or by passivity, particularly given the huge number of reviews, consultations and 'action plans' considering, in particular, the future of broadband roll out, intellectual property television news, public service broadcasting (PSB) and media ownership rules.

These activities, however, all refer to *overt* acts of government, to the decision-making actions of policymaking bodies and to the consultative work of regulatory agencies – in other words to the public exercise of 'official' power. There is, however, an entirely different approach, rooted in debates in social and political science that emerged in the 1960s (for example Bachrach and Baratz, 1962; Lukes, 2005 [1974]) that seeks to locate power in less visible arenas of decision-making and indeed to focus instead on examples of silence and non-decision-making. Instead of focusing exclusively on the evidence provided by key participants, the content of white papers and regulatory orders, the detail of draft bills and Congressional acts and the flavour of parliamentary debates, this approach examines the means by which alternative options are marginalized, conflicting values de-legitimized and rival interests de-recognized. This flows from an understanding of decision-making as an ideological process structured by unequal access to power and where power itself 'is at its most effective when least observable' (Lukes, 2005 [1974]: 1).

It is worth noting at this point that the 'liberal' narrative of non-intervention and the libertarian principles on which it is based have not disappeared in the 'hyperactivism' of recent media policy approaches. The liberal demand for non-intervention is perhaps most purely expressed by James Murdoch, the chief operating officer of 21st Century Fox, who has called for the dissolution of all rules affecting the broadcast sector: 'there is a long way to go before consumers enjoy the sovereignty that is their right. We don't need more controls to achieve that. We need a bonfire of controls. Then commerce will be free to drive out culture forward to the real "golden age" of broadcasting' (cited in Plunkett, 2005).

Indeed, this latter perspective, intimately linked to neo-liberal visions of economy and society that, as we have seen, privilege the dynamism of market forces and relegate the state to a secondary role, is far from absent in *Digital Britain*. The opening of *DB* makes it clear that meddling should be the exception and not the rule. In those cases where market forces are adequately serving customers, the 'simple position is that these sectors are working well and do not need commentary, intervention or unnecessary interference' (BIS/DCMS, 2009: 9). Furthermore, the report argues that 'an excessive focus on the sector could chill operational negotiations and decisions while participants wait to see how it all pans out' (2009: 10). Non-intervention, in other words, is the safest option, generally the default position for neo-liberal governments for whom an activist public policy approach is only needed in those areas in which 'market failure' is likely to take place, for example the provision of a universal broadband infrastructure or the delivery of regional and local news.

Now this sounds reasonable enough at first glance but, of course, it begs the question of what constitutes an example of 'market failure', who decides this, what values are brought to bear on the decision-making process and what other options are considered in achieving regulatory objectives in relation to this particular area? The central question then concerns not the rights and wrongs of intervention *per se* (as neo-liberals claim) but the reasons for intervening in *this* area and *not* in another one. Why does a particular issue become a 'problem' that is worthy of public policy attention? Why does the building of a broadband infrastructure become an urgent 'problem' for which senior government figures attempt to find a solution while the concentration of key sectors of media does not? Why is piracy such a major issue for legislators but not falling budgets for original television programming? An exclusive emphasis on immediate and observable regulatory debates misses out on this deeper level of questioning: what are the presuppositions that govern the parameters of the regulatory process and that shape what questions are asked as well as which ones are not? Who holds *definitional* as well as *organizational* power inside policy-making and regulatory networks?

This is where a focus on inaction and non-decision-making can be highly productive in terms of identifying the 'dominant values and the political myths, rituals and institutions' (Bachrach and Baratz, 1962: 950) that structure access to power in decision-making contexts. Such an approach would involve an analysis of how the policy and regulatory agenda is formulated and objectives are framed, which voices are privileged inside and which are frozen out of debates, how participation is invited or barriers erected, and from whom information (or 'evidence') is sought. It would focus on the ideological positions that are brought to bear in shaping agenda-setting questions and suggest that regulation, far from being a simple administrative act, is instead a series of highly politicized transactions in which certain preferences are celebrated and others marginalized (Freedman, 2008).

Silence, in this context, does not mean 'doing nothing' and 'inaction' does not suggest a lack of energy on the part of policymakers and regulators nor a reluctance in principle to intervene. Rather it refers to a strategic decision that the best way to promote hegemonic interests and to naturalize foundational values is through a *particular* role for the state. In recent years this has consisted of administrations assisting, as quietly as possible, in the rolling out of markets and market logic in the media sector. In helping market forces to operate 'freely' and to extend their reach, public policy, in for example the US and UK, has been crucial. From decisions on what constitutes an acceptable level of broadcast ownership concentration to their determination to secure analogue switch-off, and from their unyielding support for domestic rights holders to increase market opportunities in export ventures to their commitment to leverage existing copyright protections onto emerging distribution platforms, the British and American states have played a decisive role on behalf of key sections of their media industries. Neo-liberal commentators may call it 'deregulation' and present this as evidence of 'small government' but many critics (for example Humphreys, 1996) have long described it as re-regulation and as proof of a complex partnership between the state and private industry.

This partnership is successful to the extent that state and private elites are able to legitimize their preferences and to use their decision-making influence pre-emptively to smother any challenges (and the campaign in 2003 against the proposed relaxation of ownership rules in the US shows that this is not a foregone conclusion). Silence, in this context, refers to the options that are *not* considered, to the questions that are kept *off* the policy and regulatory agendas, to the players who are *not* invited to the bargaining table, and to the values that are seen as unrealistic or undesirable by those best able to mobilize their decision-making power. For example, in current discussions concerning the allocation of spectrum released by digital switchover, why is it that principles of remuneration dominate and why is it that the idea of handing spectrum to community and public bodies is virtually ignored? When it comes to the future of news, why is it that the idea of introducing some sort of public subsidy is anathema to policymakers, ignoring the simple fact that news production has long been subsidized in different ways by the state?

There are many other revealing absences in the policy approach articulated by initiatives like *Digital Britain*. Why does it assume that it is only the private sector that should take a lead in 'delivering the effective modern communications infrastructure we need' (BIS/DCMS, 2009: 1)? Where is the research that shows a demand for making existing radio sets redundant as part of the move to DAB? Why is there no mention at all of the possibility of industry levies on pay TV operators and mobile phone providers to fund regional television news (as proposed by Barwise, 2009 and the Institute for Public Policy Research, 2009)? Whose interests are reflected by Ed Richards, the chief executive of Ofcom, when he claims that an industry levy is unlikely 'to be an attractive proposition in the middle of a recession' (Richards, 2009): business leaders or ordinary viewers? Why does *DB* raise the spectre of siphoning off part of the television licence fee to fund public service content but promises not to interfere with the profits of private companies? As the media economist Patrick Barwise argues (2009):

> I don't think you have to be a Marxist or a paranoid schizophrenic to note that the dominant pay-TV operator is controlled by Rupert Murdoch and that this might, at the margin, influence politicians' willingness to introduce a levy on pay TV, however small and however great the benefits for the British public and the creative industries.

Here is a clear case of power being deployed inside the bargaining process not to argue against the introduction of industry levies but, far more effectively, to prevent the proposal ever being raised as a 'serious' policy option in the first place. It is the conception of the 'third face' of power raised by Lukes (2005

[1974]: 1) as being especially effective precisely because of its invisibility.

All the various proceedings, inquiries, seminars and working parties investigating ownership rules together with all the documents, reports and policies that emerge from these investigations are likely to be significant for what they exclude, for the questions they fail to pose and for the alternatives they fail to consider. It is these gaps on which those interested in securing media pluralism and diversity ought to focus.

CONCLUSION

This chapter has proposed a number of different ways of approaching questions of media ownership and regulation. It has suggested that the interested researcher may turn to empirical data to substantiate particular claims or to justify certain policy recommendations; it then turned to more normative assumptions about 'desirable' forms of ownership or institutional arrangements that may best facilitate democratic dialogue; finally, it argued that researchers should consider ownership and regulation as systems of thought that can be illuminated by treating them as inspirations for debate that reveal particular ideological positions, by focusing on the specific arguments used by elites to justify and naturalize these positions, and by revealing the lacuna in these positions – the assumptions and propositions that are marginalized and sometimes excluded from official debates.

Some readers may point to a confusion at the heart of this typology – that the chapter is encouraging us to separate out these approaches: to produce data but also to draw on conceptual models; to focus on visible instances of regulatory activity or ownership battles but simultaneously to reflect on the hidden depths of the decision-making environment; to see ownership and regulation as administrative affairs but also as ideological processes. Each of these positions may be singly justifiable but, taken together, they provide a much more productive method of communications research. The normative without the empirical tends towards the abstract while the empirical without the conceptual tends towards the tedious. By adding in the recognition that what are all too often seen as rather bureaucratic processes are in fact systems of thought and action that, in the present climate, mobilize market-oriented values and marginalize non-commercial objectives, debates concerning ownership and regulation can be seen as crucial analytical ciphers for contemporary neo-liberalism.

In conclusion, this chapter has argued for an approach to media ownership and regulation that combines a focus on structure with an emphasis on *interests*. We need to identify whose interests are best represented in policy and regulatory debates and to highlight the mechanisms both of the inclusion and exclusion of these different interests. We need to evaluate the dominant paradigms that mark the discussion of ownership and that characterize the regulatory process but also to introduce alternative frames and new starting points. Only by combining the empirical with the conceptual and the technical with the ideological, as well as analysing both existing structures and agents of change, can we be assured of capturing the real significance of debates and movements focused on media ownership and regulation.

NOTE

This chapter is based in part on 'Metrics, models and the meaning of media ownership', *International Journal of Cultural Policy*, December 2012 and on work for the author's *The Contradictions of Media Power* (Bloomsbury 2014).

REFERENCES

Associated Press (2006). 'FCC chair orders probe into why media ownership studies were destroyed', 19 September. Available at: www.foxnews.com/printer_friendly_story/0,3566,214392,00.html

Aufderheide, P. (1999). *Communications Policy and the Public Interest*. New York: Guilford.

Austin, D. (2009). Submission to Ofcom Review of Ownership Rules, 1 October. Available at: http://stakeholders.ofcom.org.uk/binaries/consultations/morr/responses/Sky.pdf

Bachrach, P. and Baratz, M. (1962). 'Two Faces of Power', *American Political Science Review*, 56(4): 947–952.

Bagdikian, B. (1990 [1983]). *The Media Monopoly*, 3rd edition. Boston: Beacon Press.

Baker, C.E. (2007). *Media Ownership and Concentration: Why Ownership Matters*. Cambridge: Cambridge University Press.

Barrett, D. (2009). Remarks to FCC Ownership Workshop, 4 November, FCC 09–182.

Barron, J. (2000). 'Structural Regulation of the Media and the Diversity Rationale', 52 *Federal Communications Law Journal*, 555–560.

Barwise, P. (2009). 'New Forms of Funding for PSB', speech to a seminar organized by the Federation of Entertainment Unions, 22 June. Available at: www.bectu.org.uk/filegrab/NewFormsofFundingforPSBbyProfessorPatrickBarwiseFEU22.6.09.doc?ref=310.

BIS (Department for Business, Innovation and Skills)/DCMS (Department for Culture, Media and Sport) (2009). *Digital Britain: Final Report*. Cm 7650. London: The Stationery Office.

Born, G. (2004). *Uncertain Vision: Birt, Dyke and the Reinvention of the BBC*. London: Secker & Warburg.

Born, G. (2006). 'Digitising Democracy', in J. Lloyd and J. Seaton (eds) *What Can Be Done? Making the Politics and Media Better*. Oxford: Blackwell, 102–123.

Braman, S. (2007). *Change of State: Information, Policy, and Power*. Cambridge, MA: MIT Press.

Cassidy, J. (2006). 'Murdoch's Game: Will he Move Left in 2008?', *New Yorker*, 16 October.

CBS (2006). Comments of CBS Corporation in the matter of the 2006 Quadrennial Review, FCC 06–121, 23 October.

Center for Social Media (2009). *Public Media 2.0: Dynamic, Engaged Publics*. Washington DC: Center for Social Media.

Chang, H-J. (2011). *23 Things They Don't Tell You About Capitalism*. London: Penguin.

Compaine, B. (2001). 'The Myths of Encroaching Global Media Ownership', openDemocracy, 6 November. Available at: www.opendemocracy.net/media-globalmediaownership/article_87.jsp

Compaine, B. and Gomery, D. (2000 [1979]). *Who Owns the Media? Competition and Concentration in the Mass Media Industry*, 3rd edition. Mahwah, NJ: Lawrence Erlbaum.

Creech, K. (2007). *Electronic Media Law and Regulation*, 5th edition. Burlington, MA: Focal Press.

Curran, J. (2002a). *Media and Power*. London: Routledge.

Curran, J. (2002b). 'Global Media Concentration: Shifting the Argument', openDemocracy, 22 May. Available at: www.opendemocracy.net/media-globalmediaownership/article_37.jsp

Curran, J., Fenton, N. and Freedman, D. (2012). *Misunderstanding the Internet*. London: Routledge.

Department for National Heritage (1995). *Media Ownership: The Government's Proposals*, white paper, CM 2872. London: HMSO.

Di Cola, P. (2006). *False Premises, False Promises: A Quantitative History of Ownership Consolidation in the Radio Industry*. Washington DC: Future of Music Coalition.

Doyle, G. (2002). *Media Ownership*. London: Sage.

Eggerton, J. (2008). 'Barack Obama's Media Agenda: An Exclusive Interview', *Broadcasting & Cable*, 16 June. Available at: http://www.broadcastingcable.com/news/programming/barack-obama's-media-agenda-exclusive-interview/28375

Elstein, D. (2002). 'Stumbling Goliaths, Dithering Davids: Unpicking the Mythology of the Media Mogul', openDemocracy, 27 February. Available at: www.opendemocracy.net/media-globalmediaownership/article_124.jsp

FCC (2003). *Report and Order and Notice of Proposed Rulemaking*, 2 June, FCC 03-127. Available at: https://apps.fcc.gov/edocs_public/attachmatch/FCC-03-127A1.pdf

Ferree, K. (2009). Remarks to FCC Ownership Workshop, 4 November. Available at: http://transition.fcc.gov/ownership/workshop-110309/ferree.pdf

Freedman, D. (2003). 'Cultural Policy-Making in the Free Trade Era: An Evaluation of the Impact of Current World Trade Organization Negotiations on Audio-Visual Industries, *International Journal of Cultural Policy*, 9(3): 285–298.

Freedman, D. (2008). *The Politics of Media Policy*. Cambridge: Polity.

Gentzkow, M. and Shapiro, J. (2006). 'Media Bias and Reputation', *Journal of Political Economy*, 114(2): 280–316.

Gibbons, T. (1998). *Regulating the Media*. London: Sweet & Maxwell.

Hart, K. (2009). 'Comcast-NBC deal finds campaign cash converging with Obama's principles', *The Hill*, 5 December.

Herman, E. and Chomsky, N. (1988). *Manufacturing Consent: The Political Economy of the Mass Media*. New York: Pantheon Books.

Herman, E. and McChesney, R. (1997). *The Global Media: The New Missionaries of Corporate Capitalism*. London: Cassell.

Hesmondhalgh, D. (2001). 'Ownership is Only Part of the Picture', openDemocracy, 28 November. Available at: www.opendemocracy.net/media-globalmediaownership/article_46.jsp

Horwitz, R. (1989). *The Irony of Regulatory Reform: The Deregulation of American Telecommunications*. Oxford: Oxford University Press.

Hoynes, W. (2007). 'Public Broadcasting for the 21st Century: Notes on an Agenda for Reform', *Critical Studies in Media Communication*, 24(4): 370–376.

Humphreys, P. (1996). *Mass Media and Media Policy in Western Europe*. Manchester: Manchester University Press.

Institute for Public Policy Research (IPPR) (2009). *Mind the Funding Gap: The Potential of Industry Levies for Continued Funding of Public Service Broadcasting*. London: IPPR.

Joint Commenters (2003). Comments of Fox, NBC and Viacom in the matter of the 2002 Biennial Regulatory Review, 02–277, 2 January.

Kelly, M., Mazzoleni, G. and McQuail, D. (eds) (2004). *The Media in Europe: The Euromedia Handbook*. London: Sage.

Lazarsfeld, P. and Merton, R. (2004 [1948]). 'Mass Communication, Popular Taste and Organized Social Action', in J.D. Peters and P. Simonson (eds) *Mass Communication and American Social Thought: Key Texts, 1919–1968*. Lanham, MD: Rowman & Littlefield, 230–242.

Leys, C. (2001). *Market-Driven Politics: Neoliberal Democracy and the Public Interest*. London: Verso.

Lukes, S. (2005 [1974]). *Power: A Radical View*, 2nd edition. Basingstoke: Macmillan.

Mago, J. (2009). Remarks to FCC Ownership Workshop, 4 November, FCC 09–182.

McChesney, R. (1999). *Rich Media, Poor Democracy*. New York: The New Press.

McChesney, R. (2001). 'Policing the Thinkable', openDemocracy, 23 October. Available at: www.opendemocracy.net/media-global mediaownership/article_56.jsp

McChesney, R. (2002). 'It's a wrap? Why media matters to democracy',openDemocracy, 9 May. Available at: https://www.opendemocracy.net/democracy-globalmediaownership/article_61.jsp

McChesney, R. (2004). *The Problem of the Media*. New York: Monthly Review Press.

McChesney, R. (2007). *Communication Revolution*. New York: The New Press.

McClintock, P. (2001). 'Nets See Regs as Best Relief', *Variety.com*, 14 October. Available at www.variety.com/article/VR1117854257.html?categoryid=14&cs=1

McQuail, D. (1992). *Media Performance*. London: Sage.

Murdoch, R. (2009). 'The Future of Journalism is More Promising Than Ever', speech to the Federal Trade Commission, 3 December. Available at: http://www.pressgazette.co.uk/node/44737

NAB (National Association of Broadcasters) (2006). Comments of the NAB in the matter of the 2006 Quadrennial Review, FCC 06–121, 23 October.

Napoli, P. (2001). *Foundations of Communications Policy*. Cresskill: Hampton Press.

Napoli, P.M. and Karaganis, J. (2007). 'Toward a Federal Data Agenda for Communications Policymaking', *CommLaw Conspectus*, 16: 53–96.

Noam, E. (2009). *Media Ownership and Concentration in America*. New York: Oxford University Press.

Obama, B. (2008). 'Barack Obama: Connecting and Empowering All Americans Through Technology and Innovation', campaign fact sheet. Available at: http://www.wired.com/images_blogs/threatlevel/2009/04/obamatechplan.pdf

Ofcom (2006). *Review of Media Ownership Rules*, 14 November. London: Ofcom.

Ofcom (2009a). 'Consumers Prefer Communications Over Celebrations in Recession: People Taking Control and Getting Good Deals in Downturn', News release,

6 August. Available at: http://media.ofcom.org.uk/news/2009/consumers-prefer-communications-over-celebrations-in-recession-people-taking-control-and-getting-good-deals-in-downturn/

Ofcom (2009b). *Communications Market Report 2009*. London: Ofcom.

Open Society Institute (2005). *Television across Europe: Regulation, Policy and Independence*. Budapest: OSI.

Overbeck, W. (2007). *Major Principles of Media Law*, 18th edition. Belmont, CA: Thomson Wadsworth.

Plunkett, J. (2005). 'Murdoch calls for "bonfire" of media regulations', *Guardian*, 22 September. Available at: http://www.theguardian.com/media/2005/sep/22/broadcasting.bskyb2

Powell, M. (2003). 'Should Limits on Broadcast Ownership Change?', *USA Today*, 21 January.

Puttnam, D. (2006). 'Nation of Fools', *New Statesman*, 4 December.

Reporters without Borders (2013). 'New broadcast media law is constitutional, Supreme Court rules', RWB, 30 October. Available at: http://en.rsf.org/argentina-new-broadcast-media-law-is-30-10-2013,45397.html

Richards, E. (2009). 'Independently Funded News Consortia', speech to the Local Media Summit, 28 April. Available at: http://media.ofcom.org.uk/speeches/2009/independently-funded-news-consortia/

Sanchez-Taberno, A. et al. (1993). *Media Concentration in Europe*. Dusseldorf: European Institute for the Media.

Schiller, D. (1999). *Digital Capitalism: Networking the Global Market System*. Cambridge, MA: MIT Press.

Seaton, J. (1998). 'A Fresh Look at Freedom of Speech', *Political Quarterly*, 69(b): 117–129.

Sheehan, S. (2003). Letter to William Safire, 3 July, private correspondence.

Streeter, T. (1996). *Selling the Air: A Critique of the Policy of Commercial Broadcasting in the United States*. Chicago: University of Chicago Press.

Thierer, A. (2005). *Media Myths: Making Sense of the Debate over Media Ownership*. Washington, DC: Progress and Freedom Foundation.

Thierer, A. and Eskelsen, G. (2008). *Media Metrics: The True State of the Modern Media Marketplace*. Available at: SSRN: www.ssrn.com/abstract=1161312

Waisbord, S. (2002). 'Media Concentration in Latin America', openDemocracy, 27 February. Available at: www.opendemocracy.net/media-globalmediaownership/article_64.jsp

Waldfogel, J. (2007). *The Tyranny of the Market*. Cambridge, MA: Harvard University Press.

Williams, R. (1968 [1962]). *Communications*. Hardmondsworth: Penguin.

2

Regulation and Ownership in the United States

Allison Perlman

Scholarship on media regulation and ownership in the US has been divided, broadly, between two traditions. The first, an administrative approach, understands communications research as an adjunct to policymaking. Deploying quantitative social scientific methods, this work presumes to seek objective answers, rooted in hard empirical data, to inform the policymaking community on topics such as media effects or the relationship between media ownership and media diversity. Its critics have argued that administrative research is no more objective than other forms of inquiry, but rather is shaped and bound by the ideological assumptions of regulators and industry, assumptions that it, by practice, normalizes and legitimates. Administrative research propels the patina of a rational, technocratic, data-driven policymaking process operating within apolitical legal frameworks. In the 1980s, Smythe and Van Dinh (1983) and Rowland (1986) would level withering critiques of the administrative approach as a form of intellectual capture, one in which academics had reduced their scholarship to instrumentalist research on behalf of business and government interests.

A critical approach to media regulation and ownership, alternatively, sees media regulation as a historically situated practice imbricated within existing social relations. Critical research in media regulation is not a unified field, but rather a predisposition towards one's object of analysis. It has encompassed research in media history, political economy, social movement theory, critical legal studies and cultural studies. Across these divergent approaches has been a rejection of the premises of administrative research and a critical engagement with the assumptions and practices within which media regulation operates. In addition, critical research frequently has desired to intervene in the policymaking process by puncturing, rather than reifying, the assumptions and definitions circulating within the regulatory community.

Significantly, these calls to arms in the 1980s for critical research in media regulation took place in the midst of substantial changes within the policy arena in the US. While advocacy for media deregulation began in Congress in the 1970s, it was implemented by the Federal Communications Commission (FCC) – along with its allies in the White House, Congress, and on the federal bench – in the 1980s. Adopting a neoliberal or 'marketplace approach to regulation' (Fowler and Brenner, 1982) the policymaking community enacted a wholesale re-evaluation of broadcast regulation, one that presumes that the 'free market', unfettered by government intervention, would best serve the communication needs of the public who expresses its will via its consumer choices. The 'marketplace approach' in practice has enabled greater levels of concentration in ownership, diminished broadcasters' public service obligations, and undercut the viability of citizen-based media reform activity. The 1996 Telecommunications Act codified into law this understanding of media regulation, along with the conflation of the public interest with marketplace efficiencies at its center (Aufderheide, 1999).

Accordingly, as this short overview suggests, critical research in media ownership and regulation consistently has responded not only to internal debates across the many disciplines from which it hails, but to concurrent regulatory paradigms and policy decisions. While encompassing a heterogeneous range of theoretical perspectives and methodological practices, critical research has been in dialogue with, and frequently has repudiated, the neoliberal framework within which media policy has operated since the 1980s.

PROLOGUE: MEDIA REGULATION AND THE CIVIL RIGHTS MOVEMENT

Beginning in the 1930s, broadcast regulation and the national commercial network system it sanctioned have been critiqued from myriad perspectives. Mass culture critics, for example, called for a publicly financed alternative to commercial broadcasting, lamenting commercial radio as insipid drivel (Kerwin, 1934). Others saw commercialism transforming radio into a ministry of propaganda for business interests, inculcating in its listeners values sympathetic to the interests of the business class while silencing dissenting views (Rorty, 1934). Free speech crusaders warned that concentration of ownership threatened the circulation of diverse views (Ernst, 1946), and economists debated the efficacy of a regulatory structure predicated on public ownership of airwaves licensed for use to private entities (Herzel, 1951; Smythe, 1952; Coase, 1959).

These early critiques of media regulation frequently invoked the 'public interest', though wavered between an economistic view of the public as consumers in a marketplace and a socio-political definition of the public as citizens in a democracy. Both definitions hinged on an abstracted and undifferentiated public, did not recognize the uneven distribution of economic and social power across publics, and saw no role for members of the public, outside their capacity as consumers or voters, to affect regulatory policy. This neutral, unitary vision of the public would come under attack in the 1960s.

The African American civil rights movement transformed on-the-ground opportunities for citizen engagement in media regulation. Racist programming practices in Jackson, Mississippi led civil rights activists to file a petition to deny the license of local station WLBT-TV. This case led to an expanded definition of 'standing', or the legal right to be heard, in license renewal decisions to include members of the public. The legal battle around WLBT also provoked concerns over regulatory capture and exposed the limiting definition of the 'public interest' that had structured broadcast regulation. The 1968 Report of the Advisory on Civil Disorders, or Kerner Report, similarly indicted the media as an instrument of white supremacy. The WLBT case and the Kerner Report both led to regulatory reforms and inspired a broadcast reform movement in which a range of identity- and issue-based groups used the petition to deny license renewal to wrangle concessions from local broadcasters regarding programming and employment practices.

Communications scholars in the 1970s, in turn, addressed the ambitions, strategies, and effects of the broadcast reform movement. Much of this work is both descriptive and celebratory, highlighting the actions of public interest groups and identifying tools at their disposal to make broadcasting more responsive to their needs (Merritt, 1974; Mills, 1974; Guimary, 1975; Lewis, 1986; Montgomery, 1989). Researchers also interrogate the efficacy of the broadcast reform movement, mapping models of the regulatory process to help maximize public impact on regulatory decisions (Chisman, 1977), or flagging potential pitfalls like the lack of cohesion across reform organizations (Branscomb and Savage, 1978), the muddied response of the FCC to a perceived usurpation of its regulatory authority by broadcast reform groups (Grundfest, 1976), or the limited impact of citizen engagement with administrative processes, deemed a 'ritualistic exercise' (Rowland, 1982, p. 2) in which reformers over time become 'captured by the theater's rules, with little hope of affecting the end of the play' (p. 41).

In addition, the policymaking process itself was put under the microscope. Institutional analyses of policymaking in the 1970s and 1980s situate the policymaking process within the localized culture of the FCC (Mosco, 1976; Cole and Oettinger, 1978) or define it as a political process in which the FCC is but one, and a comparatively not all that powerful, player (Baird, 1967; Krasnow and Longley, 1973; Baughman, 1985). While this work was part of a wider movement decrying regulatory failure and regulatory capture (see Sunstein, 1990), it also was a response to powerful criticisms of the media unveiled by reformers seeking to understand how regulation in the 'public interest' had so egregiously failed to serve the diversity of the American public.

Paradoxically, the broadcast reform movement and its criticisms of regulation helped lay the foundation for media deregulation. Horwitz astutely makes this point in *The Irony of Regulatory Reform* (1989), a field-redefining text that displaces prevailing theories of regulation with a historically grounded approach. Emerging during the Progressive Era and New Deal to balance economic stability with social and consumer safeguards, regulatory agencies displaced property and contract law as the primary mechanisms to order economic activity and operated as essentially closed systems in which economic and state interests negotiated in private. Horwitz suggests that the 1960s and 1970s brought not only new forms of regulation to address the social consequences of business behavior, but public critiques of the regulatory process itself. In response, regulated industries sought regulatory relief both from regulations of externalities and from public interest groups. The critique of regulatory capture from the left morphed into a critique of regulation from the right as an impediment to the commercial marketplace and the marketplace of ideas. This critique, along with the impact of new communication technologies and the embrace of a conservative economic agenda, led to deregulation, as well as the codification of a new interpretation of the public interest rooted in economic efficiencies and market controls.

In other words, the critique of the broadcast reform movement, as well as that of the communications scholars who used it as a starting point to analyze regulatory processes, led to its own undoing. The failure of media policy to attend to the public interest and the inefficacy of the FCC became part of the rationale to dismantle broadcast regulation, not to make it more robust and responsive to public needs. From the 1980s onwards, one of the central goals of critical media regulation research has been to challenge the premises and outcomes of the marketplace approach to regulation and to outline the hazards posed by increased media consolidation.

MEDIA OWNERSHIP

By and large, scholars in the US have addressed the issue of media ownership from two perspectives. The first ties media ownership to debates over the meaning of the First Amendment, and thus bears directly on questions of democratic self-governance. Interpretations of the First Amendment in the US have divided between free speech absolutists, who adhere to the Holmesian 'marketplace of ideas' metaphor and see the First Amendment as preventing government restrictions on speech, and free speech interventionists who, in contrast, understand the First Amendment as protecting the right of listeners, as per Meiklejohn (1960), to hear all that is worth saying. This latter interpretation thus carves an affirmative role for the government to eliminate threats posed to the speech rights of listeners. The second approach to media ownership works within a political economy tradition and asks how commercial media operate within an advanced capitalist society.

Since at least Morris Ernst's *The First Freedom* (1946), concentration of ownership has been positioned as a threat to the First Amendment. Adopting an interventionist interpretation of speech rights, Ernst argues that concentrated economic power within media industries limits public debate, the public sphere reduced to an expression of the interests of a 'narrowing group of owners and clients' (p. 180). This particular critique of media ownership would resurface in the 1960s in the context of the broadcast reform movement. While Rucker (1968) updates the argument leveled by Ernst over economic concentration and speech rights, even appropriating Ernst's title, Barron (1967) specifically attacks the 'romantic' view of speech rights, rooted in the 'marketplace of ideas' metaphor, as unsuitable to an age of mass media. Identifying civil rights nonviolent and violent protests as evidence of degraded speech conditions, and maintaining that protest has 'taken these forms because it has had nowhere else to go' (p. 1647), Barron advocates for an interventionist approach to speech rights in which the state secures speech opportunities for diverse viewpoints.

Significantly, from the 1940s through the 1970s, as Horwitz (1991) has illustrated, federal courts typically embraced an interventionist interpretation of the First Amendment in its broadcasting cases (notably in *NBC v. U.S.* [1943] and *Red Lion v. FCC* [1969]). Yet conservative appointments to the federal bench and to the FCC in the 1980s embraced an absolutist interpretation, one that privileged the 'active speaker' over the 'passive listener' (Fowler and Brenner, 1982, p. 238), and that, in turn, enfolded corporate speech rights within the sanctity of First Amendment protections. This view would find support in a spate of books published by free speech absolutists who insisted that content-based broadcast regulations and the license renewal process itself were affronts to the First Amendment (Friendly, 1976; Silber, 1980; Labunski, 1981; Krattenmaker and Powe, 1994).

To a large degree, many of the arguments regarding media concentration from the 1980s onwards have borne remarkable similarity to those of Ernst and Barron. What has been added has been detailed analyses of the scope and extent of cross-media concentration and its impact on media content (see the multiple iterations of Bagdikian's *The Media Monopoly*, 1983, 2004) and direct attacks on the political consequences of media deregulation. Baker's work exemplifies this tack (2002, 2007), as he illuminates how a neoliberal approach to regulation, and the levels of media consolidation it has produced, has resulted in a media system that harms democracy. Baker revives what had been an assumption in juridical spaces prior to the 1980s, that a nexus exists between diversity of opinion and diversity of ownership and that concentrated ownership dangerously restricts the circulation of ideas. Stein (2006) similarly takes on the neoliberal approach to speech rights ascendant in legal and policy circles by marshaling participatory democratic theory as a justification for an affirmative role of the state to assure conditions for public participation in and access to a robust public sphere.

A critical political economy approach to media ownership is sympathetic to the interventionist speech rights concerns over media ownership. Studies in the political economy of communication, as per Mosco (1996), place communication within a 'wider social totality' (p. 71). It broadly 'endeavors to connect how media and communications systems and content are shaped by ownership, market structures, commercial support, technologies, labor practices, and government policies' (McChesney, 2008, p. 12). Accordingly, it understands communications to be embedded within larger structures of political and economic power.

Since the late 1960s, political economy approaches to media ownership have emphasized how commercial media, supported and enabled by the state, circulate messages, images and values that function to maintain their power, render social critique and dissent invisible, and prop up the imperial ambitions of the military-industrial complex of which they are a part (Schiller, 1969, 1973, 1989; Tuchman, 1974; Parenti, 1986; Herman and Chomsky, 1988; Kellner, 1990). McChesney's prodigious scholarly output (1999, 2004, 2006, 2008) continues in this tradition, though it emphasizes the impact of neoliberalism on media policy, media ownership, and media power. He links media concentration and conglomeration to the deteriorating quality of journalism and public debate, the escalating presence of advertising and marketing across media, and the promotion of the neoliberal fiction that our values can be met, and our problems solved, via consumerism and the workings of a free market. McChesney consistently has advocated for a robust public media sector as an alternative source of information and perspective, suggesting that public broadcasting has become the default, but continually imperiled, public sphere in a media ecology defined by trans-national corporations and media conglomerates (Herman and McChesney, 1997; McChesney, 1999). The public has an active role to play in McChesney's schema as advocates for a democratic media system, one that privileges the rights of publics over the profit-seeking goals of corporations.

A second tradition within the political economy of media reframes the relationship between communications media and capitalism. Originating in Smythe's (1977) blind spot thesis, this line of inquiry decenters the media text as the primary commodity of media companies and suggests instead that the main commodity is the audience (Smythe) or the audience's watching time (Jhally and Livant, 1986). Leisure time for workers is transformed into a form of labor in which audiences are both sold to advertisers and cultivated as purchasers of branded consumer goods. While it was the subject of intense debate after Smythe initially published his article (see Murdock, 1978), his basic insight about the audience commodity has become common sense within studies of American television. It also finds contemporary resonance in discussions of the attention economy as a means to understand shifts in economic and labor power in digital media (Bermejo, 2009; Caraway, 2011; McGuigan, 2012).

An important backdrop to scholarship in political economy from the 1980s onwards was not only the effect of deregulation on media industries, but the ascent of cultural studies. Schiller, at the end of his *Culture, Inc.* (1989), for example, addresses technological utopianists who misunderstand new technologies for new forms of agency, in Schiller's view, mistaking 'button pushers and channel changers' (p. 147) for individuals with actual power, and attacks cultural studies. Cultural studies emerged in part as a response to the totalizing interpretation of media power offered by political economists like Schiller, one that does not make space for the instability and polysemy of media texts; conceives of power primarily along class lines; and looks at macro expressions of power, ignoring the politics of everyday life. Political economists have attacked the premises of cultural studies scholarship, suggesting it is too quick to dismiss the structuring role of capital in all other forms of inequality; it is too rosy-eyed about the potential for resistance and subversion of media content; and it is an illogical position vis-à-vis the media's role in disseminating misinformation about public affairs and foreign policy. The political economy/cultural studies debate engulfed media scholarship in the 1990s (see, for example, Garnham, 1995; Grossberg, 1995; McLaughlin, 1999). McChesney, who has become of the loudest voices articulating the threats posed to media by consolidation and conglomeration also has been one of the loudest voices denouncing cultural studies and the understandings of politics and power upon which it operates (see especially McChesney, 1996, 2006).

As John Dewey (1910) would say, we did not solve this intellectual fight, but rather got over it. As Miller (2007, pp. 4–5) notes, despite the stark battle lines drawn, numerous scholars have blended political economy and cultural studies approaches to power, his own work exemplifying this hybrid in its attention to how US television operates as a technology for cultural citizenship – one that structures and limits civic participation, political knowledges and cultural identities. Within the specific realm of media regulation and ownership, innovative scholarship has melded political economy's interest in the distribution of resources with cultural studies' emphasis on reading practices, textual meanings, multiple registers of identity and inequality, and micropolitics (as in the work of Classen, Hendershot and Noriega discussed below). Feminist media scholarship, similarly, has examined how constructions of gender are produced in and by media industries (Meehan, 1993; Meehan and Riordan, 2002) and regulatory practices (Perlman, 2007).

In addition, since the 1990s, scholars have offered new theoretical frames to examine media regulation and ownership that destabilize the sharp divide between political economy and cultural studies. Streeter (1996) applies a critical legal studies approach to broadcast regulation. This move approaches law as an expression of historically situated and socially determined beliefs, rather than fixed and stable concepts. Streeter defines commercial broadcasting as 'a kind of social philosophy in practice' (p. xii); one informed by the worldviews of the institutional actors – regulators, legislators, business managers, judges – who constitute their vision of what broadcasting should be through policy and law. In *Selling the Air*, Streeter identifies the set of ideas guiding the development of broadcasting as corporate liberal, a matrix of beliefs to emerge in the Progressive Era to square liberal ideas about free markets, individual liberty and private property with the expansion of corporate capitalism. His work stresses that policymakers occupy an 'interpretive community', and accordingly: 'What makes a ruling appear practical, a legal decision seem sound, or a procedure appear fair, is the contingent shared vision of the interpretive community itself, not simply rational policy analysis, legal reason, formal rules of process or procedure, or interest group pressures' (p. 114).

In addition, Streeter argues that media regulation is a generative act and is especially attentive to how it has *created*, rather than regulated existing, markets and property relations. Streeter's work is both a rejection of legal liberal approaches to media law and regulation and, functionally, a rejoinder to institutional analyses of policymaking that focus on the politics of broadcast regulation, but not the ideological assumptions that guide its practices. Streeter similarly has emphasized the structuring role of policy discourse, and the understandings of technology that inform them, in shaping policy decisions (Streeter, 1987). McMurria, who has one been of the most attentive critics to the historical roots and contemporary expressions of neoliberalism in media programming and practices, builds on Streeter in his advocacy for a critical cultural approach to media regulation, making the case for bringing political economy analyses in line with cultural studies by asking how regulation and law evoke 'registers of power across categories of class, race, gender, and nation' (McMurria, 2009, p. 171).

Another productive theoretical turn in this vein has been in critical cultural policy studies, or as Napoli (2008) put it, in bridging the gap between cultural policy and media policy in the US. To see media policy as part of cultural policy is to recognize it as a site for the 'production of cultural citizens' (Lewis and Miller, 2003, p. 1) as much as the regulation of media technologies. It views policy as an expression of cultural values and political power, an instrument that both distributes resources and manages populations. As per Lewis and Miller, critical cultural policy engages not only what policy does, but what it can be, a 'disciplined imagining of alternatives' (p. 3). In addition, as Napoli illustrates, thinking of media policy as cultural policy opens up our understanding of media policy as the site where decisions about cultural diversity and access, as well as the informational needs of the public, are determined. Kirkpatrick's (2011, 2012) important work on local and vernacular policymaking exemplifies this application of critical cultural policy studies analysis.

MEDIA HISTORY

Since the late 1980s, media historians have documented the structuring role of law and policy in constituting media industries. They have demonstrated how commercial broadcasting in the US resulted from legislative and regulatory decisions that instantiated a hierarchy of uses for the radio spectrum in line with business interests (Douglas, 1989; McChesney, 1993; Engelman, 1996; Slotten, 2009). They also have shown how regulatory decisions in the early years of television solidified the dominance of commercial, national networks (Boddy, 1990) and played a determinative role in shaping the contours of the relationships between the film and broadcast industries (Hilmes, 1990). The emergence of media conglomerates in the 1980s, as Holt (2011) illustrates, resulted from convergences between Reaganite economic policies, in particular the emphasis on privatization and deregulation, and the interests of media corporations seeking vertical and horizontal integration.

In addition, media historians have focused on extended case studies of past efforts at media reform. Following McChesney's (1993) seminal study of a failed reform movement to unmoor the dominance of commercial broadcasting ascendant after the Radio Act of 1927, media historians have narrated histories of media reform long erased from histories of American broadcasting. Fones-Wolf (2006) examines how labor unions, from the 1930s to the 1950s, tried to gain access to the airwaves and to diminish the stranglehold of business interests in radio, only to have their efforts reversed by the repressive political climate of the cold war. Pickard (2010a, 2010b, 2011) focuses on the 1940s as a critical juncture in which challenges to the hegemony of commercial media were derailed, and similarly gestures to the powerful corporate appropriation of cold war rhetoric to undermine efforts for more publicly responsible media. These histories of media reform are at once declension narratives and stories of roads-not-taken, and as such offer lessons for contemporary reformers as to the obstacles faced in past efforts to reimagine how broadcasting could operate in American life.

To be sure, these explorations of media history and media policy were part of a broader surge in interest in broadcasting history ascendant the US since the 1990s. In addition, they concretize how media regulation fostered the development of commercial broadcasting, displacing the fiction at the center of neoliberal policy thinking that regulation had been an impediment to, rather than safeguard of, commercial broadcasters. In addition, in recovering alternative visions of the sociopolitical function of media, they demonstrate how the structure of the US broadcasting system did not inhere to the technologies themselves or bubble up out of well-worn American principles, but resulted from the strategic victories of media corporations and their political allies, whose vision of media's political and social role displaced that of citizen groups, intellectuals and public interest advocates. Rowland's (1997a, 1997b) historical overview of how and what the 'public interest' has meant underscores the concept's close identification with the economic interests of regulated industries.

Other media historians have examined not only how media policy produces media systems, but how it reproduces political and social identities. Hendershot (1998) in her book on the regulation of children's media, demonstrates how the policymaking arena is a forum where adults enact their own constructions of childhood. Media regulation, she argues, is not primarily a zone for the protection of innocent children, but for the constitution of the child as inherently innocent. Classen (1994, 2004) and Noriega (2000) demonstrate that though media regulation is presented in the seductive guise of administrative rationality, its deck consistently has been stacked to reinforce racial and class-based hierarchies. Classen's study of African American activists and their allies in Mississippi shows how regulatory decisions like who has standing, what counts as an actionable complaint, or which forms of evidence are credible in the policymaking process had worked against communities of color in their efforts to make media responsive to their needs as citizens. Noriega, in his examination of Mexican American activists, emphasizes the role of media policy in restricting the parameters of how disfranchised groups can make claims for inclusion and reform. His analysis foregrounds how bureaucratic concessions to Chicano groups were a means to redirect claims of injury and mistreatment into administrative processes.

Accordingly, there is an important distinction in how media scholars have written the history of media policy and reform. Historians like McChesney and Pickard emphasize commercialism and the dominance of business interests within broadcasting as the primary target of reform. They interpret media policy as a tool for the consolidation of corporate power and cudgel against the interests of the public. The 'public interest' in their work is not a black box, whose meaning varies on context and deployment, but rather an identifiable set of interests associated with communities challenging corporate power. Classen, Hendershot and Noriega are less concerned with questions of regulatory capture and the primacy of corporate interests in informing policy decisions, and are more interested in how law and policy naturalize existing power relations and social categories. Emerging out of critical cultural studies, this research interrogates how policy acts on political subjectivities as much as it does on regulated institutions.

MEDIA ACTIVISM AND PUBLIC PARTICIPATION

The broadcast reform movement of the 1970s unleashed a wave of scholarship interested in media reform efforts. The advent of deregulation in the 1980s created a climate in which this form of activism was increasingly untenable, and specifically removed many of the tools in activists' toolkits to level license renewal challenges. In the subsequent years, both media reform activity and scholarship on media reform activity diminished. McChesney's (1993) recovery of early broadcast reform activism in the late 1920s thus entered an intellectual and political environment in which media reform had been fairly dormant for over a decade.

The acceleration of media conglomeration, especially in the 1990s, lent a renewed urgency to media reform. Where media scholars have split, however, has been in how they have understood the problem of the media and the correct strategies for activist intervention. Curtin (2000) and Curtin and Streeter (2001) argue that advocacy groups must understand the economic logics of television in a neo-network era, in which the stuff of media has changed from controlling communication conduits to controlling intellectual property. The product of media, in other words, is no longer the program or the film but intellectual property rights that can be leveraged as brands. This, to Curtin and Streeter, is the primary issue for media activists – not ownership or access, but the ways that intellectual property rights and assertions of corporate authorship are constraining creativity and the free flow of ideas. In the wake of a broad-based coalitional effort in 2003 to prevent the loosening of media ownership restrictions, what McChesney (2004) has labeled the 'uprising of 2003', renewed attention has been paid to the issues of ownership and access and to media advocacy groups seeking reform (Carroll and Hackett, 2006; Dunbar-Hester, 2008, 2010). Cognizant of the shifting economic arrangements of media conglomerates, this form of advocacy prioritizes citizen access, ownership diversity, and localism.

Renewed interest in media activism also has provoked interrogations of obstacles to meaningful participation, from FCC disinterest in public opinion and the powerful role of industry lobbyists and attorneys in shaping policy (Brown and Blevins, 2008) to extant interpretations of administrative law that requires agencies to seek public comment but provides them with the discretion to ignore what the public had to say (Gangadharan, 2009). However, as Brown and Blevins note, epistemic communities – in this case, public interest attorneys and leaders of established reform organizations – can reframe the stakes in policy decisions, counter neoliberal assumptions about the 'public interest', and activate community participation that can turn media attention on policy proposals typically outside the parameters of public debate. Gangadharan (2013) offers a deliberative model of public participation that prioritizes collective, inclusive discussions across stakeholders but that recognizes divergent discursive practices across communities. This model emphasizes the importance of policy translation, both in making policy issues legible for diverse publics and in rendering public positions on policy issues in a framework legible to regulators.

McMurria (2012) astutely analyzes how television activism under neoliberalism frequently has adapted to the consumer-citizen construct on which it hangs by, for example, advocating against greater media consolidation through appeals to consumer choice or via fan-based pressures to hold media companies accountable to their taste preferences. He importantly highlights how minority media rights organizations have resisted appeals based on individual consumer choice, and instead have continued to prioritize minority ownership and employment opportunities. Minority media rights activism has hinged on dual concerns over the politics of distribution and the politics of representation, of securing access but also visibility and control. McMurria rightly notes how minority rights organizations have conflicted with anti-corporate media activists, their emphasis on minority ownership and employment frequently overriding their concern over media mergers and media consolidation. This friction similarly has surfaced in debates over internet policy, and in particular over the issue of network neutrality, which a number of established civil rights organizations have seen as less important than the digital divide. This tension is worthy of further exploration, as is the continued work of minority media rights groups whose efforts have been acknowledged in media reform scholarship primarily when they sync up with the ambitions of anti-corporate media reformers.

FUTURE RESEARCH

To conclude, I would like to highlight six potential areas of future research in media regulation and ownership that address both applications of theoretical turns in media studies to this subfield and internal blind spots within it that deserve attention.

First, we should import the production studies model (Caldwell, 2008; Mayer, Banks and Caldwell, 2009; Mayer, 2011) deployed by media industries scholars to interrogate the labor practices of policymaking and the discursive practices that structure how workers within this field understand how and what they do. McMurria's critical cultural citizenship approach to regulation provides the theoretical and political justification for such analysis, and production studies scholarship provides a model and methodology to put it into practice. Such analysis should extend beyond interviews with FCC commissioners and legislators and include both the staff and clerks whose labor is critical to how regulation is imagined, established and implemented, as well local forms of regulation and policymaking.

Second, we should be attentive to the expansive historiography of conservatism in the US to emerge over the last twenty years (for a good overview, see Phillips-Fein, 2011). Scholarship on media reform, both past and present, typically has focused on progressive social movements and has neglected the comparatively successful efforts of conservative media activists and think tanks in, on the one hand, instantiating their vision of media policy and, on the other, corralling the public conversation over what constitutes the 'problem' of the media to its purported liberalism. In this vein, we need to take stock of conservative discourses on the media, including those circulated on cable news and talk radio, that yield the conditions under which neoliberalism can flourish.

Third, in imagining alternatives to commercial media conglomerates and transnational corporations, we need to stop relying on idealized notions of how public media operate over actual assessments of its practices. Scholarship on public media in the US has exposed the elitism and unexamined aesthetic hierarchies on which it has been predicated (Ouellette, 2002), its privileging of institutional insiders to the exclusion of citizens on the ground (Hoynes, 2002) as well as its history of racially discriminatory practices (Berkman, 1980), yet seems to have no bearing on the unyielding and uncritical support for public media advocated by some media regulation scholars.

Fourth, we should stop ignoring Spanish language media. As Amaya (2011) notes in his powerful call to reconceptualize access to Spanish language media as intrinsic to the citizenship rights of Latinos, US television scholars replicate the diminished space afforded Spanish language media in regulatory circles with our own dismissal of this substantial sector of US television. Discussions of media consolidation, for example, have focused nearly exclusively on the English language sector despite comparatively higher levels of concentration in the Spanish language sector. Any commitment to democratic media in the US must take stock of the linguistic diversity of the nation, and the concomitant obligations for a media system that reflects it.

Fifth, to underline McMurria's (2012) challenge, we need to be attentive to how media policy and constructions of the public, within both policy and advocacy discourse, not only replicate the conflation of the consumer with the citizen, but presume a race-less, genderless, class-less public. Since the broadcast reform movement ended, minority media activism has been either the subject of historical inquiry or has been addressed by legal scholars frantic about the devastating impact of deregulation and neoconservative racial politics on minority media rights, yet has been marginal within media regulation scholarship. While the challenges of the broadcast reform movement were hampered by the ascent of neoliberalism, we should not replicate its erasure of historical and continuing forms of inequity in our own construction of media policy and politics.

Sixth, as per Braman (2004), we need to continually reassert a definition of media policy as foundational to democratic self-governance. As Braman demonstrates, how a policy area is defined is fundamentally political, as it determines how it will be discussed, what its goals will be, and who will participate in its decision-making process. In rescuing 'media policy' from a broader, more nebulous realm of 'information policy', Braman suggests that media policy is the subset of information policy that fundamentally affects democratic process and discourses within the public sphere. As regulation continues to consider convergent media technologies and a host of questions over their governance, this emphasis on the role of communication in a democracy must be reaffirmed.

REFERENCES

Amaya, H. (2011). Television/televisión. In Kackman, M. et al. (Eds), *Flow TV: Television in the age of media convergence*, 183–197, New York: Routledge.

Aufderheide, P. (1999). *Communications policy and the public interest: The Telecommunications Act of 1996*. New York: The Guilford Press.

Bagdikian, B.H. (1983). *The media monopoly*. Boston: Beacon Press.

Bagdikian, B.H. (2004). *The new media monopoly*. Boston: Beacon Press.

Baird, F.L. (1967). Program regulation on the new frontier. *Journal of Broadcasting*, 11, 231–243.

Baker, C.E. (2002). Media concentration: Giving up on democracy. *Florida Law Review*, 54, 839–919.

Baker, C.E. (2007). *Media concentration and democracy: Why ownership matters*. New York: Cambridge University Press.

Barron, J.A. (1967). Access to the press: A new First Amendment right. *Harvard Law Review*, 80, 1641–1678.

Baughman, J.L. (1985). *Television's guardians: The FCC and the politics of programming, 1958–1967*. Knoxville: The University of Tennessee Press.

Berkman, D. (1980). Minorities in public broadcasting. *Journal of Communication*, 30, 179–188.

Bermejo, F. (2009). Audience manufacture in historical perspective: From broadcasting to Google. *New Media Society*, 11, 133–155.

Boddy, W. (1990). *Fifties television: The industry and its critics*. Urbana: University of Illinois Press.

Braman, S. (2004). Where has media policy gone? Defining the field in the twenty-first century. *Communication Law and Policy*, 9, 153–182.

Branscomb, A.W. & Savage, M. (1978). The broadcast reform movement: At the crossroads. *Journal of Communication* 28, 25–34.

Brown, D. & Blevins, J. (2008). Can the FCC still ignore the public? *Television and New Media*, 9, 447–470.

Caldwell, J.T. (2008). *Production culture: Industrial reflexivity and critical practice in film and television*. Durham: Duke University Press.

Caraway, B. (2011). Audience labor in the new media environment: A Marxian revisiting of the audience commodity. *Media, Culture & Society*, 33, 693–708.

Carroll, W.K. & Hackett, R.A. (2006). Democratic media activism through the lens of social movement theory. *Media, Culture & Society*, 28, 83–104.

Chisman, F.P. (1977). Public interest and FCC policy making. *Journal of Communications*, 27, 77–84.

Classen, S.D. (1994). Standing on unstable grounds: A reexamination of the WLBT-TV case. *Critical Studies in Mass Communication*, 11, 71–91.

Classen, S.D. (2004). *Watching Jim Crow: The struggles over Mississippi TV*, 1955–1969. Durham: Duke University Press.

Coase, R.H. (1959). The Federal Communications Commission. *Journal of Law and Economics*, 2, 1–40.

Cole, B.G. & Oettinger, M. (1978). *Reluctant regulators: The FCC and the broadcast audience*. Reading: Addison-Wesley Publishing Company.

Curtin, M. (2000). Gatekeeping in the neo-network era. In M. Suman and G. Rossman (Eds), *Advocacy Groups and the Entertainment Industry*, 65–75, Westport: Praeger.

Curtin, M. & Streeter, T. (2001). Media. In R. Maxwell (Ed.), *Culture Works: The Political Economy of Culture*, 225–249, Minneapolis: University of Minnesota Press.

Dewey, J. (1910). The influence of Darwinism on Philosophy. In. J. Dewey, *The Influence of Darwin on Philosophy and other Essays in Contemporary Thought*, 1–19, New York: Henry Holt and Company.

Douglas, S.J. (1989). *Inventing American Broadcasting*, 1899–1922. Baltimore: Johns Hopkins University Press.

Dunbar-Hester, C. (2008). Geeks, meta-geeks, and gender trouble: Activism, identity, and low-power FM radio. *Social Studies of Science*, 38, 201–232.

Dunbar-Hester, C. (2010). Beyond 'Dudecore'? Challenging gendered and 'raced' technologies through media activism. *Journal of Broadcasting & Electronic Media*, 54, 121–135.

Engelman, R. (1996). *Public radio and television in America: A political history*. Thousand Oaks: Sage Publications.

Ernst, M.L. (1946). *The first freedom*. New York: Macmillan.
Fones-Wolfe, E.A. (2006). *Waves of opposition: Labor and the struggle for democratic radio*. Urbana: University of Illinois Press.
Fowler, M.S. & Brenner, D.L. (1982). A marketplace approach to broadcast regulation. *Texas Law Review*, 60, 207–257.
Friendly, F.W. (1976). *The good guys, the bad guys, and the First Amendment: Free speech vs. fairness in broadcasting*. New York: Random House.
Gangadharan, S.P. (2009). Public participation and agency discretion in rulemaking at the Federal Communications Commission. *Journal of Communication Inquiry*, 33, 337–353.
Gangadharan, S.P. (2013). Toward a deliberative standard: Rethinking participation in policymaking. *Communication, Culture & Critique*, 6, 1–19.
Garnham, N. (1995). Political economy or cultural studies: Reconciliation or divorce? *Critical Studies in Mass Communication*, 12, 62–71.
Grossberg, L. (1995). Cultural studies vs. political economy: Is anybody else bored by this debate? *Critical Studies in Mass Communication*, 12, 72–81.
Grundfest, J.A. (1976). *Citizen participation in broadcast licensing before the FCC*. Santa Monica: Rand Corp.
Guimary, D.L. (1975). *Citizens' groups and broadcasting*. New York: Praeger.
Hendershot, H. (1998). *Saturday morning censors: Television regulation before the v-chip*. Durham: Duke University Press.
Herman, E.S. & Chomsky, N. (1988). *Manufacturing consent: The political economy of the mass media*. New York: Pantheon Books.
Herman, E.S. & McChesney, R.W. (1997). *The global media: The new missionaries of corporate capitalism*. Washington, DC: Cassell.
Herzel, L. (1951). 'Public Interest' and the market in color television regulation. *The University of Chicago Law Review*, 18, 802–816.
Hilmes, M. (1990). *Hollywood and broadcasting: From radio to cable*. Urbana: University of Illinois Press.
Holt, J. (2011). *Empires of entertainment: Media industries and the politics of deregulation, 1980–1996*. New Brunswick: Rutgers University Press.
Horwitz, R.B. (1989). *The irony of regulatory reform: The deregulation of American telecommunications*. New York: Oxford University Press.
Horwitz, R.B. (1991). The First Amendment meets some new technologies: Broadcasting, common carriers, and free speech in the 1990s. *Theory and Society*, 20, 21–72.
Hoynes, W. (2002). Political discourse and the 'New PBS'. *The International Journal of Press/Politics*, 7, 34–56.
Jhally, S. & Livant, W. (1986). Working as watching: The valorization of audience consciousness. *Journal of Communication*, 36, 124–143.
Kellner, D. (1990). *Television and the crisis of democracy*. Boulder: Westview Press.
Kerwin, J.G. (1934). *The control of radio*. Chicago: University of Chicago Press.
Kirkpatrick, B. (2011). Regulation before regulation: The local-national struggle for control of radio regulation in the 1920s. *Journal of Radio and Audio Media*, 18, 248–262.
Kirkpatrick, B. (2012). Bringing blue skies down to earth: Citizen policy making in negotiations for cable television, 1965–1975. *Television and New Media*, 13, 307–328.
Krasnow, E.G. & Longley, L.D. (1973). *The politics of broadcast regulation*. New York: St Martin's Press.
Krattenmaker, T.G. & Powe, L.A. (1994). *Regulating broadcast programming*. Cambridge: MIT Press.
Labunski, R.E. (1981). *The First Amendment under siege: The politics of broadcast regulation*. Westport: Greenwood Press.
Lewis, C.S. (1986). *Television license renewal challenges by women's groups* (unpublished doctoral dissertation). University of Minnesota: Minneapolis.
Lewis, J. & Miller, T. (Eds) (2003). *Critical cultural policy studies: A reader*. Malden: Blackwell Publishers Ltd.
Mayer, V. (2011). *Below the line: Producers and production studies in the new television economy*. Durham: Duke University Press.
Mayer, V., Banks, M.J. & Caldwell, J.T. (Eds) (2009). *Production studies: Cultural studies of media industries*. New York: Routledge.
McChesney, R.W. (1993). *Telecommunications, mass media and democracy: The battle for the control of U.S. broadcasting, 1928–1935*. New York: Oxford University Press.
McChesney, R.W. (1996). Communication for the hell of it: The triviality of U.S. broadcasting

history. *Journal of Broadcasting and Electronic Media*, 40, 540–552.

McChesney, R.W. (1999). *Rich media, poor democracy: Communication politics in dubious times*. Urbana: University of Illinois Press.

McChesney, R.W. (2004). Media policy goes to main street: The uprising of 2003. *The Communication Review*, 7, 223–258.

McChesney, R.W. (2006). *Communication revolution: Critical junctures and the future of media*. New York: New Press.

McChesney, R.W. (2008). *The political economy of media: Enduring issues, emerging dilemmas*. New York: Monthly Review Press.

McGuigan, L. (2012). Consumers: The commodity product of interactive commercial television, or, is Dallas Smythe's thesis more germane than ever? *Journal of Communication Inquiry*, 36, 288–304.

McLaughlin, L. (1999). Beyond 'Separate Spheres': Feminism, and the cultural studies/political economy debate. *Journal of Communication Inquiry*, 23, 327–354.

McMurria, J. (2009). Regulation and the law: A critical cultural citizenship approach. In Holt, J. & Perren A. (Eds), *Media industries: History, theory, and method*, 171–183, Malden: Wiley-Blackwell.

McMurria, J. (2012). Pay-for culture: Television activism in a neoliberal digital age. In Mukherjee, R. and Banet-Weiser, S. (Eds), *Commodity activism: Cultural resistance in neoliberal times*, 245–272, New York: NYU Press.

Meehan, E.R. (1993). Heads of households and ladies of the house: The political economy of gender, genre, and ratings, 1929–1990. In W.W. Solomon and R.W. McChesney (Eds), *Ruthless Criticism: New Perspectives in U.S. Communication History*, 204–221, Minneapolis: University of Minnesota Press.

Meehan, E.R. & Riordan, E. (Eds) (2002). *Sex and money: Feminism and political economy in the media*. Minneapolis: University of Minnesota Press.

Meiklejohn, A. (1960). *Political freedom: The constitutional powers of the people*. New York: Harper.

Merritt, B.D. (1974). *An historical-critical study of a pressure group in broadcasting: Black efforts for soul in television* (unpublished doctoral dissertation). The Ohio State University: Columbus.

Miller, T. (2007). *Cultural citizenship: Cosmopolitanism, consumerism, and television in a neoliberal age*. Philadelphia: Temple University Press.

Mills, K. (1974). Fighting sexism on the airwaves. *Journal of Communications*, 24, 150–155.

Montgomery, K.C. (1989). *Target: Prime time: Advocacy groups and the struggle over entertainment television*. New York: Oxford University Press.

Mosco, V. (1976). *Reforming regulation: The FCC and innovations in the broadcasting market*. Cambridge: Harvard University Program on Information Technologies and Public Policy.

Mosco, V. (1996). *The political economy of communication: Rethinking and renewal*. Thousand Oaks: Sage Publications.

Murdock, G. (1978). Blindspots about Western Marxism: A reply to Dallas Smythe. *Canadian Journal of Political and Social Theory*, 2, 109–119.

Napoli, P.M. (2008). Bridging cultural policy and media policy. *The Journal of Arts Management, Law, and Society*, 37, 311–330.

Noriega, C.A. (2000). *Shot in America: Television, the state, and the rise of Chicano cinema*. Minneapolis: University of Minnesota Press.

Ouellette, L. (2002). *Viewers like you? How public TV failed the people*. New York: Columbia University Press.

Parenti, M. (1986). *Inventing reality: The politics of mass media*. New York: St Martin's Press.

Perlman, A. (2007). Feminists in the wasteland: The National Organization for Women and television reform. *Feminist Media Studies*, 7, 413–431.

Phillips-Fein, K. (2011). Conservatism: A state of the field. *The Journal of American History*, 98, 723–743.

Pickard, V. (2010a). Reopening the postwar settlement for U.S. media: The origins and implications of the social contract between media, the state, and the polity. *Communication, Culture & Critique*, 3, 170–189.

Pickard, V. (2010b). Whether the giants should be slain or persuaded to be good: Revisiting the Hutchins Commission and the role of media in a democratic society. *Critical Studies in Media Communication*, 27, 291–411.

Pickard, V. (2011). The battle over the FCC Blue Book: Determining the role of broadcast media in a democratic society, 1945–1948. *Media, Culture & Society*, 33, 171–191.

Rorty, J. (1934). *Order on the air!* New York: John Day Co.

Rowland Jr., W.D. (1982). *The illusion of fulfillment: The broadcast reform movement.* Columbia, SC: Association for Education in Journalism and Mass Communication.

Rowland Jr., W.D. (1986). American telecommunications policy research: Its contradictory origins and influences. *Media, Culture and Society*, 8, 159–182.

Rowland Jr., W.D. (1997a). The meaning of 'the public interest' in communications policy – part I: Its origins in state and federal regulation. *Communication Law and Policy*, 2, 309–328.

Rowland Jr., W.D. (1997b). The meaning of 'the public interest' in communications policy – part II: Its implementation in early broadcast law and regulation. *Communication Law and Policy*, 2, 363–396.

Rucker, B.W. (1968). *The first freedom.* Carbondale: Southern Illinois University Press.

Schiller, H.I. (1969). *Mass communications and American empire.* New York: A.M. Kelley.

Schiller, H.I. (1973). *The mind managers.* Boston: Beacon Press.

Schiller, H.I. (1989). *Culture, Inc.: The corporate takeover of public expression.* New York: Oxford University Press.

Silber, J.S. (1980). *Broadcast regulation and the First Amendment.* Lexington: Association for Education in Journalism.

Slotten, H.R. (2009). *Radio's hidden voice: The origins of public broadcasting in the United States.* Urbana: University of Illinois Press.

Smythe, D.W. (1952). Facing facts about the broadcast business. *The University of Chicago Law Review*, 20, 96–106.

Smythe, D.W. (1977). Communications: Blindspot of Western Marxism. *Canadian Journal of Political and Social Theory*, 2, 1–27.

Smythe, D.W. & Van Dinh, T. (1983). On critical and administrative research: A critical analysis. *Journal of Communication*, 33, 117–127.

Stein, L. (2006). *Speech rights in America: The First Amendment, democracy, and the media.* Urbana: University of Illinois Press.

Streeter, T. (1987). The cable fable revisited: Discourse, policy, and the making of cable television. *Critical Studies in Media Communication*, 4, 174–200.

Streeter, T. (1996). *Selling the air: A critique of the policy of commercial broadcasting in the United States.* Chicago: University of Chicago Press.

Sunstein, C.R. (1990). *After the rights revolution: Reconceiving the regulatory state.* Cambridge: Harvard University Press.

Tuchman, G. (Ed.) (1974). *The TV establishment: Programming for power and profit.* Englewood Cliffs: Prentice Hall.

3

Television in Latin America: From Commercialism to Reform?

Martín Becerra, Guillermo Mastrini and Silvio Waisbord

This chapter offers an overview of the historical evolution of television in Latin America and discusses current trends that reflect the multi-folded transition of television systems amidst technological, economic and political changes.

In broad stokes, television systems in Latin America can be characterized as commercial systems with significant participation from governments, relatively open to foreign capital and programming, and highly uneven in terms of market size and capacity to produce domestic programming. The Latin American television system emerged and became consolidated in the early decades of the industry. It combined the runaway commercialism of US television and political patrimonialism, ubiquitous Hollywood programming and indigenous fiction (most notably telenovelas), active participation of foreign (mostly US) capital and domestic moguls with close ties to governments, and heavy doses of entertainment programming driven by commercial imperatives with politically hamstrung government-owned stations.

None of the transformation during the past decades fundamentally altered the basic characteristics of television systems. Neither neoconservative market policies nor the consolidation of democracy in the region transformed the foundations of television industries. Television systems continue to straddle commercial and political dynamics and relegate civic, public interests to a marginal role. Recent economic, political and technological changes have deepened existing tendencies. Television is a multibillion dollar industry primarily geared to the production of entertainment programming. Television ownership is highly concentrated. Some domestic markets are dominated by one corporation (Brazil), others are duopolistic (like the case of Mexico), and others are dominated by a handful of media groups (such as Argentina, Colombia and Peru). The increase of domestic production and exports in recent years reflect the consolidation of major corporations in domestic and regional markets.

Before commercialization and globalization became dominant forces in television systems worldwide starting in the 1980s, television in the region was already structured around commercial principles and linked to the dynamics of internationalization. Latin America as a region represented the first case of internationalization and commercialization of the global television industry. The push towards privatization and globalization accentuated traditional characteristics of domestic systems. As the number of channels exploded with the arrival of cable and satellite television in the 1980s, television gained a dominant position in media systems in terms of audience reach and advertising expenditures, as well as influence in politics and society.

Television systems have historically been organized around commercial principles yet they feature significant intervention by the state. Private interests have traditionally dominated both ownership and financing. Private companies have owned the vast majority of licenses and, in some cases, developed networks that dominated domestic markets. The absence of strong regulations on ownership allowed the consolidation of powerful networks, particularly in large markets such as Brazil and Mexico. In all cases, private advertising has remained the dominant form of financing. The centrality of private interests can't be understood outside the central role of the state. State policies, particularly during authoritarian periods, were responsible for allowing increased ownership concentration in private hands. The fact that television systems have been dominated by private companies is inseparable from the role of the state in promoting policies that cemented the power of the market. Also, the connivance between state and private interests was responsible for the chronic weakness of public television. Despite several experiences offering a real alternative to commercial television, public television never became consolidated. Whereas governments maintained some stations in the hand of the state, stations de facto functioned as instruments of the Executive, and were constantly subjected to political pressures. With a few exceptions, then, state-owned and -run television did not offer alternatives guided by public principles.

It remains uncertain whether technological and political changes, namely the advent of digital television and the proliferation of web-based platforms and recent reformist attempts to promote new regulatory frameworks, will effectively reorganize television systems.

THE HISTORICAL EVOLUTION OF TELEVISION

There are no systematic studies that have reconstructed the political history of television in Latin America. Curiously, the most comprehensive studies (Fox & Waisbord, 2002; Sinclair, 1999) are in English. There are plenty of studies focused on domestic industries (Gómez Orozco, 2002; Robina Bustos, 1996), as well as studies about the evolution of public television (Bolaño, 1988; Fuenzalida, 2000; Mastrini, 2005; Rincón, 2001).

Television in Latin America developed earlier than in other regions of the world. In 1950, Mexico became the sixth country in the world to launch television broadcasting, shortly followed by Brazil, Cuba, Argentina and Venezuela. Whereas postwar recovery efforts delayed the rise of television in Europe, a mix of populist politics and economic developmentalism stimulated the establishment of television in Latin America. The rise of television responded to the political ambitions of populist administrations to control a new technology for political communication coupled with the ambitions of industrial interests that dominated radio. Because governments were not interested in reshaping the pre-existing commercial structure of radio, television evolved according to the commercial logic of private ownership and advertising funding. Local media bourgeoisies played a critical role in the development of television by liaising between US technology and content providers and domestic governments. This pattern represented another case of 'associated-dependent' development (Cardoso & Faletto, 1979) in the region – that is, the association between local capital and transnational companies. By tapping into personal contacts with US corporations and domestic governments, media companies were able to assemble the necessary capital, technology and political will. Whereas US networks (CBS, NBC and ABC) provided technology, capital and programming, domestic entrepreneurs offered personal contacts with Presidents and other influential officials to ensure the viability of the new project. Furthermore, important business linkages and communication existed across the region among media moguls. One of the most telling examples was the critical role played by Cuban businessmen in the development of television in several countries in the region (Rivero, 2009).

In his comprehensive study, John Sinclair (1999) observes that close links between the state and media owners dominated television. The domination of a commercial, private system coexisted with tacit agreements of mutual advantage between business and government officials intended to reinforce the legitimacy of the political system and the media structure. Whereas television owners enjoyed great discretion to operate in unregulated markets, the State maintained informal control over content. In Argentina, for example, Brenca de Russovich and Lacroix (1985) observe that 'official privatism' has defined the industry. Elizabeth Fox (1990) has characterized it as a 'politically docile' model – that is, a system prone to contact pressures from governments. Despite these basic characteristics, there have been contradictions and conflicts between politicians and business.

Television experienced a significant expansion both geographically and commercially during the 1960s. At that time, the basic characteristics of the system became consolidated: private companies competing for audience, the centralization of program production in capital cities, and little production by state-owned television stations. With the exception of Colombia and Chile (as well as post-1959 Cuba), where public television gained modest development until the 1990s, markets featured dominant private stations and state-owned stations with chronic financial problems, low audience and strong political dependence on the Executive. The coming of military dictatorships in the 1970s and 1980s did not cause major changes. Certainly, authoritarian regimes exerted censorship, particularly on political news and so-called moral issues in fictional programming, but they left commercial and government interventions untouched.

More than the return of democracy in the 1980s, the consolidation of neoconservative policies in the majority of Latin American countries in the 1990s exacerbated commercialism. In addition to privatization in Chile and Colombia, governments promoted the relaxation of anti-concentration regulations which, in turn, facilitated the rise of large multimedia corporations. Large media groups such as Brazil's Globo, Mexico's Televisa, and Argentina's Clarin quickly became integrated in the global communication market as companies with unmatched power in their domestic markets and associated with leading companies both regionally and globally. Given their powerful position in their home markets, which are the largest audience and advertising markets in the region, these media groups have enjoyed unparalleled profits particularly compared to their counterparts in Western countries. In addition to their dominance in print media, radio and over-the-air and cable television, they consolidated their position as vertically integrated companies. They dominate the production of television and film content and wield unmatched power to determine the daily news and political agenda. Consequently, television markets are dominated by a handful of companies. On average, the four largest companies in each domestic market in the region control 80% of both the audience and advertising markets. Research on media concentration in several countries (Argentina, Bolivia, Brazil, Colombia, Chile, Ecuador, México, Paraguay, Peru, Uruguay and Venezuela) show that whereas the telephony and cable television industries are the most highly concentrated, the radio industry is the least (Becerra & Mastrini, 2009; Mastrini & Becerra, 2006).

In the past years, media corporations confronted two challenges. One challenge has been the aspirations of Telefonica and Telmex, the two dominant telecommunication groups in the region, to enter the audiovisual market. Although there are still legal restrictions to offering multi-platform services (the so-called 'triple play' in reference to offering telephony, internet and video services) in the countries dominated by media companies, the latter have faced competitors with significant economic power and deep financial pockets. Second, the coming to power of leftist and populist governments during the 2000s created tensions between large media companies and ruling administrations in several countries, a point to which we return later.

Whereas the television market in the 1950s and 1960s, in the early stages of the industry, was dominated by US-based networks (including partial ownership of leading stations), today, global corporations are influential mainly through the sale of programming and technology. Television ownership, however, remains in the hands of local companies. Television systems have become integrated in the commercial global markets through business links with US and European companies in the production of content and technology, and the development of new products and services such as satellite television. Media giants (Televisa, Globo, Cisneros and Clarín) partnered with global corporations (Sky and Direct TV) in the development of digital television as minority shareholders, and took advantage of these platforms to increase production for global markets.

Another recent development was the transition in management practices. Whereas traditionally companies were family-owned and operated according to old management principles, large companies have increasingly adopted professional management approaches to conduct business. As Sinclair (1999) indicates, although most regional companies are still family-controlled, management has shifted from traditional styles to globalized approaches linked to global capital. To gain access to global funding to finance further expansion, some companies resorted to public offerings and selling minority stock interest to global companies.

TELEVISION PROGRAMMING

The structural evolution of television provides the basic background to understanding the history of television programming in the region. During its formative years in the 1950s and 1960s, Latin American television illustrated the lopsided distribution of program production and distribution worldwide. It mostly featured Hollywood productions with a smattering of domestic and European (mostly British) programs. As a whole, the patterns of television distribution in the region represented the 'one-way street' flow of content criticized by media imperialist theorists (Nordenstreng & Varis, 1974). This situation was the subject of extensive analysis in early television studies in the region (Fox, 1990). The fact that broadcasting systems were organized around commercial principles and lacked content regulations in terms of origin of programming facilitated the importation of programs, most notably from the United States. Latin America became the first major market for the international expansion of the US networks. Private and deregulated systems in the context of economies relatively open to foreign capital offered propitious conditions for US companies to expand southward (Waisbord, 1998).

Early domestic production was largely composed of news, variety/game shows, and fiction, most notably, telenovelas – the genre that eventually became the staple of defined television production in the region. Forms of program funding that characterized US television, mainly sole sponsorship by large corporations, were also typical in the region in the 1950s and 1960s. This modality continued pre-existing trends in the radio industry, and reflected the initial timidity of major local advertisers to support television given low penetration. It was uncommon for oil, tobacco, soap and car companies to be the sole sponsors of newscasts and popular entertainment shows. Just as had happened in the US, this format was later replaced by the magazine-sponsor format, a trend that reflected both the consolidation of television companies and the realization that attracting a range of advertisers was more profitable than the single sponsorship model.

As the industry became consolidated in the 1970s and rates of television households reached over 95%, the volume of domestic production increased. This development gradually shifted the proportion of foreign/local shows in daily broadcasts. It is impossible to generalize across the region given the huge disparities in terms of audience size and the volume of advertising investments. Economies of scale in commercial, unregulated markets explain why domestic productions displaced foreign programs, particularly from the most popular time slots. Not surprisingly, the biggest producers are companies that have historically dominated the largest markets. Brazil's Globo and Mexico's Televisa have long enjoyed unparalleled access to large audience markets (130 million and 70 million in each country, respectively), earned about $3 billion and $1.5 billion respectively in advertising revenues in 2008. Certainly, the fact that both companies held quasi-monopolistic positions in their home markets for several decades gave them unique advantages. Not only did they attract some of the largest audiences worldwide, they also dominated advertising-rich markets. Companies in medium-sized markets (Argentina, Colombia, Venezuela) have also been able to churn out considerable numbers of annual hours. The situation is different in small markets (basically Central America and the Caribbean), which, despite occasional local productions, have largely relied on imports from neighboring countries and the US.

Any discussion of television programming in the region would be incomplete without discussing the telenovela genre, Latin American's staple programming, audience favorite and signature export. The origins, characteristics and success of the telenovela have been extensively discussed (Gómez Orozco, 2002; Mazzioti, 1996; Verón & Escudero, 2004). It is a genre grounded in popular literary traditions and radio programs that preceded the arrival of television. The rise of television offered a new opportunity to adapt previous media forms and narrative styles to the requirements of the new medium. Telenovelas incorporated the languages and formats of their predecessors in other media, and cultivated storytelling narratives dealing with love, hope, family, despair, and other classic themes of the genre.

As the quintessential program of the television industry, the telenovela has been both industrial product and cultural form grounded in local and regional traditions (Martín Barbero & Muñoz, 1992). As a television commodity, telenovelas are produced in a definite number of episodes (ranging from 70 to 120 annual hours), which are broadcast daily during one year. Latin American telenovelas are a predictable, easy-to-schedule product. Its flexible narrative structure allows for permanent adjustments according to events, ratings, audience demands and other factors. While following the basic premises of the genre (e.g. love stories, female heroines, social inequalities), there are distinctive styles across countries. Whereas Brazilian networks produce telenovelas that feature historical and contemporary social themes as well as political issues usually ripped from news headlines, Mexican and Venezuelan programs typically focus on conventional narratives (e.g. Cinderella, rags-to-riches, star-crossed lovers). While they follow the basic matrix of the genre, telenovelas from other countries (Argentina, Chile and Colombia) blend conventional elements with magical realism, social realism, politics and history, and headline news (Gómez Orozco & Vassallo de Lopes, 2010).

The narrative flexibility and textual openness of telenovelas have been at the center of the Latin American tradition of reception studies/audience research cultural studies. Largely in reaction to structuralist and semiological approaches that dominated the analysis of television programs (and media content) in the 1960s and 1970s, scholars have explored the process of 'mediation' in the interaction between telenovelas and audiences in the past decades. Following insights advanced by Jesús Martín Barbero (1993), studies set out to explore how people engage with telenovelas to develop and renew cultural meanings and social identities. The popularity of the genre in the region, as well as its ease in crossing national borders, attest to telenovelas' unique ability to integrate everyday stories and express a range of popular cultures in open narratives (La Pastina & McAnany, 1994; Lozano & Frankenberg 2008).

Both the intense regional traffic and popularity of telenovelas illustrate arguments about the strength of 'cultural proximity' (Straubhaar, 2007) to explain programming choices and audience preferences. Telenovelas have defined a common Latin American televisual space amidst cultural, political and economic differences (Sinclair, 2005). This space became a geo-linguistic market structured around a common language originally delineated and dominated by Mexican Televisa. Given its unique power in the largest Spanish-language market in the world, Televisa aggressively exported programs throughout the region in the 1970s and 1980s, and also pursued other business arrangements through station/network ownership as well as program distribution in the 1990s. Televisa's leading role in the formation of Univision in the United States was only one example of its strategy of regional expansion. Eventually, as production increased in other countries, other companies developed similar regional and global ambitions. During the 1990s, companies based in other countries (such as Argentina and Colombia) also became exporters, although at a smaller scale than giants Televisa and Globo. Industrial regionalization underlies the emergence of a common television market. This was not the product of official design or regional integration. Instead, it responded to the interests of leading media corporations to expand their reach, taking advantage of linguistic commonality and the absence of content regulations.

The capacity to produce considerable hours of telenovela content became the springboard for television exports. Initially, Televisa and Globo, and to a lesser extent Venezuela's Venevision, were 'export' pioneers as they successfully sold telenovelas, first in the region and later around the world. Capitalizing on their solid production and marketing structures in domestic markets, they shrewdly took advantage of the opening of television markets worldwide, initially in Asian countries and later in post-communist Eastern Europe. Consequently, Latin American telenovelas became common in countries such as Russia and the Philippines as low-budget, low-cost immensely popular programs. Simultaneously, Brazil's Globo achieved the status of the reigning power of television programming in the Portuguese-speaking world through its massive presence as an exporter and distributor of television content.

Another important aspect of the growth of television exports is the sale of program formats. In tune with trends in the global television industry, several Latin American companies have developed and exported program formats since the 1990s. The region has not been merely a 'program recipient' amidst the global boom of format television, particularly reality shows and game shows. Some large corporations and independent production companies successfully sold 'ideas' for telenovelas, children's shows and other programs in the region and elsewhere (Waisbord, 2004). This trend basically followed existing patterns of program importation and export in the region. Sinclair (2005) identifies Brazil and Mexico as 'net exporters', Colombia, Chile and Peru as 'new exporters', and Caribbean and Central American countries as 'net importers'. By the end of the first decade of the 21st century, Mexico, Brazil, Colombia and Argentina were consolidated telenovela exporters (Medina & Barrón, 2010).

In the context of highly globalized industries, patterns of program traffic and consumption show the resilience of national cultures. Whereas pay television is typically filled with US productions, over-the-air television stations commonly schedule domestic and regional shows. Local and regional programming inevitably commands audience preferences, and constitutes the bedrock of domestic businesses. The difference is that not all countries are similarly positioned to produce all types of programming, particularly content with high production values. Virtually all countries (even those with only a few million inhabitants) produce shows on shoestring budgets such as news, talk shows, game shows and other staples of low-budget television. Given the absolute primacy of market considerations, only systems with large audience and advertising markets are able to produce content such as fiction that requires higher investments.

Whereas these trends suggest the limitations of original media/cultural imperialism theses that envisioned a region dominated by US corporations and programs, they do not settle the questions of voice and diversity in television. Television programming can hardly be defined as completely uniform in terms of social, political and cultural expression. From 'high culture' to 'faith-based' (Sierra Gutierrez, 2008) programming, particularly in pay television, weekly schedules feature limited diversity. Commercial imperatives, mainly the perennial competition among large corporations for large audiences, typically elbow out considerations of voice, representativeness and democracy. Over-the-air television content basically produces content for urban, mass audiences, and is filled with vapid entertainment and sensationalism without consideration for public issues, minority voices, community issues, and a myriad of interests and demands. Pay television offers a broader yet limited menu but it hardly addresses the deficiencies of over-the-air television given the financial limitations of domestic programming and the low numbers of households. Given its inherent biases, commercial television offers a narrow view of Latin American societies.

Furthermore, the chronic weakness of public television in the region has undermined the potential to provide a civic, noncommercial, nonpartisan platform to express the rich cultural diversity of the region (Radakovich, 2005). Public television has frequently been collapsed into state-owned networks that basically function as government propaganda. Certainly, there has been no shortage of experiences and innovation in public television that were truly informed by civic ideals (Fuenzalida, 2000). Yet they have been temporary at best due to political instability, financial troubles, and weak political support. Without stable democracies, the prospects for public television managed and funded by an independent body were dim. The return and consolidation of democracy, in principle, offered promising conditions to reinvigorate public television. Also, the region's rich tradition of alternative, social and civic mobilization around media issues provided auspicious conditions to revert to the tradition of fusing public television into propaganda television, and to expand alternatives to commercialism. As discussed below, only in the last decade did such demands gain increased visibility in the policy arena, and in alliance with sympathetic administrations, civic groups were able to successfully advocate and lead the still in-process reform of television.

TELEVISION IN TRANSITION

During the past decades, television in Latin America has experienced three transitions: technological, economic and political. Whereas the technological and economic transition offers similarities with processes in other regions in the context of globalization, political changes are uniquely regional, shaped by long demands for media reform and the interest of some administrations to reform television ownership, content, and regulation. On the one hand, the consolidation of pay television and digitalization has brought industrial and technological transformations. According to 2010 data from the Latin American Multichannel Advertising Council (2011), 43.7% of television households had access to pay services at that time.

Distribution across countries largely varies, however. Whereas the proportion of pay television households is over half in some countries (Argentina, Colombia, Peru and Venezuela), the rate of penetration is 36% in Brazil and 22% in Mexico. On the other hand, government-promoted reforms, have introduced regulatory changes in ownership and content. Examples are Venezuela's 2004 Law of Social Responsibility in Radio and Television, Argentina's 2009 Law of Audiovisual Communication services and Uruguay's 2008 Law of Community Radio. Similar policy initiatives are currently under discussion in several countries, including Brazil, Ecuador, Paraguay and Chile. Whereas these reforms attracted a great deal of controversy, they also received support from some business and civic organizations. In all cases, however, social and civic organizations provided the initial push for reforms. This has been a notable development in television systems that historically excluded participation from civic society in policy, regulation, ownership and funding. Civic organizations have rallied around the demand for access to licenses and, in several cases, have vast experience in broadcasting through operating local and community stations (many have existed in a legal limbo or were persecuted as 'pirate' or 'illegal').

These simultaneous transformations are changing television systems in Latin America. The advent of pay television ushered in the modernization and globalization of channels, formats and contents. It introduced complex business relations between local companies (which operate cable services) and global corporations (that own content such as CNN, Fox or ESPN). At the same time, however, the recent push for new regulation to diversify televisions offers and ownership reflect the growing role of local social and political actors. The conflicts between industrial and political-social actors, as well as the tension between the global logic of the industry and local demands for reform and regulation, reflect the dynamics of the current television landscape in the region.

TECHNOLOGICAL CHANGES: PAY, DIGITAL AND MULTI-PLATFORM TELEVISION

The growth of pay television helped to change patterns of program importation and export. It contributed to new forms of content production, adaptation and export of programs and formats (see Arsenault & Castells, 2008; Fox & Waisbord, 2002; Sinclair, 1999; Waisbord, 2004). Simultaneously, it offered the backdrop for the rise of new businesses that produce and sell programs to television networks (traditionally the network controlled the production of their own programming, thereby minimizing the possibilities for the existence of independent companies). Consequently, several 'independent' production companies emerged and became consolidated, particularly in countries with significant domestic production. Although media powerhouses such as Televisa and Globo are the largest program producers and exporters (Medina & Barrón, 2010), other media giants, such as the case of Argentina's Clarin, are associated with independent producers like PolKa and Ideas del Sur.

Just like in the rest of the world, the implementation of terrestrial digital television has attracted a great deal of attention in Latin America in recent years (Bolaño & Brittos, 2007; Bustamante, 2008; Fuenzalida, 2010; Galperín, 2007; García Leiva, 2010; Kaplún 2008). The transition to digital television has been delayed in comparison to the US, Europe and Japan. The 'analog shutdown' is scheduled to take place between 2016 and 2020 in most countries in the region. Consequently, current digital transmissions are experimental, and the television industry continues to operate with analog technology.

After a five-year debate, most countries (with the exception of Colombia) have decided to adopt the Japanese Brazilian ISDB-T norm, over the US ATSC and the European DVB-T in the past years. Only two countries favored the European norm, and Mexico and some Central American countries opted for the US ATSC. Brazil adopted the Japanese norm after it negotiated the use the patents without additional costs as part of broad industrial policy. Unlike US and European countries, Japan exempts the levy of patent fees, which facilitates the expansion of the manufacture of set top boxes in Brazil, and eventually in other countries in the region. The Japanese-Brazilian norm allows free reception of television signals in mobile platforms, a feature that wasn't guaranteed by European and US norms. Another difference is that the norm allows the simultaneous use of mobile telephone in the same bandwidth.

The prevalence of the ISDB-T norm in the region is the outcome of economic and geopolitical considerations, namely, the economic leadership of Brazil in the region, and its promise to ensure trade benefits to countries that adopt its norm. Certainly, underlying this decision was also the political sympathy of left-wing governments with Brazil's administration. It should be mentioned that specific characteristics of the norm, such as the ease of transmitting and receiving digital signals with mobile telephony, also made it attractive.

Three factors explain the delay in the transition to digital television in the region. First, because governments have only recently defined the norms, the television industry has only recently begun to plan the process of production, storing and distribution of digital materials. Given uncertainty about policy decisions, the industry was reluctant to invest in cost and operational plans to strategize new business models. Second, persistent inequalities in the region make it unlikely that massive numbers of consumers would migrate to digital platforms. Latin America remains the region in the world with the sharpest social inequalities in the world (CEPAL, 2010). Given these conditions, for example, Argentina is planning to shut down analog services in 2019, and Mexico in 2021. Third, digital television faces tough competition from cable television in countries with better wealth distribution (such as Argentina, Chile and Uruguay), where penetration reaches around 60% – significantly higher than the average of 28% in the region. Cable television's offering of 80 channels remains a formidable competitor for digital services.

The commercial nature of over-the-air television in the region (Fox & Waisbord, 2002) and oligopolistic market structure (Becerra & Mastrini, 2009) determine the possibilities for the adoption of digital television. Systems with strong established corporations and weak public broadcasting are resistant to changes that may threaten dominant powers. Digital television may offer opportunities to provide access to actors that have been historically absent such as non-profit organizations and the creation of public, non-commercial stations and networks. Consequently, the new technology has the potential to upset the current television order as it may introduce more competition for advertising and other revenues, such as state subsidies.

Until now there have been neither clear rules about the future of the spectrum nor guidelines about the structure of digital television. Even in the few cases where the government has laid out basic guidelines, they are not enforced. In Argentina, for example, the 2009 Law of Audiovisual Communication Services determines the priority of assigning licenses to non-profit organizations, but this stipulation has not been observed yet. It is important to note that the development of digital television in the region is not led by government agencies that control television. Rather, it falls under the mandate of government offices in charge of industrial and telecommunication policies. Overlapping functions across government agencies combined with a disorganized transition to digital television and the coexistence of different priorities among key actors shape the uncertain future of economically unstable television markets.

Some governments have assigned market actors a crucial role in the transition by giving them the ability to exploit the same bandwidth that they already hold in analog television (as in Brazil), or delegating to companies that own analog signals the power to build the digital infrastructure (as in Chile). In other countries, the state remains at the center of the process of digitalization in order to ensure that decoders will be freely distributed among low-income populations (as in the case of Argentina). In addition, the process of digitalization has also taken place in cable and satellite. Although pay television remained largely analog by the end of 2010, direct-to-home satellite television has experienced fast growth in recent years.

THE POLITICS OF POLICY REFORM

In the past decade a new relationship between governments and the media has been taking shape as a consequence of the irruption of leftist and center-left presidents and political parties. This includes the administrations of Nestor Kirchner (2003–2007) and Cristina Fernández de Kirchner (2007–2011) in Argentina; Lula da Silva in Brazil (2003–2007, 2007–2011); Evo Morales (2005–2010; 2010–2015) in Bolivia; and Rafael Correa in Ecuador (2007–2011). Labeling these administrations with conventional categories is fraught with problems as they have been variously called progressive, populist and center-left. In general terms, they cultivate a political discourse that favors wealth distribution, the active role of the State in media policies, human rights and minority rights, and they are critical of US leadership in the region.

Within the so-called 'Latin American progressivism', countries can be distinguished between members of the Alianza Bolivariana para los Pueblos de Nuestra America (ALBA) (Venezuela, Bolivia, Ecuador, Nicaragua and, until recently, Honduras) and other countries that aren't members (Brazil, Uruguay, Argentina, Paraguay and Chile). All of them, however, have approached media policies, particularly television, as a central issue in their political agenda. Within this general trend there are governments that have prioritized direct communication between public officials and citizens to reduce the centrality of mainstream media. Heads of state often speak and are interviewed in programs in state-run stations and private media sympathetic to the governments. Presidents generally discuss news, and accuse large media owners and other oppositional groups. These practices reflect official attempts to bypass the mediation role of traditional news organizations, which have frequently criticized populist and leftist administrations.

What is stake here is not just a style of political communication, but, more broadly, an effort to force changes, especially the dominant position of some media corporations. Such goals are expressed in a range of policies intended to engineer important transformations in television and, more broadly, media systems. All governments have implemented regulatory policies to strengthen the presence of the state and recognize the right of non-profit actors which had been banned from owning television licenses (Becerra & Mastrini, 2009; Hallin & Papathanassopoulos, 2002). Some have implemented constitutional reforms that stipulate the right to communication (Bolivia, Ecuador). These measures are important for they reflect the recognition of the rights of civic and non-profit actors, opposition to racial discrimination, and new forms of media ownership. Administrations in Argentina and Venezuela have sanctioned new broadcasting laws that limit cross-media ownership and promote the rise of new license owners. Other governments (Brazil, Argentina, Chile, Paraguay, Ecuador, Bolivia, Venezuela) have created and supported public media in order to provide a platform for voices sympathetic to the government in order to confront private interests. Some governments (Argentina and Uruguay) have promoted community broadcasting through ensuring frequencies for non-profit associations. Other governments (Venezuela) have launched the revision of licenses. Finally, some administrations have approved legislation allowing access to public information and the regulation of official advertising.

Beyond specific differences among these policies and across countries, these initiatives reflect growing state intervention on various issues such as the active enforcement of the terms and duration of broadcasting licenses, the derogation of libel laws, the regulation of content (such as hate and racist speech), the opening of television licenses to non-profit actors, the promotion of debates about limits to media ownership, and the introduction of limitations to government intrusion in public media.

Two other developments in the regional television landscape need to be mentioned. The regional news channel Telesur was launched in 2005. Its majority owner is Venezuela, and other countries (Argentina, Uruguay, Bolivia, Ecuador, Nicaragua and Cuba) are minority partners. Its primary goal was to offer an alternative to CNN. Simultaneously, the launch of Terrestrial Digital Television was supported by 'progressive' governments and adopted the Japanese norm, following the leadership of Brazil. Despite cases of pressure from local actors to adopt other norms, governments agreed upon similar norms due to common geopolitical considerations, as discussed earlier.

Government initiatives to reform television and media systems have been supported by various business and social actors interested in gaining access to markets that have been historically dominated by large media corporations. In fact, several groups have advocated for regulatory reforms among ruling administrations. In Argentina, for example, the Coalición para una Radiodifusión Democrática (Coalition for Democratic Broadcasting) spearheaded the process that culminated in the passing of the new broadcasting law in 2009 (De Moraes, 2009). In Uruguay, civil society groups had a similar leading role in debates in society and Congress that led to the passing of community radio law (Waisbord, 2010). Both telecommunication corporations and civil organizations have actively participated in debates leading up to legislative reforms. Whereas the former successfully achieved their goals in a few cases, the latter were favored by new legislation that included them as legitimate actors and license holders. Although these dynamics offer differences across countries, this overall pattern can be widely observed.

Dominant corporations have reacted similarly to attempts to reform media systems across the region. Given the interest among governments in prioritizing media reform and making it a matter of public discussion, powerful corporations have predictably opposed any change that would limit their business interest. They have vigorously opposed attempts to change the status quo and have generally invoked the right of 'freedom of the press' to defend their business. Consequently, administrations and leading media groups have frequently clashed over policies and have waged a sustained rhetorical war over television policies (Waisbord, 2011).

The conflict between government and business actors reflects the breakdown of the traditional model of television (and broadcasting system) characterized by the paradox of weak regulation but strong political controls (Fox & Waisbord, 2002). The new regulations promoted by current administrations imply new forms of control but their impact on ownership and content still remain unknown (De Moraes, 2009; Waisbord, 2011). In Argentina, for example, there haven't been substantive changes. The combination of judicial decisions that suspended the application of the new broadcasting law, which reflect the power of media corporations opposed to new legislation, and the lack of prioritization by the Kirchner administration have slowed down the process of reform. Therefore, it remains to be seen whether recent policies will effectively implement changes and lead to major transformations in television systems.

CONCLUDING REMARKS

What can be learned from the experience of television in Latin America? What is its significance for television studies?

First, *contra* theories of media imperialism, it suggests that geo-cultural regions are not homogeneous in terms of capacity for production and export of programming.

Latin America is quite diverse in terms of the structure and size of television markets. Not all countries are similarly net producers or importers of programming. Nor are US and Western imports necessarily dominant, particularly in over-the-air television, given the strong presence of regional productions, concentrated especially in a few countries (Mexico, Venezuela, Argentina and Colombia). The prospects for domestic production also largely vary across genres. Whereas original fiction is more limited due to costs, local industries are filled with low-budget news, talk and variety shows. The growth of domestic production has allowed the rise of independent producers not affiliated with network and station ownerships.

Second, the growth of domestic production doesn't alter the fundamentals of television systems organized around private property and commercialism. In fact, ownership has become increasingly concentrated amidst the explosion of television offerings in the past decades. More production doesn't mean more diversified content, as commercial systems, predictably, prioritize 'lowest common' denominator programming to attract large audiences.

Third, runaway commercialism and globalized business do not undermine the power of governments. Even though national systems are connected with regional and global markets and business, national and state governments retain power through patrimonial and quid pro quo practices given their control of decisions affecting business calculations, programming and news. These relations should not be painted in broad strokes of collusion or opposition. Instead, governments and media corporations maintain complex and shifting relations.

Fourth, ongoing processes of reform of television systems suggest that governments, the market and civic actors play various roles in policy transformations. Some domestic television systems are in flux, not only because of technological changes, but also due to the interest of governments in spearheading changes in ownership and content. Just like in the late 1960s and early 1970s, when some governments promoted the politics of nationalization and expropriation that ended with the emergence of conservative military dictatorships, current populist administrations set out to reorganize television. Whether these processes pursue the expansion of government influence by curbing the power of large corporations, as critics suggest, or promote the democratization of television by promoting civic ownership and diversified content, as defenders argue, remains a matter of contentious debate. The outcomes of these processes remain uncertain and deserve close attention in the future for they shed light on the dynamics and impact of television reform.

REFERENCES

Arsenault, A. & Castells, M. (2008). Switching power: Rupert Murdoch and the global business of media politics: A sociological analysis. *International Sociology*, 23(4), 488–513.

Becerra, M. & Mastrini, G. (2009). *Los dueños de la palabra: Acceso, estructura y concentración de los medios en la América Latina del siglo XXI*. Buenos Aires: Prometeo.

Bolaño, C. (1988). *Mercado Brasileiro de televisão*. Aracaju: Editora da UFS.

Bolaño, C. & Brittos, V. (2007). Políticas de comunicação, governo Lula e TV digital. *Liinc em Revista*, 3(2), 91–101.

Brenca de Russovich, R. & Lacroix, M. (1985). Radio y poder en la Argentina, 1920–1953. *Cuadernos de Periodismo*, 1.

Bustamante, E. (2008). Modelos internacionales de TDT: La política y los lobbies pueden frustrar las expectativas de diversidad. *Diálogos de la Comunicación*, 77, 1–23.

Cardoso, F.H. & Faletto, E. (1979). *Dependencia y desarrollo en América Latina: Ensayo de interpretación sociológica*. México DF: Siglo XXI.

CEPAL (2010). La hora de la igualdad: Brechas por cerrar, caminos por abrir. Retrieved from www.eclac.cl/cgi-bin/getProd.asp?xml=/publicaciones/xml/0/39710/P39710.xml&xsl=/pses33/tpl/p9f.xsl&base=/pses33/tpl/top-bottom.xsl

De Moraes, D. (2009). *A batalha da mídia: Governos progresistas e politicas de comunicacao na América Latina*. Río de Janeiro: Pao e Rosas.
Fox, E. (1990) *Días de Baile: El fracaso de la reforma de la televisión de América Latina*. México DF: FELAFACS-WACC.
Fox, E. & Waisbord, S. (Eds) (2002). *Latin politics, global media*. Austin, TX: University of Texas Press.
Fuenzalida, V. (2000). *La televisión pública en América Latina: Reforma o privatización*. Santiago: FCE.
Fuenzalida, V. (2010). TVN digital. *El Mercurio*, November 2, p. A2.
Galperín, H. (2007). Digital broadcasting in the developing world: A Latin American perspective. In M. Cave & K. Nakamura (Eds), *Digital broadcasting: Policy and practice in the Americas, Europe and Japan* (pp. 39–53). Northampton, MA: Edward Elgar.
García Leiva, T. (2010). The introduction of TDT in Latin America: Politics and policies. *International Journal of Digital Television*, 1(3), 327–344.
Gómez Orozco, G. (Ed.) (2002). *Historias de la televisión en America Latina*. Barcelona: Gedisa.
Gómez Orozco, G. & Vassallo de Lopes, M.I. (2010). Observación de la ficción televisiva en ocho países iberoamericanos. *Comunicación y sociedad*, 13, 13–42.
Hallin, D.C. & Papathanassopoulos, S. (2002). Political clientelism and the media: Southern Europe and Latin America in comparative perspective. *Media, Culture & Society*, 24(2), 175–195.
Kaplún, G. (2008). Uruguay y la televisión digital: Decisiones técnicas, (in)decisiones políticas. *Diálogos de la Comunicación*, 77, 1–10.
La Pastina, A.C. & McAnany, E.G. (1994). Pesquisa sobre audiência de telenovelas na América Latina: Revisão teórica e metodológica. *Revista Brasileira de Comunicação*, 17(2), 17–37.
Latin American Multichannel Television Council. (2011). Retrieved from www.lamac.org/america-latina-ingles/pay-per-view/share-increment
Lozano, J.C. & Frankenberg, L. (2008). Enfoques teóricos y estrategias metodológicas en la investigación empírica de audiencias televisiva en América Latina: 1992–2007. *Comunicación y Sociedad*, 10, 81–110.
Martín Barbero, J. (1993). *Communication, culture and hegemony: From the media to mediations*. Thousand Oaks, CA: Sage.
Martín Barbero, J. & Muñoz, S. (Eds) (1992). *Televisión y melodrama*. Bogotá: Tercer Mundo Editores.
Mastrini, G. (Ed.) (2005). *Mucho ruido, pocas leyes*. Buenos Aires: La Crujía.
Mastrini, G. & Becerra, M. (2006). Globalización, mercado e industrias culturales: Resistencia o simulacro? *Diálogos de la comunicación*, 75, 1–7.
Mazzioti, N. (1996). *La industria de la telenovela*. Mexico: Paidos.
Medina, M. & Barrón, L. (2010). La telenovela en el mundo. *Palabra Clave*, 13(1), 77–97.
NexTVLatam (2010). *Cable digital en Latinoamérica*, 2010–2015. Retrieved from www.nextvlatam.com/Nota.aspx?IdContenido=1793
Nordenstreng, K. & Varis, T. (1974). *Television traffic – A one-way street? A survey and analysis of the international flow of television programme material*. Paris: UNESCO.
Radakovich, R. (2005). Entre espejos y espejismos: La television public en Uruguay. *Cuaderno Venezolana de Sociologia*, 14(3), 335–357.
Rincón, O. (Ed.) (2001). *Televisión pública: Del consumidor al ciudadano*. Bogotá: Convenio Andrés Bello.
Rivero, Y. (2009). Havana as a 1940s–1950s Latin American Media Capital. *Critical Studies in Media Communication*, 26(3), 275–293.
Robina Bustos, S. (1996). Televisa: De los cables subterráneos a PANAMSAT. In D. Crovi Druetta (Ed.), *Desarrollo de las industrias audiovisuales en México y Canadá*, México DF: Universidad Nacional Autónoma de México.
Sierra Gutiérrez, L.I. (2008). La Tele-Fe: Religión mediatizada. *Diálogos de la comunicación*, 77, 1–17.
Sinclair, J. (1999). *Latin American televisión: A global view*. New York, NY: Oxford University Press.
Sinclair, J. (2005). Latin American commercial television: 'Primitive capitalism'. In J. Wasko (Ed.), *A companion to television* (pp. 503–520). Malden, MA: Blackwell.
Straubhaar, J.D. (2007). *World television: From global to local. Los Angeles*, CA: Sage Publications.

Verón, E. & Escudero, L. (Eds) (2004). *Telenovela: Ficción popular y mutaciones culturales*. Barcelona: Gedisa.

Waisbord, S. (1998). Latin America. In Smith, A. (Ed.), *Television: An international history* (pp. 254–263). New York, NY: Oxford University Press.

Waisbord, S. (2004). McTV: Understanding the global popularity of television formats. *Television & New Media*, 5(4), 359–383.

Waisbord, S. (2010). The pragmatic politics of media reform: Media movements and coalition-building in Latin America. *Global Media and Communication*, 6(2), 133–153.

Waisbord, S. (2011). Between Support and Confrontation: Civic Society, Media Reform, and Populism in Latin America. *Communication, Culture & Critique*, 4(1), 97–117.

4

Ownership and Regulation of Television in Anglophone Africa[1]

Ruth Teer-Tomaselli

ESTABLISHMENT OF BROADCASTING IN AFRICA

In the British colonies of Africa, the introduction of broadcasting was a planned, rather than haphazard, affair. While some work had been done prior to the Second World War, with the Zambian Broadcasting Corporation set up under the auspices of the British Colonial Office in 1941, the end of the war brought a new urgency. Faced with a rising tide of African Nationalism, beginning first in the Gold Coast (Ghana), the Colonial Office realized that the development of broadcasting services in Africa could no longer be delayed and turned to the BBC for help. Between 1945 and 1950 radio stations, transmission networks and studios were established. In 1948 the Colonial Office dispatched a memorandum to all Colonial Governors entitled 'Broadcasting in the Colonies' (Armour, 1984: 359). Broadcasting was considered as an 'instrument of advanced government' to improve communication between governments and the governed and to enlighten and educate the masses as well as to entertain them. Initially, all the stations were responsible directly to the countries' governors, and later to the various ministries within the government. They were financed by the Colonial Office.

Thus the level of centralization was extremely high, the purpose of the radio stations clearly defined in governmental terms, and involvement of civic organizations negligible. These radio stations and the nascent broadcasting services that supported them could not be regarded as 'public service' enterprises in the way in which the British or European models are usually conceived. They were unambiguously national – or state – broadcasters with the sole purposes of supporting national governments and educating school-age children and adults.

Unlike the previous examples of broadcasting facilities set up in the erstwhile British Colonies, the Rhodesian (now Zimbabwean) and South African examples differ. Here, the status of the colonial relationship with Britain was different. South Africa was a self-governing 'Dominion' within the British Commonwealth. To all intents and purposes (Southern) Rhodesia was similarly treated, and was administered from the Dominion Office rather than the Colonial Office. After the Unilateral Declaration of Independence in 1965 through to the attainment of majority rule in 1980, the politically isolated country focused its television efforts on the white minority, while providing a series of radio stations for the black majority.

Across Africa, therefore, it is unsurprising that even after independence, the broadcast sectors were seen as part of the national ideological armoury, and were regarded as legitimate mouthpieces of governments and ruling parties. Radio was 'the exclusive preserve of the political elite who used it to communicate with one another and to issue instructions to the lesser mortals in the community' (Ahmed, 2012: 60). The continuity of the colonial autocratic tradition regarding broadcasting is one of the greatest ironies in post-colonial Africa.

TELEVISION IN THE 1960S AND 1970S – FOLLOWING THE PATTERN SET BY RADIO

Across most of Africa, the introduction of television was closely tied to independence. World War II created huge upheavals among both the colonial powers and the colonized countries; while the former were weakened substantially, the latter grew in national self-confidence. Libya was the first African state to assert itself as independent from Italy in 1951. French-colonized Tunisia and Morocco followed in 1956. Ghana was the first of the erstwhile British colonies to gain independence in 1956. The resulting 'winds of change', to use Harold Macmillan's resonant phrase, heralded both liberation and television. Newly independent African states moved rapidly to gain access to the prestige and ideological control offered by television stations. Most of the initial stations were based in the capital cities, and served only the urban areas and their immediate environments rather than the entire population (Wedell, 1986); however, the urban elites were precisely the constituency in which the new governments were interested. All television stations were directly under the control of the national broadcaster, which in turn was under the control of one or other Minister and associated department – usually a department of information, but sometimes a department of culture.

Following Ghana's independence in 1957, a commission was set up to examine the development and growth of broadcasting in an independent Ghana. Despite political independence, the Ghanaian Government did little to change the way in which broadcasting was conducted. The Ghanaian Broadcasting Corporation (GBC) remained state-owned, fully funded and controlled by the government, just as it had been under the colonial administration (Ahmed, 2012: 81). While retaining a nominal 'public service ethos', including governance by a board of directors, the instrument of incorporation of the GBC (Legislative Instrument LI 472/1965) gave the President the power to appoint and dismiss the board and determine the tenure of the board chairperson (Ahmed, 2012: 85). The Ghanaian experience set a pattern for other countries moving from a colonial to a post-colonial broadcasting regimen.

The Angolan government owned three domestic radio services and one externally beamed station. Independence from Portugal was attained in 1975, and television was established a year later in 1976 (Eribo, 1997a: 334). The new station, Televisao Popular de Angola (TPA), was directly controlled by the ruling party, the People's Movement for the Liberation of Angola (MPLA). 'The television programmes include[d] indigenous content and revolutionary messages for the mobilisation of the people in the former guerrilla army' (Eribo, 1997a: 334).

Occasionally it happened that commercial services pre-empted the establishment of the government station, and when this happened, it was soon clawed back under government control. The London Rhodesia Company (Lonrho) introduced television into the areas surrounding the copper mines of Kitwe. The enterprise was bought out by the newly independent Zambian government in 1964, amalgamated with the Zambian Broadcasting Corporation, and provided the genesis of a national television service (Kasoma, 1997: 145).

In Zaire (now Democratic Republic of the Congo) independence from Belgium was achieved in 1965, and the Ministry of Information established the first television station in 1967 (Ibelema and Onwudiwe, 1997: 313). In 1972 the Catholic Church inaugurated a private station, Tele-Star, only to be taken over by the government in 1973 and renamed Renpec (Ibelema and Onwudiwe, 1997: 313). The takeover was part of Mobute Sésé Seko's political consolidation and 'cultural reform' policy, which extended a blanket control over all organs of civil society, including the media. It also marked the period during which the state and the Catholic Church were most antagonistically opposed to each other (Callaghy, 1984: 121). The station, later called Ratelesco, was absorbed as part of the Office Zairos de Radio et Television (OZRT).

Southern Rhodesia established television in 1960, the first service in Southern Africa. From the outset, it was a commercially owned service carrying advertisements, but licensed by the Rhodesian Government with a levy of licence fees paid by viewers. Colour television was introduced in 1984. As was the case throughout Africa at the time, television reception was confined mainly to the large cities and most viewers were from the white minority, a factor that was reflected in the programme content.

South Africa, despite being one of the two large economies in Africa, was also one of the last to receive television in 1976. The reason for the delay was political, with the apartheid government wanting to retain a degree of ideological isolation from the outside world (Teer-Tomaselli, 2013). Initially, there was only one channel, divided equally between English and Afrikaans language transmission. Two further channels were introduced in 1982 and 1983 to cater for African languages. M-Net, a terrestrial encrypted channel, was licensed in 1986 and was the first company to break the monopoly of the state-owned SABC.

Nigeria, the other huge television consumer on the continent, inaugurated the Nigerian Television Authority in 1977 to co-ordinate state broadcasting. Formerly known as Nigerian television, it took on regional status in 1976 when Nigeria was under military rule. It was presented as the authentic voice of the Nigerian government.

In the 1980s television continued to take account of only the middle-class urban dwellers. In Kenya, for instance, an estimated half a million people were reached via 78,000 sets (Wedell, 1986: 171). Television was managed centrally in Nairobi, with regional boosters in other areas. Although the introduction of rural electrification schemes, and the propensity of newly wealthy elites to move back to their agricultural birthplaces helped to increase interest in television in rural areas (Wedell, 1986: 96), it remained a predominantly urban phenomenon.

In the 'first wave' of television introduction in Africa, the media, including television, continued the tradition set up in the colonial period, and served as communicators for the 'national cause'. In the majority of countries, there was a single television channel serving the nation for a few hours each day.

ALL CHANGE – THE WATERSHED YEARS OF THE 1990S

The early part of the 1990s brought together three interacting sets of circumstances: a global political shift, which saw the end of the Cold War; an economic push for neo-liberal policies spurred by growing international trade and the World Bank's structural adjustment policies; and a technological tsunami that changed the way in which communications operated, making a great deal of the old ways of doing things obsolete.

The fall of the Berlin Wall and the subsequent cessation of the Cold War, led to the end of the bipolar division of the world into blocs of American versus Soviet Union proxies. This period asserted the supremacy of democracy as the uncontested political paradigm, as well as heralding the ascendance of neo-liberal policies in all matters, including broadcasting and telecommunications (Iraki, 2010: 145).

Economically, global pressures for liberalization saw the privatization, or semi-privatization, of public utilities of all sorts, including broadcasting. The early 1990s witnessed intense debates on the nature of broadcasting across the world and a major shift in approaches to broadcasting regulation. Public service broadcasting, which had been the most dominant broadcasting model, was challenged everywhere in the world by political and economic interests, by increasing competition from commercial broadcasting, media concentration and general challenges of adapting to globalization.

Technologically, in broadcasting terms, the introduction of satellite technology, followed closely by the digital revolution, put paid to the old certainties of 'spectrum scarcity' on which previous generations of broadcasting regulation had depended.

Internally, pro-democracy movements agitated for a process of democratization and diversity of ownership and voices, while market-led forces were interested in using the media to make money. Economic liberalization also meant opening the broadcasting sector to entry by foreign media players, a movement that was fuelled by the emerging technologies of satellite and cable delivery.

A three-tier broadcasting structure comprising public service (PSB), private commercial and community broadcasters was advocated. In terms of this rather utopian vision, the hitherto national, or state broadcasters, would be 'transformed' to fulfil the mythologized roles of public service broadcasters in a free and democratic public sphere. Commercial broadcasting would satisfy the needs of advertisers and entrepreneurs, providing at the same time a varied diet of entertainment and news programming to the audience. Community broadcasting would build a web of locally produced, highly relevant grassroots stations, mainly on the platform of radio, with deep and significant relationships to the local populace. Finally, liberalization demanded the establishment of independent broadcasting regulators and the reduction of the state's role in the broadcasting sector. 'The pressures for broadcasting reform brought together the agendas of the external global actors promoting neo-liberalism and local groups campaigning for more communicative spaces during the democratic transformations' (Ndlela, 2007: 1–2).

All of this came as a rude shock to African governments and their national broadcasters, who were characterized as monolithic, lacking pluralism and deeply suspicious of external scrutiny. In Africa, there has been strong connection between broadcasting and the maintenance of political power. It is a commonplace that in the many political and military coups that have taken place on the continent the broadcast station is one of the first sites occupied. Thus, widening the net of ownership has seemed counter-productive. James Zaffiro noted wryly that:

> It is characteristic of foreign critics and consultants to assume that something other than state control (of broadcasting) is possible. It may be desirable. No African government in its right mind would willingly sacrifice direct control over broadcasting if it doesn't have to. During the crucial transitional years, when a new majority-elected regime seeks state-wide legitimation for itself, broadcasting is among its most important tools. To lose it, or even to share it with other factions, is illogical and dangerous. (Zaffiro, 2002: ix–x)

This line of thought is discernible in the words of Zimbabwean President, Robert Mugabe, who has steadfastly opposed private broadcasting, asserting that 'you don't know what a non-state radio station might broadcast' (Article 19 1995: 123). Nevertheless, the broadcasters of the continent were faced with the challenge of transforming themselves politically – fulfilling the expectations of the internal market and the global circumstances, and adapting to a rapidly changing technological milieu.

REGULATORY OVERHAUL IN THE 1990S – ARM'S LENGTH CONTROL IN BROADCASTING

During the colonial period, broadcasting typically was the preserve of the government of the day in most African countries, and there was no need for a regulatory body or agency to oversee broadcasting. This situation remained the norm in the immediate post-colonial era, and it was only after a period of liberalization and opening of the airwaves that the necessity of an arms-length regulator was recognized. The process was not immediate in all countries, nor was it uniformly enacted. Some countries opted for a converged regulator with responsibilities for all electronic communication, both broadcasting and telecommunications; while others chose to install a sector-specific body. Some countries, including for instance Zimbabwe, still do not have a separate regulator, and the responsibility for monitoring and arbitrating broadcasting, along with its ownership, remains with a government department.

The UNESCO sponsored 'Windhoek Declaration' (1991) was an important starting point for much of the regulatory reform in the Southern African Development Area Countries (SADC). 'The General Regulatory Principles' stated that the legal framework for broadcasting should include a clear statement of principles underpinning broadcast regulation, including promoting respect for freedom of expression, diversity, and the free flow of information and ideas, as well as a three-tier system for broadcasting, public service, commercial and community. The main argument was that a three-tier system would enhance diversity and pluralism in the media, elements that are central to the democratic processes (Mjwacu, 2003: 173). The document was updated a decade later (Windhoek+10) to take account of the significant changes in political and economic landscape during the intervening period.

South African broadcasting policy review and reforms have been the most extensive in Africa. Prior to 1993 the SABC, in common with many other broadcasters on the continent, was regulated, owned and controlled by the state. Dubbed 'his master's voice', the Corporation was an important part of the ideological control exercised by the apartheid government (Teer-Tomaselli, 2005). The first fully enfranchised election in South Africa took place in 1994. Ahead of the elections, the multi-party negotiating forum realized that continued control of the broadcast media by the National Party would skew the process and cast doubt on the outcome of the election. Thus, the first piece of post-apartheid legislation, the Independent Broadcasting Authority Act (1993), was actually enacted before the election to ensure that the SABC played an impartial role in the envisaged transition. In 2000, the IBA was merged with the telecommunications regulator to become the Independent Communications Authority of South Africa (ICASA) (Teer-Tomaselli, 2005). The development of the IBA/ICASA and the outcome of its policies had a profound influence on the introduction of regulators in the region.

Most of the regulators were established in the late 1990s and early part of the third millennium. Ghana was one of the earliest. The National Media Commission (NMC) was set up in 1993, and arose directly from the contents of the country's constitution. This wide-ranging body draws up and regulates all communication media – print, film, public relations and advertising, as well as the electronic media of radio and television. It sets a relatively high local content quota – 80% for the Ghanan Broadcasting Corporation (GBC) as public service provider, and 30% for all free to air commercial stations (National Media Policy, n.d., circa 2008). The National Communications Authority, set up in 2008, is responsible for licensing (National Communications Authority, n.d., circa 2008).

In Kenya, the liberalization of airwaves occurred between 1997 and 1999, ending the monopoly of the Kenya Broadcasting Corporation (KBC). The Communications Commission of Kenya (CCK) was established in 1999 as an outcome of the Communications Act (1998). The CCK was given the regulatory mandate to oversee frequency allocation and to ensure the regulatory compliance of electronic media. In Malawi, the Malawi Communications Regulatory Authority (MACRA) set up in 1998 regulates converged sectors of 'telecommunications, posts and broadcasting' directly under the Ministry of Information. In Botswana the National Broadcasting Board was established under the Broadcasting Act of 1998. The Ugandan Broadcasting Council was established under the Electronic Media Act (1996) and reaffirmed in the Ugandan Broadcasting Corporation Act (2005). The latter acts consolidated Ugandan Television and Ugandan Radio as a single entity (Hills, 2003: 105; Chibita, 2006). The Tanzania Communications Regulatory Authority (TCRA) was established in 2003, merging the Tanzania Communications Commission (TCC) and the Tanzania Broadcasting Commission (TBC).

Despite the existence of nominally independent regulators, the state still retains significant influence on broadcasting. Ghana remains in the nebulous area of a broadcaster that purports to be a 'public service' corporation, but in fact, is closer to the national broadcaster model, answering directly to the Executive through the office of the Vice-President and the Ministry of Information. A clear indication of this is its continued reliance on direct funding from the national fiscus, making is vulnerable to state interference (Ahmed, 2012: 142).

In Kenya, the close ties between the broadcasting sector and the state are emphasized by the appointment of the board of governors by the President without a public hearing. Critics have expressed concern over the independence and autonomy of the regulators: all are appointed executively, without the participation of stakeholders, parliament or any other interest groups. The President appoints the CCK chair, four of its members are permanent secretaries (again presidential appointees), and the remaining seven are ministerial appointees (Wanjiku, 2009: 5). The close ties between the broadcasting sector and the state in Malawi are emphasized by the appointment of the board of governors by the President without a public hearing. The Botswanan National Broadcasting Board was separated from the Telecoms Board, however, in order to rationalize the use of resources, the latter still provides basic secretarial services (Hills, 2003: 63). The Board is answerable to the Office of the President and the Ministry of Communications, Science and Technology (Banda 2006).

Contrary to the South African approach, policy-making in Zambia was top-down with little public input. Until 1987 broadcasting services were an integral part of the state, being placed directly under government control through the Ministry of Information. In that year the Zambia National Broadcasting Corporation (ZNBC) Act came into being. This transformed the Zambia Broadcasting Services into a nominally autonomous corporation, with greater independence and less call on state monies. However, the reforms were carried out within the one-party state establishment, and in the absence of political liberalization achieved very little in transforming the broadcasting media that remained the preserve of the ruling party. Critics have argued that these reforms were mainly cosmetic as the broadcaster remained essentially firmly in the hands of government (Kasoma, 1997; Banda, 2006).

In Zimbabwe, discussions on re-regulation of broadcasting began in 1995, but five years later, there was little to show for them. A bill catering for both broadcasting and telecommunications was introduced in1997, but the broadcasting section was withdrawn in 1999, leaving the broadcasting monopoly intact. Nonetheless, 'time' drawn from ZBC's second channel was rented out to a number of 'safe' broadcasters, discussed in the following section. Drawing on emergency legislation that allowed the government to simply promulgate a law without a consultative process, the Broadcasting Services Act was unveiled in 2001. The law allows for the three-tier system of broadcasting and the end of the monopoly of the ZBC, but under extremely tight licensing conditions, which together with a slew of legal uncertainties, make the situation unattractive to foreign investors. Thus, the national broadcaster remains to the present time (2013) a *de facto* monopoly. There is no independent regulator in Zimbabwe, and the Ministry of Information continues to own all state-owned broadcasting (radio and television) along with the majority of newspapers (Ndlela, 2007).

LIBERALIZATION AND THE INTRODUCTION OF NEW BROADCASTERS

A primary purpose of re-regulation in the 1990s was the 'opening of the airwaves' to new broadcasters and the dismantling of the state-owned broadcasting monopolies. Ghana, as might be guessed, was a first to make changes. In 1977 private television was introduced by Kwa Ansah who founded Television Africa Limited as a free-to-air station. This was the only fully Ghanaian television company for many years. The parent company was Film Africa Limited set up in 1977 with an avowedly political-cultural agenda of rectifying the perceived negative portrayal of the African continent in the international media. In 1989 GBC installed a satellite to receive CCN directly. This was followed by further installations for Deutche Welle, World Net and Canal France International (CFI) in 1997.

Elections have proved to be trigger points for democratic reforms in many areas of African social activities, including broadcasting. In Zambia, for instance, the first multi-party elections took place in 1991 and, as a result, Kenneth Kaunda was ousted in favour of the main opposition, the Movement for Multiparty Democracy (MMD) led by Fredrik Chiluba. The voting was monitored by a series of judicial proclamations that settled disputes over access, and in the process, created an enabling environment for a more liberal media policy in the country. However, despite its large election promises, Chiluba's new party was reluctant to cede any control over the airwaves. Although legislation ending the monopoly of the Zambian National Broadcast Corporation was enacted, few new licences were issued (Kasoma, 1997: 145). A limited number of frequencies were made available in each district and most of these were given to politically conservative Christian applicants. Other licences were issued to MultiChoice, the subscription satellite company discussed later in the chapter. The South African holding company owns 70% while the Zambian Government holds the remaining 30%. Zambian critics have suggested that 'in embracing a cautious deregulation of the broadcasting market, the MMD wanted to be seen to be reformist, while at the same time remaining firmly in command of broadcasting' (Banda, 2006: 2).

The South African situation also illustrates the central importance of a change of government. After the first democratic elections in 1994 and the establishment of the IBA, the broadcast sector was overhauled. The SABC began the long journey of transformation from state to public service broadcaster. It was forced to sell off some of its lucrative commercial radio stations, and a number of new licences were issued. A free-to-air television station, e-TV, began national broadcasting, and soon achieved audience figures as high as those of the SABC (Teer-Tomaselli, 2005). M-Net expanded its operations and became Direct Satellite Television (DStv), moving from an analogue to digital platform and providing the first satellite direct-to-home transmissions in Africa in 2000 (see below).

In Kenya, freedom of expression was kept in check during the entire Daniel Arap Moi and KANU Party regime, lasting from 1978 to 2007. During his presidency, newspapers perceived to be critical of the ruling party, the government or any of its allies were banned and journalists arrested and detained without trial (Wanjiku, 2009).

The first privately owned station in Kenya was Kenya Television Network (KTN), founded in 1990 by Jared Nangana under the banner of the Kenya Media Trust, which also operated a number of radio stations. Although it only served Nairobi and its environs, the station broke the monopoly of the Kenyan Broadcasting Corporation (KBC). While independent, the channel was licensed by the Ministry of Information, as was the KBC. KTN would become a model for the new wave of liberalized television stations, and a resource for the training of personnel. The channel self-consciously set itself up as having a sophisticated and aggressive news-style, seen as a vibrant antidote to the compliant KBC. By global standards, it was not radical. The core of the programming was a mixture of re-transmitted CCN and MTV programming, together with entertainment from Europe, America and Australia. A small amount of programming material from African states augmented the local news production (Heath, 1992).

The station's history is indicative of the ambiguity of the 'private' status of many commercial broadcasting stations. As with many independent television stations of the time, KTV struggled to counter political interference, particularly from the ruling KANU party. The station stopped broadcasting local news completely during 1993–1994 (Heath, 1992). Eventually, after further persecution, the Standard Group, a consortium that already owned a raft of media concerns, acquired it in unclear circumstances. Moi, party leader of KANU and then-President of Kenya, was a major shareholder of the company. Although the newspapers, television and radio stations under the Standard banner demonstrate 'relative editorial autonomy' (Iraki, 2010: 147), it could not be said that they were able to ignore the interests of the owners. The Standard Group, while nominally an independent commercial operation, acted as a quasi-government body under the KANU (the ruling party) and president of the Republic of Kenya until 2007. The overthrow of Moi and the 'second liberation' that followed inspired a wave of change that was in large part behind the African clamour for political pluralism that included a raft of new media voices. The Kenyan Constitution was revised, paving the way for multi-party politics and allowing the licensing of several new commercial broadcasters.

Royal Media, which also broadcasts about 20 FM stations in various vernacular languages, is probably the largest electronic media company in Kenya. Its television channel, Citizen Television, was able to retain greater independence than most others, and built a solid reputation for newsworthiness. However, since 2007, this has declined (Iraki, 2010: 147). At the time of writing (2013), the station is a commercially oriented entertainment channel, carrying the most recent and popular American and British television programming (Wanjiku, 2009: 6). The last of the indigenous media consortia is the Nation Media Group, which also owns a television channel among its various media holdings.

The mid-1990s were a politically tumultuous time for Zimbabwe, and by extension, for the Zimbabwean Broadcasting Corporation (ZBC). There was a significant turnover of both the management and the Board of Directors during this period. Bowing to international pressure, the Minister of Information mooted a series of liberalizations. The so-called Ibbottson Report was introduced in 1997, advocating the end of the ZBC monopoly and the 'liberalization' of the broadcasting sector. However, the process illustrated the ambiguity of the situation with the reform process attempting to reconcile two sets of incompatible interests: the desire to commercialize the ZBC ranged against the felt need to retain political control of the media (Moyo, 2004: 19). While the report recommended complete privatization, the Zimbabwean Government chose to lease the TV2 spectrum to three privately owned broadcasters, including Joy Television. Leasing meant that the licence could be withdrawn at any time, and that the station, although private, would remain under government control. 'This strong interventionist streak meant that the drive to introduce private/commercial broadcasting was never synonymous with deregulation – or even liberalization' (Moyo, 2004: 21). The politically adventurous Joy Television was taken off air in 2002.

The liberalization of television licensing in Nigeria came about rapidly. Fourteen private stations were approved in 1993, 12 of which were in the southern provinces (Eribo, 1997b: 61). In the current situation in Nigeria, radio and television fall under separate ownership, with the government-owned Federal Radio Corporation of Nigeria (FRCN) controlling radio, while the Nigerian Television Authority (NTA) has ownership of two television services. NTA1 is distributed across six 'television zones', while NTA2, the commercial, advertising-funded arm of state broadcasting, is distributed nationwide. As a consequence of the earlier regimen in which both telecommunication and broadcasting were under the direct control of the Ministry of Information, Nigerian Telecommunications (Nitel) owns a majority of the transmitters that broadcast FRCN and NTA programming. At the present time, each Nigerian state also has a broadcasting company that broadcasts one or two locally operated terrestrial stations, making up approximately fifty government-owned stations. Among the most important independent commercially owned stations are Silverbird Television, Africa Independent Television (AIT), Channels Television, and Superscreen Television. Additionally, there are a number of cable television operators as well as satellite operators (see below).

SATELLITE – THE ADVENT OF THE MULTI-CHANNEL GLOBAL BROADCASTERS

Throughout this chapter, the interconnection of the political, economic and technological aspects of broadcasting has been emphasized. Nowhere was this more evident than in the advent of the satellite transmission of broadcast programming. Hard on the heels of the political liberalization of the 1990s, satellite transmission, often provided by global players, offered the next policy and regulatory challenge for African governments. South Africa was an early adopter of the technology. The IBA (later ICASA) held public hearings in 1999. The ensuing White Paper noted that '[s]atellite broadcasting has become a key phenomenon in broadcasting globalization compelling regulatory regimes all over the world to review their policies to encompass its expansion' (quoted by Mjwacu, 2003: 178).

Three major and several smaller direct to home satellite companies operate across Africa. For the most part, these have replaced the earlier satellite feeds off the huge C-Band dishes and transponders previously found in the wealthier enclaves of African cities. In the north of Africa, Arabsat is available over the entire Sahel region. It is owned by the Panther Media Group registered in Dubai. The programmes, packed into more than a hundred channels, are sourced from countries in the Middle East and North Africa as well as the usual global news and content providers. Across the western and central part of the continent, frequently referred to as Francophone Africa, Canal Plus/ Horizon Afrique holds sway.

MultiChoice Africa, also trading at DStv (Direct Satellite Television), is the biggest provider of subscription television in Anglophone Africa. MultiChoice Africa is a wholly owned subsidiary of the South African media conglomerate, Naspers, although the African arm of the business is registered in Mauritius. In 1992 an encrypted analogue service was launched in 20 African countries. Digital satellite-delivered services were introduced in 1995, increasing the bouquet substantially. The first international joint venture was entered into with Namibia in 1993, which marked the beginning of a separate entity for the operation of the African arm of the business, now named DStv. Increased satellite capacity has enabled the service to increase both its footprint and its signal strength. In turn this has added to the services available – a bouquet of more than a hundred channels; television on demand; mobile television; and video to home delivery. The group now broadcasts in over 50 African countries (Teer-Tomaselli, Wasserman and de Beer, 2007)

MultiChoice enters into a joint venture with local entrepreneurs, sharing the investment, dividend and management strategy. Joint ventures trade under the name of MultiChoice and the country, e.g. MultiChoice Zambia or MultiChoice Nigeria, and maintain a call centre as well national and multiple regional offices in each country. Thus, unlike most direct to home satellite companies, subscribers deal with the company in their own country, in their own language and at the local level.

Like all satellite or cable companies, DStv is a both a producer and publisher, sourcing material from markets across the English-speaking world. However, much of its attraction in the African market has been its investment into especially African products, and the establishment of a raft of Africa-centric channels under the AfricaMagic brand. In doing so it has strategically partnered with local producers in Nigeria, Uganda and Kenya, as well as the huge production facilities it maintains in South Africa.

CONCLUSION

In Africa, television typically was introduced in the immediate post-colonial period, although in some countries it came much later. Television followed the pattern set by colonial-style broadcasting services, in so far as it was under the domination of the ruling party and the state. Its purpose was (and in many cases, remains) to serve the political and financial interests of the state. The global tumult of the 1990s affected many African countries, and a number of broadcasting regulators were set up during this period. However, the independence and efficacy of some of these regulators is open to question. At the same time, the licensing of commercial broadcasters broke the monopoly on state broadcasting. Some of these have close links to the state, or have been bullied into compliant submission. The third wave of television provision has seen multi-national broadcasting companies, armed with satellite digital technology and vast bouquets of global programming, offering their services to the moneyed elite in African countries. Despite this, the majority of Africa's televisual-citizens are dependent on national broadcasters that remain close to the model of their early post-colonial antecedents.

NOTE

1 A disclaimer: The chapter does not purport to be an encyclopaedic account of ownership, control and regulation of all broadcasters within the whole of Africa. Such a project would be too large for the present study. Rather, the chapter concentrates on providing a schematic outline of a periodization of ownership and regulation in a selected number of largely Anglophone countries south of the Sahara. Countries of the Sahel share an historical background with close connections to the French colonial milieu or to the Middle East, and are too different to be able to absorb these details in the limited space available in the present chapter. While the details of each country vary significantly, there is sufficient commonality to be able to make a few valid generalizations across the countries under study.

REFERENCES

Ahmed, A.S. (2012). *Broadcasting reform in Ghana: A critical analysis of broadcasting policy and regulatory change 1994–2008* (PhD thesis). Johannesburg: University of the Witwatersrand. Retrieved from http://wiredspace.wits.ac.za/handle/10539/12345

Armour, R. (1984). 'The BBC and the development of broadcasting in British Colonial Africa'. *African Affairs*, 83(332), pp. 359–402.

Banda, F. (2006). 'Key issues in public service broadcasting in sub-Saharan Africa'. Open Society Institute, London, Oct. 2006. Retrieved from http://eprints.ru.ac.za/461/

Callaghy, T.M. (1984). *The state-society struggle: Zaire in comparative perspective*. New York: Columbia State University Press.

Chibita, M. (2006). Indigenous Language Programming and Citizen Participation in Ugandan Broadcasting: An Exploratory Study. PhD thesis. Pretoria: University of South Africa.

Eribo, F. (1997a). 'Elusive press freedom in Angola'. In Festus Eribo and William Ong-Ebot (eds), *Press freedom and communication in Africa*. Asmara (Eritrea): Africa World Press. pp. 325–340.

Eribo, F. (1997b). 'Internal and external factors affecting press freedom in Nigeria'. In Festus Eribo and William Ong-Ebot (eds), *Press freedom and communication in Africa*. Asmara (Eritrea): Africa World Press. pp. 51–74.

Heath, C. (1992). 'Structural changes in Kenya's broadcasting system: A manifestation of presidential authoritarianism'. *International Communication Gazette*, 5(1), pp. 37–51.

Hills, R. (2003). *Broadcasting policy and practice in Africa*. London: Article 19.

Ibelema, M. and Onwudiwe, E. (1997). 'Congo (Zaire): Colonial legacy, autocracy and the press'. In Festus Eribo and William Ong-Ebot (eds), *Press freedom and communication in Africa*. Asmara (Eritrea): Africa World Press, pp. 303–324.

Iraki, F. (2010). 'Freedom, cross ownership and the monopolizing of public spaces in Kenya'. In M. wa Müngai and G.M. Gona (eds) *(Re) Membering Kenya: Identity, culture and freedom*. Nairobi: Twaweza Communications Limited.

Kasoma, F. (1997). 'Communication and press freedom in Zambia'. In Festus Eribo and William Ong-Ebot (eds), *Press freedom and communication in Africa*. Asmara (Eritrea): Africa World Press.

Mjwacu, T. (2003). 'Opportunities and challenges of new technologies in media and communication: The Windhoek Declaration'. In A. Zegeye and R. Harris (eds) *Media, identity and the public sphere in post-apartheid South Africa*. Leiden (Netherlands) and Boston, Mass.: Brill Publications, pp. 169–190.

Moyo, D. (2004). 'From Rhodesia to Zimbabwe: Change without change? Broadcasting policy reform and political control'. In H. Melber, D. Moyo and S Chiumbu, *Media, public discourse and political contestation in Zimbabwe*. Uppsalla: Nordic Africa Institute, pp. 12–28.

National Media Council of Ghana (n.d.). *National media policy*. Accra: NMC. Archived at www.library.fes.de/pdf-files/bueros/ghana/03699.pdf

Ndlela, N. (2007). 'Broadcasting reforms in Southern Africa: Continuity and change in the era of globalisation'. *Westminster Papers in Communication and Culture* (University of Westminster, London), 4(3), pp. 135–156.

Teer-Tomaselli, R. (2013). 'In the beginning: Politics and programmes in the first years of the SABC television'. In M. Tager and C. Chassi (eds), *Tuning In*. Cape Town: Pearsons.

Teer-Tomaselli, R., Wasserman, H. and de Beer, A.S. (2007). 'South Africa as a region media power'. In D. Thussu (ed.), *Media on the move: Contra-flow in cultural products*. London: Routledge.

Teer-Tomaselli, R.E. (2005). 'Change and Transformation in South African Television'. In J. Wasco (ed.), *Companion to Television*. London: Blackwell.

Tudesq, A-J. (1997). 'Problems of press freed in Côte d'Ivoire'. In Festus Eribo and William Ong-Ebot (ed.), *Press freedom and communication in Africa*. Asmara (Eritrea): Africa World Press, pp. 291–302.

Wanjiku, R. (2009). *Kenya Communications Amendment Act (2009) Progressive or retrogressive?* Nairobi: Association for Progressive communication. Retrieved from www5.apc.org/en/system/files/CICEWAKenya20090908_EN.pdf

Wedell, G. (ed.) (1986). *Making broadcasting useful: The development of radio and television in Africa in the 1980s*. Manchester: Manchester University Press.

Zaffiro, J. (2002). *Media and democracy in Zimbabwe, 1931–2001*. Colorado Springs, Col.: International Academic Publishers.

5

Ownership and Regulation in Europe

Stylianos Papathanassopoulos

The liberalization of the rules governing the television sector around the globe has facilitated, if not accelerated, the trend toward the creation of larger and fewer dominant groups in the audiovisual landscape and consequently in the entire media sector. As a result the media industry has become more concentrated and populated by multimedia conglomerates. This trend has raised fears of excessive concentration of ownership. Mergers, acquisitions and common shareholding, led in most instances by telecommunications groups, have created a web of common interests across the communications sector, though even here the pattern is not uniform.

The question of media ownership and its impact on media pluralism and diversity has been a major topic in media studies, since ownership affects news media and the latter play a special role in a democratic society (Baker, 2007). One the other hand, few actually understand what regulation of ownership is about. One can say that regulation 'has traditionally focused on threats arising from media ownership, resulting in narrowly tailored anti-concentration rules, in combination with specific content obligations and safeguards for editorial independence' (Valcke, 2012, p. 30). In effect, regulation of media ownership aims to prevent concentration of economic power as well as to promote diversity of ownership aiming at achieving a pluralism of voices in the field (Liu & Albarran, 2009, p. 854). Although, countries may employ similar regulatory tools, significant differences emerge in the quality of the pluralism and diversity they achieve (Ciaglia, 2013).

This chapter aims to describe the structure of ownership in the television sector, to highlight the effects of the relaxation of ownership rules, and to discuss the forms of concentration in the media field. Finally, it describes the efforts of the European Union to tackle the issue of media concentration.

THE STRUCTURE OF OWNERSHIP IN THE TELEVISION SECTOR

Compared to print media, the structure of TV ownership is characterized in most countries by a kind of 'duality', in the sense that the ownership of most, if not all, TV channels belongs to either public or private entities (see also Sussman, 2008). Public TV stations refer to those broadcasters which are funded, at least partially, by general public revenues (i.e., the traditional 'license fee' or more recently an income tax). On the other hand, private (aka commercial) TV stations are characterized by broadcasters whose financing is provided by the private sector, for example through advertising, sponsoring, etc., and whose owners are private companies, families, or groups. Public TV channels are in most cases owned by the state or managed by the state (i.e. the government of the day). In many countries, public TV channels are controlled by the government (i.e. political party) in power, or can have a charter making them relatively independent of the state or governing power (such as Britain's BBC). The break-up of state monopolies and the subsequent liberalization of the sector have restructured the television industry around the globe.

In countries such as the United States family ownership went into decline as many media companies became part of large, publicly traded companies keen to reap the profits associated with the various media sectors. In effect, media ownership regulations designed to promote competition, diversity and localism were loosened, in particular after the 1996 Telecommunications Act. But even before the Act, TV companies were controlled by major banking and financial institutions (see Bagdikian, 2004; Sussman, 2008).

In 2004, the Congress abolished all limits on the number of TV stations a single company can own, provided a single owner's stations do not exceed 39% of the US TV households. For the purposes of calculating the 'national audience reach' under this rule, TV stations on UHF channels (14 and above) count less than TV stations on VHF channels (13 and below). The Federal Communications Commission (FCC)'s 'Dual Network Ban' permits common ownership of multiple broadcast networks but prohibits mergers between the major TV networks: ABC, CBS, Fox and NBC. The 'Local TV Multiple Ownership' rule allows an entity to own up to two TV stations in the same local market (the so-called Designated Market Area – DMA) if either: (1) the signals – known as 'Grade B signal contours' – of the stations do not overlap; or (2) at least one of the stations is not ranked among the top four stations in the DMA (based on TV audiences share), and at least eight independently owned TV stations would remain in the market after the proposed combination.

Moreover, the regulation on 'Local Radio/TV Cross-Ownership' imposes restrictions based on a sliding scale that varies according to the size of the market: (1) in markets with at least 20 independently owned 'media voices' (defined as full power TV stations and radio stations, major newspapers, and the cable system in the market), an entity can own up to two TV stations and six radio stations (or one TV station and seven radio stations); (2) in markets with at least ten independently owned 'media voices' an entity can own up to two TV stations and four radio stations; and (3) in the smallest markets an entity may own two TV stations and one radio station. In all markets, an entity must comply with the local radio and TV ownership limits (see http://transition.fcc.gov/cgb/consumerfacts/reviewrules.pdf, accessed 26 August, 2014). These relaxations had an immediate influence on the status of concentration of media ownership, leading to media consolidation.

In Europe, the break-up of state monopolies and the entry of a plethora of new privately owned channels changed completely the structure and performance of the industry. Broadly speaking, privately owned TV channels have become dominant in most countries (with a few exceptions such as China) all over the world. The restructuring of the Western European television systems has brought about an increase in the number of private channels in operation and, consequently, has changed the relationship between the private and the public broadcasting sector. For example, at the end of 2010 the number of channels in Europe was 9,893 compared to 220 in 1996 and less than 90 in 1989. Moreover, only 12% of the TV channels are public (EAO, 2011).

THE EFFECTS OF OWNERSHIP RELAXATIONS

In the last two decades various waves of mergers and acquisitions in the media sector and subsequently in the TV landscape have also been seen (Sanchez-Tabernero, 1993; Albarran, 2002; Bagdikian, 2004; McChesney, 2004). According to the European Audiovisual Observatory (EAO), 62% of the market was controlled by 15 companies in 2006-2009 (Deutche Telekom, Sony, Telefónica, Microsoft Inc, Vodafone Group, Groupe France Télécom, Nokia, Hutchinson Whampoa Ltd, Bouygues, Apple Inc., Telecom Italia, Vivendi, News Corporation, Walt Disney, BT Group PLC). The percentage increased further, up to 67% in 2010, indicating a growing concentration of ownership at the global level (EAO, 2011).

In the US, the consolidation in the sector was facilitated, as noted, by the Telecommunications Act, but even before the Act, non-media businesses, such as General Electric and Westinghouse, were allowed to purchase two of the major television networks, RCA and CBS. Although there is a small number of television stations that are owned independently or by families, the 1996 merger between Time Warner and America Online ($156 billion) signifies a landmark for the consolidation of the communications industry, while at the same time paving the way for new acquisitions and mergers in Europe (Papathanassopoulos, 2002).

In Europe, the merger between Bertelsmann, Germany's largest media company and Compagnie Luxembourgeoise de Télédiffusion (CLT) in 1996 created CLT-Ufa, a major player in the increasingly competitive German market. Using Bertelsmann's firm footing in free television, CLT-Ufa was able to exert greater pressure on its main rival, the Kirch group. CLT-Ufa also has stakes in RTL4 and 5 in the Netherlands, M6 and TMC in France, Premiere in Germany and Channel 5 in Britain. Among the largest deals in this period was Vivendi's $46.7 billion take-over of Seagram's Universal in 2000, forming Vivendi Universal, followed by France Télécom's $40.1 billion purchase of British cellular company, Orange. Moreover, the number of smaller deals increased.

According to the Euromedia Research Group, in recent years the emergence of a small number of larger western European media companies alongside US-based global giants Disney, Time Warner and News Corporation could be perceived. German-based Bertelsmann AG is present in about 50 countries, and Vivendi, which had its origins in France, is present in more than 70. These companies, like their US counterparts, conduct diversified activities, are engaged in all types of media production and are present in regional/national media markets across the globe (Vartanova, Bergés-Saura, Steemers & Papathanassopoulos, 2011, p. 173). Moreover, a number of large media companies can also be observed (such as, among others, Lagardère (France), Pearson (UK), Reed Elsevier (Netherlands), VNU (Netherlands), ProSiebenSat1 Media AG (Germany), Sanoma (Finland), Bonnier (Sweden), Mediaset (Italy)), which are predominantly oriented towards the regional markets. On the other hand, these players as well as other European media players have yet to establish a global player in the online domain, with Google, Yahoo and Facebook all originating in the US (Vartanova et al., 2011, pp. 173–174).

THE CONCERN OVER FOREIGN OWNERSHIP

Another important issue has been the impact of foreign media ownership on national media systems. In the US the relevant restrictions reflected the potential political impact of news outlets in both domestic and international affairs, and policymakers' concerns about whether foreign owners might incorporate propaganda. Australian media mogul Rupert Murdoch became a US citizen in 1985 primarily to circumvent FCC restrictions on foreign ownership of broadcast licenses.

In Europe, there is also some concern regarding the impact of US-based transnational conglomerates on European media industries (Iosifidis, Steemers & Wheeler 2005, p. 83), including the impact of foreign investment – both European and non-European – on national media systems. The issue of foreign media ownership within Europe is in fact multi-faceted, increasingly transnational, multi-sectorial and complex (Vartanova et al., 2011, p. 173). In Europe two trends can be noted: the first is that leading European companies have come under non-European ownership. For example, the 'French' group Vivendi is 90% owned by US registered stakeholders; the 'British' BSkyB, the 'Italian' Sky Italia, Premiere Germany and Premiere Austria come under the control of US-based News Corporation; and even the Bertelsmann Group is owned by the Bertelsmann foundation (76.9%), while the founding Mohn family owns 23.1% (Vartanova et al., 2011, p. 174). The second is that leading European TV companies like the RTL have already developed a pan-European strategy, usually launched in league with local partners, such as RTL in Germany, M6 in France, Channel 5 in the UK, Antenna 3 TV in Spain, Alpha TV in Greece, RTL4 in the Netherlands and RTL Klub in Hungary (Papathanassopoulos & Negrine, 2011, pp. 30–31).

One also has to add the changes in the media industries and the ownership status affected by social change in Central/Eastern Europe and the former Soviet Union. As the Euromedia Research Group points out media ownership has become a lucrative area for national politicians and foreign media investors. The latter enter a 'virgin' market, while for politicians media ownership brings a combination of political influence and business advantages (Vartanova et al., 2011, p. 174).

THE ISSUE OF MEDIA CONCENTRATION

Media ownership regulation is designed, in most cases, to control or limit developments that distort competition and generate imbalances in the marketplace. As Robert Picard notes: 'A significant challenge in dealing with issues of media concentration is that conflicting economic policies affect media consolidation and concentration because policy-makers often pursue different objectives simultaneously through various administrative and regulatory agencies' (2011, p. 359).

In the US, for example, after the passage of the Telecommunications Act of 1996, there was substantial consolidation across most forms of media sectors (McChesney, 2003). Currently, most US television stations are either owned by or affiliated to the large networks (ABC, CBS, NBC, Fox, Viacom and Time Warner), both in over-the-air broadcasting and cable, constituting an oligopoly. Other countries, at least in the Western world, have followed a similar pattern of consolidation. In Australia Rupert Murdoch's New-York-based News Corporation controls some 70% of the country's press. Similarly, in Canada, the relaxation of ownership rules led to a round of mergers and acquisitions resulting in a tight consolidation of television station ownership across the country (Pitts, 2002). Europe has also experienced, as noted, an expansion of commercial television and the media sector overall.

The study of media concentration remains an important concern among media economists and policymakers (Albarran, 2006; Rice, 2007; Picard, 2011). In effect, whether media concentration constitutes a condition that violates the public interest is one of the most debated issues (Napoli, 2006). There are various approaches used in media research, encompassing both quantitative and qualitative perspectives (Garcia & Surles, 2007). By and large, there are both liberal and the critical perspectives regarding media concentration (see also Mastrini & Becerra, 2008).

In the liberal perspective, the concentration processes are not questioned apart from the case of monopoly. This supposes that concentration is problematic as a measure of the vigor of a marketplace of ideas (Noam, 2009). On the contrary, it provides media companies with synergies and economies of scale, as well as the possibility to improve distribution and lower costs, while the homogeneity of audiovisual services has more to do with competition than with the concentrated structure of the industry (Bustamante, 2004).

In the critical perspective, media concentration is seen as one of the main tools capitalism uses to maintain its legitimacy. It views the concentration of ownership as a tool of the media companies (owners) to promote their own values and interests, either in the media business sector or in the content, directly through synergies or indirectly though the influence of editors and self-censorship. Also, it is considered that media concentration tends to constrain the expression of voices critical of the system (McChesney, 2004).

By and large, concentration refers to the process by which similar entities are progressively grouped together, while diversification refers to the process by which different entities are brought together under one centre control (Murdock & Golding, 1999). The former is sometimes referred to as *horizontal integration* whilst the latter is more commonly known as *vertical integration*. But, there is also a third kind, the *cross-ownership* of media. This takes places when a person or company owns outlets in more than one medium in the same geographical market. It is a business strategy driven by advances in technology and it is aimed at achieving economies of scope across multiple media. Costs may be reduced through the 'synergy' of sharing staff and content between outlets in different media, and revenues may be increased through the sale of multimedia advertising packages. For example, the French company Vivendi owns the pay TV channel Canal Plus, cinema studios, theme parks, cable channels, record labels, a video games division, a telecommunications group, and other businesses. Time Warner Inc., as noted, owns AOL and other computer-based enterprises, one of the biggest cable operations in the US with the Turner Broadcasting System (CNN and other cable broadcasting companies and the Atlanta Braves baseball franchise), New Line Cinema, Warner Bros. Pictures and the Warner Bros. Television Group, Home Box Office operations, dozens of Time Inc. magazines, comic books, and still other media holdings (McChesney, 2003, 2004).

The proponents of media consolidation or concentration argue that profits increase as middlemen are eliminated and risk is reduced in that the possibility of a producer not being able to sell a product (vertical integration) is lessened. They also argue that horizontal integration can achieve economies of scale by spreading fixed costs over a larger number of stations in the case of television. They also argue that horizontal integration can improve the ability of a company (such as a production company) to shift unwanted material (such as movies), by packaging it with the hits. It can also improve the terms on which channels buy programs.

THE REALITY OF CONCENTRATION

The reasons why media companies seek alliances and mergers are simple: to maintain and/or gain market strength and supremacy (Sanchez-Tabernero & Carvajal, 2002; Doyle, 2002). In attempting to gain these advantages, entrepreneurs and corporations engage in mergers and acquisitions, which have given them greater control over specific sectors or combinations of sectors. In effect, they have extended long running trends in the media industry: first, the long-turn towards concentration; second, a tendency towards diversification, which in turn gives conglomerates the advantage of having, if not control, several stakes in the communication industry.

At the same time, diversification increases the control which individual entrepreneurs or media organizations can wield over a larger part of the whole linkage between content production and distribution. In the media sector in the last decades there has been a gigantic redeployment marked in the field by acquisitions, transnational alliances and megamergers. This creates obvious problems in the media sector since it unleashes pressures for companies to diversify, merge or expand. Size has become an important factor. The trends described above (such as rationalization, expansion, growth, increased competition, larger markets, etc.) may be the unintended consequences of a policy for competition and the process of restructuring the European communications sector. But, as Murdock and Golding (1999, pp. 117–118) argue, in the last decades two parallel movements in the field of policy of both the European Union and its major member states can be observed: the first is the ascendancy of *marketization* policies and the second the *convergence* of the communication sector. In effect, 'corporate thinking presents the new spaces of private enterprises opened up by convergence as a logical, and therefore, largely incontestable, consequence of digital technologies' (Murdock & Golding, 1999, p. 118).

Murdock and Golding (1999, p. 118) consider marketization as 'all those policy interventions designed to increase the freedom of action of private corporations and to institute corporate goals and organizational procedures'. In fact, many countries have become less willing to write off the perceived economic opportunity costs associated with restricting domestic media and cross-media ownership. Murdock and Golding (1999, pp. 118–119) provide five basic dimensions to this general process:

- *Privatization*, the sale of public communication assets to private investors.
- *Liberalization*, the introduction of competition into monopoly markets or its extension into markets with limited competition.
- *Corporatization*, encouraging or compelling organizations still within the public sector to pursue market opportunities.
- A move from *license to auctions.*
- *Reorienting the regulatory system*, i.e., a reform of the regulatory system which will increase corporate access to markets and enhance their freedom of action within it.

The strategy of media companies in merging or vertically integrating is clear: the advantage of owning a distribution system means that a TV service provider will have a secure outlet for programs and/or TV channels. This allows them to dominate over the small and medium companies, 'both established and new, over the mass and niche markets both global and local' (Bustamante, 2004, p. 813). On the other hand, these moves are not the inherent 'logic' of digital technology, but, through marketization policies, the ambitions and interests of the major corporate players within the communication system. As Murdock and Golding (1999, p. 120) point out: 'They [corporate players] deploy appeals to technological inevitability to provide justifications for continuing and extending marketization to their maximum advantage'.

OWNERSHIP CONCENTRATION REGULATION AT THE EUROPEAN UNION LEVEL

Although most of the member states of the European Union (EU) impose some special restrictions on ownership of the media (Doyle, 2002, p. 148), large firms, especially after the Maastricht Treaty, have become integral players in the policy formation process, participating directly as private actors, or collectively through new loose cross-border alliances (Papathanassopoulos, 2002).

Since the mid-1980s the European Parliament has been calling on the Commission to take an initiative in order to control, if not impede, the on-going concentration of ownership in the European media sector. In 1992, the Commission published its Green Paper on *Pluralism and Media Concentrations in the Internal Market* (CEC, 1992). This consultative document was heavily influenced by industrial imperatives and argued that the European media industry was hindered by extensively different ownership rules in each member state, whereas media companies were attempting to pursue their activities and investments. The media owners, like News Corporation and Fininvest, were against any EU initiative to harmonize legislation on media ownership at the European level. In the meantime, within the Commission there was growing debate on the matter, in particular over whether the Commission was competent to pursue policies aimed at safeguarding pluralism (Papathanassopoulos, 2002).

In 1994, the Commission published a follow up to the 1992 Green paper, in which it presented and evaluated the outcome of the consultation process (CEC, 1994). This second round of consultations produced exactly the same results as the first one, i.e., a lack of consensus concerning any attempt to harmonize media ownership policy at the EU level. After a two-year period of consultations, debates and considerable lobbying from major international and national media groupings, Commissioner Mario Monti aimed to present a first draft for a Directive on Media Pluralism to his colleagues. The main proposals of this draft in July 1996 were as follows (Papathanassopoulos, 2002, pp. 112–113):

- A 30% upper limit on 'monomedia' ownership: one broadcaster could not control another (new or existing) if that equaled 30% or more in designated market 'zones';
- A 'multimedia threshold': an upper limit of 10% for 'multimedia' concentration (i.e., ownership for a combination of different media (television, radio and/or newspapers)). A venture already based in one media could not own a different media (new or existing) if the total audience share of its media equaled 10% or more in the area concerned;
- All market shares would be based on multiple audience measures within the area in question.
- Member states could exclude public broadcasters from these upper limits.

Once again these proposals raised new controversies. The main one focused on the level of diversity of ownership appropriate for different market sizes (at the local, regional or national level). The Commission put forward a new version in March 1997. In this new 'Exploratory Memorandum', the title was changed from 'Concentrations and Pluralism' to 'Media Ownership' in the Internal Market. This signaled a 'flexibility' to deflect the focus away from pluralism (where the Commission's competence would be in question) toward the aim of removing obstacles to the Internal Market. As Doyle (2002: 164) points out, it was clear that in practice, the 'flexibility clause' would allow member states to maintain whatever upper restrictions on ownership were affordable – either economically or politically – in their own territories.

As Radaelli (2000, p. 36) points out: 'the EU media concentration policy process has witnessed a persistent attempt of the Commission to finalize a draft proposal', but it has become rather clear that any attempt to tackle the issue of media ownership and concentration at a European level is unrealistic. The industrial imperatives of the information society and the new opportunities, stakes and interests of the convergence of the media, telecommunications and information society make it difficult to tackle this thorny problem. In effect, 'convergence has already revealed the inadequacy of European initiatives to harmonize sectorial ownership regulation' (Iosifidis, 1999, p. 10). Added to this is the inadequacy of the tools used to measure and evaluate such things as market concentration and market influence.

In short, the EU seemed powerless to regulate the issue of concentration, apart from scrutinizing the mergers and acquisitions under the competition law (see Doyle, 2002; Papathanassopoulos, 2002; Harcourt, 2005). This is not an easy task: it is obvious that any merger, acquisition or joint merger aims to eliminate competition and at the same time to strengthen its position in the market. In practice, it seems that the European

Commission has favored the growth of larger European groups with media interests in many member states but it has prevented concentration in national markets (Harcourt, 2005, p. 117). It is unlikely that the trends toward concentration in the wider communications environment will cease since every dominant player will attempt to extend its dominant position into new or neighboring markets. However, the Commission still aims to ensure that citizens have access to a variety of information sources, allowing them to form opinions (CEC, 2007). Recently, after pressure from members of the European Parliament, the Commission established a High-Level Group on Media Freedom and Pluralism to make 'recommendations for the respect, protection, support and promotion of pluralism and freedom of the media in Europe'. There have been some talks concerning the creation of a 'media pluralism monitor' (Valcke, 2012) or even the establishment of independent National Regulatory Authorities to be responsible for media freedom and media pluralism (CMPF, 2013). As with other policy areas, it is easier to point to the risks and the dangers than to actually create mechanisms to support those aims and objectives. Nevertheless, the result is that the situation, up to now, creates a regulatory uncertainty rather than a coherent competitive environment (Papathanassopoulos & Negrine, 2011).

CONCLUDING REMARKS

In the digital era the growth of vertical integration strategies could be interpreted as 'defensive moments in the face of uncertainty and convulsions brought about by digital networks and their challenge to the power distribution' (Bustamante, 2004, p. 813). Concentration of ownership, on the one hand, was due to the strategy of large media groups to ensure that they form part of the competitive content and service offering, and on the other, it was a response to new and powerful competitors that emerged outside the content sectors.

This coupling of media giants, characterized by 'a strong sense of continuity from the analogue era' (Iosifidis et al., 2005, p. 3), is certain to cause new problems in terms of media ownership, concentration, media diversity and quality (Barnett, 2004). At the same time, the economic recession in the last years and the downturn in advertising seem to drive toward a new consolidation in the media sector. In the traditional television sector, media pundits predict that that there would only be five major players in European commercial TV and only two in pay-tv, a picture that resembles to some extent the US. It remains to be seen, but the financial crisis in the economy will certainly affect the media sector in general and TV in particular, leading to further consolidation and vertical integration.

Broadly speaking, the issue of concentration will remain at the center of heated debate in the first half of the 21st century. Past experience has shown that neither intergovernmental organizations like the European Union nor the individual states, either larger or smaller, in the era of globalization, convergence and digitalization of the media seem able to regulate ownership and concentration in the sector since there are too many risks involved. In fact, due to the technological developments, concentration of ownership needs to be reconsidered not only with respect to specific media sources (such as television), but across different media and in relation to distribution channels, whether for traditional media or for new media. On the other hand, new players such as search engines, news aggregators, social networks and app stores increase their presence and scope in the communications sector. The increasingly important role they play in either improving or restricting media pluralism should be considered, especially as they start producing content. However, care must be taken to distinguish between media that publish original work directly, and services that allow users to republish or link to other people's work.

REFERENCES

Albarran, A.B. (2002). *Media Economics: Understanding Markets Industries and Concepts* (2nd edn). Ames: Iowa State University Press.

Albarran, A.B. (2006). Historical Trends and Patterns in Media Management Research. In A.B. Albarran, S.M. Chan-Olmsted & M.O. Wirth (Eds) *The Handbook of Media Management and Economics* (pp. 3–22). Mahwah, NJ: Lawrence Erlbaum.

Bagdikian, B. (2004). *The New Media Monopoly*. Boston, MA: Beacon Press.

Baker, E.C. (2007). *Media Concentration and Democracy: Why Ownership Matters*. New York: Cambridge University Press.

Barnett, S. (2004). Media ownership policies: Pressures for change and implications. *Pacific Journalism Review*, 10 (2): 9–20.

Bustamante, E. (2004). Cultural Industries in the Digital Age: Some Provisional Conclusions. *Media, Culture and Society*, 26(6): 803–820.

CEC (Commission of the European Communities) (1992). Pluralism and Media Concentration in the Internal Market: An Assessment of the Need for Community Action. COM (92) 480 Final, Brussels: 23 December.

CEC (1994). Communication to Parliament and Council: Follow-up to the Consultation Process Relating to the Green Paper on Pluralism and Media Concentration in the Internal Market – An Assessment of the Need for Community Action. COM (94) 353 Final, Brussels: 5 October.

CEC (1997). *Explanatory Memorandum* (Media Ownership in the Internal Market). DG XV, Brussels: February.

CEC (2004). Guidelines on the Assessment of Horizontal Mergers under the Council Regulation on the Control of Concentrations between Undertakings. *Official Journal of the European Communities*, 31(5).

CEC (2007). Directive 2007/65/EC of The European Parliament and of the Council of 11 December 2007 amending Council Directive 89/552/EEC on the coordination of certain provisions laid down by law, regulation or administrative action in Member States concerning the pursuit of television broadcasting activities, *Official Journal of the European Union*, L. 332/27, 18/12.

Centre for Media Pluralism and Media Freedom (CMPF) (2013). *European Union Competencies in Respect of Media Pluralism and Media Freedom – Policy Report.* Robert Schuman Centre for Advances Studies/European University Institute/European Union.

Ciaglia, A. (2013). Pluralism of the System, Pluralism in the System: Assessing the Nature of Media Diversity in Two European Countries. *International Communication Gazette*, 75(4): 410–426.

Doyle, G. (2002). *Media Ownership: The Economics and Politics of Convergence and Concentration in the UK and European Media*. London: Sage.

EAO (European Audiovisual Observatory) (2011). *Television in 36 European States,* Vol. 2. Strasbourg: EAO.

Garcia, L.D. & Surles, E. (2007). Media Ownership and Communications: Enriching the Research Agenda. *Telecommunications Policy*, 31: 473–492.

Harcourt, A. (2005). *The European Union and the Regulation of Media Markets.* Manchester: Manchester University Press.

Iosifidis, P, Steemers, J. & Wheeler, M. (2005). *European Television Industries*. London: BFI.

Iosifidis, P. (1999). Diversity versus concentration in the deregulated mass media domain. *Journalism & Mass Communication Quarterly*, 76 (1): 162–162.

Liu, F. & Albarran, A.B. (2009). Media Economics and Ownership. In W.F. Eadie (Ed.), *21st Century Communication: A Reference Handbook* (pp. 851–858). London: SAGE Publications.

Mastrini, G. & Becerra, M. (2008). Concentration in Media Systems. In D. Wolfgang (Ed.), *The International Encyclopedia of Communication*. London: Blackwell Publishing. Retrieved from: www.communicationencyclopedia.com/subscriber/tocnode?id=g9781405131997_chunk_g97814051319958_ss120-1 (accessed 8 November 2008).

McChesney, R.W. (2003). 'The Nine Firms That Dominate the World', Global Policy Forum. Retrieved from: www.globalpolicy.org/component/content/article/162/27621.html (accessed 27 May 2013).

McChesney, R.W. (2004). *The Problem of the Media: US Communication Politics in the 21st Century*. New York: Monthly Review Press.

Murdock, G. & Golding, P. (1999). Corporate Ambitions and Communication Trends in the UK and Europe. *Journal of Media Economics*, 12(2): 117–132.

Napoli, P.M. (2006). 'Issues in Media Management and the Public Interest'. In A.B. Albarran, Chan-Olmsted, S.M. & Wirth, S.O. (Eds), *Handbook of Media Management and Economics* (pp. 275–296). Mahwah, NJ: Lawrence Erlbaum.

Noam, E. (2009). *Media Ownership and Concentration in America*. New York: Oxford University Press.

Papathanassopoulos, S. (2002). *European Television in the Digital Age: Issues, Dynamics and Realities*. Cambridge: Polity.

Papathanassopoulos, S. & Negrine, R. (2011). *European Media: Structures, Policies and Identities*. Cambridge: Polity.

Picard, R.G. (2011). 'Economic Approaches to Media Policy'. In R. Mansell and M. Raboy (Eds), *The Handbook of Global Media and Communication Policy* (pp. 355–365). Chichester, West Sussex: Wiley-Blackwell.

Pitts, G. (2002). *Kings of Convergence: The Fight for Control of Canada's Media*. Toronto: Doubleday.

Radaelli, C.M. (2000). Policy Transfer in the European Union: Institutional Isomorphism as a Source of Legitimacy. *Governance: An International Journal of Policy and Administration*, 13(1): 25–43.

Rice, R. (Ed.) (2007). *Media Ownership: Research and Regulation*. Cresskill, NJ: Hampton Press.

Sanchez-Tabernero, A. (1993). *Media Concentration in Europe: Commercial Enterprise and the Public Interest*. Düsseldorf: European Institute for the Media.

Sanchez-Tabernero, A. & Carvajal, M. (2002). *Media Concentration in the European Market: New Trends and Challenges*. Media Markets Monograph. Pamplona, Spain: Servicio de Publicciones de la Universidad de Navarra.

Sussman, G. (2008). 'Ownership in the Media'. *The International Encyclopedia of Communication*. In D. Wolfgang (Ed.), The International Encyclopedia of Communication. London: Blackwell Publishing. Retrieved from: www.communicationencyclopedia.com/subscriber/tocnode?id=g9781405131995_chunk_g978140513199520_ss31-1 (accessed 2 January 2009).

Valcke, P. (2012). Challenges of Regulating Media Pluralism in the European Union: the Potential of Risk-Based Regulation. *Quaderns del CAC 38*, XV(1): 25–35.

Vartanova, E., Bergés-Saura, L. Steemers, J. & Papathanassopoulos, S. (2011). Media Industries: Ownership, Copyright and Regulation. In J. Trappel, W.A. Meier, L. D'Haenens, J. Steemers & B. Thomass (Eds), *Media in Europe Today* (pp.167–183). Bristol: Intellect.

6

International Regulation and Organizations

Paschal Preston and Roderick Flynn

INITIAL INSTITUTIONALIZATION: INTERNATIONAL REGULATION OF NATIONAL MEDIA

The relative prominence of a series of international organizations with regard to television since the early decades of the 20th century both reflect the changing conceptions of what television is but also the capacity of at least some of those organizations to influence those conceptions. That these conceptions were fluid from the outset is illustrated by George Bernard Shaw's 1922 play *Back to Methuselah.* Set in the year 2170, one scene depicts the head of the British government in conference with his cabinet ministers, who are several hundred miles distant. The Prime Minister operates a number of keys on a switchboard and life-size pictures of the people he is speaking to appear on a silver screen, their voices simultaneously transmitted (Laufer, 1928).

Although the Shaw example reminds a 21st-century reader that there was nothing inherent in television technology that demanded it be developed as a broadcasting technology, once the decision to exploit it in that fashion was taken, the institutional structures initially devised for radio broadcasting were, for the most part, simply extended to television in the 1930s and 1940s. As Raymond Williams (1974) has noted, these structures reflected the initial assumption that broadcasting institutions would operate within national boundaries addressing only domestic audiences. (There were exceptions to this – Radio Luxembourg, the precursor of RTL, pioneered international commercial broadcasting, commencing experimental transmissions aimed at Britain and Ireland as early as 1932.) Nonetheless, broadcasting's reliance on the electromagnetic frequency spectrum, which paid no heed to national boundaries, meant that even a primarily national medium required some international regulation, if only to prevent one national broadcaster's signal interfering with those of neighbouring countries. The key framework for this regulation in the early years of broadcasting was the International Telegraph Union (ITU). The ITU's very existence reflected the inherently international nature of electrical and electronic communications – as national telegraph systems began to breach their national borders the ITU constituted the mechanism for agreeing and setting international standards for telegraphic apparatus and use. When radio emerged in the 1890s, conceived as a substitute for wired telecommunications, it was unsurprising that the ITU should become the locus for regulating wireless communications; first at the 1906 International Radiotelegraph Conference in Berlin and then the subsequent conference in London in 1912. However, the first ITU Radiotelegraph conference to comprehensively address radio as a broadcast medium was not held until 1927 in Washington.

The Washington Conference established a number of themes which would recur through the history of international discussions regarding radio and television broadcasting. Not least amongst these was the conception of how broadcasting should be run. European delegates assumed that broadcasting was an element of the broader state-owned communication apparatus that included telegraphy and telephony. In the USA, by contrast, the exploitation of radio waves was assumed to be a primarily commercial enterprise. Thus although European delegates argued that 'the dignity and authority of the committee would be impaired if representatives of private enterprises were accorded the full rights and powers granted to representatives of the administrations [governments]' (Stewart, 1928, 46), the USA adopted the view that such a position would entirely exclude the USA since USA broadcasting was dominated by private enterprises. Thus the conference permitted private companies to participate whenever 'a country is not represented by an administration' (Stewart, 1928, 46). This was a significant concession because although the USA delegation was led by the USA Secretary of Commerce Herbert Hoover (who also presided over the Conference), it permitted private commercial interests, and in particular, USA commercial interests to exercise a degree of influence over the conference proceedings and outcomes.

In this respect it might be argued that the key outcome of the Washington Radio Conference was the decision to adopt the USA proposal for allocating frequency spectrum for all services carried by 'Hertzian waves' (including radio and later television). European delegates advanced the argument that the spectrum should be allocated on national lines. However, for the USA this raised the prospect that access to a scarce resource would be wasted on nations not yet capable of exploiting it economically or technologically. Thus the USA proposed that that frequency be divided on the basis of service types (two-way radio, broadcasting, etc.) so as to reward those making technological innovations with larger portions of the spectrum. In other words:

> the American system meant that the spectrum would be allocated into portions for each useful application, or service, achievable through radio technology and not on the basis of a certain section of the spectrum set aside for each nation. (Schwoch, 1990, 77)

The effect of this was to place future control of the spectrum in the hands of those (soon to be largely US-based) major radio corporations with the R&D capacity to drive technological innovation. Furthermore, the Conference effectively enshrined the 'first come, first served' principle with regard to spectrum access: when a country wished to establish a new broadcasting service they needed to work around the frequencies already occupied by neighbouring countries and which had been previously notified to the ITU's International Bureau. Thus, by definition early-industrializing nations were favoured and lesser developed nations (and in particular post-colonial nations in the post-World War II period) would struggle to access workable frequencies when they did eventually acquire the wherewithal to establish indigenous broadcasting services later in the

20th century. This codifying of unequal access effectively became an unquestioned element of the ITU's DNA. When satellite broadcasting began to take off in the 1960s following the launch of Syncom and the first series of Intelsat satellites (followed from 1971 by Intersputnik serving Eastern Bloc countries), concerns were expressed about the advantages accruing to those nations in a position to first place satellites in favourable geostationary orbital slots. Although the Outer Space Treaty of 1967 explicitly described the geostationary orbit as 'the province of all mankind' and 'not subject to national appropriation by claim of sovereignty, by means of use or occupation, or by any other means' (Rothblatt, 1982, 59), in practice, ownership of satellite slots was secured by dint of early notification to the ITU's International Frequency Registration Bureau creating a 'clear incentive to move into new frequency bands as quickly as is technologically possible' (Rothblatt, 1982, 60).

INTERNATIONALIZATION OF TELEVISION CONSUMPTION AND DISTRIBUTION

Regardless, this was the extant international regulatory framework when television arrived as 'radio with pictures' in the 1930s and 1940s. The explosive success of the medium in the USA in the late 1940s and early 1950s was followed by the more delayed diffusion of the medium elsewhere. According to UNESCO there were just five countries regularly broadcasting television in 1950 but by 1960 this had leapt to 63. Nor was this early development limited to what were then termed 'First World' countries: of the 63, four were on the African continent and 12 in Asia (UNESCO, 1963, 19).

Although not initially reflected by international organizational structures, the transnational nature of television became evident in global consumption patterns from the 1950s onwards. In the core capitalist countries, the diffusion of television was taken as a key symbol (alongside the motor car) of a new era of full employment, raising real standards of living and mass consumerism or 'affluence'.

Much like radio, the distribution of television was initially organized on a national basis. The sheer cost of television production, however, soon meant that even in the wealthier countries of the North, there was a reliance on the importation of programmes primarily produced for consumption in other markets. Not surprisingly, given the dominant role of Hollywood-based studios in the film sector since the early 1920s, the USA was key amongst these (Garnham, 1990).

If in the 1940s and 1950s much television was live, the increasing incidence of filmed material produced by Hollywood for television and the arrival of video recording in the late 1950s meant it became more feasible to sell programmes to overseas markets. Thus when Independent Television was launched in the UK in 1955 its schedules included more than a smattering of USA fiction content. By the 1960s international trade in television content was taken for granted and was a virtual prerequisite for broadcasting institutions established in smaller countries. Even if audiences preferred local content over imported material (the so-called 'cultural discount') the cost of buying imported material was usually a fraction of the cost of producing domestic content because the price of television imports was not predicated on production costs but on estimates of what the importing market could bear. Thus even for public service broadcasters addressing local audiences conceived of, at least in part, as citizens, the economic logic of purchasing imports was compelling.

It should be acknowledged, however, that the USA was not the only source of international television sales. By the start of the 1970s the UK was second only to the USA as a distributor of television programmes, France exported over 15,000 hours of material annually (the bulk of it to former African colonies) and West Germany sold over 5,000 hours (mainly to Latin America and Asia).

Furthermore, although operating on a smaller scale, Mexico exported over 1,300 hours of material to Latin America in 1970 while the Lebanon and Egypt exported around 2,000 hours of material to the Middle East in the same year.

INTERNATIONAL JOINT/ COLLABORATIVE PRODUCTIONS

Furthermore, the economic logic of television production also incentivized the establishment of broadcasting unions to spread production costs. The prototype was the OIRT (Organisation Internationale de Radiodiffusion et de Télévision) established by broadcasters from 26 European states in 1946 to exchange broadcasting information. In 1950, however, following the rise of Cold War tensions, most of the Western European members defected to the newly established European Broadcasting Union (EBU). In addition to serving as a vehicle for technical cooperation (a function it continues to serve), the EBU sought to encourage programme sharing through the establishment of Eurovision. (The OIRT would adopt a similar approach with the creation of Intervision in 1960). By the end of the 1960s, Eurovision was producing approximately 1,000 hours of material each year for shared use by its members. Although the highest profile programme was the annual 'Eurovision' song contest, the bulk of the material was constituted by news and sports content. Similar rationales drove the establishment of the Union of National Radio and Television Organisations of Africa (1962), the Asian (-Pacific) Broadcasting Union (1964), the Arab States Broadcasting Union (1969) and the Caribbean Broadcasting Union (1970).

For some theorists and governments, the scale of international television sales was a cause for celebration. Modernization theorists such as Daniel Lerner and Wilbur Schramm identified television programming as part of the electronic media which could have a dynamic impact on national development goals. They argued that the 'free flow' of television contributed to the development of the 'mobile personality' and other mental attitudes deemed to be prerequisites for developing nations to progress by encouraging an open attitude to new influences, insights and innovations. A decade later, the image of an unfolding 'global village' was ecstatically preached by one Marshall McLuhan. International programme sales or the increasing incidence of live global transmissions extended 'our central nervous system in a global embrace, abolishing both space and time as far as our planet is concerned' (McLuhan, 1964, 4) and creating an aspiration for 'wholeness, empathy and depth of awareness ... a faith that concerns the ultimate harmony of all being' (McLuhan, 1964, 5–6).

GLOBAL FLOWS OF TELEVISION: FREE FLOW VS CULTURAL IMPERIALISM

However, as the 1960s progressed, elements within the international organization charged with promoting peace and security through encouraging international education, scientific and cultural collaboration began to voice concerns about the nature of these televisual flows. Early UNESCO statistics on television concentrated on capturing the number of transmitters and television sets rather than on content. Unsurprisingly, even though broadcasters had begun to emerge on all five continents by 1960, there were huge disparities in levels of access: in 1960 there were just 0.5 television sets per 1,000 population in Africa as opposed to 231 in North America. However, by the late 1960s, UNESCO, driven in particular by states from the Non-Aligned Movement began to focus on the geopolitical implications of global imbalances in international flows of television and other media products. This concern was informed by the assumption that international political cooperation could only occur in a context where all nations had a proper understanding of other peoples and nations: television and other mass media occupied a key role in shaping this understanding.

On one level the US espousal of the 'free flow of information' principle as a key element of post-war foreign policy appeared to support the UNESCO perspective: starting from the premise that access to information was, ipso facto, a good thing, the advocacy of 'free flow of information' argued for removing artificial barriers (including trade barriers) to the international flow of information goods (including television).

However, at a UNESCO meeting in Montreal in 1969, delegates expressed the concern that 'what has come to be known as the 'free flow of information' at the present time is often in fact a 'one-way' flow rather than a true exchange of information' (UNESCO, 1970, 27). This concern was apparently validated with the 1974 publication by UNESCO of the Kaarle Nordenstreng and Tapio Varis-prepared report *Television traffic – a one-way street*. Although the report found that the USA and China were 'highly self-supporting television countries and Japan and Russia 'mainly' used their own programmes, it nonetheless concluded that 'in all parts of the world there are countries which are heavily dependent on foreign imports in their programming' (UNESCO, 1974, 12).

Some of the results were stark: in South America, the Near East and Africa, more than half of the countries surveyed relied on imported content for the bulk of their television. This pattern was even evident in more peripheral European countries: in both Iceland and Ireland domestic content accounted for only a minority of all programming. Furthermore, it was clear that in the West and in non-aligned nations imported material was disproportionately sourced from the USA with the UK a distant second. Nor would new distribution technologies such as satellites address the imbalances: 'If anything, satellites have only confirmed the traditional patterns of flow' (Hulten, 1973, cited in UNESCO, 1974, 52)

A symposium on the Report held at the University of Tampere in May 1973, stressed the need to reformulate national and international policies in order to 'redress the imbalance of resources which presently characterizes the international flow and direction of information among nations'. It concluded, in words that would ultimately produce a strongly negative response from the nations which dominated television production, that culturally weaker nations 'may require special assistance, subsidies or protection to permit them an enlarged role in the communications process' (UNESCO, 1974, 59).

This analysis – which clearly countered the arguments, claims and much of the optimism of modernization theorists – encouraged the Non-Aligned Movement within the UN and UNESCO to call for a New World Information and Communication Order (NWICO) in 1976. At UNESCO's 1976 General Conference in Nairobi, the Directorate General established the 16-person International Commission for the Study of Communications Problems chaired by Irish jurist Sean MacBride. Four years later MacBride produced a substantial report suggesting that a NWICO could be based upon, inter alia, the following principles:

- that the negative effects of certain communications monopolies, public or private and other excessive concentrations of communication power should be eliminated;
- that obstacles to a free flow and *wider and better balanced* dissemination of information and ideas should be removed;
- that journalists and other media and communications professionals should be at liberty to do their jobs but that they should also take seriously the responsibility that came with that freedom;
- that developing countries should be facilitated in developing their own communications infrastructures and personnel in a manner suitable to their needs and aspirations.

The reaction of the United States (and more broadly though with less vehemence other Western nations) to the implicit and explicit NWICO assertion that Western dominance of information flows constituted 'cultural imperialism' was immediately hostile. Even before the publication of the MacBride report in 1980 a coalition of Western governments, media companies and other private sector interest groups had begun to wage a campaign against the direction UNESCO was taking, seeing the focus on the global dominance of Western media as nothing less than an attack on capitalism. The USA position held that NWICO equated to state censorship and control of media. For example, they criticized UNESCO's support for a system to protect journalists in the conduct of their work through state licencing. Eventually in 1984 the USA withdrew entirely from UNESCO (to be followed a year later by the UK).

This withdrawal dealt a fatal blow to the NWICO vision since by definition it required a degree of cooperation from the two countries where some of the largest corporate producers of media content were based – even if both the USA and UK have since returned to UNESO (in 2003 and 1997 respectively). Over the course of the 1980s, in the face of much diplomatic and financial pressures, the UNESCO secretariat (if not the majority of the nation states who constituted its membership) gradually adopted policy concepts and language much closer to that of the USA.

REGIME CHANGE: NEO-LIBERALISM RESHAPES THE INTERNATIONAL TELEVISION ORDER

If the two successive hegemonic powers (UK and USA) lost a few skirmishes with governments of the non-aligned movement and their allies in the debates over international structures of television and news flows in the 1970s' UNESCO setting, they were far from defeated. Indeed they soon showed that they still possessed the material, political and symbolic resources required to totally transform the conceptual and value frames underpinning those debates and, indeed, to construct entirely new policy fora and organizations to deal with the ever-growing internationalization of television. In sum, they possessed and exercised the considerable power required to introduce all-round change in the international organization of television and other media structures and flows.

Since the 1940s at least, the USA's defence of 'free flow of information' principles advocated allowing market logic to regulate the distribution and consumption of information. By the late 1970s, the capacity to advance arguments against NWICO on these grounds reflected a fundamental shift in the political economy in the Western powers which was becoming more manifest precisely during the very period in which the MacBride Group was deliberating. At the start of the 1970s the Keynesian consensus which had been the political economic orthodoxy of the West since the 1950s seemed reasonably solid. By the end of that decade, however, the effect of two oil crises and the apparent ineffectiveness of traditional Keynesian remedies in addressing economic recession created a space in which a new orthodoxy – neo-liberalism – could come to fore.

This shift was symbolized by the near simultaneous rise to power of two new political leaders in the heartlands of the two most powerful imperial powers of the modern era: Margaret Thatcher in the UK in 1979 and Ronald Reagan in the USA in 1980. The two new leaders were not only enthusiastic missionaries of neo-liberalism and champions of both 'free markets' and 'free flow' who happened to be located in highly influential national nodes within the international system setting. Furthermore, they both shared a strategy of national economic renewal and industrial development which placed a strong emphasis on 'making a business of information' (Preston, 2001).

Central to neo-liberalism is the conviction that, almost by definition, state-run industries were less efficient than their private sector counterparts, a conviction that found expression in moves to privatize state-owned industries or to deregulate state monopolies to allow competition from the private sector.

These pressures were felt in television broadcasting as elsewhere: by the early 1980s, there were increasing pressures from various interests seeking a radical restructuring of television and radio broadcasting in Europe and other countries which had not initially followed the predominantly commercial and market-driven approach adopted in the USA.

These pressures originated from industrial interests, including advertisers seeking lower-cost advertising platforms and investors seeking an opportunity to invest in new areas, as well as from politicians and film/programme makers – including those seeking a wider diversity of channels and greater scope for 'independent' programming. Technological developments also played a role, not least because they were perceived to overcome the limits imposed by spectrum scarcity, creating scope for multiple channel service provision and helping to reduce the costs of both television production and distribution.

Even if multi-channel cable television which had expanded rapidly in the USA in the 1970s was slower to take-off in many of the larger European countries, nevertheless the 1980s marked a significant turning point in the structures and characteristics of the television sector globally. The number of national and regional terrestrial channels as well as satellite-based services grew rapidly within the EU area, from approximately 50 in 1980 to over 200 by the end of that decade. This shift was driven by another key international (if not supranational) organization: the European Commission's efforts to construct an increasingly integrated 'single market' embraced an ever greater number of countries and industrial sectors as well as policy functions. As this overall integration project was developed, there were increasing pressures to construct a common regulatory framework for the television and film sectors within the EU area.

The rationale behind the Commission's push for an integrated EU 'audio-visual space' mirrored the economic logics applied to all other service industries: thus Europe's failure to compete with the USA in global audio-visual markets was ascribed to, firstly, the insufficiently market-led nature of state broadcasters and, secondly, the absence of a large-scale European 'home market' (to parallel that of the USA). The response was the introduction of the 'Television without frontiers' (TWF) policies introduced in 1988 which represented a shift away from the hitherto strongly national orientation of media regulation in European countries. In an era when direct-to-home satellite broadcasting was on the cusp of becoming a major player in television distribution, TWF encouraged the development of a single European broadcasting market by prohibiting EU member states from blocking television broadcasts originating in another member state. At the same time, in a bid to undermine the dominance of established state-owned public service broadcasters, TWF obliged all EU broadcasters to set aside 10% of their transmission time for works produced by independent production companies. However, conscious that these rules potentially created a market opportunity for television content holders from outside the EU, TWF also insisted that EU broadcasters should reserve 51% of transmission time for EU works.

As such, the European Commission created a somewhat equivocal context. On the one hand it clearly sought to create new opportunities for the expanded role of large integrated cross-media and transnational corporations in the television sector within the EU area; on the other hand it raised a barrier against precisely such corporations if they were based beyond the EU. From an external perspective this might have been regarded as protectionism, an approach at odds with the broad sweep of neo-liberal policies.

And, ironically, it was precisely the move to introduce a market-driven logic to the European audio-visual industry that allowed a call for greater external access to EU audio-visual markets to dominate debates held under the auspices of one new, but increasingly significant player amongst the international organizations considered by this chapter – the World Trade Organization (WTO) and its predecessor the General Agreement on Trade and Tariffs (GATT).

GATT/WTO

From 1948 onwards, the first seven rounds of GATT negotiations concentrated on reducing artificial barriers (tariffs/quotas on imports, subsidies to local industries) to trade in agricultural and industrial goods. At the eighth round launched in Punta Del Este in Uruguay in 1986, however, trade in services was added to the liberalization agenda. This included trade in audio-visual goods such as television programmes. As early as the 1960s the USA had asserted that 'European limitations on the import of television programmes contravened GATT provisions' (Pauwels and Loisen, 2003, 293). European governments had then rejected this assertion on the grounds that, as cultural artefacts, television programmes could not be treated in the same fashion as industrial commodities.

However, the TWF directive made this commodity/cultural artefact distinction harder to maintain, as did the wave of deregulation then transforming the formerly state-dominated European broadcasting sector. Thus although the initial years of the Uruguay Round focused on the question of agricultural subsidies, from 1993 the spotlight was turned to the audio-visual sector. The USA specifically objected to the existence of local content quotas (as provided for in TWF), the subsidization of national screen industries and the imposition of restrictions on foreign ownership of the press. For its part, the EU pointed out that, regardless of these supports, the USA already enjoyed substantial access to EU markets. In 1993, the EU imported $3.7bn worth of audio-visual material from the USA whilst the USA imported just under $300m worth from the EU (Miller, 1993, 76).

Although in the closing weeks of the Uruguay Round the conflict was framed as a USA/EU (and in particular French) dispute, USA attempts to remove support for national audio-visual industries was almost universally opposed. India, Canada, Japan, Australia and most Third World nations sought to defend those industries on the grounds of cultural sovereignty.

That the dispute over liberalizing trade in audio-visual goods went down to the wire (and almost threatened the collapse of the entire GATT agreement) was a result of the very different premises informing the USA and EU positions. As Miller (1993, 79) put it, the USA position sought an end to all screen subsidies and quotas 'arguing from the precepts of neoclassical economics for untrammelled play of comparative advantage inside laws of supply and demand'. In other words monopolistic competition could be justified on the grounds of the sovereign consumer, 'a fully formed subject making a rational choice in favour of entertainment and distraction'.

By contrast the EU position argued that cultural products were public as well as private goods with a historical and national significance which couldn't be captured within economic formulas and as such should be entirely excluded from the trade in services negotiations. In this instance, state support was justified on the grounds of the 'sovereign citizen … an insufficiently knowing subject in need of education in civics' (Miller, 1993, 79).

In the end, rather than see the entire GATT agreement collapse, the USA and the EU agreed to park the issue. As the USA Trade Representative Mickey Kantor put it 'we can best advance the interests of our artists, performers and producers – and the free flow of information around the world – by reserving all our legal rights to respond to policies that discriminate in these areas' (Cited in Schlesinger, 2000, 98)

Yet despite the absence of any specific commitments on audio-visual services, the sector was still included within the general framework of the Uruguay Round of trade negotiations under GATT (specifically the GATS – General Agreement on Trade in Services). Thus it would be subject to the rules laid down for international trade generally, particularly in connection with transparency and settlement of trade conflicts at GATT level. This also meant that, in future, the audio-visual services sector would come under the dispute resolution mechanisms of the WTO, the permanent successor to GATT.

It was anticipated that the USA-based industry, in particular, would seek to exploit the WTO framework to bring liberalization to the audio-visual sector. In 2000 the USA advanced the argument that the increasing prevalence of digital delivery of audio-visual content meant that television programmes and films delivered in this way could be treated as 'virtual goods' and thus subject to more stringent GATT rules on liberalization rather than in the GATS framework. In practice this argument gained little traction with other WTO members. And in the agreement theoretically most directly associated with television – the GATS – there has been relatively little progress towards liberalization. As the WTO itself acknowledged in 2010: 'Audiovisual services is one of the sectors where the number of WTO members with commitments is the lowest' (WTO, 2010).

INTERNATIONAL CONGLOMERATES AS INTERNATIONAL ORGANIZATIONS

Nonetheless, the establishment of the WTO, an organization which takes for granted the commodified nature of television and other informational product, as the dominant force in the international regulation of trade in television, is reflective of the huge increase in the scale, operations and influence of a handful of international, commercial corporations now active in the television services sector. Though subject to the regulatory regime of the WTO, these corporations constitute in themselves significant international organizations.

Their dominance is the outcome of the innovations in policy and regulatory regimes introduced in the 1980s in the wake of the neo-liberal turn. Several of the largest are direct corporate descendants of the USA-based 'majors' which have played such a dominant role in the global audio-visual sector since the 1920s (Garnham, 1990). A few other members of this exclusive club of major international organizations in the contemporary television services space comprise relatively new corporate empires led by ambitious leaders, of which the News International group, headed by one Rupert Murdoch, is a key exemplar. Still others have accumulated significant capital and power resources based on distinctive ICT-related functions or technological capabilities (e.g. Google, currently planning to charge for some content on its YouTube arm) or through ownership and control of distribution networks (e.g. cable TV or telecoms services corporations). These are the most influential and powerful organizational actors active in the television sector and policymaking arenas in many countries in most regions of the world. When it comes to initiating or shaping changes in public policy and regulations affecting the television sector, or in their implementation, these major corporate actors can rely on fulsome professional support from the clusters of small-scale firms specializing in lobbying, public relations, IPRs and other relevant fields, many of whom are conveniently located in crucial cities such as Washington and Brussels. As before, one of the key functions of that supporting cast of actors is to ensure that, above all, the core issues and stakes involved in the reshaping of policy or regulation directly affecting the media of public communication never quite enter the supposed domain of public information or become the subject of informed debate amongst citizens (Preston, 2009).

The neo-liberal regime has also created novel opportunities for new 'independent' firms to emerge and engage in programme making. Some have notched up success that is truly (figuratively and literally) exceptional, although the majority operate in highly competitive, and often highly precarious, economic conditions and remain structurally dependent on those larger firms which own and control access to a distribution or broadcasting network. Certainly such small-scale or 'independent' firms possess little by way of collective clout when it comes to the shaping and implementation of policies impacting on the television services sector.

The potential of the micro, small or independent segments of the television (or 'creative') services sector has also been frequently emphasized, indeed celebrated, in much of the recent literature advancing certain visions of an emerging new economy. For Manuel Castells they are characteristic of his definition of the contemporary moment in terms of 'network society' or 'information society' theories (Castells, 1996, 2009). He proposes that the contemporary economy and society, and its media sector, is marked by the growing scope for collaborative, networking relationships between actors (generally presented or presumed to be equals) and a concomitant decline in the role and influence of large-scale, hierarchical organizations.

Castells' stress on the fact that novel electronic networking systems (such as the internet, Web 2.0 and social networking), which comprise leading technologies of our time, will only serve to further amplify and enhance the potential for collaborative networking-type of social relationships over those based on hierarchical organizations may not seem immediately relevant for television. However, as medium, industry and technological form, television has demonstrated a considerable capacity to appropriate and harness successive waves of technological (and major political and economic) innovations of recent times. In line with that, it has become more and more intricately interwoven with other, mature and new, media services. Television screens and video clips, and related cultural forms or formats, now comprise ever more prominent features on the websites of 'old' media such as national newspaper organizations. Likewise, much of the content on the 'new' digital platforms of the internet, Web 2.0 or social media comprise television forms and formats, available in various modes and forms, from 'on-demand', pay-per-view to user-created content (e.g. Netflix, and VOD services; YouTube). In sum, it is now becoming increasingly difficult to distinguish or separate a discrete television sector from a growing array of neighbouring media fields. The implication here is that television may be equally subject to the (presumed) decentralizing, non-hierarchical impulses of the net. By extension, Castells' argument suggests that, as a medium, television is unlikely to remain exclusively dominated by a handful of international corporations.

Accepting this conclusion is only possible, however, if one concurs with Castells' account of the influence of networks on organizing social, economic or political relationships. In particular, it requires an acceptance of his portrayal of 'networking' as affording, encouraging and facilitating collaborative, peer-to-peer relations between presumed equals. Castells presents the image of a fuzzy and complex system of overlapping networks involving multiple actors where power differentials seem to disappear in a radical new pluralism based on the plurality of technicalist or network-related functions. However, this approach downplays the continuing and distinctive role, power and salience of concentration and centralization of capital (and attendant power relations) in major multimedia conglomerates. This is not to suggest that questions of corporate power are entirely absent from his account, but his techno-centric model places them in a very secondary, more minor role compared to the standards of the analytical apparatus developed in the critical political economy of communication. Many of the key words and concepts of critical political economy are flagged and mentioned by Castells, but in ways whereby they lose much of their usual analytical and political purchase. As a consequence, Castells' analysis does not offer a convincing basis on which to anticipate the demise (or even substantial weakening) of the power exercised by corporations in international television.

CONCLUSION

In sum then the configurations of key international organizations relevant to the television services sector have undergone radical change since the mid-20th century.

Commencing with a medium structurally constituted by a relatively small number of state-centred organizations committed to one vision or other of television as a public service, the role of international organizations was initially limited to regulation of frequencies. However, as the technological potentials of television were extended and the commercial imperatives of spreading the high cost of television production became manifest, television became, de facto, an international medium, first at the point of consumption (via sales of taped programmes) and later – via satellite – at the level of distribution. In an increasingly post-colonial world, the dominance of Western television content across the world's screens raised fears of 'cultural imperialism' which found its most expression in the forums of UNESCO. The basis upon which NWICO principles were defeated – that television content and other informational media should be regarded as tradable commodities, primarily regulated by the invisible hand of the market – presaged the transition to the final key organization, the ascendancy of the WTO amid increasingly global media corporations.

FURTHER READING

Bignall, J. & A. Fickers (Eds) (2008). *A European Television History*. Oxford: Wiley-Blackwell.

Chakravartty, P. & Y. Zhao (Eds) (2008). *Global Communications: Toward a Transcultural Political Economy*. Lanham, MD: Rowman and Littlefield.

Guback, T. (1992). 'Television Without Frontiers, 2: What The Quota Really Means'. *Television Quarterly*, 24(3): 81–89.

Hall, P. and P. Preston (1988). *The Carrier Wave*. London: Unwin Hyman.

Howkins, J. (2002). *The Creative Economy – How People Make Money from Ideas*. London: Penguin.

IPTS (2012). *European Television in the New Media Landscape*. Seville: IPTS.

Klaehn, J. (2010). *The Political Economy of Media and Power*. New York: Peter Lang Publishing, Inc.

Lowe, G.F. & J. Steemers [for RIPE] (Eds) (2012). *Regaining the Initiative for Public Service Media*. Göteborg: Nordicom.

Tunstall, J. (2008). *The Media Were American*. Oxford and New York: Oxford University Press.

Wasko, J., G. Murdock & H. Sousa (Eds) (2011). *The Handbook of Political Economy of Communication*. Malden, MA: Wiley-Blackwell.

Winseck, D. & D.Y. Jin (Eds) (2011). *Political Economies of the Media: The Transformation of the Global Media*. London: Bloomsbury Academic, pp. 3–48.

REFERENCES

Castells, M. (1996). *The Information Age: Economy, Society and Culture. Volume I. The Rise of the Network Society*. Malden, MA: Blackwell Publishers.

Castells, M. (2009). *Communication Power*. Oxford: Oxford University Press.

Garnham, N. (1990). *Capitalism and Communication*. London: Sage.

Laufer, B. (1928). 'The Prehistory of Television'. *The Scientific Monthly*, 27(5): 455–459.

McLuhan, M. (1964). *Understanding Media: The Extensions of Man*. London: Ark.

Miller, T. (1993). 'The Crime of Monsieur Lang: GATT, the Screen and the New International Division of Cultural Labour' in A. Moran (ed.) *Film Policy: International, National and Regional Perspectives*. London: Routledge.

Pauwels, C. and J. Loisen (2003). 'The WTO and the Audiovisual Sector: Economic Free Trade vs Cultural Horse Trading?' *European Journal of Communication*, 18(3): 291–314.

Preston, P. (2001). *Reshaping Communications : Technology, Information and Social Change*. London and Thousand Oaks, California: Sage.

Preston, P. (2009). *Making the News : Contemporary Journalism Practices and News Cultures in Europe*. London: Routledge.

Rothblatt, M. (1982). 'Satellite Communication and Spectrum Allocation' *The American Journal of International Law*, 76(1): 56–77.

Schwoch, J. (1990). *The American Radio Industry and its Latin America Activities 1900–1939)*. Champaign, IL: University of Illinois Press.

Schlesinger, P (2000). 'From Cultural Protection to Political Culture? Media Policy and the European Union' in Cedarman, L.E. (ed.) *Constructing Europe's Identity: The External Dimension.*

Stewart, I. (1928). 'The International Radio Telegraph Conference of Washington'. *The American Journal of International Law*, 22(1): 28–49.

UNESCO (1963). *Statistical Reports on Radio and Television 1950–1960*. Paris: UNESCO.

UNESCO (1970). *Mass Media in Society: The Need of Research* (UNESCO Reports and Paperson Mass Communication No. 59). Paris: UNESCO.

UNESCO (1974). *Television Traffic – A One-Way Street? A Survey and Analysis of the International Flow of Television Programme Material*. Paris: UNESCO.

UNESCO (2006). *Trends in Audiovisual Markets – Regional Perspectives from the South*. Paris: UNESCO.

Williams, R. (1974). *Television: Technology and Cultural Form*. London: Routledge.

WTO (2010). 'Audiovisual Services', www.wto.org/english/tratop_e/serv_e/audiovisual_e/audiovisual_e.htm (accessed 27 April 2013).

7

Television in India: Ideas, Institutions and Practices

Arvind Rajagopal

NON-WESTERN TELEVISION: COEVAL TEMPORALITIES OF NATION AND TELEVISION

The story usually told of television is of a technology from the West, exported to other regions. This story has both presumed and produced universalist narratives of the modular character of social development enabled by the spread of technology and capitalism. The best-known recent example of an account along these lines was about the growth of print markets, in Benedict Anderson's (2006) argument about the creation of reading publics. Anderson argued that 'print capitalism' stabilized linguistic identities and made possible claims of cultural uniqueness and political sovereignty across territorial boundaries. But the implication of mass media as an inexorable force has also produced critiques centered around Americanization and media-led consumerism as aspects of cultural imperialism, and have expressed resistance to it (Horkheimer and Adorno, 1972; Mattelart and Siegelaub, 1979–83; Gitlin, 1980; Miller, 1998).

Even if they disagree about mass media or the uses it is put to, advocates of both perspectives point to a historically new set of media effects arising with television. If print induced the sense of anonymous camaraderie in shared linguistic markets, television conveyed the senses of intimacy and simultaneity, and generated quotidian participation on a new scale. Unlike the textual focus in cinema studies, therefore, scholars of television have pointed to the medium's everyday flows within and alongside daily life as constituting its distinctness (Williams, 2003). If print came to be identified with the formation of modern publics, television pointed to the possibility of an even deeper identification, although critical scholarship argued that mass manipulation became more likely than mass emancipation (Horkheimer and Adorno, 1972).

What is noticeable in scholarly writing is therefore a media teleology that often associates shifts internal to the technological forms of mass media with broader sociopolitical conjunctures. For example, electronic media is correlated with the rise of bureaucratized soft power and institutions for the management of cultural aspiration and political ideology. Hence media histories often get collapsed into received historical accounts of Europe and North America. Such arguments have their *locus classicus* where electronic media arrived subsequent to the development of infrastructures that were a long time in the making. For example, scholars of France and Britain have described the process of making peasants into Frenchmen or proletarians into Britons that was already more than a century or two old before the arrival of modern mass media (Weber, 1976; Colley, 1994). The standardization of spoken language, the institution of a common educational system, the building of communications infrastructure across localities and regions, and campaigns in and against foreign nations, went alongside the development of print capitalism and the formation of a national culture, but comprised distinct processes in themselves. The national identity that resulted from such processes took on a life of its own, as something that appeared to be more than the sum of its parts, and prior to them.

We can recall Raymond Williams' argument that television fulfilled the need for a unified information uptake for society: the technology was developed only when such a need came to be expressed, though the means for realizing it had already existed for some time. Electronic media came to be established after a substantial degree of social integration had occurred in these nations. In other words, Williams understood the entry of television as subsequent to the formation of the modern public in Western society.

The analytical frame critics tend to use for assessing television arises from historical contexts where techno-social developments are pegged to broader ideological shifts associated with capitalist modernity, and with the rise of the modern nation-state. However, in the global South, cinema, radio and television have developed simultaneously and alongside the process of nation building. Rather than an incremental development of media technology, it is their simultaneity, and the heterogeneous modes of address that result, that require analysis. An important explanation for this phenomenon lies in the fact that substantial populations in such regions could not read or write, making mass media an essential element of nation building for the majority.[1] Indeed it is the case today that though Euro-American experiences dominate accounts of television, it is in the global South that television has had maximum exposure.[2]

Thus most non-Western countries witnessed television's growth when nation building had only recently begun, and the gap between elite and mass culture was still wide.

New, i.e., developing nations, by contrast, usually had a thin crust of official nationalism on top, and multilingual, multi-ethnic cultures underneath, that were seldom ready to embrace each other. Infrastructure was usually limited where it existed, and the political consensus required to make the nation's parts cohere was precarious, and had to be backed by more explicit and frequent threats of violence than in most Western nations. The euphoria of winning national independence from colonial powers could not compensate for underdeveloped institutions and for the challenges of securing a nationalist project in an increasingly globalized world.

States in non-Western countries planned to use television to assist in the creation of modern publics where they did not exist before: developmentalism, rather than the sudden apprehension of a global village, or of a society of the spectacle, defines the first phase of the institutionalization of television. Television was assumed to be a technology for making mass education cheaper, and since market forces would turn the medium into a means of entertaining the well to do, granting them entry would mean abdicating the state's responsibility. In this understanding, technologies had to be used in the order of their development; print-based education had come first, and so television should impart print-based literacy skills where audiences lacked them, experts assumed (Agrawal, 1977).

Broadcasting emerged as a utility in Euro-American contexts with relatively well-developed civil societies, but for developing nations it usually required technological assistance from abroad and a state strong enough not only to engage in meaningful program production but also to mediate mass media reception in socially constructive ways. Techno-mediation was supposed to help develop new nations, but in fact it would remake them in unpredictable ways, just as the *ancien régime* was taken by surprise by the effects of print. Perhaps anticipating such an outcome, developing nations retained state control over broadcasting in most cases. National programming found it difficult to stray very far from the bounds of official nationalism, and therefore had limited reach.

Worldwide shifts in the 1980s, associated with neoliberalism and media-market monopoly in the West, has also impacted the non-West, but differently: here, a profusion of media expressions, and the opening up of state-controlled media has been the norm, from Al Jazeera, to Beijing-opera-influenced action thrillers in East Asia, and Indian soap operas. This moment offers us a way of rethinking media history and its teleologies, but in ways that are both more global and more critical.

The way television grew in most countries contradicted the modernist teleology of communications media. This teleology assumed a linear succession of technologies that enable a transition from mythic to realist modes of perception, the latter being exemplified by state reason. The market was assumed to be a vanishing mediator in this process, necessary but subservient to enlightenment-style thought and its designated inheritor, the bureaucracy. However, with the spread of television in the global South, information uptake did not occur as policymakers expected, either on the part of the broadcaster or the audience, until the state ceded control and the market, which was usually transnational, took over. Since the state was believed to be the indispensable means of brokering modernization, such an event was unforeseen. It raised questions about many assumptions underwriting investment in the technology by developing nation-states.

What could be noticed, for example, was the co-existence of multiple mediatic forms; rather than older media always becoming the content of new media as McLuhan had predicted, the power of older media forms could be evident in newer media such as television. For example, the advent of Hindu tele-epics that rendered Indian television into a nationwide phenomenon, adopted a mode of presentation that combined the aesthetics of calendar art, early cinema and mythological films, together with the shot-reverse shot editing style of the soap opera. The varying epistemic status of 'new media' in this context was more characteristic of market pluralism than of the more prescriptive character of state reason, which television had been envisaged as promoting.

In the West, liberal democratic norms usually provided the critical lens to view the transformation of mass media into purveyors of 'infotainment', as a perversion of the norms of enlightened civil society. A fuller expression of these developments occurred in non-Western societies, which can be summarized as market-led modernization. The most striking characteristic of market-led modernization is its extensive reliance on electronic communication, and the creation of an increasing sense of simultaneity with events in the rest of the world, without necessarily being accompanied by many of the institutional features and resources associated with modernization. Television is inseparable from the very different public sphere formations which arise in such contexts that in effect implode the temporal lag between advanced and backward regions of the world. Western television has already been parochialized in this context, since it assumes no need to acknowledge or comprehend the non-Western majority, whereas direct or indirect consciousness of the West has grown more acute precisely due to television. It is in the non-Western world that we can more clearly obtain a sense of the global character of television, therefore.[3]

TELEVISION IN INDIA: A BRIEF OVERVIEW

Contemporary Indian television in India presents a case of market-centric cultural regulation arising in largely state-free space, after decades of state regulation effective mainly in expanding communications hardware. It thus provides an exemplary site for understanding the centrality of media to emerging forms of (Asian) globalization. On the one hand, there is an extraordinary explosion of public communication with the expansion of television alongside growing print markets in national and regional languages. On the other hand, this more liberal regime of communication is accompanied not by the abandonment of punitive colonial-era censorship but by its transformation into unofficial but strong forms of cultural regulation, often backed by the threat of violence from non-state organizations, sometimes with tacit official support (Mazzarella and Kaur, 2009). For example, caste-based, regional and religio-political groups and parties increasingly dictate what can be seen and said publicly, using criteria of cultural authenticity rather than impartial norms of the public good as such. Market-led modernization appears fully to accommodate this development, even if the corporate character of media entails an overwhelming focus on the elite segment of the market. A number of paradoxes result, some of which I list below:

1. First, while culture is being upheld as deep-rooted and unchanging, it is reshaped by various forces, and especially by the fastest growing segment of the economy, *viz.*, the media industry.
2. Second, democratization grows not around or despite but through particularistic identities, as indicated in the salience of regional and caste parties of late.
3. Third, the expansion of modern infrastructure, of which media industries are one component, is accompanied by the exclusion of growing numbers of the poor as the state diminishes its welfare functions. Interestingly, as people are pushed outside what passes for a state safety net, new mechanisms of surveillance and control arise, whether from government efforts to establish an identity database of the entire population, or due to the organization of an increasing segment of people's attention in terms of audience viewing time.
4. This leads to a fourth, and possibly an overarching paradox. With the media growing in reach and visibility, it is an increasingly important stakeholder in culture and politics; but it is unregulated and its influence is poorly understood.

Though studies of television tend to provide nation-specific examples, or assemble data organized according to national media systems, its growth and development are embedded in global forces. In India, international agencies are prominent at the very inception. In the Cold War era, developmental concerns on the part of the West (initially, via grants from Philips and from UNESCO, and later in the stationing of NASA's ATS-6 satellite over India for a year) led to the introduction and promotion of television. State television eventually came to be established, and achieved a degree of success in attracting audiences, but the period of meaningful state influence over television has been all too brief. Nationwide broadcasting began only in 1982, more than two decades after the introduction of television in 1959, in order to televise the Ninth Asian Games to an audience that was more foreign than domestic.

By 1991, a mere nine years later, and two years after the Cold War ended, satellite channels from abroad made their entry. When the Gulf War began in January 1991, some hotels installed private dish antennas to receive CNN news reports. Apart from interest in the conflict itself, a large number of Indian nationals were stranded in Kuwait, at the time occupied by Iraqi forces. This led to more demand for news than domestic sources could supply. Although some private companies had begun producing weekly tapes of video news, these were obviously restricted in their reach.[4]

The breakdown of the Cold War dispensation was thus the beginning of the end for the state's monopoly. International events acquired an urgency that could not necessarily be accommodated by the cautious proceduralism of state television news.

This was soon followed by the privatization of television and its enormous expansion, fuelled by foreign investment. Foreign satellite channels had already been operating for four years when a Supreme Court judgment in 1995 declared that broadcasting could not be a state monopoly, tacitly legimitizing existing and new private initiatives. Less than a third of the population had access to television in their homes at this time, but this ratio doubled in a decade (Rajagopal, 2009, appendix). By then, the earlier developmental agenda was abandoned in favor of a turn to cultural authenticity, and ultimately, programming favoring a Hindu majority.

Scholars writing on television have until relatively recently focused on the dominance of the state, but for most of the period of state control, audiences formed quite a small percentage of the population.[5] It is with the relaxation of state regulations that broadcasting grew most rapidly and today television is tantamount to a utility for the middle classes. One Indian state government has even declared that free color television sets should be available to all, and is close to achieving this goal.[6] Meanwhile, commercial firms control 95% of the existing channels at the time of writing, and dominate the urban television landscape.[7] Although precise figures are not available, there is extensive foreign presence in television after years of proudly asserting state sovereignty: 100% foreign ownership is allowed for 'entertainment' channels, and up to 26% for news channels.[8] Government reports, however, celebrate the growth of the media industry, without acknowledging that its decades-old policy of using broadcasting to promote 'education, information and entertainment' for the nation has unobtrusively been replaced by one of revenue maximization via private content providers. A full, nationwide analysis is not possible at this time, since the finances of most television channels are opaque; for instance it is not clear how 700 or 800 channels survive when hardly one or two appear to make a profit.[9] In many states, TV channels are financed by the leading political parties, or they represent a 'convergence' of industry, political families and media houses. The growth of media has thus provided state power new channels for its exercise, while often hiding in plain sight. The days of state monopoly over the airwaves seem quaint and ineffective by comparison with the 24/7 emotional churning of the channels today, where the same message can be reinforced across dozens of media outlets simultaneously.

This more recent, and more international phase, is characterized by largely unregulated programming, including partisan use of news media by private parties in the guise of objectivity, and a plethora of soap operas, game shows and reality TV. Competitive individualism is inculcated and rewarded on game shows and on reality TV, and characterizes most of the protagonists in these programs. News channels take up issue-based campaigns for justice and periodically champion popular mobilization, but within strict limits. The Anna Hazare campaign against political corruption, for example, dominated national news in 2011 until the major news media decided the story was over. The media dependence of the movement, which had seemed irresistible until then, thereafter became evident (Rajagopal, 2011).

The academic literature acknowledges the interpretive challenge posed by the quiet withdrawal of the state as an overt presence and the sudden appearance of an ostensibly state-free televisual space in the context of market-led globalization, and notes the importance of national and regional identity. Identity provides categories that help organize the plethora of content both for broadcasters and for viewers, although such categories are not necessarily the same as they used to be.

Previously, national and regional identities were commonsense referents within a closed domestic economy, which the government regulated under an import substitution industrialization regime. Audiences today experience programs from the US or UK recast in domestic formats, as well as shows that acknowledge traditional and conservative Indian values albeit in upscale, consumerist settings (Kumar, 2006). With the opening of the economy and the exponential growth in its technological mediation, both regional and national identity appear, ironically, as the aim and outcome of global capital and the talent it employs. Culture had hitherto been assumed to be pre-existing, infusing all aspects of society, and unchanging, in a conceptual scheme that itself arose from the colonial encounter, according to which the West had modernity and the rest had tradition. In this binary model, 'tradition' signaled culture, while modernity was an open category standing for universal reason.[10] Now culture in the shape of regional and national identity became dynamic and politicized, and a battleground for competing interests, in addition to comprising a sector that increasingly appeared to be driving the rest of the economy (Joshi, 1989).[11]

In the rest of this essay I briefly discuss three historical phases that can be used to analytically distinguish topical and theoretical emphases in the literature on Indian television: development communication; an intensified focus on national identity and the nation-state; and finally, globalization, which is more extensively explored here. In the first phase, the state was the prime mover, spearheading the growth of television. In the second, when commercial sponsorship of programming commenced, the state adopted the role of regulator of the activity it had initiated, becoming the ringmaster of programming, deciding what shows would be sanctioned, and thus what cultural fare millions of Indians consumed every day. In the third phase, the state shifted from the role of ringmaster to being a participant and an enabler, seeking to share in the profits like any other business, and facilitating further opportunities for revenue. Its regulatory role continues, but mainly to assist in the growth of the industry. State television channels continue to be operated at considerable expense, but little discussion of their content can be noticed either in the urban media or in scholarly literature. Certainly the effect of a cultural policy could be felt in the first and especially in the second phase, but by the third phase, the state ceded cultural policy, which had always been neglected in any case, to the market (Swati Chaturvedi, interview with Ambika Soni, July 26, 2009). This involved a 'change of mind-set', to use a phrase favored by senior government officials, treating private business as a means of growing the economy, and acknowledging limits on the state's knowledge and capacity. Due to the unrestricted entry of market forces, however, state monopoly over television was exchanged for a scenario where corporate monopoly looks set to take over instead. As I have noted above, the social effects of such media liberalization are emergent, and require further empirical study.

ESTABLISHING THE INFRASTRUCTURE FOR MASS MEDIA: SITE TO DOORDARSHAN

When electronic media became available, the experts of the time deemed them to be important in educating the masses, communicating appropriate information to enhance national development. Developmentalism was thus the broad rubric under which communications media developed, and it was viewed as capable of achieving consensus without visible coercion (Lerner, 1959). But people's needs were too local and varied, and mass communication not sufficiently context-sensitive or attuned to audience responses for much success in these efforts.

In India, for example, the state television system, Doordarshan, provided the sole option for a growing number of Indians between 1959 and 1991, and was responsible for developing programming formats that remain influential today (Awasthy, 1965; Masani, 1976; Chatterji, 1991; Jeffery, 2006). The audience for this system is now so much smaller and poorer that few if any of the numerous recent reports on television take much note of it, and virtually all of the profuse news coverage focuses on the private channels alone. However Doordarshan indexed the first phase of India's televisual modernity, which can be extended from the inception of the medium until the time a significant corpus of television programming actually began to emerge in the mid-1980s.

The bulk of the literature during this period addressed the hypothetical potential of television, and was oriented towards development communication, although critiques of this perspective were also made. A significant section of the literature centers on the Satellite Instructional Television Experiment (SITE) conducted in 1975–76, analyzing the results of studies carried out with educational programming in rural areas, with the help of a satellite whose services were loaned by NASA to the Indian government, specifically via the Space Applications Centre in Ahmedabad, for a year.

In the case of SITE, the bulk of the funding went to hardware installation, and relatively little was spent on software testing and implementation. Like much of the development communication literature, studies of SITE are loquacious on objectives and laconic on outcomes. However, it is clear that broadcasting programs were envisaging without some means for clarifying and reinforcing desired audience responses, making it unlikely that the planners' intentions would bear fruit. One former broadcasting official observed, 'Much time is spent in putting up transmitters, building and equipping studios … but no thought is given to the programs which are to be broadcast from these centers. The entire machinery is upside down' (Awasthy, 1965, p. 28).

Such an operational model, whether upside down or not, suggested an unreflective reliance on a conveyor belt model of pedagogy that assumed messages stayed relatively unchanged from sender to receiver. However, as Umberto Eco (1965) has pointed out, aberrant interpretation is the rule, not the exception, in mass communication. Ultimately, villagers were reported to prefer the feature films that were aired to the educational programs, which they found unsuited to their needs. Revealingly, one study reported that villagers referred to television as *tamasha*, a Hindustani word for spectacle or drama, or as *sanima* (cinema) (Agrawal and Rai, 1980, pp. 83–4). Small wonder that the Minister of Information and Broadcasting, when asked about the predominance of movies on Doordarshan, protested that if the government stopped screening films no one would watch television.[12] The task of socializing the population into TV viewership was, therefore, the self-appointed task of the state, and this was understood to be a modernizing activity *in itself*, regardless of what people viewed. One SITE study noted, '[M]ass media exposure is a great force in making man modern', citing Daniel Lerner (Agrawal and Rai, 1980, p. 296). The extension of media infrastructure was understood as the state's mission, alongside creating railroads and canals, setting up heavy industries and designing new townships, laying the basis for modern attitudes that would undergird the state's nation-building. The outcomes of the SITE experiment on viewers were unclear at best, but the expansion of television enlarged the state's reach, and expressed the sense, if not of what James Scott has called 'seeing like a state', then of looking like one (Scott, 1998; Pierce, 2006, pp. 48, 887–914).

Although the expansion of communications was part of a plan to enlarge state infrastructure, it was not foreseen that political authority would itself fundamentally transform in the process, from state to non-state agents and sites, and from bureaucratic planning to corporate strategy. This occurred with the government opening the door to commercial programming, initially within state television and later, outside it. Although developmentalism had always presumed that state communication achieved the best results, the commitment to a rigid model of control and the inability to manage the heterogeneity of its audience, together with the success of market actors in securing and organizing people's attention would lead to a sea change that is as yet only very partially understood.

IMAGINING AN AUDIENCE FOR TELEVISION

The second phase of India's televisual history spans the growth of commercial television under the aegis of state television, which occurred roughly between the mid-1980s and the mid-1990s. A wider array of perspectives can be found in this phase, ranging from developmentalism to critiques of the forms of nationalism represented on Doordarshan. A tacit assumption in this period concerned an instrumental relationship to television: if only television was properly used, national development might be advanced, and the cultural and political abuses of the technology eliminated. A distinctive shift in this period concerned Doordarshan's experiment with entertainment genres, although developmental programming aimed for example at farmers, women and children dominated, together with a limited amount of news coverage.

This much-discussed experiment in what were described as 'pro-development' soap operas combined entertainment and education to both attract audiences and improve them. It was inaugurated by an energetic bureaucrat, S.S. Gill, Secretary to the Information and Broadcasting (I&B) Ministry during the critical phase of television's expansion in the early 1980s. The prospect of making profits while fulfilling development objectives was enticing; some proponents deemed it a 'win-win' proposition (Singhal et al., 1993). The prototype was *Hum Log* (We People), written by a well-known Hindi novelist and journalist, Manohar Shyam Joshi, which ran for 17 months. Developmental themes were found to depress audience size however, and sponsors withdrew until such messages faded out. It was as a family melodrama that the serial eventually succeeded.[13]

As the government's emphasis shifted from building infrastructure to developing programming ideas and content, demands for economic growth increasingly came into conflict with assertions of cultural continuity, creating a dilemma for those making the serials. The problem lay in seeking overt messages of change, although the strategy of national development had in fact been to rely on an assertion of cultural continuity, while assuming tacitly that cultural change would follow from economic growth. It was the makers of commercial entertainment who mastered deniability as a tactic, providing narratives that overtly endorsed traditional cultural codes while subtly inserting new ones of individual aspiration and achievement.

It was the unexpectedly successful serialization of the Hindu epics on television, beginning in 1987 that brought home the political dividends of claiming that the culture was in fact unchanging, despite all the evidence of social change.[14] Bureaucrats had been skeptical that the Ramayan and Mahabharat would win substantial audiences, and that these epics could successfully fit with their developmentalist mission. They could not have been more wrong. To the bureaucrats' surprise, the telecasts brought urban India to a standstill and decisively distinguished the new medium from the morally more ambiguous cinema, giving it respectability. The values upheld in these epics, of devotion and sacrifice, were widely seen as belonging to the past, and as unavailable in present-day society (Rajagopal, 2001). The collective desire to commemorate the absence of such Indian values confirmed the existence of a winning programming formula. The winds of change were blowing, but television could provide a shelter or a buffer rather than a mirror to the storm, with nostalgia for an upper caste Hindu past rendered as historical commonsense.

Doordarshan eventually succeeded (viewership grew from 1.2% to 18.2% of the population between 1981 and 1991), but at the cost of displacing its earlier mission of development and embracing instead something closer to a market-friendly Hindu orthodoxy (Rajagopal, 2009, p. 326). An Indian personality took shape for television not so much by advocacy as by denial of change, and tactful acknowledgment of a new dispensation that was assumed to be a *fait accompli*.[15]

TELEVISION UNDER LIBERALIZATION: BETWEEN DEREGULATION AND LOCALIZATION

Until the early 1980s at least, it was assumed that television was an unaffordable luxury for a poor country (Pal, 1983). As late as 1978, the B.G. Verghese Committee Report estimated that by 2000 there would be between 12 and 18 million TV sets in the country. That figure was reached in just over a decade, and by 2000 there were over 70 million television sets. In fact by 1996 bureaucratic perspectives on television had changed – the Nitish Sengupta Report declared it to be a 'misconception … that television is a luxury' and recommended acknowledging 'the immense potential of television as an instrument of rapid socioeconomic transformation' (Ministry of Information and Broadcasting, Government of India, 1996, paragraph 2.16). Within two more decades television had become a utility, for example in the state of Tamil Nadu where more than 16 million free color TV sets were distributed in a few years at a cost of nearly $900m.[16]

By 2010, TV had the highest penetration among all mass media in India at 56%, while 90% of those surveyed in a media industry study reported that it was their preferred entertainment.[17] Meanwhile, market reforms led to the increase of TV channels from a handful to 450 or more in less than 20 years. A 2009 study reported that there were 110 million TV households in India, out of which 70 million homes subscribed to cable and satellite, while Doordarshan alone served the remaining 40 million homes.[18]

The speed of developments, if nothing else, has ensured that the lacunae in the scholarship on Indian television are numerous. Ownership data on the industry are scarce, due to mostly private control, scanty regulations, and a state record-keeping system that prefers secrecy to disclosure.[19] Television channels, whether public or private, have not always welcomed outside observers, and no public archives are available of television programming; hence studies have concentrated on content analysis more than on televisual sound and image. Given the enormous volume of output, studies thus far are understandably limited in scope.[20] In what follows I offer some preliminary notes in understanding the nature of the changes affecting television, and the effects that these changes in turn had on emerging cultural trends.

Now when economic reforms were carried out, the task of the private sector was to advance the growth of the economy. What was the task of the media industry once it was granted freedom by the state, besides growing and making money? What was to be communicated, and what outcomes were desired, when state regulations were limited at best, and focused on containing overt violence and obscenity? As it turned out, the government wanted very little more than revenue growth; issues of political censorship could be informally managed and usually required little overt coercion. Meanwhile, after decades of warning against corrosive Western values, foreign investment flooded into the Indian media industry. The result, however, was in some respects the opposite of what had been feared: *Indian language programming expressing conservative values and upholding traditional roles became among the most widely watched programs*. As it turned out, cultural identity was not only the basis for expanding the media market, but also in a way, its outcome. Corporations sought to brand goods so as to encourage old and new consumers to imagine intimacy with these goods. Cultural identity defined in regional, religious or national terms served this purpose. Identities necessarily changed in the process, into something more public and stereotyped, but reflexive and individuated as well; that is, they became harder externally and more plastic internally. While a measure of turbulence resulted from these new expressions of identity, the political order itself revealed itself to be surprisingly resilient.

This resilience was in part due to the fact that despite ostentatious ideological differences, the major political parties, except for the Communist Parties, have been in agreement on the necessity of economic reforms and market liberalization. With the media industry loudly championing economic reforms, debates over the relevant policies have been confined to the opinion pages of daily newspapers rather than spilling out onto the streets. Television as itself a site where economic reforms are being carried out is a subject that has largely been ignored, both by economists and by other scholars. This is surely something that requires analysis across the board and through attention to the various aspects of the televisual apparatus (including satellite uplink agencies, broadcasters, content providers, televisual content, cable operators, advertising and market research companies, and audiences). Instead, economists have at best focused on the media's rate of growth while bracketing the media industry itself. Meanwhile other scholars have discussed the media industry while assuming reforms as a backdrop.

Since the media consume an increasing share of the working day, and since media companies are stakeholders in the process, issues of media production and reception are relevant to discussions about both the changing structure of the media industry and on the manner in which media legitimize the new context of neoliberalism. Existing commentary tends to be dominated by industry analysts, for whom economic growth is an end in itself and state intervention is mainly perceptible as either pro- or anti-business, however. Official commentary itself tends to adhere to such an instrumental perspective, and expresses reluctance to intervene amidst rapid market growth regardless of the fears of monopoly.[21]

What would it mean to think about the growing relevance of television under liberalization? In what follows, I offer a brief discussion of state deregulation and its outcomes, which can be summarized as follows: (1) Liberalization of India's economy has produced a far more deterritorialized form of media control due to extensive investments coming in from abroad. (2) Paradoxically, this is accompanied by an intensive localization of television programming, with an explosion of Indian-language news and soap operas whose message of overt patriarchal dominance and covert individual aspiration, amidst the proliferation of game shows and other imported formats, was indisputably a homegrown product.

State Deregulation and Deterritorialization: The Changing Structure of Televisual Control

Litigation is often cited as the proximate cause for relaxing state control over the airwaves. In 1995, India's Supreme Court ruled, in response to a petition arguing that the airwaves were public property and not the property of the state.[22] But corporations and entrepreneurs had already been serving Indian audiences for some years, beginning with the Gulf War in January 1991, when some hotels installed private dish antennas to receive CNN news reports. Apart from interest in the conflict itself, a large number of Indian nationals were stranded in Kuwait, at the time occupied by Iraqi forces. This led to more demand for news than domestic sources could supply. Although some private companies had begun producing weekly tapes of video news, these were obviously restricted in their reach.[23]

A few months later, Star TV began to broadcast *The Bold and the Beautiful* in the region. Television channels could tape-record their shows and courier them abroad, as well as uplinking from Russia, the Philippines, Hong Kong or Singapore, as in fact broadcasters like JAIN, Sun TV and Asianet,

Zee TV, Sony, ATN and BITV had been doing since 1990, depriving the government of a substantial amount in revenues (Ministry of Information and Broadcasting, Government of India, 1996, article 7.6).

Bypassing terrestrial authority highlighted at least three factors:

1. It showed that new technological developments were altering the ground on which many aspects of state control were based. Technologies such as video, dish antennas and the internet required intensive policing on the one hand, and sophisticated expertise on the other, if the government monopoly over what people watched was to be retained. The reluctance of the state to commit scarce resources for such measures was compounded by the lack of political will to do so.[24]
2. Audiences were increasingly knowledgeable enough to locate and use these technologies, which meant that the artificial information scarcity the government sought to preserve would be harder to maintain.
3. New media technologies and their ability to create audience demand entailed not only the *de facto* end of the monopoly in broadcasting, but also a loss of legitimacy for the government, which was propounding laws that were unenforceable and out of date, while also depriving it of much revenue. Broadcasting monopoly was leading not to persuasion, but to popular skepticism about government propaganda. The result of excluding private initiatives was therefore counter-productive (Ministry of Information and Broadcasting, Government of India, 1996, articles 2.36, 7.6). The urban middle classes wanted news and entertainment, and businesses were able to cater to them. The government could stand in their way, at considerable expense to itself, or it could yield to the pressure, and reap the benefits.

Pressures to liberalize access to television were enabled by the fact that the potential for revenue dominated government policy objectives. Frequently, the means of determining and implementing these objectives was short-sighted, and oriented towards preserving state control rather than allowing non-state parties to gain influence in the name of development. But as liberalization gained momentum, revenue enhancement by itself became an overarching policy objective (Ahluwalia, 2002). Practices that yielded high revenue growth therefore implicitly justified themselves. Ministerial pronouncements on television, for example, dwelt on the growth of the sector without acknowledging that it came at the expense of state broadcast media, whose pioneering efforts in establishing television were aimed above all at promoting not-for-profit education, whereas entertainment was overwhelmingly the objective of commercial media.

Meanwhile, cable operators all over the country had for years been operating satellite dish antennas to receive signals, which they then transmitted through informal cable connections to residential neighborhoods, in a completely unregulated process. The Cable Television Networks (Regulation) Act of 1995 belatedly sought to register these operators and obtain their consent to a code of conduct to guard against concerns about mores, taste and security. But the code is a voluntary one, and although government monitors a limited number of channels, there is no systematic oversight of television programming in India.[25] Thus the Minister for Information and Broadcasting, Ambika Soni, when asked why her ministry was not exercising its authority in this regard, replied, 'Self-regulation is the proper way for educated people' (Chaturvedi, 2009). Invariably, discussions about the lack of government regulation were diverted to appreciating the media industry's rapid expansion, implying that such a lack resulted in increasing growth and was thereby justified (Newstrack India, 2010).[26]

By 2006, a UNESCO study would observe that India had one of the *least restrictive* policies regarding the entry of foreign media companies, and that available government records tended to obscure the extent of foreign presence in the sector. This was the result of a gradual devolution of government regulation. Though government-appointed observers were invariably critical of commercial entertainment in broadcasting, and urged the reinforcement of public service components in the field, they seemed aware that their recommendations could be unpersuasive.

For example, the Sengupta Report, after a ringing denunciation of 'alien entertainment', suggested the establishment of one public service channel each in TV and radio, aware that to expect government broadcasting as a whole to embrace the public service ethos, even while commercial broadcasting was expanding rapidly, would be unrealistic (Ministry of Information and Broadcasting, Government of India, 1996, articles 5.2 and 5.4). Ironically, liberalization, rather than being viewed as an acknowledgment of regulation's limits, was increasingly viewed as a default option. For example, on the subject of foreign investment in the news media, one information and broadcasting secretary told reporters, 'We have all along said that we will be happy to take any step to facilitate the growth of the industry but the industry hasn't demanded it'.[27] The erstwhile roles of the state and the media were flipped.

Region and Deregulation

Given the plethora of channels, no comprehensive map of televisual dynamics is possible. However, a brief discussion of television in one region can help in a symptomatic reading of the emerging media field. One of the largest television companies in the country is the Sun TV group owned by Sumangali Cablevision, which dominates the television market in Tamil Nadu, and has channels in all four southern states. The other leading television company in Tamil Nadu is Jaya TV, named for its founder J. Jayalalitha, leader of the one of the two major Dravidian political parties in the state, the AIADMK. Competition in television is an extension of political party competition, and viewers understand that television is partisan; hence it is necessary to move between channels to get alternative views on important issues. Due to its size and influence, Sumangali Cablevision is the more significant of the two by far, however. The company was founded as an arm of the erstwhile ruling party in Tamil Nadu, the DMK, and its owners are close relatives of the leader of the DMK. The DMK itself was founded on a lower caste, anti-brahmin platform in 1949, and came to power in 1967 with a campaign based on linguistic identity through its extensive grass-roots cadre, and its astute use of mass media in its campaigns; the AIADMK was a breakaway faction.[28]

Sun TV's own success was the result of a deft reading of the caste landscape of Tamil Nadu. Although Tamil Brahmins probably formed 2% or 3% of the population, and lacked political power, Brahmin norms remained influential in the domestic sphere, and thus presented a gateway that programmers had to cross to reach mass viewership. Accordingly, for the first several years of its existence, Sun TV's programming was entirely given over to Brahmin producers, such as K. Balachander, Y.G. Parthasarathy, Y.G. Mahendra, 'Crazy' Mohan, Mowli, and S.V. Sekar, who produced Brahmin-dominated serials. Other networks followed suit. In the words of one observer, a TV viewer until the late 1990s could have been forgiven for thinking that only Brahmins lived in Tamil Nadu (Bala Kailasam, Creative Head, SRM Group Television, 2011). Once audiences appeared to be hooked, Brahmin producers and stories were phased out. Serials switched focus instead to non-Brahmin characters and settings, and proceeded to garner much larger audiences. In following such a route to winning viewers, Sun TV showed the subtlety of a party that had won political power from the Brahmins, but knew that they remained culturally hegemonic. In other words, winning audiences required broadcasters to read both viewer preferences and contextually defined social preferences, and finding a way to negotiate between the two.

Since the starting point for mass media in India, as in most other countries, was a situation of minority cultural dominance, there tended to be a wide disparity between these sets of preferences. In south, representations of the caste elite were deliberately used as a Trojan horse to lower the threshold of acceptance with mass audiences. Such a clear strategy was premised on the more than half-century-old political polarization of lower caste identities, an event that was, however, not accompanied by a social revolution that significantly improved lower caste status in general.

In north India where the political ascendancy of lower castes was still contested, lower caste identity was not available as a vehicle for viewers. Here, mass programming strategy took its cues from events that were themselves enmeshed in television, with art imitating art imitating life. The success of the Hindu mythological serials in the late 1980s and early 1990s made the Hindu supremacist Bharatiya Janata Party (BJP) aware of the mobilizational potential of Hindu symbols and mythological themes. Until then, the conventional wisdom had been that Hindus could not be unified politically on a religious platform, due to their divisions across caste and sect. The BJP reaped the electoral dividends of using such themes and eventually came to power at the Center in 1998, an event that had been unthinkable barely a decade earlier. The successive triumphs of the tele-epics and of Hindu cultural nationalism set the stage for televisual formulas in which overtly upper caste Hindu orthodoxy became the backdrop, medium and boundary for dramas.

In both north and south India however, acquisitive individualism was a subversive element galvanizing the social context, with the familiarity of the melodramatic form softening the nature and extent of the shift being attempted.

ASPIRATIONS, IDENTITIES, AND CHANGING SOCIAL RELATIONS

The reversal was most striking in the case of films. It was an historical irony that the long-disparaged film industry provided the cultural resources to incubate programming for an Indian audience. The Indian film industry, probably the most prolific in the world, was granted industry status in 1998, nearly a century after its inception. By and large, filmmakers had to win market share without the subsidies other industries obtained.[29] But as a result, they found themselves free to address their audience without rehearsing the dogmas of official nationalism. The narratives they provided rendered the law of the state as, typically, external to the customary laws of the community when dealing with all conflict including crime. The police invariably arrived after the trouble was over and the culprit confessed, to provide symbolic closure to a method of conflict resolution regulated by the community itself (Prasad, 1999). This formula emerged outside the matrix of developmental institutions, but, with liberalization, came to infuse television as well. This was a noteworthy departure from an earlier period when it was necessary to sharply distinguish films from other mass media. Similarly, film producers for many years treated television as merely a convenient vehicle for films, but industrial consolidation created an increasing continuity of both form and content across media platforms. With new regulations allowing 100% foreign ownership in entertainment media including the film industry, Bollywood became a convenient resource for expanding the media market, leading to a general Bollywoodization of mass media: pictures of films and of film stars and sounds from film song tracks became fairly ubiquitous in radio, television and print as well as mobile phones.[30]

Bollywood filmmakers produced not only the Hindu mythological epics for Doordarshan. The templates they created were used also for the so-called pro-development soap operas. Serials like *Hum Log* appealed to audiences that had been exposed to social realist cinema from filmmakers like Guru Dutt and Bimal Roy, and later Basu Chatterjee, Gulzar and Hrishikesh Mukerjee. In such films, the rhythms of ordinary life were treated as adequate for cinema, with melodrama confined to the domestic sphere rather than extending to more sensational topics. The mythological epics helped extend this framework of television entertainment, and, in situating it within the bounds of Hindu conservatism, sanctioned the investment as morally appropriate. It was from here that soap operas began anew in the liberalization period.

Prominently, these were serials revolving around mother-in-law and daughter-in-law tensions within joint family households, popularly known as *saas-bahu* serials, and initiated in the show *Kyunki Saas Bhi Kabhi Bahu Thi* (Because the Mother-in-Law Was Once a Daughter-in-Law) produced by Balaji Telefilms, which set the trend. The family setting was invariably extremely wealthy, but domestic rivalries, livened by sensational crimes such as kidnapping, rape and murder, created fraught scenarios for the family members. In effect, the greater explicitness and intensity of lower-class domestic conflict was transposed to upper-class communities, thereby providing something for everyone.

Exaggerated audio-visual effects provided interpretive cues, for example different characters would be signaled by distinct audio tracks. A respected elder could come before the camera to the sound of devotional music, and the arrival of a fashionable young man could be signaled by reggae or hip-hop tracks. Abnormal events would be replayed to signal distress or surprise, for example a slap of rebuke administered by a husband or a parent; swish pans would be used to render scenes blurry or make them spin, to suggest the loss of balance and perspective induced by shock. These and other elements of melodramatic form could provide aesthetic props creating a sense of continuity in narratives that were themselves changing.

Thus although soap operas centered on the family, the family was no longer portrayed in quite the same way. For example, portraying romantic love, which was at the center of most film stories, involved a delicate negotiation between reassuring elders of their importance and allowing fulfillment to young lovers, and much of the drama turned on how these competing tensions were resolved. Frequently, filmmakers exercised creativity in endorsing traditional authority while allowing romance to be requited. What shifted was the way in which the individual's relationships with the community were resolved; earlier they were, with rare exceptions, harmonious. In television currently, there is a premium placed on individualism, and seldom a respite to antagonism. One viewer posted a remark about Tamil soap operas that could be applied to Indian soaps more generally:

> [Do you want to know] how to drive your sister-in-law out of the house? How to plan for looting property from your father-in-law? How to take revenge on those who spoiled your family? How to deal with the illegal affairs of your wife/husband? One point solution for all your queries, watch Mega Serials on TV. (Srinivasan, 2006)

Remarkably, however, *saas-bahu* soaps tend to be filled with pious references to scripture and tradition. Scenes of prayer and worship are frequent, as are claims of upholding custom and observing ritual prescriptions. The fact that viewers choose to watch shows with overtly feudal and patriarchal values while market reforms are dissolving many existing certitudes is often seen as proof of deep-rooted traditionalism. A more nuanced diagnosis is obviously necessary. Resurgent traditionalism alone could not explain the multiple currents present together in society (Shailaja Kejriwal, personal interview in 2004). Increasingly stereotyped invocations of tradition are accompanied by subtle changes in social relations, especially of caste, gender, and other forms of embedded inequality. Women portrayed in soap operas, for example, are increasingly assertive, even if in the name of custom; when sacrificial behavior is depicted, it appears increasingly unmoored from the support systems of the community, and more personalized. Such individualized attitudes are clearly compatible with a media system premised on promoting the consumption of goods.

We can ask what kind of media regulation might be possible in such a context. Heavy-handed *dirigisme* is not favored any longer, and the subtlety required to address social change is assumed to reside in the industry itself. Meanwhile, the reliable old standby of 'national identity' has been so thoroughly reworked that some of the more popular presentations of Indian cultural themes often emerge from foreign-owned channels, embracing cosmopolitanism while abundantly signaling rootedness. What is interesting, however, is that if the earlier state regime was criticized for being narrow in scope, uniform in content and conformist in spirit, similar tendencies are visible even with hundreds of private channels.

Although the promotion of markets is supposed to guarantee competition, uncertainty about how to address an emerging market of television viewers has led to conservative programming choices and zealous imitation of market leaders. As a result, the formula of paying lip service to convention and tradition while covertly espousing aggressive individualism has become the norm. One example will suffice to convey the point. For the moment on Indian television, even such well-established Western genres as law and order shows or hospital-based dramas, where individualism is subordinated to the demands of cooperation and professional discipline, are glaring in their absence. This moment may pass soon. Even so, there are lessons to be drawn here. The collective thrust of the media corporations appears to be to delegitimize traditional structures of loyalty and trust, understanding them as backward, cumbersome to the individual, and above all hostile to their own goal of championing consumer choice and enlarging the market. Religious and familial orthodoxy and the rule of elders are assumed as conditions against which overt challenge is inadmissible. No doubt these messages are meant to speak across ideological divides. What is interesting is that neither side comes off well: if the established order is cruel, its adversaries are cunning or dishonest.

Overwhelmingly, the events that these soap operas portray correspond to events in the private domain, but the public sphere offers a privileged space for their enactment and resolution. Since the progress of social reforms stalled after national independence, citizens had to cope with traditional social structures that could not easily mediate ongoing processes of modernization; hence such an outcome is not surprising. The collective urgency attending private concerns seeking expression far outweighs the felt need of public sphere issues as such. What we have therefore is a disappearing public, one that is swallowed up by an expanding private sphere.

Thus, televisual exposure has provided a sanctified space for social and moral experimentation, for example in the guise of reality shows. Popular reality shows such as *Bigg Boss* (*sic*), which is said to have 114 million viewers in India and 100 million more abroad, make individuals on the show and off it confront cultural difference in highly contrived settings.[31] They stage what would normally be private or intimate forms of behavior, on the part of ordinary persons, in public. This leads to a certain homogenization of conduct on the one hand, and an intense awareness of the power of publicity on the other. In this way television is socializing audiences through shock and public spectacle.

Here is a specific method of education. If the reality shows are brazenly sensationalist (*Bigg Boss* has featured a woman who was formerly a bandit, a seven-foot-tall professional wrestler, a cross-dresser, and so on) they succeed because they satisfy not only curiosity about the bizarre, but give individuality to it, and so underline the drama of detaching oneself from the community and standing out as an individual, albeit in stereotyped ways. Such modes of performance allow expression of tendencies not sanctioned in daily life. They allow vicarious participation for audiences who can learn to assess the potential of their own behavior as a spectacle to be managed in an identifiable manner.

Orthodoxy appears fascinated, not threatened, by such portrayals, and at present cannot conceive that such programming might seriously undermine it. This is surely an error. Indian business is already being nudged by foreign corporate capital to adopt a more socially liberal attitude on questions of caste, class and gender. An example would be the Oscar-winning *Slumdog Millionaire* (dir. Danny Boyle, 2008), released by Rupert Murdoch's Fox Productions, where the sub-human treatment of a poor Muslim boy in Mumbai is clinically portrayed, while sentimentality is reserved for his love story.

INDIAN TELEVISION TODAY

No categorical position can be adopted, therefore, on the effects of inflow of foreign capital on Indian television, but the manner and scale of its deployment, as well as the paradoxical social reactions to it – from renewed commitments to social orthodoxy, to the embrace of individuated models of pleasure and consumption – deserves notice. According to one reckoning, between 40% and 60% of the investment in Indian media and entertainment industry comes from foreign investors.[32] At the time of writing, there exists a limit of 26% foreign direct investment (FDI) in news channels, and 76% in 'non-news' channels, but various factors compromise this *cordon sanitaire* around 'news'. For example, up to 100% FDI is possible in advertising, market research and public relations, and together these industries dominate strategy and content choices across the media. The 'readership' surveys done by market research agencies together with television ratings influence the decisions of advertising agencies as well as the news and entertainment media. Taken together with the importance of media planning and public relations, the customary role of media professionals is defined less by journalistic norms and more by market concerns. In 2007, advertising expenditure amounted to 120 billion rupees, at least three-quarters of which go to TV, radio and print media. The five leading advertising agencies, mostly controlled by foreign conglomerates that in some cases own all the equity in these firms, account for well over half of advertising business in the country (Vasanti, 2008a).

It is interesting and perhaps symptomatic that in such a context, no precise accounting exists of the extent of foreign investment in Indian media. For example, the UNESCO report cited earlier noted that the Indian government did not record foreign private equity or venture capital investments as foreign direct investment (FDI), nor did it include the retained earnings from foreign subsidiaries under such a heading. This led to underestimating foreign investment, the report noted. For example, in 2002 Rupert Murdoch's News Corp invested about $200 million in its Indian cable television operations (Star TV), but although the amount equaled all the rest of FDI that year, it was not included in the Indian FDI statistics (UNESCO, 2006).

It is clear that in the government's view, regulation and controls over foreign investment, although performed in the national interest, had for years acted as a drag on economic growth, and gave power to bureaucrats who neither understood the technicalities of what they were regulating nor were effective in their interventions. It is the wholesale character of the shift that is striking, from a state-centered approach to one that is led by the market's intuitions and desires. Public service television should have its place, state officials agree. But it seems that for the present, that place is mainly by the side of rural and poor viewers, who watch Doordarshan if they cannot get something better.[33]

The Cable Television Networks (Regulation) Amendment Bill 2011, which makes full digitization of cable television across the country mandatory, has led to a wave of media mergers. Corporate houses envision that greater economies of scale for business promoters will lead to greater profits, and far more viewing options for those who can afford them. Affordability for the wider public is the subject that remains a big question. One immediate result will be that consumers can no longer negotiate rates and viewing options with their local cable operators (LCOs). Currently, broadcasters claim they receive only about a quarter of the fees that viewers pay for their programs, because

LCOs under-report subscribers and gain disproportionate revenues (Rajagopal, 2012).

A comparison with the US context is instructive. In the case of US media, the emergence of media monopoly led to diminishing content from local and from foreign sources in favor of more generic, syndicated national features (Bagdikian, 1983). Meanwhile the growing political conservatism of the media, if it was perceived at all by the public, appeared as a status quo orientation, with little hint of political scandal. In India by contrast, both local and foreign content increase, with Western program formats often receiving a markedly indigenous coloring. And the growth of media corporations in India occurs through a corrupt state-industry nexus that is readily perceived through the numerous television channels owned and operated by members of political parties, or is otherwise made public through leaks, sting operations and other increasingly familiar genres of scandal.

Even where domestic ownership exists, since many major media are privately held, information is not available regarding the nature of their control. We do know, however, that in at least ten states, major television channels belong to political families; the real numbers maybe higher (Vasanti, 2009). A further hint of concentration is suggested by the fact that in 2008 11 cable distribution companies provided nearly all of the 400 television channels then present in the country (Vasanti, 2008b). The largest companies in the electronic media segment are Star India, New Delhi Television (NDTV), Sony, Essel group, India Today, Sun Network and Television Eighteen. Recently, the Walt Disney group has emerged as leader of the children's entertainment segment. Since there are no particular rules and laws to limit cross-media ownership, almost all major newspaper groups have entered the electronic media market. The fast-paced growth of the media business has forced these groups to issue public equity or invite FDI to survive in the market. Consequently, privately owned media are being reshaping in a corporatized structure, in a departure from the traditional family-owned business model that deserves investigation.

DIRECTIONS FOR FUTURE RESEARCH

In lieu of a conclusion, let me end this survey of Indian television with some suggestions for further research.

1. The growth of television has been accompanied by the growth of cross-media ownership, and by the movement of content across platforms such as film, radio, print, television and cellular telephones. Earlier assumptions that took for granted a degree of medium insularity are untenable in such a context. Furthermore, where political independence bestowed citizenship rights before social reforms took root,[34] cultural norms are multivalent and contextually defined, and need to be distinguished not simply in terms of what people believe, but in terms of what is assumed to be permissible and what is not. What is screened on television is part of a chain of events in the domestic sphere, the political public sphere, and the mass media more generally. Exclusive attention to television is liable to be misleading in such contexts. These factors demand a decentering of television, and an attention to the medium as part of a broader canvas. We can recall here a remark by Oskar Negt (1978) 'A critical theory of the media cannot have the media as its center'.
2. While it is a commonplace that sensationalism has proliferated with the growth of television, the multivalent character of cultural norms entail a reversal of the public-private opposition in some crucial ways that clarify the forms of prevailing sensationalism. Processes of self-fashioning and subject formation that occurred in an era when print was dominant, are necessarily far more public today, especially where for the majority of the population, literacy is less than a generation old. We may decry the drowning of properly public issues by sensationally rendered private matters, but the imperative to represent and in some measure adjudicate questions of identity formation that neither the domestic sphere nor the political public sphere can cope with, can hardly be overstated. The dynamics of this distinct space where the public can disappear into

the private sphere, require to be studied with a more expansive historical imagination than is sometimes brought to bear on it.

3. If private concerns take over public space, what happens to norms for communicating properly public issues? The norms of news reportage are not those of the received journalistic canon, in this context. Objectivity and neutrality appear to matter less than emotional resonance and ethical appropriateness, and partisan attitudes are more prevalent; viewers moving between channels may have to negotiate a plurality of opinions that may be mutually irreconcilable (Hallin and Mancini, 2004). Unlike the Habermasian model, whose parliamentary decorum and rule-governed rational criticism relegates the violence of politics outside the public sphere, *violence is internal to the public sphere* in such contexts. How it is managed in televisual news reportage is a topic that deserves study.
4. Although communication requires language, communication theories tend to be parsimonious in their assumptions about language, and presume at best monolingual conditions. Few societies today fulfill this condition. What are the dynamics of multilingualism in historical contexts where the dominance of a colonial language gives way to indigenous vernaculars as the latter gain in significance with the growth of markets and media? What kind of cross-pollination occurs between linguistic domains, as English words and syntax are borrowed in indigenous languages, and metaphor, vocabulary and cultural styles from native languages infuse the former language of command?[35]

NOTES

1 They were not the only relevant mediators, to be sure; there were also canals, roads and railways, as well as institutions of the state, etc.

2 For example, the combined population of North America and Europe constituted an estimated 1.38 billion in 2010, for example, when the transcontinental countries of Russia and Turkey are included, or about 23% of the world's population. Meanwhile the number of television sets in the world is approximately 1.4 billion, of which 285 million are in North America. No estimates are at hand for the number of TV sets for all of Europe, but if we tentatively assume 250 million sets across the 842 million population of Europe (UN figures for 2005), more than 900million television sets exist outside Europe and North America, or 64% of all television sets. (Department of Social and Economic Affairs, UNPD. [Graph illustration of population by continent and world's ten most populated countries.] *Wikipedia: The Free Encyclopedia.* Retrieved from www.en.wikipedia.org/wiki/File:World_Population_by_Continent_and_10_Most_Populated_Countries.png).

3 In this essay, I move between arguments about the media in general and about television more specifically; my arguments do not apply exclusively to television except where I indicate that they do.

4 Started between 1988 and 1991, *Hindustan Times*' Eyewitness, *India Today*'s Newstrack, the Observer Group's Observer News Channel (ONC), and Independent Television's India View were the most prominent. Their combined circulation was probably in the tens of thousands (Tanna, n.d.).

5 The potential audience in 1982 was under 2% of the population, for example, and in 1991, when foreign channels made their entrance, less than a fifth of the population had television in their homes (Rajagopal, 2009).

6 In the state of Tamil Nadu more than 16.2 million free color TV sets have been distributed in four years for nearly $900 m. A million more are promised (see Times News Network, 2011).

7 Of the more than 450 television channels estimated to be operating according to a 2009 report, the state television network, Doordarshan, controls 31. (See PriceWaterhouseCoopers, 2009.)

8 About 40% of foreign direct investment in India originates from companies based in Mauritius. Singapore, the US and the UK, respectively, follow with single digit figures in comparison (Department of Industrial Policy and Promotion, Government of India, 2010).

9 See Sucheta Dalal (2011), cited in Sandeep Bhushan (2013).

10 For a useful discussion of this conceptual scheme and its implications in the Cold War era, see Pletsch (1981).

11 The Entertainment and Media industry ('E&M') grew at a cumulative rate of 16.6% between 2004 and 2008 (PriceWaterhouseCoopers, 2009, 16). The Gross Domestic Product in India, by comparison, has grown at a rate between 7% and 9% during those years. One media professional has remarked, '[Earlier] the market was driving media, by the late '80s media started driving the market' (G. Krishnan, cited in Kohli-Khandekar, 2003, 28).

12 H.K.L. Bhagat quoted in *The Indian Express*, 22 June 1983.

13 The prototype of the pro-development soap opera was derived from Peru, with a serial called *Simple-*

mente Maria (Simply Mary), showing how a working-class girl, with hard work and discipline, could overcome her status as a poor single mother and achieve social mobility. It initially aired on Panamericana Televisión Channel 5 and achieved high ratings in the United States, Argentina, Mexico, Colombia, as well as other countries (see Singhal et al., 1994; see also Rajagopal, 2001, 79).

14 This period was marked more broadly by the beginning of liberalization together with the rise of increasingly powerful political competition to the ruling Congress party in the form of politicized religious and caste identities. Television became a key place for observing both the anxieties and the possibilities of this conjuncture. See in this context Mankekar (1999) and Rajagopal (2001).

15 This is a reference to the title of the widely cited P.C. Joshi Committee Report on Doordarshan. See Ministry of Information and Broadcasting, Government of India (1985).

16 See footnote 6 above.

17 These are the results of a survey by Deloitte India of 2,000 Indians living in cities, and between the ages of 14 and 75, published in a report titled *The State of the Media Democracy*, released in 2010, cited in Kohli-Khandekar (2010b).

18 Figures from *Indian media and entertainment outlook* (PriceWaterhouseCoopers, 2009). The figure given in Wikipedia for the number of television channels at the time of writing, Jan 2011, is 550.

19 The Right to Information Act, passed in 2005, allows a targeted disclosure of information, and has thus opened a crack in the wall of state secrecy. Information can now be selectively sought and obtained, but only as part of a campaign where many bureaucrats understand the public to be beneficiaries rather than masters; in this sense colonial-era state culture is slow to change as yet.

20 Noteworthy amongst recent studies is Munshi (2010).

21 A characteristic note is struck in the Nitish Sengupta Committee Report of 1996: 'Thus, there is no point being ostrich-like by refusing to grant such permission and not facing the real issue which is that it may be a better option to grant permission and having satellite channels under some discipline rather than letting them operate as free uncontrolled agents. On the other hand, despite the proliferation of the satellite channels, and the government controlled electronic media losing their earlier monopoly, paradoxically, the need for autonomy for electronic media remains stronger. They have to compete successfully with new challengers who have entered into the arena and at the same time successfully reflect the true non-political facets of our nation and its rich and variegated cultural heritage' (Ministry of Information and Broadcasting, Government of India, 1996, Article 1.8).

22 The case involved the attempt by the Cricket Association of Bengal (CAB) to ensure that a six-nation cricket match it was hosting would be telecast by Doordarshan on mutually agreeable terms. Doordarshan's refusal to parlay led to CAB obtaining an agreement with Trans World International (TWI) to generate the broadcasts. TWI is the 'event arm' of New York City based IMG (formerly International Management Group), a global media and sports business. However the agreement required Doordarshan cooperation in hosting the signals, which was not forthcoming. When the case came to court, the lawyers for the I&B Ministry contended, among other things, that the CAB was a trade organization, whereas Doordarshan was empowered to serve the public interest. The Supreme Court eventually ruled that trade organizations too could serve the public interest, and that no monopoly over the airwaves could be allowed to the government (*The Secretary, Ministry of Information and Broadcasting vs. Cricket Association of Bengal & Anr.* dated Feb. 9, 1995. 1995 AIR 1236; SCC (2) 161).

23 Started between 1988 and 1991, *Hindustan Times*' Eyewitness, *India Today*'s Newstrack, the Observer Group's Observer News Channel (ONC), and Independent Television's India View were the most prominent. Their combined circulation was probably in the tens of thousands (Tanna, n.d.).

24 Thus police may be unwilling to 'raid' a cable operator who is merely 'showing' a film without the necessary permission. However, if there is the prospect of media publicity, police initiative immediately becomes more likely. See the discussion with an erstwhile cable entrepreneur, Purushottam Samraj of Showtime Advertisement (Pvt) Ltd., who claims to have introduced cable to Delhi. See Media Researchers @ Sarai (2005). See below for further discussion.

25 The Secretary to the I&B Ministry disclosed in 2010 that 100 channels were being monitored on a round-the-clock basis, with a plan to extend surveillance to more channels. (Interview with Asha Swarup, Secretary, I&B Ministry, quoted in Sruthijith, 2008).

26 At the time of writing, about 30,000 cable operators are estimated to cater to 83 million cable homes between them. There are also several multi-system operators who receive satellite signals and convey them to cable operators, obtaining a carriage fee from TV channels of ten or twenty million rupees a year to carry their content. However, they form a relatively small section of the market as a whole. Local-level

cable operators retain the bulk of subscription revenue, therefore, and it is estimated that only 10–15 per cent of it reaches the cable companies (Kohli-Khandekar, 2003).

27 The remark was in response to a specific question about print media (Sruthijith, 2008).

28 The DMK, or Dravida Munnetra Kazhagam (Dravidian Progressive Party) led by C.N. Annadurai ('Anna') was itself a breakaway from the Dravida Kazhagam (Dravidian Party). The latter was founded by E.V. Ramasami Naicker, originally as the Justice Party, in 1916, and came to power in Madras in the first General Elections in 1920. The AIADMK (All India Anna DMK) was formed by a leading Tamil cine star, M.G. Ramachandra ('MGR'), whose erstwhile friend and colleague Jayalalitha now heads the party.

29 The difference is one of degree; no field was free of state influence. The Films Division and the National Film Development Corporation provided a limited amount of funding for select films, enabling the so-called 'parallel cinema' to emerge; in addition, documentaries were funded by these organizations. Censorship provided another avenue of influence: with politics and overt romance being forbidden, filmmakers had to find creative means of circumventing these obstacles. See Virdi (2003).

30 However, with the growth of the middle classes and development of overseas audiences, Bollywood itself became more upmarket, while regional film industries such as the Bhojpuri film industry addressed more of the urban working class audiences in Northern India. See e.g., Tripathy & Verma (n.d.). My thanks to Ratnakar Tripathy for sharing this paper with me.

31 The show results from a collaboration between Network 18 in India and Viacom, the media corporation; its host is the actor Salman Khan (Timmons, 2011).

32 Government figures indicate that $2.1 bn in foreign investment went to 'information and broadcasting' between 2000 and 2010, while in the same period, the service sector attracted $26.2 bn. These represented 1.7% and 21% respectively of total FDI for those years. The figures for the service sector include but do not isolate advertising, marketing and public relations, which are all dominated by foreign firms (Department of Industrial Policy and Promotion, Government of India, 2010).

33 About one-third of the 119 million television households did not have cable television in 2009, according to a recent report (PriceWaterhouse Coopers, 2009). I have not been able to obtain precise figures for Doordarshan audiences for recent years; my account is drawn from interviews with industry personnel.

34 I am thinking here of the stagist process of growth envisaged, e.g., by sociologists such as T.H. Marshall (see Marshall, 1950).

35 Typically, even the most highly rated top Hindi news channels are only able to charge about a third of the rate for advertisements that English language news channels get, reproducing a difference that has long operated in print media. While the ratios may vary, the story is true for news television as well. This rate differential does not apply to entertainment TV (Kohli-Khandekar, 2010a).

REFERENCES

Agrawal, Binod C. (1977). Satellite instructional television experiment: Social evaluation: impact on adults. Bangalore: Indian Space research Organization.

Agrawal, Binod & Rai, Kumkum (1980). *Women, television and rural development: An evaluative study of SITE in a Rajasthan village*. Ahmedabad: Space Applications Centre.

Ahluwalia, Montek Singh (2002). Economic reforms in India since 1991: Has gradualism worked? *Journal of Economic Perspectives*, 16(3). Reprinted in Mukerji, Rahul (Ed.) (2007). *India's Eeconomic economic transition: The politics of reforms*. New Delhi: Oxford University Press.

Anderson, Benedict (2006). Imagined communities: Reflections on the origin and spread of nationalism (revised edition). London and New York: Verso Books.

Awasthy, G.C. (1965). *Broadcasting in India*. Bombay: Allied Publishers.

Bagdikian, Ben (1983). *The Media Monopoly*. Boston: Beacon Press.

Bhushan, Sandeep (2013). Manufacturing consent. *Economic & Political Weekly*, June 8, 68(23), 12–15.

Chatterji, P.C. (1991). *Broadcasting in India*. New Delhi: Sage.

Chaturvedi, Swati (July 26, 2009). Interview with Information and Broadcasting Minister Ambika Soni. There's no place for censorship in a democracy. Zee news. Retrieved from www.zeenews.com/news550249.html

Colley, Linda (1994). *Britons: Forging the modern nation 1707–1837*. New Haven, CT: Yale University Press.

Dalal, Sucheta (2011). Glamour stocks turn ugly. *Money Life*, August 25. www.moneylife.in (accessed June 5, 2013).

Department of Industrial Policy and Promotion, Government of India (2010). *India FDI fact sheet November 2010*. Retrieved from http://dipp.nic.in/fdi_statistics/india_fdi_index.htm

Eco, Umberto (1965). Towards a semiotic enquiry into the television message. In J. Corner & J. Hawthorn (Eds), *Communication studies: An introductory reader* (131–150). London: Edward Arnold.

Gitlin, Todd (1980). *The whole world is watching: Mass media in the making and unmaking of the new left*. Berkeley, CA: University of California Press.

Hallin, Daniel C. & Mancini, Paolo (2004). *Comparing media systems: Three models of media and politics*. Cambridge, UK: Cambridge University Press.

Horkheimer, Max & Adorno, Theodor W (1972). The culture industry: Enlightenment as mass deception. In M. Horkheimer & T.W. Adorno (Eds), *Dialectic of enlightenment* (120–167). New York: Continuum.

Jeffrey, Robin (2006). The Mahatma didn't like the movies and why it matters: Indian broadcasting policy, 1920s–1990s. *Global Media and Communication*, 2(2), 204–224.

Joshi, Manohar Shyam (1989). Manohar Shyam Joshi, novelist and television script-writer, New Delhi, personal interview, August.

Kailasam, Bala (2011). Bala Kailasam, Creative Head, SRM Group Television, Chennai, telephone interview, January 8.

Kejriwal, Shailaja (2004). Shailaja Kejriwal, STAR TV creative head, Mumbai, personal interview, September.

Kohli-Khandekar, Vanita (2003). *The Indian media business*. New Delhi: Response Books, Sage Publications.

Kohli-Khandekar, Vanita (2010a). What is regional media? *Business Standard*, April 13.

Kohli-Khandekar, Vanita (2010b). Who says mass media is dead? *Business Standard*, 16 June.

Kumar, Shanti (2006). *Gandhi meets primetime: Globalization and nationalism in India television*. Champaign-Urbana, IL: University of Illinois Press.

Lerner, Daniel (1959). *The passing of traditional society: Modernizing the Middle East*. Glencoe, IL: The Free Press.

Mankekar, Purnima (1999). *Screening culture, viewing politics: An ethnography of television, womanhood, and nation in postcolonial India*. Durham, NC: Duke University Press.

Marshall, T.H. (1950). *Citizenship and social class and other essays*. Cambridge, UK: Cambridge University Press.

Masani, Mehra (1976). *Broadcasting and the people*. New Delhi: National Book Trust.

Mattlelart, Armand & Siegelaub, Seth (Eds) (1979–83). *Communication and class struggle* (Vols 1–2). New York: International General.

Mazzarella, William & Kaur, Raminder (2009). Between sedition and seduction: Thinking censorship in South Asia, in R. Kaur and W. Mazzarella (Eds), *Censorship in South Asia: Cultural regulation from sedition to seduction* (1–29). Bloomington and Indianapolis, IN: Indiana University Press.

Media Researchers @ Sarai (2005). Complicating the city: Media itineraries. *Bare acts: Sarai reader* (270–271). Retrieved from www.sarai.net

Miller, Toby (1998). *Technologies of truth: Cultural citizenship and the popular media*. Minneapolis, MN: University of Minnesota Press.

Ministry of Information and Broadcasting, Government of India (1978). *Akash Bharati: Report of the Working Group on Autonomy for Akashvani and Doordarshan* (B.G. Verghese Committee Report). New Delhi.

Ministry of Information and Broadcasting, Government of India (1985). *An Indian personality for television: Report of the working group on software for Doordarshan, Volumes 1 and 2*. New Delhi

Ministry of Information and Broadcasting, Government of India (1996). *Report on Prasar Bharati (Nitish Sengupta Committee Report)*. New Delhi.

Munshi, Shoma (2010). *Prime time soap operas on Indian television*. New Delhi and London: Routledge.

Negt, Oskar (1978). Mass media: Tools of domination or instruments of liberation? Aspects of the Frankfurt School's communication analysis. *New German Critique*, 14, 63.

Newstrack India (2010). Regulatory Act will resolve broadcast media issue: Ambika Soni.

Newstrack India, March 30, New Delhi. Retrieved from www.newstrackindia.com/newsdetails/156455

Pal, Yash (1983). Thoughts on communication. *Seminar*, December, 292, 36.

Pierce, Steven (2006). Looking like a state: Colonialism and the discourse of corruption in northern Nigeria. *Comparative Studies in Society and History*, 48(4): 887–914.

Pletsch, Carl (1981). The three worlds, or the division of social scientific labor, circa 1950–1975. *Comparative Studies in Society and History*, 23(4), 565–90.

Prasad, Madhava (1999). *The ideology of the Hindi film*. New Delhi: Oxford University Press.

PriceWaterhouseCoopers (2009). *Indian entertainment and media outlook 2009*. Retrieved from www.pwc.com/in/en/forms/indian-entertainment-and-media-outlook-2009.jhc-ctml (accessed Sept. 10, 2010).

Punathambekar, Aswin (2013). From Bombay to Bollywood : The making of a global media industry. New York: NYU Press.

Rajagopal, Arvind (2001). Politics after Television: Hindu nationalism and the reshaping of the public in India. Cambridge, UK: Cambridge University Press.

Rajagopal, Arvind (2009). Literacy, print, radio and television growth, 1941–2006. In Rajagopal, Arvind (Ed.), *The Indian public sphere: Readings in media history* (320–326). New Delhi: Oxford University Press.

Rajagopal, Arvind (2011). Visibility as a trap in the Anna Hazare campaign. *Economic and Political Weekly*, November 19, 19–21.

Rajagopal, Arvind (2012). 'Wanted, a comunications policy'. *The Hindu*, January 24.

Scott, James (1998). *Seeing like a state: How certain schemes to improve the human condition have failed*. New Haven, CT: Yale University Press.

Singhal, A., Obregon, R. & Rogers, E. (1994). Reconstructing the story of Simplemente Maria, the most popular soap opera of all time. *Gazette*, 54, 1–15.

Singhal, A., Rogers, E.M. & Brown, W. J. (1993). Entertainment telenovelas for development: Lessons learned. In A. Fadul (Ed.), *Serial fiction in the Latin American telenovelas with an annotated bibliography of Brazilian telenovelas* (158). Sao Paulo: Scholl of Communication and Arts.

Srinivasan, Jitendra (2006). Mega headaches … Polambals [gripes] on TV and radio, April 2. Retrieved from www.ksr.sweetcircles.com/category/polambals-on-tv-and-radio/

Sruthijith, K.K. (2008). There has been no demand from the print media to raise FDI cap. *Livemint*, June 19. Retrieved from www.livemint.com/2008/06/19224000/There-has-been-no-demand-from.html

Tanna, Ketan (n.d.). Video wars: Can the new video magazines survive in the cutthroat market? A special report. *TV& Video World Magazine*. Retrieved from www.ketan.net/art7_4.html

Times News Network (2011). 10 lakh more colour TV sets for free distribution. *Times of India*, January 10.

Timmons, Heather (2011). In India, reality TV catches on, with some qualms. *New York Times*, January 10, pp. B6.

Tripathy, Ratnakar & Verma, Jitendra (n.d.). *Identities in ferment: Reflections on the predicament of Bhojpuri cinema, music and language in Bihar*. Asian Development Research Institute, Patna, India.

UNESCO (2006). *Trends in Audio-Visual Markets: Regional Perspectives from the South*. Paris. 306–7.

Vasanti, P.N. (2008a). FDI tightening its grip over Indian media space. *Livemint*, August 8. Retrieved from www.livemint.com/Articles/2008/08/08002512/FDI-tightening-its-grip-over-I.html

Vasanti, P.N. (2008b). Need for a debate on cross-media ownership rules. *Livemint*, December 12. Retrieved from www.livemint.com/Companies/4jForQqXitdzFhUvGzC7sO/Need-for-a-debate-on-crossmedia-ownership-rules.html

Vasanti, P.N. (2009). Political parties and their media mouthpieces. *Livemint*, October 8. Retrieved from www.livemint.com/2009/10/08224021/Political-parties-and-their-me.html

Virdi, Jyotika (2003). *The cinematic imagination: Indian popular films as social history*. New Brunswick, NJ: Rutgers University Press.

Weber, Eugen (1976). *Peasants into Frenchmen: The modernization of rural France 1870–191*. Stanford, CA: Stanford University Press.

Williams, Raymond (2003). *Television: Technology and cultural form*. New York and London: Routledge.

8

Mexican Research on TV: A Tradition Framed by a Powerful Quasi-Monopolistic TV System

Guillermo Orozco

Writing about research on TV today, but thinking about its future, at a time of great global transformation and of mass media and its interactions with its audiences, and doing so from Mexico as the stage for a peculiar industrial and cultural TV development: 'private-commercial-non-competitive TV', leads us to set out an open future as the horizon for research on TV, one that is 'free' as much as possible from the ideological burden of its historical condition, which does not mean an idealistic or strong-willed horizon but rather one with great empirical realism that would favor research on TV refocused on the audiences themselves, but not just as viewers, net surfers or consumers but as communication-producing citizens who are undergoing a transition to other audio-visual entertainment stages, both the existing ones and those to be constructed and to other means of logging on, getting information, producing knowledge, communicating and of course, enjoying TV.

It is precisely this multiple, complex, constructive transition, not to one but to several audio-visual scenes, along with the 'implosive and explosive' decentralization of mass TV as the hegemonic mass medium of its audiences' every-day lives, that appears to be timely and coherent to propose as the major object of study for research on TV at present and in the future. The account of TV and its research in Mexico is an illustrative example of this future need and priority, and it presupposes saying something about several topics that have become interconnected throughout its history and present, not only because TV and its research intertwine with them (politics, culture, the media, the market) but also and above all because television itself is 'many things at the same time': an institution, a medium, a language, a technology, a business…

AN OVERVIEW OF MEXICAN TV

Once Mexican TV and its research have been located within a specific 'media geography' that Sinclair and Straubhaar (2013) have classified as Latin American due to its specific commercial geo-political-linguistic coincidences, the Mexican TV system, dominated by Televisa, the big Mexican TV company, has developed with TV, in addition to a large market, a unique, effective 'entertainment regime'.[1] A regime, not just a mere style or perspective that has managed to place itself in the audiences' programming preference; a regime because the TV model has entailed a great deal of political and commercial imposition, a great deal of surveillance from the sponsors and other powerful institutions, a great deal of seduction and contempt for its audiences,[2] who were educated as viewers at will by that regime, although it has also gratified them, kept them informed and entertained and it has even given them some shows that allowed them to grow as persons, such as those telenovelas from the 1970s with a pro-social message that managed to make a great impact (Garnica & Sabido, 2011), which ended up by legitimizing right in front of 'their own audiences' eyes.

These issues, among other more circumstantial ingredients, have *overshadowed* the program relevance and the parallel existence in the country of other types of TV, which are neither commercial nor private, which have been around since 1958, which despite their permanence, geographic multiplication in different regions in the country and even a certain degree of popular support attained among 'learned' audiences, have always been options for minorities; first for those that have managed to pick up their open signals around Mexico City and then for those who seek to enjoy, albeit sporadically, what is usually referred to as 'cultural' and/or proximity TV. These channels, 'this other TV' that is not exactly public, as proven and argued by Ortega (2006), has been bedazzled by the glitter of the 'star system' that has always accompanied programming at Televisa, that has been one of the hooks it uses to snare its audiences, thus achieving the highest ratings among its viewers (OBITEL, 2013).

Televisa's success has therefore been at once a cultural and a marketing success. It has been based on five points essentially. The first has been a 'quasi-total and permanent' political alliance with the Mexican State throughout the administration of the moment, especially the 'Priistas',[3] only temporarily with another party, whose complicity has protected and benefitted TV companies exceptionally throughout its history while it has not only permitted, but also promoted, what has been the key to their success: the *permanent abolition of competition*, favoring its monopolistic, oligopolistic development as the only and then the main national TV companies.

This abolition –the second point – was perpetrated in four moments: in 1955, when *Telesistema Mexicano* was born by fusing the first three channels granted to the pioneering TV business people into one single company. The second moment was in 1973, when the then President facilitated the creation of what is known today as *Televisa*, settling the growing differences between *Telesistema Mexicano*, based in the capital, and *Canal 8*, from the city of Monterey, sponsored by the growing industry to the north of the country and its managers who were not accomplices of the presidential power (Hernández & Orozco, 2007). The third moment of the abolition of competition was 1993, when after a schism of the so-called 'Revolutionary Family' (during which the PRI government in office first set up its own 'national TV system' with IMEVISIÓN), they decided to get rid of it because it was unproductive, selling it to what is known today as TV AZTECA. This company, after a period of disruption with Televisa and after a brief period of competition in news and telenovelas programming, finally decided to share 'the TV cake' in friendly terms with Televisa, forming thus the current 'duopoly' and concentrating in their hands around 90% of the total TV production and scope in Mexico. At present Televisa controls 70%, TV Azteca 20% and the remaining 10% is distributed among the different cultural and regional channels sponsored by the government (OBITEL, 2013).

This alliance between the government and the TV companies also included an agreement (third point) that was never made official; it was rather tacit, among the government, Televisa and TV AZTECA with respect to programming. The commercial companies kept the production and distribution of fiction while the other non-profit cultural channels were in charge of producing and broadcasting news, documentaries, educational programming and the fine arts. That is, they focused on that 'high-quality cultural' programming, which, although it gained international recognition, like that of the UNESCO in 1996, awarded to Channel 22 when it was recognized as the 'cultural channel from Mexico', it never had more than 4 rating points as opposed to the 35 that the Televisa telenovelas usually enjoyed (OBITEL, 2007).

The fourth point is the media development and exploitation of melodrama first as a radio and then as a TV genre. This has been the favorite genre among Mexican and Latin American audiences, since as Martín-Barbero & Muñoz (1992) point out, these audiences are prone to 'retell' over again what is happening on the screen to the others who are enjoying the melodrama. Thus the melodrama goes from being a story told to being one retold and it expands into life itself involving its audiences. Televisa had then the ability to make big TV business out of popular culture (Orozco, 2002) by producing and offering the telenovelas[4] on its screen, a format that has been highly sellable and has gone around the world, even becoming the main source of income of the TV consortium and the emblematic format of Latin American TV (Vassallo & Orozco, 2010).

The fifth point is that the novel Mexican TV system, from the technological and commercial points of view, found the political platform and knowhow necessary to take advantage of the favorable circumstances of the postwar world market, where the strengthening of capitalism implied an ever-growing involvement with the media above all through the expansion of publicity on its screens. Likewise, their proximity to the US helped the media business people to exert effective pressure for carrying on with a strong commercial TV system in Mexico projected as its 'partner' (Sinclair & Straubhaar, 2013).

CONSEQUENCES FOR THE MEXICAN TV RESEARCH

In the previous context Mexican TV research in general has differed from other research traditions in their understanding of what TV is, since in contrast to the research undertaken in the framework of Anglo-Saxon geographies, where TV has been assumed in research rather as a 'medium' in its *mediacity*,[5] in Mexico and the Latin American universe it has been understood more or almost exclusively in its 'institutionalism', more often than not related with power, as a source producing ideological, political and publicity contents, contents that are considered powerful by many researchers, with a high potential for affecting, influencing and 'educating the viewers on their own' (Sanchez-Ruiz, 1992), at times even without taking into account the formal and technical aspects of the contents that make it possible for them to be viewed and broadcast on and from the screens.

In keeping with this vision, research here has also been *bipolar* more so than in other media geographies because it has been the result of stands that are on one hand critical and on the other, administrative about TV itself and about its research; these perspectives have usually developed in parallel, without one recognizing the other's legitimacy, both in what respects the conceptions of TV as object of research and the methodological strategies to tackle it. Academic research, which is mostly qualitative and critical, has sponsored studies that aim at understanding specific TV objects: contents, audiences, mediations, TV companies' capital, etc., while corporate research is quantitative and it intends to measure and count events and regularities, especially those related to rating and programming consumption, and this way define favorite tendencies, consumptions and TV times. It was but less than a decade ago that some points of mutual connection and feedback between both research traditions were found (Jara & Garnica, 2007).

Upon reviewing the studies on TV, which are mostly academic, and therefore can be consulted publicly, it is possible to see that the results of much of the research undertaken, above all that on TV and the audiences, have shown the most criticizable side in this relationship and these subjects since it has exposed a TV full of questionable messages and an ignorant audience, one that is vulnerable, faithful, docile and conformist with a banal, low-quality programming. Thus, the effects of TV which have been unveiled by the research have mostly been *negative effects*, above all in terms of the viewers' nationalism and behavior (Montoya & Rebeil, 1987). It is noteworthy that there are no positive effects of TV recorded in the Mexican studies, which must be understood from a triple dimension: firstly, due to the hegemonic weight of the Mexican TV system and its complicity with power, which has been present in the researchers' critical thinking; secondly, because in academic investigative practice on TV ideology has often prevailed over empirical results (Lozano & Frankenberg, 2008); and thirdly, due to the geo-cultural 'macro mediation' pointed out by Martín-Barbero (1997) when he says that TV has been 'the intellectuals' evil eye' in Latin America.

BACK TO THE FUTURE WITH THE MEXICAN RESEARCH ON TV

The first record of studies on TV in Mexico dates back to 1966,[6] it was conducted by Antonio Castro, and it is titled: *El pueblo de México espera. Estudio sobre la radio y la televisión* (*Mexican people wait. Study about Radio and TV*) (Castro, 1966). This is a text that presents a critical analysis of the first Radio and Television Federal Law, issued in 1960, whose most controversial point, the one with the greatest social impact, was the change of the constitutional framework of the radio and television industry from being a *public service* industry to being *public interest* one. This change might seem subtle, but it has profound economic, legal and political implications, which paved the way for unprecedented private and commercial development in audio-visual media in Mexico, eliminating almost completely its social responsibility in terms of the citizenry and its commitment to the public's education. These characteristics were assumed by 'other television networks,' which, without becoming Televisa's rivals, have managed to co-exist, frequently even in a dignified manner (Hernández & Orozco, 2007).

Unlike what happened in other Latin American countries, where there were TV systems made up of private and public channels, in Mexico a TV system was founded and consolidated in the first eight years with only one type of channel: private, even though the government, not the society at large, claimed the right to a small share of the air time (12.5%) to broadcast their own programming.[7] This unique characteristic of the Mexican TV system seems to have served as a 'citizens' distracter', since it made it possible for Mexican society to sit back quietly knowing that ultimately TV control was in the hands of the State and not in those of the media companies, trusting that this State would safeguard their 'cultural citizenship' in front of the screen. Nevertheless, what no one could imagine, and it was proved in 2003, was the armed attack on the Channel 40 facilities – a TV signal threatening to compete with the news programming of the established networks, Televisa and TV AZTECA – this event made it clear that the alleged pact entered into by the government of the moment and the Mexican TV duopoly must have been like that pact by means of which a super police agency, in order to defend its 'always legitimate' interests, assigns almost impossible missions to Agent 007, who was granted a 'licence to kill'. Thus, the then Mexican government allowed Channel 40 to 'be killed and die', with which the reign of monopoly remained once again intact in the Mexican TV system.

Going back to research, in 1968 *La televisión y el alumno de secundaria del Distrito Federal* (*TV and the secondary school student in México City*) was published by Raúl Cremoux, an author who now directs the 'award-winning' TV Channel 22. This was perhaps the first research in Mexico to undertake an empirical approach on the use of TV in the pioneering Mexican educational system called: *Telesecundarias*, and the educational effects of commercial TV among its student body, comparing both types of effects – the intended, desirable ones through *Telesecundarias* and the unwanted mostly undesirable effects resulting from the students' exposure to commercial TV (Cremoux, 1968).

These two studies have pioneered research on TV in Mexico and they may be considered *exemplary* inasmuch as they inaugurate and reflect the 'classical', permanent, central interests of media researchers in the country. These studies have marked two priority lines of research up to the present. On the one hand, to expose different parts of that 'great monster' that the Mexican TV system has come to be, showing its alliances with the national and transnational political power, and on the other, to explore the educational impact that the programming of such a TV system has had on its public (Orozco, 2002). To this last end, researchers have used perspectives from Agenda Setting, Encoding-Decoding, Mediations, Cultivation Analysis, Cultural Studies, Discourse Analysis, Media Effects and Uses and Gratifications Analysis, among others.

The following can be cited as examples of the first research interest, the TV system: *La respondabilidad de los medios de comunicacion en México* by Fernández-Christlieb (2000) (*The responsibility of mass media in Mexico*) and those works by Raul Trejo dealing with the relation between the media and power, specifically the role played by Televisa: *Televisa el quinto poder* (Trejo, 1985) (*Televisa, the fifth power*) and *Las redes de Televisa* (Trejo, 1988) (*The Televisa networks*) that combined chapters by the most renowned TV academic researchers in Mexico.

There have been several theoretical perspectives in the different studies related to Televisa, among them that of Cultural Imperialism, the Theory of the Elites, the Theory of Dependence and its variable of 'Asymmetric Interdependence' proposed by Straubhaar (1991), as well as Political Economy variants with approaches that ranged from an emphasis on the role of the State and politics to the role of the market (Gómez & Sosa, 2008).

More recently and in the context of media reform in Mexico, Esteinou and Alva (2009) coordinated a multiple study on the notorious law that Televisa intended to impose in 2006 in the media reform at the threshold of the transfer into the digital world: *La 'Ley Televisa' y la lucha por el poder en México (The Televisa Law and the struggle for power in Mexico*, 2009), which is a vast publishing effort, comparable to that of a Handbook, since its 400 pages review almost all the topics concerning TV research written by its most representative critical researchers.

Along with the research on the TV system and companies, interest in the relation between *education* and television has also occupied a central position in research on TV in Mexico. One way or another almost one-third of the research, especially that conducted between the 1960s and the 1990s, had the educational aspect linked to TV as the raison d'être of its undertakings (Orozco & Padilla, 2005). The educational aspect manifested first as the stage of non-educational TV's affectation. What was at stake, then, was the proposal of 'bad education' in the audiences, as demonstrated by an important study conducted by Sánchez-Ruiz in the 1980s (1989): *Teleadicción infantil, ¿mito o realidad?* (*TV addiction: Myth or reality*?) Beginning in the 1990s there occurs a sort of turning point in research on TV and education, since the objective becomes teaching the audiences how to deal with TV, and the exploration of communicative intervention strategies in their reception processes (Charles & Orozco, 1990; Crovi, 2006).

The perspective that inspired some of the most significant studies along these lines was the model of Múltiple Mediación devised by Orozco (2001), which was in turn inspired in the perspective of Mediaciones, by Martín-Barbero, but landing on different manifestations of them that could be observed empirically.

In addition to education, the telenovela has been in the spotlight of research on TV in Mexico, but it has also been approached from two very different and even opposing perspectives, which confirms this aforementioned investigative 'bipolarity'. First, in the 1980s and 1990s the telenovela was researched from a culturalist approach, centered on its 'retellings' by different sectors of their audiences; that is, focused on its reception and interpretation, not on its production. Much of this research was inspired in the conceptual-methodological proposal by González (1998) who developed a variant of Cultural Studies centered on the analytical category of Cultural Fronts where the telenovela, rather than an object of analysis as such, was a pretext or catalyst to explore the people's greatest cultural process (Orozco, 1994).

Since 2005 the telenovela has been approached from an integral perspective developed by the OBITEL: The Ibero American Observatory of TV Fiction, which, from the production itself of TV fiction as well as its distribution, seeks to explain in qualitative and quantitative terms the characteristics, rating, changes, contexts, markets and tendencies of TV fiction, and with it that of the Ibero American TV systems in which fiction is considered its epicenter (OBITEL, 2007). From this investigative program, fiction has been understood and approached as 'another way of telling reality', one that perhaps has a stronger impact on their audiences due to the dramatization it entails and the receptiveness it finds in Latin American geopolitics.

TV fiction is not just a very popular genre in Latin America, but also its telenovela, at least in Mexico, has become, since 2005, 'the big format', whose narratives aim at maintaining and expanding the contemporary political, commercial hegemony since it has turned into the privileged field for naturalized or integrated publicity of both brands and products and political propaganda, above all in electoral times (Orozco & Franco, 2012). This is an important challenge for any democracy and for the strengthening of the citizenry since audiences are surprised with political messages because, without previous warning, the 'agenda is loaded' on the very plots of their favorite fiction shows. In this new situation, TV news reports gradually lose their importance in TV's ideological framing (Orozco et al., 2011).

At the same time, since TV fiction is the programming with the highest ratings in Mexico, always accounting for 20– 30%, TV has a long life ahead and, as claimed by Carlón (2012), it will continue to maintain its validity as a programmer of social life. This does not mean that the media side of the growing communication imposes itself on the personal side through the audiences' connectivity on the social networks in the digital world; it means that TV and its audiences will continue to be 'linked' albeit in different ways, since television invades other screens and favors communicative experiences on the internet, YouTube, Facebook and many other virtual sites. And, as Miller claims (2012) as long as TV maintains that characteristic of never demanding too much effort from their audiences, in the time to come there will be more TV, not less TV; now, there will be different types of TV, not just the ones known at present, and there will be larger audiences who will stay longer around several screens.

TV WITHIN A 'SELF-MASS COMMUNICATION' ENVIRONMENT: A MAJOR RESEARCH CHALLENGE ABOUT TV AND ITS AUDIENCES

Having mentioned some of the major elements that make up the Mexican TV system and outlining some of the characteristics that make it possible for the prevailing type of research to be distinguished, but also looking beyond towards the ambivalent global stage where economic differences widen among the countries and within them, as well as their *digital divide* where contrasting, sharp situations are experienced in the political, economic, cultural and communicative fields, what aspects of TV are worthwhile researching?

After reading the Mexican case, I would dare put forward a big *why – what for* concerning research on TV centered on its audiences and their *audience-making processes (audienciación)*. In terms of theoretical-conceptual supports for new research, it would be convenient to think about *cognitive reasoning logics* other than the hypothetical-deductive ones of the scientific method. The point is not to fall into mere ideologizing either. No. For one thing, the model of indications (Guinzburg, 1999) would have to be incorporated as it has been developed from historiography, seeking to uncover the past to explain and understand the present, as well as that of *abductive inference*, proposed by Pierce and rescued by Jensen (1998) for communication studies. Abduction intends to explain what the event must have been by inferring it from its consequences or indications, understood under the theorization related with the object approached. This 'alibi' makes it possible for us to unveil the power that is not usually transparent, and it makes it easier to understand from where it is being exerted and by means of what strategies and frameworks it intends to impose itself on the TV audiences. Likewise, abduction would be useful in making evident unwritten rules in TV practices and in the TV systems in specific media geographies (Orozco & González, 2012).

It would be convenient also to design proactive research that promotes and proposes new paths to 'empower' everyone as citizens, but from their specific role as audiences, which is estimated to be the prevailing one in the future. This is said without underestimating the occasional debate that has emerged in this respect to abolishing the category of audience under the dazzle of the participation of new 'entities': users, net surfers, consumers or *prosumers*. In many of these debates it has been assumed that the category of the audience was born with TV (Rincón, 2013) and therefore, it is concluded that it is soon to die just like TV itself is dying.

But TV is not dying and the audiences were not born with TV either! TV, for its part, is undergoing a transformation in a Darwinian manner so as not to lose its rating, producing *transmedia fiction* and *web telenovelas*, among other innovations, and it continues to enjoy their new audiences' preference. As to the audiences, it must be underscored that they had existed for a long time, although maybe they became more popular with TV. It must be remembered that audiences were not always as we have known them in front of the TV. According to what Butsch (2000) explains in his book, *The Making of American Audiences*, audiences underwent a transformation with the modern mass media, the radio, the cinema and the TV, which moved them into a position of passiveness, or rather induced them to think and act in a certain manner or to consume what was advertised on screen, etc., something pursued mainly by commercial TV, which has reigned supreme in most Western countries. As the making of the audiences progressed, Butsch maintains, the audiences gradually stopped being *active reactionary publics*, demanding a public who was capable of having the authors alter the scripts and having the directors modify their productions, or of debating in the written press or producing their own versions, slowly allowing themselves to be turned into essentially screen spectators.

Being the audience means becoming involved with the information, with the others and the other 'stuff' in a way that is always mediated, not directly. That is why *audience-making* has been the characteristic that defines and distinguishes exchange with contemporary societies, where the involvement tends to be ever more mediated and technologized. And as long as audiences remain like this, no matter how interactive or even hyperactive the result, the term will continue to be applicable and the audience-making process will continue to expand.

In addition, the more the audience-making process grows and expands, the more distinct it becomes as an element to identify the current socio-cultural-political economic exchange. Being the audience is then equivalent to what at one time it meant to be working class, which distinguished being fundamentally part of the majority of the citizens during the industrial revolution. Being the audience is then a very rich and at the same time complex category, which must be understood in each entertainment regime. Audience is also a dynamic category, always in a process of re-composition, and it is an elastic category that 'stretches' from audiences who are apparently passive to others who are hyperactive, immersed in the virtual world. The concept of audience entails two dimensions that make it possible for its meaning to be understood. It is possible to 'be' the audience in a way, aware of what is going on, without that preventing us from 'being' or staying as part of the audience in a different way, carefree, relaxed where what is watched on TV serves as a means to escape, not to reflect. There is no contradiction between these different states among the audiences during their exchange with the screens.

The greatest problem does not lie in the dissonance between being one or another kind of audience, but rather on what I have conceptualized as a contemporary *communicative condition*. This condition consists precisely in the fact that even having the growing possibility of interactivity, of creating and distributing one's own messages horizontally, not being an audience defined only by reception, vast audience sectors, including the young, do not make full use of their potential, or at least do not always do so (Orozco, 2011). Therefore, the challenge faced by committed research is to understand the particular historical and contemporary processes, as well as the potentials and dangers that audience-making have in different media geographies, first to 'intuit' and then search, check or understand better the particulars of these processes and the meanings the audiences assume about it. The applicable methodological strategies must be invented or readapted to unveil 'routes to audience-making'[8] and their mediations. That can be achieved by knowing audiences' stories with media and by making what is not evident *evident* for the audiences, as well as their constitution as such, and therefore their areas of opportunity and weakness to move as citizens in the TV scenarios. It can be done using the Framing and the Agenda Building models together with Oral History focused on exploring the ways audiences have been addressed by media stories.

Accompanying the audiences in their reinvention as active subjects of communication is a pending undertaking for research on TV. If fiction in general is so well accepted everywhere, a concrete alternative is that of promoting and provoking *fan fictions* (Jenkins, 2008; Scolari, 2013) as reconversion exercises to help audiences to become active and creative, and from there hopefully involved with local and global interlocution with the others. What is becoming clear is that if the audiences have more or less kept their role as simple spectators during the reign of mass commercial TV, it has been the 'authoritarian way of addressing them' of such TV systems as the Mexican one, that mainly kept them in that condition. Then, finding most horizontal forms of TV audience's interlocution seems to be one of the major challenges to know about, with research on TV yet to be undertaken.

NOTES

1 The author's own concept, produced from Padilla's (2013) research to refer to and eventually research within the cultural, the concrete ways of spending free time and having fun, which are evermore a vital objective among viewers and net surfers.
2 As businessman Azcárraga Milmo, the owner of Televisa, has declared: 'we produce television for the poor, for those without education,

who are the majority of the Mexicans ... the rich can afford to enjoy themselves somehow else' (*Proceso*, 1997: 13).

3 The dominant permanence of the PRI (Institutional Revolutionary Party) for 71 years to the year 2000, when it was referred to as the 'Perfect Dictatorship' by Literature Nobel-Prize winner Vargas Llosa, has mostly been possible thanks to its alliance with this other 'TV dictatorship', the Televisa Consortium (Orozco, 2002).

4 Although it is common in academic literature to equal the American *soap opera* with the *telenovela*, in reality, there are differentiating characteristics both in genre and in format, such as length and sequencing (in addition to historical themes), that make them completely different TV products framed by cultural processes of consumption that have little in common.

5 The author's concept denoting the specific media dimension of a mass medium (Orozco, 2001).

6 The database of the Catálogo de Documentación en Ciencias de la Comunicación (CC-DOC) was used, this catalogue offers a systemized access to the references – and in many cases complete documents – of academic research on communication in Mexico. There are 1,111 references on TV studies. They are not the only ones that deal with the topic of TV, but they are the ones that tackle it in a central manner as an object of research. There are others where TV is researched among other mass media, which have not been considered for this study.

7 A measure made in 1973, that was eventually eliminated by PAN President Vicente Fox in 2003 due to the pressures exerted by Televisa, in what is known as the 'Decretazo': a presidential decree that eliminates by decree a previous decree.

8 It is the author's term and it refers to the audiences' stories with the media; it collects their tastes, times, customs, disappointments and changes with a medium or programming genre, and recovers its context and their memories about their interaction with that medium.

REFERENCES

Butsch, R. (2000). *The making of American audiences: From stage to television 1750–1990*. Boston: Cambridge University Press.

Carlón, M. (2012). Una reflexión sobre los debates anglosajón y latinoamericano sobre el fin de la TV. In Orozco, G. (Ed.) *Tvmorfosis, La televisión abierta hacia la sociedad de redes*. Guadalajara: Tintable.

Castro, A. (1966). *El pueblo de México espera. Estudio sobre la radio y la televisión*. Mexico: Cuadernos Americanos.

Charles, M. & Orozco, G. (Eds) (1990). *Educación para la recepción. Hacia una lectura crítica de los medios*. Mexico: Trillas.

Cremoux, R. (1968). *La televisión y el alumno de secundaria del Distrito Federal*. Mexico: Centro de Estudios Educativos.

Crovi, D. (2006). *Educar en la era de las redes: una mirada desde la comunicación*. México: UNAM.

Esteinou, J. & Alva, R. (Eds) (2009). *La 'Ley Televisa' y la lucha por el poder en México*. Mexico: UAM.

Fernández-Christlieb, F. (2000). *La responsabilidad de los medios de comunicación en México*. México: Paidós.

Garnica, A. & Sabido. M. (2011). Las 'tele – visiones' de un visionario. In R. Jara (Ed.), *Telenovelas en México. Nuestras íntimas extrañas*. México: Delphi.

Gómez, R. & Sosa, G. (2008). Reforma a la legislación de radio, televisión y telecomunicaciones en México (2005–2007). In Vega, A., Portillo. M. & Repoll, J. (Eds), *Las claves necesarias de una comunicación para la democracia*. Mexico: AMIC–UJAT.

González, J. (1998). *La cofradía de las emociones interminables*. Guadalajara: Universidad de Guadalajara.

Guinzburg, C. (1999). *El queso y los gusanos*. Barcelona: Muchnik.

Hernández, F. & Orozco, G. (2007). *Televisiones en México. Un recuento histórico*. Guadalajara: Universidad de Guadalajara.

Jara, R. & Garnica, A. (2007). *¿Cómo la ves?*. Mexico: IBOPE–AGB.

Jenkins, H. (2008). *Convergence culture. La cultura de la convergencia de los medios de comunicación*. Barcelona: Paidós.

Jensen, K.B. (1998). *The semiotics of mass communication*. London: Routledge.

Lozano, J.C. & Frankenberg, L. (2008). Enfoques teóricos y estrategias metodológicos en la investigación empírica de audiencias televisivas en América Latina: 1992–2007. *Comunicación y Sociedad*, (10), 81–110.

Martín-Barbero, J. (1997). La televisión o el mal de ojo de los intelectuales. *Comunicación y Sociedad*, (29), 14–27.

Martín-Barbero, J. & Muñoz, S. (Eds) (1992). *Televisión y melodrama. Géneros y lecturas de la telenovela en Colombia*. Bogota: Tercer Mundo Ediciones.

Miller, T. (2012). El Ahora y el futuro de la televisión. In Orozco, G. (Ed.), *Tvmorfosis, la televisión abierta hacia la sociedad de redes*. Guadalajara: Tintable/Universidad de Guadalajara.

Montoya, A. & Rebeil, M.A. (Eds) (1987). *Televisión y desnacionalización*. Colima: Universidad de Colima/Asociación Mexicana de Investigadores de la Comunicación.

OBITEL (2007). *Anuario Obitel 2007*. Sao Paulo: Globo Universidade.

OBITEL (2013). *Anuario Obitel 2013*. Sao Paulo: Globo Universidade.

Orozco, G. (1994). *Televisión y producción de significados: Tres ensayos*. Mexico: Universidad de Guadalajara.

Orozco, G. (2001). *Televisión, audiencias y educación*. Buenos Aires: Norma.

Orozco, G. (Ed.) (2002). *Historias de la TV en America Latina*. Barcelona: Gedisa.

Orozco, G. (2011). La condición comunicacional contemporánea. Desafíos latinoamericanos de la investigación de las interacciones en la sociedad red. In Jacks, N. (Ed.), *Análisis de la recepción en América, Latina. Un recuento histórico con perspectivas a futuro* (45–67). Quito: CIESPAL.

Orozco, G. (2014). TV: Causa y efecto de sí misma. In Carlón, M. & Scolari, C. (Eds), *El fin de los medios. El comienzo de un debate* (34–47). Buenos Aires: La Crujía.

Orozco, G. & Franco, D. (2012). Telenovelas 'con causa política'. Zócalo, (151), 22–24.

Orozco, G. & González, R. (2012). *Una coartada metodológica. Abordajes cualitativos en la investigación en comunicación, medios y audiencias*. Mexico: Tintable.

Orozco, G. & Padilla, R. (2005). Los estudios de recepción en México. Un Itinerario. In Lozano, J.C. (Ed.), *La Comunicación en México: diagnósticos, balances y retos* (147–166). Mexico: CONEICC/Instituto Tecnológico de Estudios Superiores de Monterrey.

Orozco, G. et al. (2011). México: la investigación de la recepción y sus audiencias. Hallazgos recientes y perspectivas. In Jacks, N. (Ed.), *Análisis de recepción en América Latina: un recuento histórico con perspectivas al futuro* (227–266). Quito: CIESPAL.

Ortega, P. (2006). *La otra televisión. Por qué no tenemos televisión pública*. México: UAM.

Padilla de la Torre, M.R. (2013). *Geografías ciudadanas y mediáticas*. Mexico: UAG.

Proceso (1997). Entrevista a Emilio Azcárraga Milmo (1216). *Proceso*, 3–15.

Rincón, O. (2013). La televisión, la máquina popular en América Latina. In Rincón, O. (Ed.), *Zapping TV, el paisaje de la tele latina* (5–16). Bogota: Friedrich Ebert Stifftung.

Sánchez-Ruiz, E. (1989). *Teleadicción Infantil: Mito o realidad?* Guadalajara: Universidad de Guadalajara.

Sánchez-Ruiz, E. (1992). *Tendencias en la Investigación sobre televisión en México (1950–1990)*. Guadalajara: Universidad de Guadalajara.

Scolari, C. (2013). *Narrativas transmedia. Cuando todos los medios cuentan*. Bilbao: Universidad de Deusto.

Sinclair, J. & Straubhaar, J.D. (2013). *Latin american television industries*. London: BFI-Palgrave Macmillan.

Straubhaar, J.D. (1991). Beyond media imperialism: Asymmetrical interdependence and cultural proximity. *Critical Studies in Mass Communication*, (8), 38–59.

Trejo, R. (Ed.) (1985). *Televisa el quinto poder.* Mexico: Claves latinoamericanas.

Trejo, R. (1988). *Las redes de Televisa*. Mexico: Claves latinoamericanas.

Vassallo, M.I. & Orozco, G. (2010). Convergencias y transmediación de la ficción televisiva; síntesis comparativa de los países Obitel en 2009. In Vassallo, M.I. & Orozco, G., *Anuario Obitel 2010*. Sao Paulo: Globo Universidade.

PART II

Makers and Making

9

How to Study Makers and Making

Miranda J. Banks

In an era of tell-all autobiographies, making-of documentaries, online production blogs, behind-the-scenes web videos, and DVD outtakes, scholars and viewers have more access to the makers of television and the processes of production than ever before. Unearthing the stories of television production is quite easy, if these mediated texts are taken at their word. To assume that these ancillary narratives of makers and making provide an honest, unmitigated view of the making of television would be at best naïve, most likely misguided. These produced co-narratives are presented to audiences with the same economically-driven agenda as the original television product. Thus, this chapter maps methods and models of how to study the makers and making of television using approaches that aim to reveal more nuanced, ideally more authentic representations of what happens behind the scenes and beyond the screen.

Two central questions arise almost immediately though: to whom do we refer when we say 'makers' of television, and what aspect of the making of television do we wish to examine? These questions get to the very heart of studies of media production. Is the maker of television a writer, a network mogul, an agent, a showrunner, an actor, an editor, a camera operator, or a practitioner honing his or her craft? Would *Sex and the City* have been the same series without Patricia Fields as its costume designer? Would *All in the Family*, *Mary Tyler Moore*, *Newhart*, *M*A*S*H** and *The Carol Burnett Show* have been such enormous ratings hits on CBS if Fred Silverman had not organized them into a programming block? Are Fields or Silverman arguably television 'makers'?

In a collaborative medium like television, the talent and skills of many practitioners are integral to the production of a series or program. Most often when scholars use the term 'maker' to describe someone, they are a key member of the production staff – often a series creator, head writer, executive producer, or showrunner. Also, when academics explore the making of television, often they focus on the production process, from the inception of an idea for a television series through to the airing of the series on television. And yet there are many other ways in which television-making occurs: from the building of a television set, to the production of industrial videos, to the meaning constructed of television by audiences. Thus, when academics enter into a study of television makers and making, rather than simply focusing in on its most elite or popular forms, scholarly pursuits explore at the margins as well.

The choices of subjects scholars explore in their studies of television production influence our perception of the medium and of production practices. The privileging of certain aspects of media-making within some academic writing reinforces the hierarchies often seen within media production, without questing categorizations and ranking of labor. As this chapter details, production studies scholarship explores the processes of production and the intricate webs of collaboration that are far more complex than those suggested by the terms 'media-maker' or 'media-making'. Contemporary production studies research explores the work of practitioners at every level of the television-making process.

This chapter maps two guiding methodologies central to the study of television production, explaining why this particular area of television scholarship often necessitates an interdisciplinary approach to both research and analysis, and how scholars have used this interdisciplinarity to their advantage. The second part of the chapter details four themes that have held particular interest to scholars of television production and continue to inform contemporary and continuing scholarship in television production studies.

METHODS

The necessity for television scholarship to address its subject on multiple registers is inherent in the design of the medium itself. Television – or for that matter film, streaming media, or video games – is often at once, a combination of a commercial product, an art form, a form of mass communication, a technology, an industrial product, a cultural product, and a text created through the work of both maker and reader.[1] The multidimensionality of media makes it a slippery subject in our hands.

When studying television, academics engage with some combination of methods mapped out by such diverse scholarly fields as communication, history, sociology, cultural studies, political economy, geography, statistics, engineering, marketing, language and literature, and anthropology. What kind of methodology allows for such flow between seemingly different academic paradigms? How do we draw the line between these fields when we are studying media? Should we draw such a line? These are the problems researchers interested in using multidisciplinary methods face when they study the texts produced by media industries. But the solutions they have found – and the research that has come out of crossing these registers – have shifted media scholarship and helped to define this complex relationship between media and maker, between industry and text, between writer and reader.

When pared down to their essence, two central methods are employed in studies of television makers and making: those emerging from within the disciplines of the social sciences and those traditionally practiced within the humanities. Granted, defined as such, these are quite broad methodological categories for exploration. But in order to understand the history of this type of media scholarship and how television scholars might move forward in their research of producers and production, articulating the nuanced relationship between these two disciplinary methods inherent in most studies of production proves particularly useful. In the end, what we often find is that the difference between scientific and humanistic methods, are more a matter of degree than of kind.

Methodological trends within television scholarship that emerge through the social sciences generally involve a structuring approach that attempts to quantify social and industrial relationships inherent in media production. Scholarship is evidence based and often employs scientific methods of quantitative analysis (content analysis, data gathering, media effects research, surveys). There is, as Jennifer Holt and Alisa Perren (2009) deftly argue in their book *Media Industries*, an inherent linearity within these scholarly traditions (p. 4). Often economists, anthropologists, sociologists and communications researchers studying television production endeavor to uncover the social phenomena of this creative industry using scientific method, laying bare what can be seen as objective knowledge and definitively proving truths about media producers and trends within media production. As defined by sociologists Richard Peterson and Narasimhan Anand (2004), the production of culture perspective takes as its guiding principal that 'the symbolic elements of culture are shaped by the systems within which they are created, distributed, evaluated, taught, and preserved' (p. 311).

Television scholarship that has emerged out of humanities traditions tends to focus more on the individual human condition, interrogating core assumptions about creative labor, meaning making, and the television text through qualitative, critical analyses. In many ways, the narrative is central to this scholarship: not necessarily the narrative of a series, but perhaps the narrative of creative thought, personal achievement, or industrial production. Here scholars have trained their analysis on the plan and design of media production, exploring with critical tools questions of art, authorship, creativity, identity, collectivity, history and culture.

While I have separated them here, in reality, few scholars engage with materials or issues using methods that rest solely from within one of these two disciplinary registers. In many ways, it is the structure of the academy has kept the industry and the text apart: social scientists look at media through the lens of policy, economy, government, geography, communication and sociology, and find the focus of their attention on the structure of the industry. Across the campus, humanities scholars examine media using methods established within the fields of literature, language, art and history, and ask questions that speak to the cultural and aesthetic meanings found within media. While I am generalizing here, it is important to see how media scholarship has been segregated across disciplines, departments, and scholarly traditions. These unique disciplines approach media studies from their own domains – so much so that many media scholars are often unaware of the diversity of scholarship coming out of other fields that could be in dialogue with their own research.

The study of producers and production, by virtue of the nature of knowledge enabled and engaged through their research, almost demands that scholars undertake an integrated analysis. The most engaging scholarship tends to stitch together a research method that fits the author's particular area of inquiry rather than conforming to the requirements of a set scholarly sub-discipline. Granted, few scholars have the necessary expertise within multiple disciplines to completely blur these intellectual boundaries, and yet, this is precisely why those who study makers and making must rely on the work of past and contemporary scholars who have already begun to conceptualize a vision of makers and making that intersects at unique methodological points.

Two defining texts that highlight the use of interdisciplinary methods within media studies are David Bordwell, Janet Staiger and Kristin Thompson's (1985) *The Classical Hollywood Cinema: Film Style and Mode of Production* and Julie D'Acci's (1994) *Defining Women*. While Bordwell, Staiger and Thompson's book covers American cinema, its significance as a work of cross-disciplinary media analysis makes it a key publication within media studies as a whole. Examining Hollywood film as an artistic and economic mode of film practice, the authors examine cinematic style and industrial conditions in relation to one another. In outlining their book project, the authors' preface reads almost like a manifesto, defining their project as not just a history of Hollywood cinema, but as a new theoretical approach to film history that emphasizes the nature of Classical Hollywood film practice as 'a coherent system whereby aesthetic norms and the mode of film production reinforced one another' (Bordwell et al., 1985, p. xiv). Bordwell, Thompson and Staiger trace the history of this deeply structured, bounded system of production that institutionalized structures of narrative, form and style. *Classical Hollywood Cinema* uncovers the complex web of relations between politics, economics, aesthetics and culture. Though this work is typically catalogued as film history, it is the kind of interdisciplinary scholarship that defines industrial studies of media production. Julie D'Acci's work combines an archival-based industrial history with feminist textual analysis, locating these texts within the specific social history of the women's movement in order to better define the industrial struggles the creators and network faced, as well as a program's reception.

Rather than separate media production scholarship based on traditional scholarly or departmental boundaries, many contemporary scholars of television production and media producers have begun defining our work under the interdisciplinary heading of production studies. Production studies is grounded in the assumption that there is a unique and significant relationship between an individual media artifact and the production of that product, the industry from which it was created, as well as the governmental policies and socio-economic conditions that were in existence during its historical moment of creation. In other words, understanding the modes of production, as well as the culture of production help scholars define the meaning and role of a media object. Inherently interdisciplinary, this scholarship pulls from a wide variety of scholarly fields. For example, some scholars apply Richard Johnson's (1986) circuit of culture model, which examines the cross-flow of institutional hierarchies, production flows, and audience analysis within a particular work. This type of work provides a theoretical intervention to traditional humanities-based, text-centered analyses by incorporating a theorization of the material conditions of labor within the context of a specific industry history. In her article, 'Cultural Studies, Television Studies, and the Crisis in the Humanities' Julie D'Acci (2004) calls precisely for this interdisciplinarity as set out by Johnson. For her, television studies has suffered from its close ties with cultural studies; rather it should pull back and realign itself equally with historical, industrial, institutional and aesthetics frameworks. What this demands is an understanding of the interconnectedness of these different registers and their interrelations: from production, to reception, to a grounded reading of socio-historical contexts to an examination of the text as a cultural artifact (D'Acci, 2004, p. 432). This circuit of culture model opens up scholarship to show the connections organic to television. By exploring these disparate points of convergence in production, consumption, and the discursive practices at each of these registers provides a heuristic, integrated vision of media.

What this requires of the scholar is an interdisciplinary approach: applying research, skill sets, methodological tools, and theoretical traditions across scholarly boundaries. Though the methods might be pulled from multiple theoretical registers, it is important to note that production studies work itself is empirical and disciplined. Recent work in production studies provides cultural, textual and contextual analysis organized by industrial theory grounded in social-science-based ethnographic field research and data gathering. And yet, when looking at these examples of industrial studies of the media text, the research itself, as read, feels methodologically organic. And that makes sense: the maker of television, the processes of production, the text, and the industry could not exist without one another – they are fundamentally connected. By nature of the processes of production, they are inseparable. These embedded industrial theorizations of the cultures of production and production culture are harvested by scholars from practitioners' experiences at every level in the production process. While the fruits of this scholarship are rewarding for its readers, having the interest and the means to conduct a study of media across these registers takes both creativity and fortitude. The following section will further detail the usefulness of methodological integration in research of makers and making in television studies while exploring some of the prevalent themes within this work.

THEMES

This section maps four themes common within production studies research that transcend disciplinary and methodological boundaries. Many of these authors employ hybrid methods to ask key questions central to understanding the nature of television producers and production. These themes are not hermetically sealed, and the boundaries between much of this scholarship is fluid. There were other ways that I could have organized this chapter: chronologically, via the primary discipline of the author, by country or production community, or by the primary research method or mode of analysis. Instead, I find focusing on patterns and common constructions of production studies knowledge significantly more useful for understanding the possible approaches a scholar might adopt in a study of the makers or making of television. The following four themes touch on essential questions – location, insider/outsider status, authorship and self-theorization – that help media scholars examine the lived realities of people working in production and articulate common or compelling trends.

Location

The specific geographic location of a production work space dramatically affects – and perhaps even determines – the possibilities for which practitioners will be involved in a project, the processes of production, and what kind of television can be made. Knowledge of local media industries, both business practices and economic models, and governmental policy toward media production will determine even further the nature of media production and the players involved. Location within production studies scholarship, as I am defining it here, refers both to a broad sense of a geographic place (at the level of nation, state, region, city) and an extremely localized understanding of a specific space (district, studio, sound stage). Different scholars have mapped out these production locales, providing insight not only into the specificities of a particular work world, but also offering scholars a model for work to be continued in the future.

Early production studies research within the United States trained its eye on Hollywood, even before television entered millions of American homes and television production moved out West (see Rosten, 1941; Powdermaker, 1950). Subsequently, a substantial number of Western film and television studies of makers and making has held Los Angeles as a key site of exploration and discovery. The work of Allen J. Scott (2005) in *On Hollywood* is particularly cogent and useful for scholars studying local media spaces and the agglomeration of talent within media capitals. A geographer, Scott traces the shift of television production from New York to Los Angeles, as well as the conglomeration of the media industries, tracking not just the spread of media production through the city, but the rise of ancillary support services (sound stages, prop houses, talent agents) and concentrated labor pools that serve as local resources for the production community. Scott's research covers both film and television production, but his chapter 'The Other Hollywood', narrows in on the unique situation in Los Angeles, arguing that because US television is not nationally subsidized, as it is in other countries, television production has become increasingly corporatized and co-dependent on other entertainment media based in Los Angeles, namely the film industry. For Scott, media production cannot be dissociated from its locale; in mapping the space of the city, Scott's argument about the agglomeration of media production and its subsequent susceptibility to erosion has begun to occur (p. 9). What Scott offers to television scholars is a skillful mapping of the culture of television production that can serve as an informative insight to those scholars working within the same geographic space – or as a model for much research that could be followed into other spaces of media production.

Narrowing down further from place to space, the single television set can offer a comprehensive view of the inner workings of practitioners: a view into production culture that can, with some adjustment, apply as a model for scholarship into other comparable work spaces. Barry Dornfeld's (1998) *Producing Public Television, Producing Public Culture*, Laura Grindstaff's (2002) *The Money Shot* and Elana Levine's (2001) 'Toward a Paradigm for Media Production Research', focus their attention on unique sites of media production, closely tracking the production process through case studies of a single television program or genre. In the process of researching their subjects, these authors immersed themselves into the world of media production, and in articulating their findings they apply arguments steeped in a language that they have pulled from both the social sciences and the humanities. These authors cross methodological boundaries in order to understand the nature of a particular media production as it tracks the production environment across the localized spaces of the boardroom to the sound stage.

Given the nature of television production today, oftentimes the boardroom and the sound stage are thousands of miles apart. A number of scholars have examined the cultural and economic meanings of place in production labor given the growing number of regional production communities developing outside media capitals as well as the increased globalization of media production, for example Eva Novrup Redvall's (2013) *Writing and Producing Television Drama in Denmark*. Serra Tinic's (2005) *On Location: Canada's Television Industry in a Global Market* and her chapter 'Borders of Production Research', examine regional production, using Vancouver's bustling media industry as a way of exploring the intersection between cultural identity, regional economics, national politics and global culture. For her subjects, the question of Canadian identity is a defining characteristic for media production and media producers. And yet, when so much of their output must be globally relevant, for Tinic's practitioners, the meaning of home or heartland is more of an idea than a specific place. As Tinic says:

> In an era of intensified global production and distribution strategies, it is worth questioning the value of continuing the cultural homogenization debate (when read as 'Americanization') and to, instead, begin exploring the possibility that an imaginary 'heartland' may actually be the transnational referent for producer in search of international audiences. (2009, p. 172)

As Tinic shows, describing the distinct nature of a production community and exploring their contribution on a regional and global level is important to understanding the shifting nature of regional production within both a national and global context.

Certain scholars' deftness in boundary crossing can be encouraging to scholars new to the area of production studies, or who might be interested in exploring regions of the globe that have been less examined. Michael Curtin's (2004, 2007) work crosses methodological borders – being media industry history and production studies – in the process applying materials from scholarship originating in economics, geography, history, cultural studies, ethnic studies and media studies, among others. First in his book chapter 'Media Capitals' (2004), and then with *Playing to the World's Biggest Audience* (2007), his book on the screen industries in China, Hong Kong, Singapore and Taiwan,

Curtin documents the history and culture of media production within local, national and global landscapes – examining the specificities of local production communities and the history of their interaction alongside the constantly changing intricate relationships of these locales to the increased scattering of media capitals such as these across the globe. In 'Media Capitals', Curtin employs the history of early television production in Chicago as a cautionary tale for the future of media production, both at the national and global level – for Hollywood and for Hong Kong. This skillful study opens the door for future scholars to continue an exchange of production knowledge, both contextualizing their scholarship within arenas already familiar to some, as well as pushing research further to explore the unique specificity of local production cultures.

The work of authors who have explored television makers and making at local, regional and national levels across the globe offers models of the possibilities of future production studies scholarship. As media capitals scatter and agglomerate our understanding of production practices and practitioners will need to be informed by scholars willing to take the time and give their attention to work being done not just at the central sites of production, but also on the margins. This understanding of the shifting relationship between the centers and the margins of media production is, in its own right, another theme that pervades scholarship on television makers and making.

Insiders/Outsiders

The conditions of television production dramatically affect the agency of individual media workers. But precisely how these production practices shape the lived experience of practitioners is transformed through the unique specificities any individual production and the makers attach to it. The relationship between gender dynamics, professional hierarchies, and amateur and professional labor have all been explored by scholars interested in the culture of television production. In different ways, these studies explain the dynamics of media production and the cultural significance of holding a privileged position within a particular production system.

The use of the terms 'insider' and 'outsider' not only applies to a person's position in relation to the major television industries. Scholars within the field of production studies have theorized these terms of labor stratification, learning in the process the complex hierarchies of community and culture within television production. John Caldwell (2008) distinguishes between the rituals of production as expressed by those above and below the line in his book *Production Culture*. The distinction between 'above-the-line' and 'below-the-line' work within the industry itself point to just one of the ways in which certain media practitioners are in positions of greater professional and cultural advantage – or have 'insider' status into the major media productions and the culture of power and celebrity. Above-the-line and below-the-line are professional terms derived from a particular worker's position in relation to a bold horizontal line on a standard production budget sheet between creative and technical costs, establishing a hierarchy that stratifies levels of creative and craft labor. According to David Hesmondhalgh (2007) in *The Cultural Industries*, those practitioners who work above-the-line are responsible for generating symbolic meanings, but this reification of 'the symbolic', and of those in the foreground, speaks to hierarchies of cultural value that get compensated at rates significantly higher than below-the-line wage labor.

As television scholars have examined, questions of discrimination by sex and the gendering of professional roles are at the crux of these questions of hierarchies and status within the television work world. Through ethnographic study and field research, scholars have found that gender dynamics can become a guiding principal in the daily lives of television workers. Andrew Painter's

(1996) 'The Telerepresentation of Gender in Japan', provides a somewhat disturbing account of the troubled gender dynamics of a Japanese television studio, and how the patriarchal environment of the company were often reflected in the narratives of the series that the company produced. The women who served as managers dressed in an androgynous style, attempting to play down their sexual difference from the rest of the management team and, as Painter observes:

> As a consequence of their attempts to deflect attention from gender within the production department, the women did not protest the ways in which they were treated inside the station or gender was depicted on TV. Intent on achieving harmony with their male workmates, women directors let many things pass that they probably objected to personally. (1996, p. 51)

The notion that behind the scenes narratives of production culture might mirror public discourse is not unique to Painter's study. Carolina Acosta-Alzuru's (2003) 'Tackling the Issues: Meaning Making in a *Telenovela*' examines storylines related to gender-based violence and the representation of sexual dynamics. Her study suggests that the series' writers and actors negotiated a discourse on these topics that contributed to larger public debates on these controversial social issues. Masculinity can play a role in television production as well, and as Vicki Mayer's (2008) study on video pornographers shows, while the subject of women's sexuality is front-and-center with amateur pornography cameramen, discourses surrounding masculinity are performed in their behind-the-scenes interactions that prove to be equally manipulative, complex, and hierarchical – and worthy of scholarly examination.

The distinction between professional and amateur production is gaining increasing importance as the nature of television production transforms. Some early research has been initiated on questions of film school production cultures and the professionalization of students (see Seiter and Banks, 2007; Caldwell, 2008). Laura Grindstaff's (2002, 2009) work on television talk show and Reality TV production explores the role of the amateur performer within television production and how their non-professionalism, or as Grindstaff refers to it, their 'ordinariness' is being used by media makers as the content from which to produce series. In her work, she challenges readers to confront ideas of who within a series can be a producer. Grindstaff notes that distinctions between actor and producer, as well as between above- and below-the-line labor do not allow for the nuances of power hierarchies within these different communities of media makers (Grindstaff, 2009, p. 75). This in many ways gets at the root of questions that categorize another theme within production studies scholarship, that of authorship within television.

Authorship

As discussed briefly in the introduction, the idea of a television maker must be interrogated. Terms like 'producer' or 'maker', identify a specific category of media practitioner that is in a position of power and creative control. The use of these terms elevates the significance of some practitioners while marginalizing other practitioners in a production process that is by nature collaborative. Vicki Mayer, John Caldwell and I (2009) note in our introduction to *Production Studies*, 'grounded in organizational hierarchies of media labor, and reinforced by "auteur" studies of film and television producers as the "authors" of their creative projects, the notion of authorship is both one of subjective identification and outsider objectification' (p. 7). The idea of an individual author of a text speaks as much to the production process as it does to the desire for the author or scholar to identify and attribute talent and distinctive work to an individual practitioner.

Theories of authorship of a media text, or auteurism, can be traced back to the work of young scholars, filmmakers and cinephiles in post-war France. As auteurism developed, especially as filmmakers and scholars turned their attention to American cinema, the auteur came to define the kind of filmmaker whose creative vision and inspired cinematic techniques were manifest in spite of the restrictive, bounded nature of the studio system. In other words, the auteur's command over the text overrides the collaborative nature of film production, let alone the industrial conditions of production. Auteurism is still a central theory in film criticism and theory. While it has had many critics and in many ways it has been debunked, scholars continue to valorize those auteur-worthy television producers, television showrunners, even television programmers whose artistry and skill transcends the conditions of the industry within which they work. Susan Boyd-Bowman's (1990) work contrasts the differences between French and British television production, explaining how the 'pride of place' within British television's credit sequences has privileged the producer over the director (p. 62). Horace Newcomb and Robert Alley's (1983) *The Producer's Medium* applied auteur theory to American television producers. Through a series of essays and interviews, Alley and Newcomb demonstrate that there are distinctive similarities between the work of the television producer and cinema's director, in that, in television, it is the producer who is the primary person responsible for deciding on consistency and character, as well as tone, style and content – and that a great producer could establish an oeuvre as individual as that of the cinema auteur. In defining the role of the producer as an 'artist' within the systemic constraints of American commercial television, they write:

> Generally, our cultural vision of the 'creative artist' is closely bound to the image of the solitary, hungry, alienated genius ... Many of the interviews recorded here indicate a struggle for autonomy, freedom, power. They indicate definitions of success measured primarily in terms of radically individual distinction and they lament the necessity for resemblance and repetition. .... Instead of seeking to define creative television production in terms of distinction alone, then, we seek to define creativity in the context of the choric. (1983, pp. 37–38)

Thus Newcomb and Alley even further ground their definition of the auteur in relation to an economic system that often undermines their creative power in favor of resemblance, repetition, or flow. Thus, within the chorus of television, the producer-auteur might stand out with less flair as simply – and yet powerfully – distinctive or recognizable. A case study of corporate auteurism is presented in Jane Feuer, Paul Kerr and Tise Vahimagi's (1984) *MTM: 'Quality Television'*. By outlining the industrial, material and formal qualities of the product produced out of the independent Hollywood studio MTM, Feuer, Kerr and Vahimagi make a convincing argument for how one studio's individual artistry and creative vision stands out for its representation within the texts produced out of the studio of a distinctive tone, style, content and character, later characterized as 'quality television' (p. 56).

Understanding the cult of the auteur provides a perspective on professional hierarchies within the media world – a world which denies the collaborative nature of television, not just from the above-the-line creative types, but also the below-the-line technicians, craftspeople, or investors who all work together to produce the media text. Though sometimes below-the-line craftspeople are acknowledged, for the most part the work of the majority of television's makers slips by on the end credits – and sometimes they are not even listed there. These below-the-line laborers (technicians, craftspeople, designers) play necessary, if often unglamorous, roles in the production of television. While much of this work is anonymous because of the sheer number of crew members on a set or workers within the network and studio offices virtually prohibit acknowledgement of their individual labor, there are particular professions that are purposefully or systematically under-acknowledged – or left unacknowledged. To do otherwise would not just upset the 'magic' of the medium, but would in fact upset the balance of the highly manufactured image that a series, studio, or press-packet is designed to sell to audiences about television makers. Furthermore, celebrating the auteur is central fan discourse: and therefore auteurism owes as much to the adoring fan as it does – by default – to the industry.

The idea of the producer as artist is not necessarily the norm, especially outside the American context. Jeremy Tunstall's (1993) *Television Producers* tracks the history of the profession within the BBC to three origins: the radio producer, the film producer and the civil servant (p. 3). In Tunstall's research, producing entails honed skills, management experience, and economic responsibility. What many scholars who study makers have done is define their terms. Grindstaff (2009), like Tunstall, explicates the precise meanings behind a number of types – or hierarchies – of producers (pp. 74–75). Vicki Mayer's (2011) *Below the Line* examines four tropes of television production that often carry assumed positions of power or agency as seen through the work of a particular television practitioner: creative/creativity (set assemblers in the Brazilian Amazon); professional/professionalism (soft-core Reality videographers); sponsor/sponsorship (Reality show casters); and regulator/regulation (public access citizens). Each case study turns on its head our understanding of these positions of centrality and power, forcing scholars to reconsider the meaning of terms like maker and making.

Self-theorization

As discussed throughout this chapter, the study of makers and makings demands an integration of methods and theoretical insights pulled from the social sciences and humanities. But also, a scholar's research findings must include the language, lay theories, and self-assessment of the makers themselves. Scholars examining the makers and making of television must ground their work in a deep understanding of the lived realities of practitioners, which includes not just empirical data, but an examination of the self-theorization of makers about their own creative and craft labor. Studies of makers and the routines and rituals of production must map the economic, political and cultural forces that influence media-making and balance them against the ways in which media producers represent themselves and their professional processes.

This theme within production studies scholarship emerges in part from the work of anthropologist and ethnographer Clifford Geertz (1973), as laid out in his essay 'Thick Description: Toward an Interpretive Theory of Culture'. Geertz argues that one cannot comprehend human behavior without understanding the context within which the behavior is expressed. In the case of television studies, the language and practices of media makers cannot be deciphered unless the researcher can also read the culture of production. As Todd Gitlin (1983) says in the introduction to his groundbreaking study, *Inside Prime Time*:

> I want to convey not only how and why I think the networks do what they do, but a sense of the ambiance and texture of the industry's life-as-it-is-lived. For this reason I recount a number of stories at some length, and often in the words of the people telling them. Anecdote is the style of industry speech, dialogue its body, and narrative its structure ... I think these stories, once scrutinized and interrogated, are the royal road to the industry's working. (p. 14)

As can be seen in this quotation, this kind of research demands proximity to the subject of inquiry – and, ironically, a process of distancing. The vast numbers of making-of documentaries, DVD commentaries and journalistic interviews offer access in a pinch, but ideally production studies scholars enter into the spaces and places of production themselves so that this type of thick description and analysis can occur.

The opportunities for insight and understanding proliferate with this grounded, empirical scholarship. John Caldwell (2008) in *Production Culture* examines film and video practitioners' deep industrial practices and industrial reflexivity, in other words their reflexive expressions of identity, critique, ritual and routine. Caldwell harvests these embedded industrial theorizations of the cultures of production and production culture from media practitioners at every stage of the production process. What Caldwell shows so expertly in his scholarship is an ability to tease out the lay theories of production, pointing to the varied and contested nature of self-representation and self-theorization (see Glaser, 1998; Seiter, 1999). Caldwell does not take his subjects at their word, but rather explores their words and examines the use of language and the meaning of particular terms to a specific production community. The aim of this work is not to uncover an authentic truth about the processes of production, but rather to examine the cultural and industrial theorizations derived from practitioners' observational, insider perspectives. A subject's experiential knowledge provides a corrective testing of academic, research-based assessments. Production studies both privilege and interrogate the subject interview, placing the researcher in dialogue with their subjects.

Notions of the desired audience circulate among industrial practitioners as part of their self-theorization, guiding their process of making a media text. This happens twice: both as they design their project and attempt to target their desired viewer, and retrospectively, when they assess the success or failure at grasping the audience and their viewing behaviors and interests. Jane Shattuc (1997) in *The Talking Cure* argues how the production process of the television talk show attempts to construct a specific type of femininity that will engage with their programs. And yet, as Shattuc explains, an actual audience of women is not a part of this theorization:

> The concept of a knowable audience of women is ultimately a fiction used by producers and advertisers to support their self-ascribed ability to attract specific audiences for specific producers and to charge the advertisers of those products accordingly ... James Ettema and Charles Whitney point out: 'the measured audience is produced by research companies, sold to the media channels and bought by the advertisers.' From this perspective, the female audience is a self-perpetuating industry construct that really needs no real women. (p. 48)

This industrial theorization of the audience is of particular interest to production studies scholars who attempt to engage with their audience and speak to their interests and concerns through issue-oriented content and the building of a fan base. Using Nick Couldry's (2003) *Media Rituals*, Candace Moore (2009) explores how media producers attempt to choreograph lesbian fan experience of the television series *The L Word*, incorporating them into the production process. This integration of the fan helps media makers frame their plan for a series. What Moore argues is that this ritual of controlled interaction and observation of group viewing allows producers to conceive of a supportive spectatorship and control fan activity. This kind of fan-approved surveillance serves the needs of both the makers and the fans who wish to gain access – or at least proximity – to the production process. This theme of self-theorization within the work of production studies reveals precisely how important it is to analyze not just the relationship between maker and audience, but between practitioner and scholar.

CONCLUSIONS AND CONTINUATIONS

Much of this chapter has focused on the necessity of disciplinary integration and a series of boundary crossings: between theoretical methods, production locales, insiders and outsiders, scholars and practitioners. One boundary not explored as of yet within this discussion is that of the medium of television itself. The technologies of television are in a state of extraordinary flux. In an era of media industry conglomeration, convergence culture and transmedia storytelling, not only is articulating medium specificity increasingly difficult, but the nature of media production is transforming in dramatic ways as well. It is the work of future media scholars to start tracking and delineating the way these technological changes have affected the lived experience, working conditions, and production processes of media makers.

The potential for interactivity in convergent media that is seen as empowering for users has produced an enormous amount of uncertainty and anxiety for media professionals around the globe. With the economic downturn that started in the fall of 2008, there has only been an increase in what scholars have already seen as a rise in contract labor and outsourcing of production, changes in employment patterns that speak to a division of labor that does not just cross media borders but international boundaries as well (Miller et al., 2005; Hesmondhalgh, 2007). The globalization of production has not been as empowering to workers as it has been to the users of media.

These shifts in television production complicate our view of the makers and making of television. Each one of the themes identified in this chapter is, in many ways, turned on its head with current transitions in television and media convergence. When media is produced across international boundaries or when labor is outsourced regionally or globally, what then can we define as a national production? Our understanding of media production, with the rise of global franchising of television and the work scholars have done to track the translation processes of media producer, is still quite new (see Parks and Kumar, 2002). But how are these cultural adaptations complicated by multi-platform storytelling? Where does a production exist if the workers who create it are spread across many locations? The terms insider and outsider are also made less clear by this globalization. The patterns of insider and outsider status have always been complex, and John Caldwell (2008) has shown in his work that insiders pose as outsiders as much as outsiders are invited in during trade rituals to access insiders. Given the nature of media convergence, fan production, how will these boundaries continue to blur – or redefine themselves – in the light of these new paradigms of what can be considered media work? Notions of authorship and proprietary labor are becoming increasingly difficult to discern as viewers are regularly urged to become authors of fan-based digital media, too. Henry Jenkins, Sam Ford and Joshua Green (2013) explore the relationship between producers and consumers and the rise of fan participation in the production process. These digital spaces of simultaneous co-production with television have shifted our understanding not just of cultural production, but the economies of media production as well. With all of these shifts, the language of media production will inevitably continue to change. A new lexicon, not just of technologies, but perhaps even of job descriptions, will continue to appear as media shifts and changes over the coming decade. Mark Deuze's (2007) *Media Work* proves helpful in tracking the contemporary terrain of production professionals, but as labor practices are revised given the changing state of technology and the economy, we will need to study the ways in which media practitioners imagine and describe their labor and their experiences within this rapidly transforming work world.

The bibliography that follows provides a list of references on makers and making in television. These books and articles offer readers not just findings specific to their subjects, but also methods that can be applied to other topics, and themes that could open up explorations of new subjects. While crossing over these academic divides as we explore media is theoretically and methodologically complicated, the insight we can gain through this process is inevitably worth the struggle. As the object of television transforms before our eyes, the possibilities for research topics expand and diversify. In mapping the study of makers and making, we see both the complexities of the production process, but also must track its evolution: geographically, creatively, technologically, industrially and professionally.

NOTE

1 Semioticians were the first scholars to talk about a piece of literature, art or media not as a 'work',

but rather as a 'text'. This choice to use the term text adds prestige to media often considered popular or low-brow, but also addresses ideas that the object of study constantly shifts in relation to its production and reception. See Roland Barthes, 'From Work to Text' (1977, pp. 155–164), as well as Barthes' discussion of 'readerly' and 'writerly' texts in *S/Z* (1975).

FURTHER READING

Bielby, William T. and Denise D. Bielby. (1999). 'Organizational Media of Project-Based Labor Markets: Talent Agencies and the Careers of Screenwriters'. *American Sociological Review* 64(1): 64–85.

Born, Georgina. (2005). *Uncertain Vision: Birt, Dyke and the Reinvention of the BBC*. London: Vintage.

Caldwell, John Thornton. (2006). 'Cultural Studies of Media Production: Critical Industrial Practices'. *Questions of Method in Cultural Studies*. Eds Mimi White and James Schwoch, 109–153. Malden, MA: Blackwell.

Caldwell, John Thornton. (1995). *Televisuality: Style, Crisis, and Authority in American Television*. New Brunswick, NJ: Rutgers University Press.

Cantor, Muriel G. (1984). 'Review: The Perils of Prime-Time Politics'. *Contemporary Sociology* 13(4): 417–419.

Cantor, Muriel G. (1971). *The Hollywood TV Producer: His Work and His Audience*. New York: Basic Books.

Cantor, Muriel G. and Cheryl Zollars, eds. (1993). *Current Research on Occupations and Professions: A Research Annual*, Vol. 8. Greenwich, CT: JAI Press.

Couldry, Nick. (2000). *The Place of Media Power: Pilgrims and Witnesses of the Media Age*. New York and London: Routledge.

D'Acci, Julie. (1997). 'Nobody's Woman? Honey West and the New Sexuality'. *The Revolution Wasn't Televised: Sixties Television and Social Conflict*. Eds. Lynn Spigel and Michael Curtin, 72–93. New York: Routledge.

Elliott, Philip. (1972). *The Making of a Television Series: A Case Study in the Sociology of Culture*. Ed. Jeremy Tunstall. London: Constable.

Ettema, James S. and D. Charles Whitney, eds. (1994). *Audiencemaking: How the Media Create the Audience*. Thousand Oaks, CA: Sage Publications.

Gamson, Joshua. (1998). *Freaks Talk Back: Tabloid Talk Shows and Sexual Nonconformity*. Chicago: University of Chicago Press.

Ginsberg, Faye. (2006). 'Ethnography and American Studies', *Cultural Anthropology* 21(3): 487–495.

Ginsberg, Faye, Lila Abu-Lughod and Brian Larkin, eds. (2002). *Media Worlds: Anthropology on New Terrain*. Berkeley and Los Angeles: University of California Press.

Gitlin, Todd. (1979). 'Prime Time Ideology: The Hegemonic Process in Television Entertainment'. *Social Problems* 26: 251–266.

Grindstaff, Laura and Joseph Turow. (2006). 'Video Cultures: Television Sociology in the "New TV" Age'. *Annual Review of Sociology* 32: 103–125.

Hartley, John, ed. (2005). *Creative Industries*. Malden, MA: Blackwell.

Havens, Timothy. (2007). 'Universal Childhood: The Global Trade in Children's Television and Changing Ideals of Childhood'. *Global Media Journal* 6(10).

Havens, Timothy, Amanda D. Lotz, and Serra Tinic. (2009). 'Critical Media Industry Studies: A Research Approach'. *Communication, Culture & Critique* 2(2): 234–253.

Henderson, Felicia D. (2011). 'The Culture Behind Closed Doors: Issues of Gender and Race in the Writers' Room'. *Cinema Journal* 50(20): 145–152.

Hesmondhalgh, David. (2009). 'Politics, Theory and Method in Media Industries Research'. *Media Industries: History, Theory, and Methods*. Eds Jennifer Holt and Alisa Perren. Malden MA: Blackwell.

Jenkins, Henry. (2006). *Convergence Culture: Where Old and New Media Collide*. New York and London: New York University Press.

Jenkins, Henry, Tara McPherson and Jane Shattuc, eds. (2002). *Hop on Pop: The Politics and Pleasures of Popular Culture*. Durham, NC: Duke University Press.

Jensen, Joli. (1984). 'An Interpretive Approach to Culture Production'. *Interpreting Television: Current Perspectives*. Eds William Rowland and Bruce Watkins, 98–118. Beverly Hills: Sage.

Johnson, Catherine. (2005). *Telefantasy*. London: British Film Institute.

Khalil, Joe and Marwan M. Kraidy. (2009). *Arab Television Industries*. London: BFI.

Kraszewski, Jon. (2010). *The New Entrepreneurs: And Institutional History of Television Anthology Writers*. Middletown, CT: Wesleyan University Press.

Lent, John. (2001). 'Overseas Animation Production in Asia'. *Animation in Asia and the Pacific*. Ed. John Lent, 239–245. Bloomington: Indiana University Press.

Levine, Elana. (2009). 'National Television, Global Market: Canada's Degrassi: The Next Generation'. *Media, Culture & Society* 31(4): 515–531.

Levinson, Richard and William Link. (1981). *Stay Tuned: An Inside Look at the Making of Prime-Time Television*. New York: St Martin's Press.

Lotz, Amanda D. (2007). 'The Promotional Role of the Network Upfront Presentations in the Production of Culture'. *Television & New Media* 8(1): 3–24.

Lotz, Amanda D. (2007). *The Television Will Be Revolutionized*. New York: New York University Press.

Marc, David and Robert J. Thompson. (1992). *Prime Time, Prime Movers: From I Love Lucy to L.A. Law – America's Greatest TV Shows and the People Who Created Them*. Boston and London: Little, Brown and Company.

Mazzarella, William. (2003). *Shoveling Smoke: Advertising and Globalization in Contemporary India*. Durham: Duke University Press.

Newcomb, Horace and Amanda D. Lotz. (2002). 'The Production of Media Fiction'. *A Handbook of Media and Communication Research*. Ed. Klaus Bruhn Jensen, 62–77. New York: Routledge.

Newman, Michael Z. and Elana Levine. (2011). *Legitimating Television: Media Convergence and Cultural Status*, 38–58. New York: Routledge.

Paul, Alan and Achie Kleingartner. (1994). 'Flexible Production and the Transformation of Industrial Relations in the Motion Picture and Television Industry'. *Industrial and Labor Relations Review* 47(4): 663–678.

Paul, Alan and Achie Kleingartner. (1996). 'The Transformation of Industrial Relations in the Motion Picture and Television Industries: The Talent Sector'. *Under the Stars: Essays on Labor Relations in Arts and Entertainment*. Ed. Lois Gray and Ronald Seeber, 156–180. Ithaca: ILR Press.

Raphael, Chad. (1997). 'Political Economy of Reali-TV'. *Jump Cut* 41: 102–109.

Rifkin, Jeremy. (2005). 'When Markets Give Way to Networks', *The Creative Industries*, Ed. John Hartley, 361–373. Malden, MA: Blackwell.

Sito, Tom. (2006). *Drawing the Line: The Untold Story of the Animation Unions from Bosko to Bart Simpson*. Lexington, KY: University Press of Kentucky.

Sullivan, John L. (2007). 'Marketing Creative Labor: Hollywood "Making of" Documentary Features'. *Knowledge Workers in the Information Society*. Ed. Catherine McKercher and Vincent Mosco, 69–83. Lanham, MD: Lexington Books.

Turow, Joseph. (1997). *Media Systems in Society: Understanding Industries, Strategies, and Power*, 2nd edn. New York: A.B. Longman.

Wasko, Janet. (2003). *How Hollywood Works*. London: Sage.

Whitney, D. Charles and James S. Ettema. (2003). 'Media Production: Individuals, Organizations, Institutions'. *A Companion to Media Studies*. Ed. Angharad Valdivia, 157–186. Oxford: Blackwell.

REFERENCES

Acosta-Alzuru, Carolina. (2003). 'Tackling the Issues: Meaning Making in a *Telenovela*'. *Popular Communication* 1(4): 193–215.

Barthes, Roland. (1977). 'From Work to Text', *Image-Music-Text*. Trans. Stephen Heath, 155–164. New York: Noonday Press.

Barthes, Roland. (1975). *S/Z*. Trans. Richard Miller. New York: Noonday.

Bordwell, David, Janet Staiger and Kristin Thompson. (1985). *The Classical Hollywood Cinema: Film Style and Mode of Production to 1960*. New York: Columbia University Press.

Boyd-Bowman, Susan. (1990). 'Television Authorship in France: Le Réalisateur'. *Making Television: Authorship and the Production Process*. Ed. Robert J. Thompson and Gary Burns. New York: Praeger.

Caldwell, John Thornton. (2008). *Production Culture: Industrial Reflexivity and Critical Practice in Film and Television*. Durham, NC: Duke University Press.

Couldry, Nick. (2003). *Media Rituals: A Critical Approach*. New York and London: Routledge.

Curtin, Michael. (2004). 'Media Capitals: Cultural Geographies of Global TV'. *Television After TV: Essays on a Medium in Transition*. Eds Lynn Spigel and Jan Olsson, 270–302. Durham: Duke University Press.

Curtin, Michael. (2007). *Playing to the World's Biggest Audience: The Globalization of Chinese Film and TV*. Berkeley, CA: University of California Press.

D'Acci, Julie. (2004). 'Cultural Studies, Television Studies, and the Crisis in the Humanities'. *Television After TV*. Eds Lynn Spigel and Jan Olsson, 418–446. Durham: Duke University Press.

D'Acci, Julie. (1994). *Defining Women: Television and the Case of Cagney & Lacey*. Chapel Hill, NC: University of North Carolina Press.

Deuze, Mark. (2007). *Media Work*. Malden, MA: Polity.

Dornfeld, Barry. (1998). *Producing Public Television, Producing Public Culture*. Princeton: Princeton University Press.

Feuer, Jane, Paul Kerr and Tise Vahimagi, eds. (1984). *MTM: 'Quality Television'*. London: British Film Institute.

Geertz, Clifford. (1973). 'Thick Description: Toward an Interpretive Theory of Culture'. *The Interpretation of Cultures: Selected Essays*, 3–30. New York: Basic Books.

Gitlin, Todd. (1983). *Inside Prime Time*. New York: Pantheon Books.

Glaser, Barney G. (1998). *Doing Grounded Theory: Issues and Discussions*. Mill Valley, CA: Sociology Press.

Grindstaff, Laura. (2002). *The Money Shot: Trash, Class, and the Making of TV Talk Shows*. Chicago: University of Chicago Press.

Grindstaff, Laura. (2009). 'Self-Serve Celebrity: The Production of Ordinariness and the Odinariness of Production in Reality Television'. *Production Studies: Cultural Studies of Media Industries*. Eds. Vicki Mayer, Miranda J. Banks, and John T. Caldwell, 72–86. New York: Routledge.

Hesmondhalgh, David. (2007). *The Cultural Industries*. London: Sage.

Holt, Jennifer and Alisa Perren, eds. (2009). *Media Industries: History, Theory and Method*. Malden, MA: Wiley Blackwell.

Jenkins, Henry, Sam Ford and Joshua Green. (2013). *Spreadable Media: Creating Value and Meaning in a Networked Culture*. New York: NYU Press.

Johnson, Richard. (1986). 'What is Cultural Studies Anyway?' *Social Text* 16: 38–80.

Levine, Elana. (2001). 'Toward a Paradigm for Media Production Research: Behind the Scenes at General Hospital'. *Critical Studies in Media Communication* 18: 66–82.

Mayer, Vicki. (2011). *Below the Line: Producers and Production Studies in the New Television Economy*. Durham: Duke University Press.

Mayer, Vicki. (2008). 'Guys Gone Wild? Soft-Core Video Professionalism and New Realities in Television Production'. *Cinema Journal* 47(2): 97–116.

Mayer, Vicki, Miranda Banks, John Caldwell, eds. (2009). *Production Studies: Cultural Studies of Media Industries*. London: Routledge.

Miller, Tony, Nitin Govil, John McMurria and Ting Wang. (2005). *Global Hollywood 2*, London: British Film Institute.

Moore, Candace. (2007). 'Having it All Ways: The Tourist, the Traveler, and the Local in The L Word'. *Cinema Journal* 46(4): 3–23.

Moore, Candace. (2009). 'Liminal Places and Spaces: Public/Private Considerations'. Production Studies: Cultural Studies of Media Industries. Eds. Vicki Mayer, Miranda J. Banks, and John T. Caldwell. New York: Routledge.

Newcomb, Horace and Robert S. Alley. (1983). *The Producer's Medium: Conversations with Creators of American TV*. New York: Oxford University Press.

Painter, Andrew. (1996). 'The Telerepresentation of Gender in Japan'. *Re-imagining Japanese Women*. Ed. Anne E. Imamura, 46–72. Berkeley: University of California Press.

Parks, Lisa and Shanti Kumar, eds. (2002). *Planet TV: A Global Television Reader*. New York: New York University Press.

Peterson, Richard A. and Narasimhan Anand. (2004). 'The Production of Culture Perspective'. *Annual Review of Sociology* 30: 311–334.

Powdermaker, Hortense. (1950). *Hollywood the Dream Factory*. New York: Little, Brown and Company.

Redvall, Eva Novrup. (2013). *Writing and Producing Television Drama in Denmark: From The Kingdom to The Killing*. New York: Palgrave Macmillan.

Rosten, Leo C. (1941). *Hollywood: The Movie Colony, the Movie Makers*. New York: Harcourt Brace.

Scott, Allen J. (2005). *On Hollywood: The Place, The Industry*. Princeton, NJ: Princeton University Press.

Seiter, Ellen. (1999). *Television and New Media Audiences*. Oxford: Oxford University Press.

Seiter, Ellen and Miranda Banks. (2007). 'Spoilers at the Digital Utopia Party: The WGA and Students Now'. *Flow* 4, December.

Shattuc, Jane. (1997). *The Talking Cure: TV Talk Shows and Women*. New York: Routledge.

Tinic, Serra. (2009). 'Borders of Production Research'. *Production Studies: Cultural Studies of Media Industries*. Eds Vicki Mayer, Miranda Banks and John Caldwell, 167–172. London: Routledge.

Tinic, Serra. (2005). *On Location: Canada's Television Industry in a Global Market*. Toronto: University of Toronto Press.

Tunstall, Jeremy. (1993). *Television Producers*. London: Routledge.

10

The Division of Labor in Television

Michael Curtin and Kevin Sanson

Division of labor is a concept that rarely appears in the literature of television studies. Indeed, labor itself receives comparatively little attention. Perhaps this is because many critics see television as a social phenomenon, thinking less about where TV comes from than what role it plays in politics, culture and home life. Moreover, the humanities tradition of television studies is steeped in textual criticism and tends to pay more attention to what is on the screen than what is behind it. To the extent that scholars consider labor issues, they tend to focus on the creativity of actors, directors, showrunners, journalists and musicians. Entertainment news and mainstream media likewise focus on the personal lives and creative activities of celebrity talent. Those with less exalted jobs, such as videographers, costume designers and lighting technicians, operate outside the spotlight, but they nevertheless tend to be portrayed as a creative class rather than as hourly laborers toiling away in a factory or a sweatshop setting. Both scholars and the general public tend to see television employees as a labor elite, working in artistic, sometimes glamorous settings, where they earn top wages and enjoy high levels of job satisfaction. Tellingly, that's an illusion that major television studios have always tried to promote: television is both fun to watch and fun to make. Why else would so many applicants vie for the very few jobs that the industry has to offer?

Yet the conditions of labor in the television industry have changed dramatically since the late 1980s and today many feel that their work is increasingly burdensome, bureaucratized, and poorly compensated. Their jobs also are insecure, with most workers experiencing regular bouts of unemployment or underemployment throughout their careers. Numerous factors are driving these changes, but most prominent among them are corporate conglomeration, globalization and technological innovation. Each will be addressed in due course, but first we will define what we mean by the 'division of labor', connecting our discussion of television labor to more general theories about the social relations of production. Then we will show how conditions today in the television industries are tied to a longer trajectory of labor relations reaching back to the early years of radio and cinema. We will begin with a discussion of US media as a case example and then turn to the globalization of television labor, considering its implications for workers and cultures around the world.

Thousands of years ago, humans began to divvy up labor within their families, clans and communities. No doubt the delegation of tasks was based on the perceived abilities of various workers, but it was also interwoven with the power relations among members of each group. Men might plow the fields, children plant seeds, and women harvest. In such cases, the relations of production, however unequal, were expressive of other social relations and were readily apparent to all involved. At a broader level, divisions of labor also existed between communities. Some villages might specialize in particular crafts or products that could be traded with other communities. Certain cities emerged as centers of trade and as places where diverse craftspeople might cluster (Jennings, 2011). During the 19th century, sociologist Emile Durkheim (1893) went so far as to say that the division of labor was fundamental to human existence, but he, like many others, believed it was undergoing a dramatic transformation in the modern era, especially with the expansion of factory labor.

Writing in the early days of the Industrial Revolution, philosopher Adam Smith (1776) noted that the specialization and coordination of workers within a factory was crucial to the improvement of productivity and profitability. Smith saw factories as an ingenious investment of capital that blended human labor and machinery within an organized space, resulting in an integrated and elaborate manufacturing technology. This was for him emblematic of capitalism's invisible hand, forever spurring efficiencies and improvements. Smith also contended that some locations enjoyed comparative advantages that made them more efficient producers of particular products, encouraging expanded networks of trade and increasing complexity in the relations of production. Yet Smith also expressed concern that highly specialized and repetitive factory labor could dull the minds of workers.

Karl Marx (2007) took up this latter concern, arguing that routine and specialized labor isolated workers from each other, undermining their sense of social connection and eroding their knowledge of the overall production process. Reduced to unskilled, hourly wage laborers, they traded their time for meager pay, no longer able to sustain themselves outside the factory setting and no longer able to recognize the fruits of their labor in the finished products that rolled out the factory door. This alienation of labor and the relentless downward pressure on wages was, according to Marx, due to the *process* of capital accumulation. What Smith saw as the ingenuity of capitalism, Marx saw as the unquenchable thirst of industrial capital to seek ever-greater returns by pitting the interests of an investor class against those of the workers. Capital used factories to increase efficiencies but also to conceal the relations among workers and thereby insure their compliance to a system that did not serve their interests. The division of labor secured both the economic and social dominance of the capitalist class. This trend would grow during the 20th century as entrepreneurial capitalism gave way to corporate capitalism, which enhanced the role of a professional managerial class that refined the tools of planning, surveillance and control over the productive apparatus (Braverman 1974; Chandler, 1993).

Thinking beyond the shop floor, Adam Smith's (and David Ricardo's (1817)) notion of comparative advantage among producers was reassessed by Marx and later Vladimir Lenin to account for the rise of transnational enterprises that both extracted raw materials from distant lands and exploited local labor, thereby creating spatially extended but nevertheless integrated chains of production, linking colonial societies to the manufacturing operations of the industrializing West. This international division of labor (IDL) prevailed for over a century, but it began to unwind as factory workers organized, demanding better wages and working conditions. In response, capital developed new transportation and communication technologies that made it possible to shift manufacturing to the 'developing world' where wages and other costs were much lower.

The transnational deployment and integration of finance, design and manufacturing operations began to take off during the 1970s, creating what some economists refer to as a new international division of labor (NIDL). 'Flexibility' and mobility are key features of the NIDL, as huge corporations scour the globe in search of efficiencies and cost advantages in manufacturing, and increasingly in service industries as well (Frobel, Heinrichs and Kreye 1980; Harvey, 1990). Miller et al. (2005) extend this analysis to show that media now operate under a New International Division of *Cultural* Labor (NICL), in which Hollywood continues to act as a dominant force, but major media conglomerates exploit their newfound mobility to refigure the geography of creative labor, thereby economizing on production costs and undermining the power of labor organizations.

Thus, the division of labor in television refers both to the social relations of the workplace but also the relations of labor worldwide. It can furthermore be thought about as delimiting a broader set of social and cultural relations within societies, defining who speaks, who watches, when, where and about what.[1] It follows then that studying the division of labor necessarily requires attention be paid to political-economic structures and everyday interactions amongst employers and employees, teasing out the ways in which market forces, institutional dynamics, social ideologies and the professional practices of television workers shape labor relations in any given context. Such questions have been taken up across a range of disciplines, from organizational and management studies to political economy, cultural studies and the emergent field of critical media industry studies. Yet, the object is an elusive one, often written about in limited or indirect ways with little regard for historical antecedents, medium specificity, or a comprehensive grasp of the particular power relations that structure the production and distribution of television content. Moreover, some of the recent, and in our opinion, strongest accounts (for example, Hesmondhalgh and Baker, 2011) don't fully consider in their projects how globalization has both dismantled and reconstituted the division of labor in myriad ways, oftentimes reshaping center–periphery relationships amongst competing locations of media production.

Although the history of the division of labor in US television is a limited case example, we'll discuss it first before considering how the division of labor plays out across a broader geographic terrain. Our rationale is that US commercial media were early innovators in technology, programming, production and labor organizing. They established patterns and protocols that have been adopted or adapted by others around the world, especially since the 1980s as the neo-liberal wave of privatization swept through national economies in other parts of the globe. Today, most media institutions are either capitalist enterprises or they are public institutions that have been pressured to adopt many of the commercial, stylistic and professional practices of their peers.

Interestingly, the workplace routines of US television first took hold more than twenty years before the medium arrived. During the 1920s, American radio broadcasters established a successful business model that television networks later adopted wholesale: broadcasters provided regularly scheduled programming to a national audience via a network of local television affiliates, and did so in order to 'deliver' a critical mass of viewers to the sponsors and advertising agencies that funded the cost of production. Furthermore, radio – inspired by vaudeville and theater – developed a durable set of programming models, focused especially on the live broadcast of dramas and variety shows from New York City. Radio also offered television its first cadre of celebrities, as many radio stars were lured to the new medium by the promise of higher salaries and greater national exposure.

Live radio proved remarkably popular but also exceptionally demanding, as the voracious appetite for novel performances, storylines, gags and music meant that performers and production staff were forced to come up with seemingly fresh material on a weekly if not daily or even hourly basis. Although much of the work was creative or at least craft-like, the pressures to produce were not unlike factory conditions: the assembly line rarely stopped and due to fierce competition, workers were constantly pressed to increase both their efficiency and creativity (Barnouw 2003; Thurber, 1948). Efficiency required standardization via genres and formats, while creativity demanded variations that could distinguish a program, hopefully engendering reliable if not singular popularity with listeners and advertisers. The programs were carefully monitored by ratings agencies and advertising firms that assigned standards of excellence according to the size and enthusiasm of each program's audience. In the eyes of the workers, the output was a show well made, but in the eyes of sponsors the core concern was the size of the audience that any given program could assemble. That is what sponsors paid for, so, in essence, the audience became a commodity, as the social relations between performers and listeners were converted into abstract numbers (ratings), which became the foundational unit of exchange between radio broadcasters and advertisers. Moreover, listeners became part of the productive apparatus, purchasing their own receivers in order to listen to programming supposedly at their leisure but ultimately at the discretion of advertisers. Audience labor therefore became twinned with the creative labor of broadcast employees (Meehan, 1990; Smythe, 1981). Consequently, creativity, popularity and entertainment were all reconstituted through an industrial and managerial order that divided and organized the labor to insure efficiency and obscure the social relations among workers, and between workers and their audiences. This division of labor allowed corporate owners to control the workforce and extract the surplus value of their labor.

Although the division of labor primarily served capitalist interests, it also was embraced by workers who sought to distinguish themselves as creative artists. On-air talent usually joined craft guilds in their areas of expertise, eventually forming alliances with their West Coast counterparts in Hollywood. Scriptwriters affiliated with the Writers Guild and performers with the American Federation of Radio Artists (later AFTRA). These workers usually negotiated with program producers that were either employed by advertising agencies or working under contract with them. Producers worked with sponsors to cultivate the concept for a show and then assembled the performers, writers and directors to execute that vision. The performance troupe therefore had a separate existence outside the network corporation, so when it came to labor negotiations this creative elite sought to distinguish itself from stagehands or clerical assistants, negotiating contracts that usually set minimum compensation standards that could be parlayed into even higher pay packages by especially talented or renowned individuals. The guilds that represented them also thought it important to negotiate separately, since many of their members were hired for specific programs and would therefore be terminated if the show were cancelled. Guilds tried to provide benefits that could tide their members over during inevitable bouts of unemployment and also tried to set boundaries as to who could lay claim to the title of writer, actor or director. Their contracts therefore looked considerably different from those of staff members at a network studio who might spend an entire career at the same facility, servicing whatever shows were scheduled for broadcast.

Although sponsors and their ad agencies produced most radio programs, the shows were performed and broadcast out of network studios staffed by sound editors, effects artists, electricians, and other skilled employees. Some of these workers joined craft guilds, but most became members of industry-wide labor organizations, such as the International Brotherhood of Electrical Workers (IBEW), the National Association of Broadcast Employees and Technicians (NABET), and the United Electrical, Radio, and Machine Workers of America (UE), bringing employees together in collective bargaining units that wielded power along the lines of the Congress of Industrial Organizations (CIO). Sound mixers, stagehands and clerical staff negotiated as a combined unit, seeing their collective influence over the production infrastructure as providing more negotiating leverage than their specific job skills. This form of collective bargaining was very similar to the labor strategies used in the automobile, garment and steel industries. Interestingly, these unions often connected radio workers to counterparts outside the broadcast industry. For example, UE represented some radio station employees as well as factory workers who manufactured radio equipment. Moreover, organizing at major network facilities – where most of the program production took place – in turn led to organizing drives at local stations, bringing together workers from many parts of the radio industry but also making it more challenging to represent their diverse interests.

The intensive demand for hours of programming each day necessitated a detailed division of tasks amongst all involved in a radio production, which was further solidified by union work rules that restricted employees to their respective areas of specialization. Although this in many ways limited a worker's range of creativity, it also ensured against attempts by management to curb costs by eroding the status and pay of skilled workers. By the end of the 1930s, the division of tasks was clearly defined and labor groups secured respectable compensation, due in large part to the robust prosperity of the industry and a general surge in union influence throughout the US economy at that time. Both network and station managers seemed well aware that labor unrest could be especially costly for an industry that faced insatiable production demands and one that furthermore rested its reputation on projecting a spirit of communal engagement. Radio invited itself into the home and therefore needed to be wary that it should not appear to be a source of turmoil or unease.

Labor relations in the radio industry were therefore relatively stable until commercial television upset the applecart during the 1950s, growing by mid-decade to become the most popular and profitable mass medium in the US. This transition was a site of immense contestation and negotiation among key players. Networks, sponsors, advertising agencies, producers and labor organizations battled over issues that ranged from creative control to the most appropriate textual forms for television's mass audience.[2] In the process, they also established a distinct set of production practices and power relations that set television broadcasting apart from its forbearers.

Perhaps the most decisive change in the broadcasting industry was the introduction of telefilm recordings, which eventually moved the center of gravity for primetime program production to the West Coast and into the embrace of Hollywood producers. But the transition took time, as some major studio executives expressed reluctance about involving themselves in what they saw as a competing medium and others held back because the majors were at the time going through a period of intensive reorganization in the wake of a 1948 anti-trust ruling that ordered the break-up of the vertically integrated Paramount Studios. At the very moment that television was taking off, major Hollywood studios were required to sell their theater chains, which in turn sparked staff lay-offs, capital equipment reductions, and the outsourcing of many production tasks to small, independent enterprises that specialized in such services as editing, make-up and set construction (Christopherson and Storper, 1989).

Therefore independent producers and minor studios were first to warm to the new medium, seeing it as a market niche left open by their larger competitors. Yet, much of this early telefilm production – underfunded, uncoordinated and largely speculative – rarely made it onto the broadcast networks' schedules. This was because sponsors wielded an enormous amount of power to dictate the terms and conditions of telefilm production during the early 1950s, much as they did in the radio model, and they were leery of the inconsistent output from unknown telefilm producers. They therefore were reluctant to finance telefilm series, especially since live broadcasts remained a highly popular and less expensive alternative (Anderson, 1994, 59). When they did opt for telefilm, sponsors offered to cover only a portion of production costs in exchange for network or first-run syndication rights, while the producer retained subsequent distribution rights for domestic and foreign sales. Ultimately, deficit financing – still very much a common financial arrangement in television production today – placed a serious financial burden on producers who had to hope that a series could be exploited in secondary markets after the network run.

The enormous financial success of *I Love Lucy* (1951–1957) became the aspirational benchmark for all telefilm producers. It also helped to draw the major studios into the telefilm business and it caught the attention of network executives who had previously showed little interest in participating in the risky business of entertainment program production. Networks began to take an ownership interest in studio productions and established cozy working relationships with key suppliers. Successful producers flourished by wagering higher and higher budgets (and thus deficits) to compete for coveted spots on primetime schedules. This inflated production costs and marginalized undercapitalized competitors. Telefilm production simply became too expensive and technologically complex for producers on the margins, especially those who lacked the financial reserves and human capital necessary to sustain budget deficits, deferred profits and relentless production schedules. It also posed a significant financial burden to sponsors. Single sponsored series quickly disappeared, as the multiple-sponsor model became the norm for television. Broadcasters successfully wrestled scheduling control away from sponsors and advertising agencies who could no longer afford the risk, and used their newfound power to create a cohesive primetime schedule (encapsulated by the newly created position of programming chief) as a tactic to maximize their appeal to viewers and affiliates, and thus increase revenues (Schatz, 1990, 125–126).

By the middle of the decade, the major film studios joined established independents, like Desilu and Four Star Productions, as the primary producers of the networks' entire primetime schedules. For the studios, telefilm production was an opportunity to overcome the postwar economic woes that emerged from the demise of vertical integration and shifts in consumer spending. Telefilm production was a cost-effective way to revive activity on empty backlots and a new source of income to offset the financial risks associated with blockbuster production strategies. This period saw a return to the factory-like efficiencies and outsized profits of Hollywood cinema's golden age, and a workforce generally compliant with the restrictive labor relations that helped ensure the studios' confidence in their television ventures (Anderson, 1994).

After the wave of studio downsizing during the early part of the decade, workers embraced the move to telefilm, even if it meant onerous working conditions, as was the case at Warner Bros. where the mode of production was overseen by a few key producers who coordinated all creative and operational aspects of telefilm production. Such tight administrative control – maintained through restrictive labor contracts and other policies, like formulaic scripts and standardized writing, directing and editing – kept costs at a minimum and streamlined production practices. It also placed an enormous burden on talent, craft and crew, who worked long hours for below-standard pay, turning out the equivalent of an episode per day. Moreover, the studio fully owned all of the series it produced, effectively barring producers from claiming original ideas and making the issue of residual payments moot. All profit was studio profit. In fact, the studio demonstrated an astonishing degree of shortsightedness in relation to the economics of television, valuing the network license fee as a source of immediate income instead of looking to the syndication market as a form of long-term investment. In short, this was a culture of risk aversion and cost containment that disproportionally taxed the studio's workforce to benefit its bottom line, an early episode in the history of corporate encroachment upon creative endeavor that foreshadowed future events.

Of course, not all producers organized their labor in the same way. Four Star, for example, was more open to engaging independent writer-producers, granting them a significant degree of creative autonomy to develop and oversee their own productions and share in a portion of the studio's profits (Anderson, 1994, 260–262). This strategy helped boost production at Four Star, and proved equally lucrative for writers who penned successful series. Nevertheless, it didn't stop the trend toward disintegration. Term contracts remained the norm, which meant employees were hired by a show and released when it was cancelled. Regular bouts of unemployment became common and therefore refocused the efforts of labor organizations. Unions and guilds placed greater emphasis on negotiating residuals (for above-the-line) and benefits (for below-the-line) to meet the needs of workers who were no longer under long-term contract with major studios and thus required the means to subsist when recurring work wasn't available. They also focused more diligently on the representation of workers within particular specializations who might float more easily between productions. Ultimately, such factors helped sustain the division of labor and normalize the divergent (and unequal) working conditions for different job categories and functions, further obscuring the social relations of production as both natural and inherent aspects of workplace distinctions (Stahl, 2009).

With the shift away from live broadcasts to telefilm production firmly established, and the programming pipeline successfully transplanted to Hollywood, the networks had fully eliminated the sponsor and advertising agency from production and programming decisions by the early sixties. Networks further solidified control over the schedules by taking an ownership stake in the programs they commissioned, ending up with vested interests in nearly all of their primetime programming by the end of the decade. Eventually, in 1970, the FCC formulated the Financial Interest and Syndication Rules (finsyn) in response to mounting criticism, effectively barring the networks from owning the programs they aired or syndicated and thereby creating opportunities for a new crop of independent producers like Norman Lear and Carsey-Werner. While a major boon for producers operating outside major studios, the regulatory intervention did little to restructure the division of labor – studios, producers, and labor organizations kept intact the production practices and labor relations that they had solidified nearly two decades earlier.

Not until the emergence of cable in the eighties did the rather efficient and complacent labor relations encounter significant challenges. Specifically, the proliferation of distribution outlets, including cable, satellite and upstart network Fox, increased competition and chipped away at the broadcast networks' dominance of the national television audience (and thus their economic stability vis-à-vis declining ad revenue). In response to the economic crisis, producers and networks embraced two countervailing programming tendencies, and, in an era of deregulation and consolidation, sought to internalize those production practices into increasingly large media conglomerates that owned both studios and television distribution systems across broadcast, satellite and cable. First was a turn to more elaborate production and programming, and representational strategies – high production values and stylistic excesses that were designed to distinguish the networks from their emergent competition and reassert their key position in an increasingly fragmented marketplace (Caldwell, 1995). This strategy had little immediate impact on the division of labor except for the few instances when new digital production technologies blurred job distinctions or replaced certain work functions. In fact, the most significant threat posed to labor relations was not television's new-found penchant for elaborate style, but the inflated production costs required to sustain it – what was once considered a necessary investment to secure competitive advantage quickly turned into a financial liability for both producers and networks and a source of increasing insecurity for union labor in the television industry.

Now part of much larger conglomerate companies with diverse media holdings, network executives faced mounting pressure to curb operating and programming costs without sacrificing output, thereby obliging producers to contend with ever greater demands for content but with less financial resources to get it done. In addition to cutting staff (in some instances, by as much as 30%), production companies and studios gravitated towards low-cost reality television genres, like prime-time news magazines and tabloid television shows, because of their shoestring budgets and high resale value in syndication markets; yet, low-cost reality television even more radically challenged the existing division of labor by tapping new forms of talent onscreen (many of them amateurs) and employing non-union labor off screen in locations outside union jurisdiction (Raphael, 1997). Once again we find the creative workforce, largely below-the-line workers, forced to bear the brunt of these cost-cutting strategies and at the mercy of their corporate overlords. A series of strikes by the craft unions and guilds throughout the eighties, which culminated in the 1988 Writers Strike, was an unprecedented display of labor unrest. Yet, common logic at the time characterized the rallying cries as futile, the final death gasps of organized labor in Hollywood that could do little to dismantle a system increasingly at odds with its best interests (Caldwell, 1995, 293).

Indeed, that the writers' strike delayed the fall schedule by 22 weeks only cemented the privileged position unscripted reality programming enjoyed across broadcast and cable, a position that has arguably remained relatively stable in the intervening years. Today, reality television is often a welcomed opportunity to the legions of young workers who flock to Hollywood more concerned about getting work than they are worried about union protections (Perren, 2011). Moreover, the rise of reality television and its formal elements, like amateur talent, repurposed home videos and hidden camera footage, has also raised larger questions about the boundaries of creative labor in the television industry, drawing much needed attention to the creative work done by individuals outside the professional confines of film sets and studio back lots (see for example, Mayer, 2011). Alongside the advent of digital technologies that not only make the means of production more accessible but also more easily integrate users' interactive, participatory labor into the productive apparatus itself, there remain urgent questions about the role individual or amateur creativity plays in the process of capital accumulation in the television industry (Deuze, 2009). In fact, the rise of user-generated content has even prompted content providers, advertisers and creators to experiment with the same digital tools as marketing and branding mechanisms. Unfortunately, this experimentation has increased the demands on writers and directors, who now find themselves producing for websites and social media pages without receiving additional compensation for their time and creative output (Curtin, Holt and Sanson, 2014).

For much of the 20th century, the most influential forms of broadcasting were national and therefore the most significant division of labor was at that level. Yet, there were countervailing forces at work as well, and it's equally important to think through these issues from an international perspective. The prolific output of the major studios engendered very significant libraries of telefilm content that, having already recouped their original production costs in the US, could then be marketed internationally at elastic prices. Although the content often failed to resonate with the cultural preferences of audiences in overseas markets, Hollywood distributors offered their programming at prices that proved irresistible, in many cases undermining the motivation to produce original content at the national or local level. This consolidated Hollywood's status as the world's leading producer, ushering in a new international division of labor that positioned the major US studios at the center of television drama and scripted comedy production, while national systems increasingly focused their attention on such genres as news, talk, game and variety programming.

Furthermore, deregulated broadcasting markets and rapid privatization disproportionately benefited US-based media conglomerates whose access to new delivery technologies, like cable and satellite, allowed them to conceive of audiences at a global, rather than national, scale. While the regionalization and localization of both staff and content proved an integral tactic for success – what Morley and Robins (1995, 16) call 'learn[ing] to reconcile global ambitions with local complexity' – the emergence of a new global media order posed a fundamental challenge to the regulatory regimes in countries with a history of public service. As the political pressures of neoliberalism mounted, the 'enlightened' ethos of public service broadcasting appeared esoteric and out of sync with the persistent logic of competition and consumer choice. Soon, public service broadcasters found state subsidies a matter of political debate, and there was a growing call for them to adopt a mindset much more familiar to their commercial counterparts.

Yet, it would be a mistake to assume Western media companies were the only ones upending national broadcasting systems via new technologies and trade liberalization. As the commercial service Zee TV prospered throughout India, for example, it also developed satellite services for South Asia, the Middle East and Europe. These trends further industrialized program production in India, expanded the number of television jobs, and set the stage for new relations of production and new labor organizing initiatives. It also put in place new forms of infrastructure owned by commercial companies that exerted pressure on public service broadcasters to marketize their operations (Thussu, 2005). In Canada, similar trends led to cuts at the Canadian Broadcasting Corporation, which made broadcasters scramble for ways to utilize their productive capacity in the face of diminishing state support (Tinic, 2005). Toronto and Vancouver opened their doors to Hollywood, leading to a rush across the border, as producers sought cost savings, compliant unions and favorable exchange rates. This became a model for media markets in other parts of the world hoping to attract 'footloose' producers with tax rebates and other incentives.

Accordingly, global media scholars like Miller et al. (2005) are right to point out the emergence of a New International Division of Cultural Labor from the growing collusion among international financial institutions and regulatory bodies, national and municipal governments, and Hollywood. Criticisms along these lines reassert the hegemony of US-based media empires, highlighting their abilities to discipline a global workforce by exploiting geographical mobility. At the same time, however, this new mobility has enabled television companies in other parts of the world to ramp up the scale of their activities in other locations. It has, for example, made it possible for companies in Latin America to migrate production to Miami, and for Arab satellite televisions channels to take flight in Beirut and Dubai where they operate outside the restrictive constraints of national TV systems in Saudi Arabia and Egypt. In fact, these multi-dimensional spatial flows have enabled the formation of alternative media capitals, which not only realign center–periphery relationships, but also constitute cultural and financial operating logics of their own (Curtin, 2003). Labor relations, then, are not exclusively dismantled and reassembled in the shadow of Hollywood. Rather, an increasingly flexible production apparatus can facilitate knowledge transfer, enable skills training, and build capacity, as much as it can reaffirm prevailing power structures.

Television's division of labor is significant at a number of levels. At the most local level it structures the unit of production and the conditions under which labor organizing takes place. It also operates at a transnational level, ordering the relations between television institutions worldwide and consolidating a hierarchy of cultural power that is embedded in the programming, practices and values of the television industry. Furthermore, history structures the division of labor as a pattern of contest and concession, pitting production efficiencies and corporate ethos against creative autonomy and empowerment. Ultimately, then, television's division of labor is an effect of rationalization as much as it is social relations and power struggles. It is an unresolved, constantly shifting conflict that remains a relatively untold story in the history of television and its related media sectors.

NOTES

1 The division of labor also functions as an effect of gender and racial ideologies, which has made certain work functions more or less available to one group of people over another. See Banks (2009).
2 The 'red scare' was one of the weapons in the battle for the control of broadcasting and an especially important tool for restraining the influence of unions, guilds and politically progressive employees. See Prindle (1988).

REFERENCES

Anderson, C. (1994). *Hollywood TV: The studio system in the fifties.* Austin: University of Texas Press.

Banks, M. (2009). Gender below-the-line: Defining feminist production studies. In V. Mayer, M.J. Banks and J.T. Caldwell (Eds), *Production studies: Cultural studies of media industries* (87–98). New York: Routledge.

Barnouw, E. (2003). *The sponsor: Notes on modern potentates*. Piscataway: Transaction Publishers.

Braverman, H. (1974). *Labor and monopoly capital: The degradation of work in the twentieth century*. New York: Monthly Review Press.

Caldwell, J.T. (1995). *Televisuality: Style, crisis, and authority in American television*. New Brunswick: Rutgers University Press.

Chandler, A.D. Jr. (1993). *The visible hand: The managerial revolution in American business*. Cambridge: Belknap Press of Harvard University Press.

Christopher, A. (1994). *Hollywood TV: The studio system in the fifties*. Austin: University of Texas Press.

Christopherson, S. and Storper, M. (1989). The effects of flexible specialization on industrial politics and the labor market: The motion picture industry. *Industrial and Labor Relations Review*, 42(3), 331–347.

Curtin, M. (2003). Media capital: Towards the study of spatial flows. *International Journal of Cultural Studies*, 6, 202–228.

Curtin, M., Holt, J. and Sanson, K. (2014). *Distribution revolution: Conversations about the digital future of film and television*. Berkeley: University of California Press.

Deuze, M. (2009). Convergence culture and media work. In A. Perren and J. Holt (Eds), *Media industries: History, theory, method* (144–156). Malden, MA: Wiley-Blackwell.

Durkheim, E. (1893). *The division of labor in society*. New York: The Free Press.

Frobel, F., Heinrichs, J. and Kreye, O. (1980). *The new international division of labour: Structural unemployment in industrialised countries and industrialisation in developing countries*. Cambridge: Cambridge University Press.

Harvey, D. (1990). *The condition of postmodernity: An enquiry into the origins of cultural change*. Malden: Blackwell Publishers Inc.

Hesmondhalgh, D. and Baker, S. (2011). *Creative labour: Media work in three cultural industries*. New York: Routledge.

Jennings, J. (2011). *Globalizations and the ancient world*. Cambridge: Cambridge University Press.

Marx, K. (2007). *Economic and philosophic manuscripts of 1844*. Mineola: Dover.

Mayer, V. (2011). *Below the line: Producers and production studies in the new television economy*. Durham: Duke University Press.

Meehan, E. (1990). Why don't we count? The commodity audience. In P. Mellencamp (Ed.), *Logics of television: Essays in cultural criticism* (117–137). Bloomington: Indiana University Press.

Miller, T., Govil, N., McMurria, J., Wang, T. and Maxwell, R. (2005). *Global Hollywood 2*. New York: BFI.

Morley, D. and Robins, K. (1995). *Spaces of identity: Global media, electronic landscapes, and cultural boundaries*. New York: Routledge.

Perren, A. (2011). Producing filmed entertainment. In M. Deuze (Ed.), *Managing Media Work* (155–165). Thousand Oaks: SAGE.

Prindle, D. (1988). *The politics of glamour: Ideology and democracy in the Screen Actors Guild*. Madison: University of Wisconsin Press.

Raphael, C. (1997). The political economy of reali-TV. *Jump Cut*, 41, 102–109.

Ricardo, D. (1817). *On the principles of political economy and taxation*. London: John Murray.

Schatz, T. (1990). Desilu: I Love Lucy, and the rise of network TV. In R.J. Thompson and G. Burns (Eds), *Making television: Authorship and the production process* (117–136). New York: Praeger.

Smith, A. (1776). *An inquiry into the nature and causes of the wealth of nations*. Chicago: University of Chicago.

Smythe, D. (1981). *Dependency road: Communications, capitalism, consciousness, and Canada*. New York: Praeger.

Stahl, M. (2009). Privilege and distinction in production worlds: Copyright, collective bargaining, and working conditions in media making. In V. Mayer, M.J. Banks and J.T. Caldwell (Eds), *Production studies: Cultural studies of media industries* (54–68). New York: Routledge.

Thurber, J. (1948). Soapland I – O pioneers. *The New Yorker*, May 15, p. 34.

Thussu, D.K. (2005). The transnationalization of television: The Indian experience. In J. Chalaby (Ed.), *Transnational television worldwide: Towards a new media order* (156–172). New York: I.B. Tauris & Co.

Tinic, S. (2005). *On location: Canada's television industry in a global market*. Toronto: University of Toronto Press.

11

From Network to Post-Network Age of US Television News

Oliver Boyd-Barrett

INTRODUCTION

Some scholars have pondered the approaching 'death' of television (Katz and Scannell, 2009), taking their cue from the rise and fall of US terrestrial television network audiences – from a peak of 50 million homes in 1980, falling to 22 million in 2009. Yet this audience considerably exceeded the total reached by any one cable television channel. Pronouncements of the 'death' of television are perhaps a way of marking the decided transition from the exclusive networks of old to a 'post-network' society of abundant television, anywhere and at any time. Factors contributing to this transition have included: (1) proliferating competition of supply across platforms (including terrestrial, cable, satellite, wireless delivery to computer, phone and other mobile devices); within a framework (2) of industry concentration, corporatization and conglomeration; in addition to (3) a combination of technological advance and diminishing human resources; and (4) apparent abandonment of public service criteria in favor of market orientation and infotainment.

Numbers of Television Journalists

According to US Census and the Bureau of Labor Statistics, quoted in Weaver et al. (2007), the full-time editorial workforce in US news media in 2002 was 116,148, with the largest concentration (58,769 or 50.6%) in newspapers, followed by weekly newspapers (21,908 or 18.9%), television (20,288 or 17.5%), radio (13,393 or 11.5%), news magazines (1,152 or 1%) and news services (638 or 0.5%). Television news staffs were spread across approximately 1,300 stations nationwide. In the period 1971–2002, the total size of the full-time editorial workforce in the traditional mainstream US news had grown 67%. In absolute numbers there had been growth in all but two categories: news magazines shrank slightly, from 1,900 to 1,152 and news agency services had shrunk dramatically from 3,300 to 638. Numbers of television journalists had increased the most dramatically of all, by 190% for the period as a whole, but slipping in the 2000s.

News networks each employed 1,000–1,400 staff in 2010. The 2004 Pew report on the State of the News Media noted general declines in investment. Numbers of free-to-air (ABC, CBS, NBC) network news correspondents were down by a *third* from the 1980s, the number of foreign news bureaus had fallen by *half*, while correspondent workload was up by 30%. In local television, almost 60% of news directors reported either budget or staff cuts. Cutbacks were noted again in the 2006 report, with older journalists being replaced by younger staff. Broadcast network news had experienced a total 10% reduction of expenditure in the period 2002–2006. In cable, Fox expenditure was up 11%, CNN up by 5% while MSNBC was still cutting. Staff in local television news experienced an average size increase of 36% in this period, although journalists were expected to do more: hours of news on local television stood at a record high (3.8 hours a day). ABC News reduced its worldwide news staff from 1,500 to 1,100 in 2010. CBS reduced its staff by 70. Only one of the free-to-air networks, NBC, was robustly profitable ($400m annually), the only one of the three linked to a cable news channel, MSNBC, with which it shared costs and from which it benefited from advertising and subscription revenues (Stelter and Carter, 2010)

NETWORK HISTORY

Origins

Over several decades three principal US networks delivered audiovisual news to US citizens: they were ABC, CBS and NBC. A contender, the DuMont network, ceased broadcasting in 1956, lacking the safety-net afforded its rivals from radio network revenues. DuMont assets contributed to the assembly of the Metropolitan Broadcasting Corporation, a small network renamed Metromedia in 1961, then sold to News Corporation and Twentieth Century Fox Film Corporation in 1985, forming the Fox Broadcasting Company in 1986. This grew to become, in effect, the nation's fourth network.

Together the three majors reached almost the entire television population, constructing a 'national' portfolio of news that was highly attractive to advertisers, who provided almost all the revenue. Then, in reaction to a series of 1970s FCC deregulatory measures that favored cable, an increasing number of households converted to receiving the traditional networks within their basic-tier cable or satellite portfolios, where the networks constituted an important but tiny cluster within 100–500+ channel universes. One-time 'free-to-air' services were now increasingly bundled into cable or satellite packages that by the 2000s cost subscribers in the range of $30–$80 monthly. In this way, cable operators easily recouped the payments they made for the right to distribute network services. For audiences, the trade-off brought higher quality signals and more channels. In the early 2000s, less than one in five households continued to receive networks by rooftop antennae as opposed to coaxial cable or satellite dish delivery. The federally mandated switch from analogue to digital transmission in 2009 obliged those audiences to purchase (discounted) set-top converter boxes (sometimes requiring new antennae) if they wished to receive the digital versions of the old 'free to air' services. Many elected to switch to cable or satellite providers.

ABC

The three 'terrestrial' channels emerged under the umbrella of parent radio organizations in the 1940s, transmitting first in black-and-white and, by the mid-1960s, in color. In 1940, the FCC broke up the radio duopoly headed by CBS and RCA's NBC, obliging

RCA to abandon one of its chains. RCA's 'Blue Network', acquired by retail magnate Edward Noble in 1945, was renamed the American Broadcasting Company. Its first evening news anchors were H.R. Baukhage and Jim Gibbons (1948–1951). The ABC Television Network, broadcasting from 1948, was acquired in 1953 by United Paramount Theatres to form American Broadcasting – Paramount Theatres Inc. (shortened to American Broadcasting Companies Inc. in 1968). The new company fostered strong relations with Hollywood studios, notably Disney, and in the 1970s set up a theatrical division, ABC Pictures, producing both for cinema and television. ABC introduced *Good Morning America* in 1975. ABC was acquired in 1985 by Capital Cities Communications to form Capital Cities/ABC. In 1996, Capital Cities was taken over by Disney, which renamed its broadcasting group ABC, Inc. In 2004, it started a 24/7 channel, ABC News Now for online and mobile phone consumers. Principal news shows in 2009 included *America This Morning*, *Good Morning America*, *Good Morning America Weekend Edition*, *This Week with George Stephanopoulos*, *ABC World News with Charles Gibson*, *ABC World News Saturday & Sunday*, *20/20*, *Nightline*, *ABC World News Now*, *ABC News Brief* and *Primetime* (Auletta, 1992; Barkin, 2002; Barnouw, 1968; Epstein 2000; Goldenson and Wolf, 1991).

CBS

CBS Broadcasting Inc, previously known as the Columbia Broadcasting System, a radio network, was established in 1928. It had only one television station in 1950 but grew steadily during the following decade. *CBS TV News* was introduced in 1948, extending from 15 to 30 minutes in 1963 (a move soon imitated by NBC and ABC). The first evening news anchor was Douglas Edwards (1948–1962). Walter Cronkite began as anchor in 1952 and was succeeded by Dan Rather in 1981. Edward Murrow, with producer Fred Friendly, converted CBS's *Hear it Now* to *See it Now* from 1951.CBS launched *60 Minutes* in 1968. The company was acquired by Westinghouse Electric Corporation in 1995, later coming under the control of Viacom in 2000. Viacom had started as a spin-off of CBS in 1971. In the late 2000s, Viacom's principal owner, Sumner Redstone, split the company into two: CBS Corporation, with the CBS television network at its core, and the new Viacom. But both companies continued to be controlled by Sumner Redstone through National Amusements. News shows in 2009 included *CBS Morning News*, *The Early Show*, *The Saturday Early Show*, *CBS News Sunday Morning*, *Face the Nation with Bob Schieffer*, *CBS Evening News with Katie Couric*, *CBS Evening News with Jeff Glor* (Saturday edition), *CBS Evening News with Russ Mitchell* (Sunday edition), *60 Minutes*, *48 Hours Mystery*, *Up to the Minute*, *CBS NewsBreak* (twice-daily 90-second daytime broadcast) and *CBS News* (Auletta, 1992; Barkin, 2002; Barnouw, 1968; Epstein, 2000; Paper, 1987).

NBC

The National Broadcasting Company (NBC) was formed in 1926 by Radio Corporation of America (RCA), which owned 50%, and its partners General Electric (30%) and Westinghouse (20%). The first NBC network television broadcasts began in 1939. The first evening news anchor was John Cameron Swayze (1948–1956), succeeded by Chet Huntley and David Brinkley (1956–1970). *NBC Today*, the first television morning news magazine show, began in 1952. Ownership of NBC passed to General Electric in 1986 when General Electric purchased RCA. GE sold off the NBC Radio Network. By the late 2000s, amidst a gathering crisis in the news industry, there was evidence of growing disenchantment with NBC within GE, and negotiations commenced with the nation's largest cable company, Comcast, to purchase a major part of GE's shares in NBC. Principal news programs in 2009 were *Dateline NBC*, *Early Today*, *Meet the Press*, *NBC Nightly News* and *Today* (Auletta, 1992; Barnouw, 1968; Barkin, 2002; Epstein, 2000).

The major terrestrial networks (and Fox, established in 1986 and comparable in scope, by the 2000s, to the Big Three) are not the only players. Smaller networks, such as CW Network (established 2006, and owned by CBS and Warner Bros.), typically comprise small numbers of affiliate stations that may receive network programming for only parts of the day. They include Spanish-language or Hispanic networks (e.g. Univision) catering to the growing size and purchasing power of the Latino/a population.

The Big Three Model

The Big Three sustained the traditions of their radio predecessors that had been established by the 1934 Communication Act, and transitioned from radio to television on the basis of those earlier structures. These comprised nuclei of network owned-and-operated stations based in the nation's principal markets, and larger clusters of affiliate stations, independently owned or owned by chains, that took most of their programming and advertising from the networks while adding local news (and advertising) and some other content. Local television news in 2008 was watched regularly by over half of the US population. A typical late newscast reached 12% of viewers watching television in a given market (but down from 21% in 1998) (Stelter, 2008). News represented about 40% of the average local station's revenue, contributing – at least until the economic downturn of 2008 – to the very high rates of annual return (sometimes 50%) on television investment, and accounting for the competitive influence of local television news on television news gathering generally.

Together the networks represented a model of broadcasting ownership, control and operation that was commercial, plural (though tending towards oligopolistic) and dependent for almost all revenue on advertising. In this way, the constitution of US broadcasting policy represented a diametrically different approach to that of most of Europe, Asia, Africa and the Middle East (Latin America was mixed), where, for the first few decades of broadcasting, governments adopted one of two models: (1) outright state-controlled and often state-financed monopolistic systems, or (2) 'public' broadcasting systems, as in Western Europe, financed indirectly by the state, by license fee or by some other, generally non-commercial, measure intended to sustain a 'Chinese wall' between the state and the broadcaster. These systems were generally not dependent, at least initially, on advertising. In many such countries there has been increasing convergence since 1990 towards the US model – namely, towards a plurality of commercial stations, diminishing state involvement, and increased dependence on advertising.

From the opposite direction, and less dramatically, there has been a measure of convergence in the US toward the 'public' and, we might say 'public sphere', European model. First, there were stipulations built into the 1934 Communications Act and subsequent legislation that broadcasters should serve (an ill-defined) 'public interest'. Criteria of 'public interest' were established on a case-law basis. The news divisions of the major networks provided a public spirited face to an otherwise highly commercial operation. Their primetime evening newscasts were typically fronted by patriarchal white male anchors whose performances projected authoritative but avuncular, judicious impartiality within a framework of loyalty to nation and the flag. Morning shows were generally lighter mixes of information and entertainment, featuring more teamwork. The first evening woman co-anchor was Barbara Walters, paired with Harry Reasoner on ABC in 1976; Connie Chung was appointed co-anchor to Dan Rather on CBS in 1993. Neither of these matches was happy. Katie Couric was the first female appointed as sole anchor, in 2006, for CBS. She was joined by

Diane Sawyer for ABC in 2009, leaving one male, Brian Williams, on NBC.

The 1996 Telecommunications Act in effect redefined public interest in terms of market competition, which it stimulated primarily by permitting access for telephony operators into television distribution and vice versa. Regulatory developments from the 1980s up to and including the 1996 Telecommunications Act, removed many 'public interest' conditions that had been imposed subsequent to the 1934 Communications Act. These included the 'fairness doctrine', whose purpose was to ensure a balance of viewpoint in matters of controversy. (Surviving 'equal time' regulations require that the voices of competing electoral candidates be heard in the measure to which their parties had gained voters' support in previous comparable elections.).The 'syn-fin' rules restricted the freedom of networks to syndicate their original programming. Other regulations had restricted: (1) the share of the national market that any network could capture with the stations that it owned; (2) the number of television stations any network could own outright (network owned-and-operated stations constitute small proportions, less than 20%, of the entire networks, and are concentrated in the large markets); and (3) multiple ownership of media properties, and cross-ownership of print and broadcasting outlets in the same markets. The 1996 Act lifted such restraints more than a notch, although later FCC attempts at further relaxation were successfully resisted by Congress (Alexander et al., 2003; Croteau and Hoynes, 2005).

US Public Broadcasting

The US 'public' broadcasting tradition was established by successful passage of the Public Broadcasting Act of 1967, which set up the Corporation for Public Broadcasting (CPB). Supported mainly from state and nonprofit local sources, the CPB developed, but did not own, a public broadcasting network in radio (including National Public Radio, NPR) and television (Public Broadcasting Service, PBS). PBS, launched in 1970, had grown to 354 affiliate stations by the 2000s (larger than any of the Big Three or Fox). These exercise collective ownership of the network.

PBS owns no stations. Many PBS stations are nonprofit enterprises attached to such entities as state, health or educational (university) institutions. Unlike commercial networks whose affiliates depend heavily on network programming, PBS stations take most of their content from third party suppliers. Stations pay for both PBS and third party feeds. A PBS policy of 'common carriage' requires most member stations to adhere to the national primetime programs on a common schedule. All content is created by or in conjunction with third parties, including member stations, although PBS retains television distribution rights. CPB, which part funds both local stations and suppliers, obtains 15–20% of its annual operating revenue from federal sources, 25–29% from state and local taxes, and 53–60% from pledge drives, donations and grants (including a substantial corporate sponsorship). In 2005, CPB received $464m from Congress and $650 in donations from individual Americans.

The 1967 Act required strict PBS adherence to objectivity and balance, and prohibited federal interference in or control of programming content. Who should define what these principles mean in practice, or how government could guarantee them without simultaneously contravening them, are unresolved. CPB chair Kenneth Tomlinson (2003–2005) attempted to use CPB resources to eliminate what he perceived to be a liberal bias. He forced the resignation in 2005 of Bill Moyers, celebrated anchor of *Now with Bill Moyers*, although Moyers soon returned for the *Bill Moyers Journal*. Already countering the weight of Bill Moyers, however, was conservative commentator Tucker Carlson and, for a while, the *Journal Editorial Report*, anchored by an editor of the Wall Street Journal's editorial page (who subsequently moved to the right-wing Fox News channel).

Leading primetime public affairs programming from PBS includes *Frontline*, *NOW on PBS*, *The Newshour with Jim Lehrer*, *The Charlie Rose Interview show*, and the *Nightly Business Report* (Engelman, 1996).

CNN

The development of cable and satellite television has had major implications for both the business and content of television news. Cable News Network (CNN), the world's first 24-hour television news channel, was founded in 1980 by Atlanta station owner Ted Turner who in 1976 had uploaded his local Atlanta station (WTCG-TV-Channel 17) to satellite, making it available for multiple downloads by cable operators. This occurred a decade following the launch of the world's first commercial satellite, Early Bird, by COMSAT in 1965, one year after HBO became the first nationally-syndicated cable channel, and three years ahead of C-SPAN, ESPN and Nickelodeon. At the time of writing, in 2014, CNN was owned by the Turner Broadcasting Systems division of the media conglomerate Time Warner. Its two major products are CNN for domestic news consumers and CNNI for international consumers. Alongside the main domestic channel, Turner launched the Headline News channel in 1982, originally under the title CNN2. This provided a 24-hour cycle of 30-minute news broadcasts. From 2005 this became Headline Prime, featuring primetime opinion anchors such as Glenn Beck and Nancy Grace. CNN facilitated the development of round-the-clock coverage of high intensity, high ratings news stories, such as the Shuttle Challenger disaster of 1986, the Jessica McClure toddler-in-the-well story of 1987, the 1991 Gulf War (an incomparable boost for CNN domestically and world-wide, offering the only television coverage of the war from inside Baghdad), the Los Angeles riots of 1992 and the O.J. Simpson trial for murder in 1995. Where CNN went, the networks followed. ABC, CBS and NBC coverage of Simpson eclipsed coverage of significant foreign events such as the war in Bosnia (Kellner, 2003). The business model for television news had transitioned from one based on high-minded public interest principles to a market model driven by consumer interest, ratings and advertising. This transformation was stimulated principally by competing news sources through cable and satellite, and the discovery by local television stations during the 1980s of new wealth through the tabloidization of local news.

High visibility CNN news programs in 2009 included *American Morning*, *CNN Newsroom*, *The Situation Room* (with Wolf Blitzer), *Lou Dobbs' Tonight*, *Campbell Brown*, *Larry King Live* and *Anderson Cooper 360*. The website CNN.com was launched in 1995. Media watchdogs such as Fair and Media Matters (Liberal), AIM and MRC (conservative) inevitably disagree as to whether CNN leans left or right. A 'CNN effect' (Cottle, 2008; Volkmer, 2005) has been postulated to denote the possible impact of CNN on pressuring governments into humanitarian responses to catastrophes such as the abandonment of Shiites in Iraq in 1991, and Western interventions in Somalia 1992 and Kosovo 1995. This may have lent a 'liberal' aura to CNN that may be true only by contrast to the predominantly conservative tone of most mainstream US television news. Thussu (2000) examined CNN coverage of NATO's bombing in 1999 – an unprecedented NATO intervention in the domestic affairs of a sovereign state. He found the bombing was reported uncritically as 'humanitarian intervention', and tended to follow the agenda set by the US military. Ackerman (2001) found that CNN was only slightly more likely than other US networks to use the word 'occupation' with reference to the Israeli occupation of Palestine. More than 90% of US network TV reporting on the occupied territories failed to report that the territories were occupied. The percentage was closer to 80% in the case of CNN. Voorhees, Vick and Perkins (2007) suggest that in coverage of Hurricane Katrina, CNN represented minorities in a negative light, ultimately reinforcing inequalities. Minorities were disproportionately cast as victims, passive or in need of help, and rarely in positions of power or expertise. Eke (2008) cites findings that coverage of the Darfur Genocide in 2005 by CNN, MSNBC and Fox News were quickly overwhelmed by competing, domestic infotainment stories.

CNN's response to the competitive threat of Fox News (which in 2009 enjoyed higher ratings than CNN but fewer unique viewers per month) has been to give a higher profile to opinionated anchors and, in the judgment of this author, move rightwards on political-economic, though not necessarily on social, issues. With occasional exceptions. such as Anderson Cooper's gentle outrage against Federal fecklessness in its response to Hurricane Katrina, the line-up of celebrity anchors could scarcely be described as 'progressive'. Some have been associated with conservative positions (e.g. Wolf Blitzer on the Israeli-Palestinian conflict; Lou Dobbs on immigration) or feature conservative anchors (e.g. Glenn Beck until he moved to Fox). Notwithstanding its coverage from Baghdad during the first Gulf War, including reports of US-inflicted carnage of civilians (e.g. the US bombing of hundreds of civilians in the el-Amiriyah air-raid shelter), it cannot be argued that CNN or CNNI has significantly opposed or critiqued US foreign policy interests. In 1999 CNN had fired foreign correspondent Peter Arnett (AP veteran of 1960s Saigon and CNN correspondent from Baghdad in Gulf War One, 1991), following the network's retraction of an Arnett report that investigated whether US forces had used nerve gas in Laos during the Vietnam War (Grey, 1999)

FOX NEWS

After CNN's debut in 1980, and the increasing carriage by cable operators of both free-to-air national networks and local stations, the next significant arrival to cable was Fox News (Whittemore, 1990). The Fox Network was established in 1986 when Rupert Murdoch (for News Corporation) and Marvin Davis (co-owner with Murdoch of 20th Century Fox Film Corp.) agreed to buy six television stations that had previously belonged to Metromedia. This was the same year that Murdoch completed his purchase of 50% of Fox Filmed Entertainment, the parent company of 20th Century Fox Film Corporation. Ten years later, in 1996, Murdoch announced the creation of a 24/7 cable news channel, Fox News. He appointed Roger Ailes to head it up. Ailes, previously with *America's Talking* (later part of MSNBC), had been a senior Republican advisor. Belying the new channel's 'fair and balanced' slogan, Ailes demonstrated that through strongly opinionated evening talk shows (including *The O'Reilly Report*, *The Crier Report*, *Hannity and Colmes*, and later shows featuring such anchors as Greta Van Susteren, Shepard Smith, Neil Cavuto, Glenn Beck) he could shift the center of television news gravity significantly rightwards and in doing so realize enviable profits and political impact. The right-wing character of Fox news was amply demonstrated by Glen Greenwald's 2004 documentary *Outfoxed* and Al Franken's book: *Lies and the lying liars who tell them*, and in copious research (e.g. Jaramillo, 2009; Thussu, 2007). Channel ratings were boosted by spectacularly gung-ho coverage of the invasion of Iraq in 2003, and coverage of the Republican National Convention in 2004. For most of the 2000s, Fox led the ratings among cable news channels (though not in cumulative audience). In October 2009, White House Communications Director Anita Dun told CNN that 'Fox News often operates almost as either the research arm or the communications arm of the Republican Party' (quoted by Melber, 2009).

Fox News was the first channel, following the demise of the 'fairness doctrine' in 1986, to leverage the ensuing opportunities for different ways of 'doing' news. The disappearance of the fairness doctrine removed even the necessity for *pretence* at being 'balanced'. The move by President Reagan's FCC to kill the fairness doctrine presumed that with the onset of digitization, cable and satellite, the age of television 'scarcity' was over. This was merely one part of the Reagan Administration's effective abandonment of the principle of media regulation for the public interest. Television could now be openly propagandistic, even if claiming not to be. Fox News's seizure of the *zeitgeist* occurred only a few years ahead of the collapse of the Soviet Union and Eastern Europe (1989–1990), the state capitalism 'turn' of China in the 1990s, and a frenzy of capitalist expansion, globalization, boom (and then bust).

Fox pushed other networks, both cable and terrestrial, in a similar though not as extreme a direction. The political center as a whole shifted (further) rightwards, assisted by 'liberal media's' uncritical buy-into neo-liberal ideology. In terms of a few social or lifestyle issues such as tolerance for racial diversity, sexual orientation and abortion, there may have been substance to the Republican attempt to paint all mainstream media as 'liberal'. But for the charge to have carried real weight, at least a part of that mainstream would have had to have been identifiably, persistently and aggressively against war and the defense industries and sympathetic to labor, progressive taxation, and measures of wealth redistribution such as free universal health care, affordable university education, the regulation of Wall Street and the dismemberment of oligopolies. I am aware of no evidence that such was the case. During most years of the Bush presidency, from 2000 to at least 2006 (and, in most respects, during the subsequent Obama administration), indeed, there was nothing remotely 'left wing' about any part of the television news establishment, therefore no hope or expectation anywhere that the Bush administration would be rigorously, persistently and in timely fashion held to account for its many proven crimes and abuses. These included but were certainly not limited to the invasions and occupations (in whole or in part) of two sovereign countries (Afghanistan, Iraq), threats of invasion against a third (Iran) on the basis of dubious evidence, the destabilization of Pakistan, the use of unmanned drones as tools for assassination, the removal of a democratically elected regime in Haiti, connivance at or participation in an attempted *coup d'état* against a democratically elected regime in Venezuela, illegal spying on US citizens, removal of habeus corpus protections, and wide-scale rendition, torture and human rights abuses against both US and non-US citizens.

MSNBC

Nothing remotely 'left-wing,' that is, until a curious 'liberal' turn at MSNBC in 2006, which belatedly propelled anchorman Keith Olbermann (*Countdown with Keith Olbermann*) to resurrect a more astringent version of CBS's Edward R. Murrow (who in 1954 had helped puncture the sleazy red-baiting of Senator Joseph McCarthy). Olbermann finally gave television voice – a literate, rational and passionate voice – to the accumulated outrage that opinion polls suggest was felt by at least half of the US population against the extreme corporatism of the Bush Administration, an Administration that had hurled the country back to levels of social inequality not experienced since the 1900s, struck a dangerously cavalier pose in the face of global environmental catastrophe and the threat of nuclear war, and embarked on an imperial rampage through the Middle East and Central Asia that it ingenuously described as a 'war on terror'. Other 'liberal' faces of the network included Chris Matthews, David Gregory and Rachel Maddow. The channel's morning show, however, was anchored from 2007 by former Republican congressman (1994–2001) and 'traditional' conservative, Joe Scarborough. Olbermann himself was forced out of MSNBC in 2010 and transferred to Al Gore's new minority liberal channel, Current TV.

MSNBC, like Fox News, was established in 1996, in a partnership between Microsoft and General Electic's NBC (now NBC

Universal). NBC Universal parents one of the three major 'free-to-air' news networks. In 2005, NBC Universal purchased a majority stake in the channel, leaving Microsoft with 18%. The two companies maintained a 50–50 partnership in their website, msnbc.com. In comparison with CNN, MSNBC stories were longer and more detailed but not evidently more 'liberal'. Conservative icon Ann Coulter appeared on one of the channel's first shows. One of its first legal analysts was Barbara Olson, who told John Dean that she hated the Clintons and used Republican talking points in her campaign for the impeachment of President Clinton. In 2003, the channel hired right-wing talk radio star Michael Savage to host a weekend show, but fired him for scandalous language. That same year, NBC and MSNBC fired Peter Arnett for saying in an interview with Iraqi Television that the US war plan had 'failed'. And in another incident that year, veteran talk show host Phil Donahue was fired as host of a nightly debate program because he was allowing 'too many' antiwar voices on the air. Then MSNBC's top rated show, Donahue aired in the same time slot as Fox News's *Bill O'Reilly*. Donahue explained that MSNBC had ordered there would have to be two conservatives for every liberal, and that he himself counted as two liberals (Moyers, 2007). Another incident occurred in 2007 when right-wing talk radio star Don Imus was simulcast on MSNBC and passed scandalous comment on Rutgers University Women's Basketball Team, prompting several advertisers to withdraw support. Additionally, a study of covert lobbying practices on network television by *The Nation* (Jones, 2010) found that MSNBC was 'the cable network with the most egregious instances of airing guests with conflicts of interest'.

MSNBC's so-called left-leaning turn corresponds with the arrival of Phil Griffin, who assumed executive oversight in 2006 (and became the network's president in 2008). His policy seems to have been to turn MSNBC into the antithesis of Fox News. That it should have taken the network a decade to formulate such an obvious response smacks of undependable, corporate desperation, not principle. Main MSNBC news shows in 2009 included *Morning Joe*, *Andrea Mitchell Reports*, *Hardball with Chris Matthews*, *The Ed Show* (with Ed Schultz), *Countdown with Keith Olbermann* and *Rachel Maddow*.

A QUESTION OF COMPLICITY

The age of network news progressively entered into decline as cable and satellite delivery overtook terrestrial television, the remote control facilitated consumer choice of what to view, the number of channels available to most homes multiplied from a handful to several hundred, and television content could be downloaded to computers and mobile devices.

It had been a worthy age, inspired by a public service ethos. While also patrician and authoritarian, it was likely preferable to some of the opinionated, partisan slanging matches that pass for television news in the 2000s on cable news networks. But it was also routinely partisan on behalf of the power elite. Rarely did it question US foreign policy until too late to make much difference. It faithfully beat the drums for (often dubious and/or illegal) imperial war through Korea, Vietnam, Panama, the two Gulf Wars, Kosovo, Afghanistan, Iraq, Iran and Pakistan, and remained studiously innocent if not ignorant through the many covert US acts of regime destabilization including US direct responsibility for, or participation in, the military dictatorship of Greece 1967–74, the overthrow of the democratically elected leader of Guatemala, Jacobo Arbenz in 1954, the overthrow of the democratically elected leader of Iran, Mohammed Mossadaq in 1953, the overthrow of the popular leader of Iraq, Karim Kassem, in 1963 (in favor, eventually, of Saddam Hussein), the overthrow of the democratically elected leader of Chile, Salvador Allende, in 1973, continuing onwards through to the overthrow of Haiti's first democratically elected leader, Jean-Bertrand Aristidt, in 2000, and support for the attempted overthrow of the democratically elected leader of Venezuela, Hugo Chavez, in 2002 (Blum, 2008).

Contrary to the Pentagon post-war rationalization, television did not 'lose' the Vietnam War. Certainly the release of iconic, revealing photos and footage of war atrocities such as the 'napalm girl' in 1972, the street execution of a guerrilla fighter in 1968, the My Lai massacre in 1968 and Walter Cronkite's gloomy prognostication after the 1968 Tet offensive likely had a profound and critical impact on many Americans. But the media broadly supported the Vietnam War. As Herman and Chomsky (2002) argue, US news media may have questioned the *strategy* for war, but failed to hold America's leaders accountable for a war in which America had no legal presence and which it fought on the basis of an ever-shifting series of false or contestable pretexts, extending their transgressions covertly into Laos and Cambodia, at hideous cost in terms of regime destabilization, human life and suffering. America's media, led by the broadcast networks failed to expose anything like the full fraud of the war until after Daniel Ellsberg's leak of the Pentagon Papers in 1971, by which time public opinion was turning solidly against the war.

Media complicity with the initial invasion and *de facto* occupation of Vietnam was the stimulus for Herman and Chomsky's celebrated propaganda model (see the first chapter of their 1988 book *Manufacturing Consent*), and for the works of Daniel Hallin (1986) and Norman Solomon (2007), among many others. Similar critical and compelling analyses are available of almost every US overseas military engagement (for example, see Douglas Kellner (1992, 2003) on both the first and second Gulf Wars). Scholarly dissection of network television coverage of war, and of media coverage of war more generally, is copious (among many additional sources see Allan and Zelizer (2004), Hoskins (2004) and Thussu and Freedman (2003)). The work of Douglas Kellner (1992, 2003) on network coverage of both the first and the second Gulf Wars has been particularly insightful, and frequently scathing.

Kellner's (1992) study of US network coverage of the first Gulf War confirms and exemplifies many of the observations of Hallin's (1986), and Herman and Chomsky's 1989 study of media coverage of Vietnam. As in Vietnam there were obfuscations as to the *causes* of war. Was it really about Iraq's threat to Kuwait (which had once been part of Iraq)? This was a threat whose importance US diplomats had seemed to downplay in communications with Hussein. Was it about cutting Iraq down to size following the defeat of Iran in the Iraq-Iran war (the US had egged on Saddam and played both sides), softening it up for an eventual return of Western oil interests to Iraq? Rather than adhering to skeptical, questioning, historically aware self-distancing from US authority that the mythology of an independent press might suggest as appropriate, network anchors and journalists predictably undertook the role of cheerleading on behalf of US forces and the Administration. Networks uncritically endorsed most of the Administration's 'Big Lies'. These included: the allegation that Hussein was preparing to invade Saudi Arabia (probably hatched to convince the Saudis to allow US troops on Saudi soil); the 'babies taken out of incubators' invention of Hill and Knowlton, a public relations company representing the Kuwaiti government in exile; the claim that the Iraqi air force and its scud missiles were annihilated on the first night of attack; assertions as to the efficiency of the very inefficient Patriot missiles; the claims that the bombing atrocities of a baby milk plant and a civilian bomb shelter in Baghdad had been in fact legitimate targets; the myth of 'precision bombings' that were in fact generally imprecise and accounted for only a small proportion of all bombing; and the claim that oil spills in the Gulf were effected by Hussein when they were in fact the result of US bombing.

Kellner attributed to networks the function of 'conduiting' – uncritical, unquestioning and frequent repetition of official Administration, Pentagon claims. It is demonstrated in Kellner's analysis of network coverage of both the 1991 and 2003 Gulf wars and of the war in Afghanistan (Kellner 1992, 2003) and in many other studies (including Hallin (1986) on Vietnam and Jaramillo (2009) on Iraq). The networks afforded scant attention to sources that favored alternatives to war. The antiwar movement was rarely seen or heard on television, even though in practice it was well organized, large and vocal. The war was personalized for mass consumption as being a battle of wills between President Bush and Hussein (who was more often referred to by his first name, Saddam, faintly resonant of 'Satan', while the president was rarely, if ever, called 'George'). Hussein was regularly compared with Hitler, a far graver and more powerful threat to the Western world. There was very little contextualization: the media did not ask why the US had not stepped in when Iran invaded Iraq in the eight-year Iran-Iraqi war; why it had not intervened when Israel invaded the Lebanon in 1988; why it tolerated Israeli occupation of Palestine; why it tolerated Israeli nuclear weaponry.

Most journalists who covered the war were organized into press 'pools' where selected correspondents reported the war on behalf of groups of media. They were largely dependent on the military for access and transportation and were subject to the guidance of 'minders'. Stories deemed problematic could be subject to lengthy delays while reviewed by military censors. Infotainment frames were adopted from pleasure-inducing Hollywood narratives and other dramatic strategies. Networks relied primarily on pro-Administration sources, including veteran military experts – an earlier manifestation of the phenomenon that came to prominence in the second Gulf War, namely retired generals masquerading as 'Television pundits', but, unknown to the audience, pre-approved and pre-briefed by the Pentagon, and sometimes doubling as defense consultants. The networks worked to create the illusion of real-time presence, while obscuring the realities of censorship and technology. Truth was often lost in (misleading) numbers of sorties, hits, troops killed and wounded, etc. The enemy was often accused of crimes of which the Americans were guilty, such as violations of the Geneva Convention. The horror of war was sanitized both through control of the image and through language: the enemy was 'engaged', bombs were 'dropped ordnance', civilian deaths were 'collateral damage', targets were 'assets' and 'assets' were 'visited', 'acquired', 'taken out', 'suppressed'. War planes were 'force packages' whose routine engagements in wholesale and cruel slaughter over large areas were sanitized as 'carpet bombing'.

Many of these issues resurfaced in network coverage of the invasion of Afghanistan in 2001 and of Iraq in 2003, where the mildly complicit 'press pool' ceded to egregious military stage-management of coverage through the 'embed' system (Jaramillo, 2009). The scandal of 'Pentagon television pundits' persisted, along with evidence of routine appearance of lobbyists, the identity of their employers hidden from viewers. Barstow (2008) chronicled how, before the US invasion of Iraq in 2003, the Pentagon nurtured retired generals as 'expert' consultants to television networks. The generals benefited from access to top sources and expense-free travel that they sometimes used in association with work for defense companies to whom they were also consultants. Television networks seemed incurious about the practice, even though their use of such sources dated back at least to the first Gulf War (Kellner, 1992). A former CNN chief news executive acknowledged in 2003 that CNN allowed the Pentagon to vet its military analysts. A four-month investigation by *The Nation* (Jones 2010) found evidence of egregious abuse of public trust by the cable networks who routinely invited highly-paid lobbyists on to programs without disclosing their affiliation. Sometimes these lobbyists parleyed the opportunity to interview news sources into an occasion for pitching their corporate 'sell'.

> Since 2007 at least seventy-five registered lobbyists, public relations representatives and corporate officials – people paid by companies and trade groups to manage their public image and promote their financial and political interests – have appeared on MSNBC, Fox News, CNN, CNBC and Fox Business Network with no disclosure of the corporate interests that had paid them. Many have been regulars on more than one of the cable networks, turning in dozens – and in some cases hundreds – of appearances. ... Since then, guests with serious conflicts of interest have popped up with alarming regularity on every network. ... Frequent television news commentators are also often given access to policy-makers, who may find that they are meeting with not just a TV pundit but also a paid lobbyist. (Jones, 2010)

These lapses suggest a collapse of rudimentary professional ethics and signify routine, unaccountable complicity between the networks and corporate, military and political interests. Whereas it was once possible to argue (Meyrowitz, 2009) that the arrival of television demystified the Presidency and the political process, we might conclude that in the 2000s, television has obfuscated the nature and methods of real power in society. Many sources have chronicled and critiqued network failure to adequately present to American audiences an account of recent American wars. I will suggest only four, namely Chossudovsky (2005) on the Taleban, Griffin (2008, 2009) on 9/11 and Bin Laden, Scott (2008) on the war on terror, and the BBC documentaries of Adam Curtis (particularly *The Power of Nightmares*) (2004). From these and other sources, we can reasonably argue, among many other things, that the networks failed to investigate: the Administration's history of relations with the Taleban; the extent to which Afghanistan had been reconfigured by American and Pakistani support of the mujahadeen during the 1980s war against the Soviets; US corporate energy interests in oil and gas pipelines across Afghanistan from the Caspian; the complex relationship between US, Pakistani and Saudi intelligence, on the one hand, and Bin Laden, multiple Afghan warlords, and the partly-fictitious entity of Al Qaeda, on the other. They failed to interrogate the many unanswered and unsettling questions that surrounded 9/11. They failed to hold the Administration properly to account for the Bush Administration's startling 2002 doctrine of pre-emptive warfare. They failed to interrogate and expose in timely fashion the Administration's deliberate manipulation of media and public opinion to falsely leverage 9/11 as the justification for a long prepared attack on Iraq, and its peddling, on the basis of highly dubious sources, of a mythology of 'weapons of mass destruction'. They played along with the Administration's staged public relations events such as Bush's 'Mission Accomplished' speech, the destruction of the Saddam Hussein statue, and the 'rescue' of Private Jessica Lynch. They reported some of the worst atrocities of their nation's history with all the bravery of liveried courtiers before their King: these included the looting of historical artifacts, Bremer's reckless disbandment of the Iraqi army, the brutal detention and killing of large numbers of Iraqis on flimsy or nonexistent pretexts, illegal rendition, the tortures of Abu Ghraib, Baghram Air Base and Guantanamo, the 'friendly fire' shooting death of sports hero Pat Tillman, the deaths of US servicemen from faulty electrical wiring at the hands of Halliburton, and the reckless killings of civilians by mercenary contractors such as Blackwater. Media frequently and misleadingly presented the US occupiers as a neutral hand protecting different Iraqi sects from one another. For far too long they maintained the façade of a puppet Iraqi regime as an 'independent' government. Patriots seeking to oust the occupiers were routinely described as 'insurgents'. Media failed to expose or give voice to legitimate public outrage against war profiteering, beginning with Bush family involvement with defense contractor Carlyle, through the obscene profits of no-bid hand-outs to contractors like KBR, to the CIA-linked, mercenary regime of Blackwater (later Xe, now Academi). In short, they shied timidly from the central controversies of their times. In Afghanistan even into 2009 they continued to support without qualification the debatable proposition (Griffin, 2009) that Osama Bin Laden was a main culprit behind 9/11 (it is Khalid Sheikh Mohammed, after years of torture in US captivity, who is the one actually charged with being the mastermind and whose trial, in 2014, still appeared to be far from completion) or that he was still alive (the USA claims to have killed him in 2011), and media struggled fecklessly to find meaning in a war in which, even on the US military admission, Al Qaeda was no longer a significant presence (Reid, 2009).

TRENDS OF CHANGE IN THE NEW MILLENNIUM

Growth of Infotainment

The US television broadcast and cable networks bear a unique responsibility for the development of market-driven 'infotainment' television and its export around the world, a process that includes: physical export and dissemination of television documentary products (e.g. CNNI, The Discovery Channel, the History Channel, the proliferation of 'reality' television shows); sale of program 'formats'; multinational advertising that pushes channel producers and mangers towards infotainment by enhancing the importance of ratings; US investment in the ownership of television channels and programs; and, increasingly, local imitation of US entertainment business practices.

Thussu (2007) has reviewed the rise of television news infotainment. He quotes a 1998 Project for Excellence in Journalism study of US mass media over two decades, tracking a shift towards 'lifestyle, celebrity, entertainment and celebrity crime/scandal', and away from government and foreign affairs. Stories about scandals rose from just 0.015% in 1977 to 15% in 1997. Human interest and quality of life stories doubled from 8% in 1977 to 16% of all stories in 1997. Thussu quotes Kovach and Rosenstiel's (1999) identification of four factors behind the rise of infotainment and its 'corrosive' impact on journalism: the 24/7 news cycle and its appetite for 'live' news; the proliferation of many new networks; the growing sophistication of news management and 'spin'; and the intensification of a ratings mind-set that prioritized the dramatic and the entertaining

Pew Project for Excellence in Journalism Annual Reports

The gathering crisis of the 2000s is nowhere referenced in more detail than in the annual reports on the state of news media by the Pew Project for Excellence in Journalism. Beginning in 2004, each traces key trends of the previous year across different news media. The 2004 report set a tone for the decade. It noted the paradox of growing diversity of news outlets while audiences were shrinking, amidst evidence of increasing press manipulation (including the Jessica Lynch story) and investors' fetish for news dissemination over news collection. The outlook for many outlets was bleak. The report doubted whether online news sites would prove a durable or sufficient compensation for loss of advertising in traditional media. Convergence was less of a threat.

The 2005 report noted a growth in available models of doing journalism, some of them partisan ('journalism of affirmation'), seemingly at odds with what most Americans said they wanted. The report proposed that journalism needed more transparency and expertise, through the roles of 'authenticator' or 'referee' – a 'show me' approach very different to the 'trust me' authority of the network age, some of whose doyens (e.g. Dan Rather or Peter Jennings) had recently died or retired. Mainstream media were investing too cautiously in building (even online) new audiences, leaving real innovation to Google, YouTube and blogs. Profitability increasingly depended on cost-cutting tactics such as cheap unscripted news programs.

The 2006 report noted the paradox of more journalistic outlets yielding fewer stories, sameness of style, modest resources and smaller audiences. Authorities could easily divert journalists away from real news. In newsrooms, 'accountants' had won their long battle against 'idealists'. Notions of 'public interest' were derided. 'Fragmentation' of product and platform, argued the 2007 report, was a response to too many organizations essentially doing the same thing – but not enough of the really important things, as demonstrated by deteriorating coverage of local government. The industry was becoming more aggressive in redeveloping its economic model, moving away from dependence on an advertising industry that was outgrowing its need for media. The 2007 report ruminated whether the investment community would look beyond news as a declining industry and redefine it as one in transition. Bravely, it asked whether the public corporation was a suitable framework for the kind of transition that newsrooms needed and whether private capital might be more effective. In content, the report noted a move from a journalistic culture of 'argument' to one of 'answers', offering solutions, certainty, crusades, yet not committed to giving the 'other side' equal play.

Content

The 2004 report noted a growing range of topics covered, but lighter treatment, more self-branding and commercials. There was pressure to run with stories more quickly, and concentration on 'blockbuster' stories. Cable news favored live interviews and reporter stand-ups rather than written and edited packages.

The 2005 report concentrated on media bias in coverage of Iraq and the 2004 election. Across *all* media, 25% of coverage was deemed negative towards US involvement in Iraq, 20% positive, 35% neutral. The three nightly newscasts and PBS were slightly more negative. Such recompense for media failure to expose false pretexts for war came too late, I would argue, to make a difference and ignored the ultimate 'success' of the war – the return of Western majors to Iraqi oil development. Fox was twice as likely to be positive. CNN and MSNBC were evenly split. A high percentage (60%) of selected foreign news stories used anonymous sources, as against 50% of morning TV and 29% of newspaper front page stories, but only 9% of cable. In contrast to the free-to-air networks, cable news was more thinly reported, using fewer sources. Cable was more one-sided and opinionated, most notably so in the case of Fox. The broadcast evening newscasts on the other hand provided greater depth, through more edited packages (involving more planning and script).

The 2006 report observed that most media news was transitory, lacking in long-term consequence, with few top stories continuing to be covered two or three days after first appearance, too often drawing on a small number of sources. Some stories on national television, the report noted, 'essentially relied on a single source, sometimes the same one on every channel'. There was excessive repetition within the 24-hour news cycle.

News coverage in 2007, as in 2004, was dominated by Iraq and election topics consumed one third of the total news hole, thus contributing to a notable narrowing of the news agenda. Overall in 2007 US-related foreign affairs accounted for 17% of coverage, followed by US election and political coverage at 13%. Non-US-related foreign news coverage accounted for 11%. Other categories were crime 7%; government 6%; disasters 5%; health 5%; economics 4%; lifestyle 4%; and business 3%. There had been a precipitous decline in government coverage from 16% on the broadcast networks in 2003 to 6% in 2007; and from 29% on cable in 2003 to 7% in 2007. Important domestic subjects given only superficial attention included education (1% of the domestic news hole), transportation (0.8%), religion (0.7%), court/legal system (0.4%) and development/sprawl (0.2%). These topics, said the report, 'bend', i.e. require constant coverage so that consumers will understand the significance of incremental changes.

'World' coverage was not really about the world but about 'some US interests abroad'. Only three countries received notable coverage: Iraq, Iran and Pakistan. Regions receiving scant attention included, surprisingly, Afghanistan, North Korea, Darfur, Russia, China and the Lebanon. Israel-Palestine accounted for 0.5% of the foreign news whole; nuclear negotiations with North Korea, 0.4%; violence in Darfur 0.2%; deteriorating relations with Russia 0.2%; China 0.5%. The report lamented the 'remarkably short attention span' of what it called 'drive by' coverage – the brief flooding of a hot news zone with instant coverage, followed by next to nothing. Also called 'hit-and-run', this was particularly notable on cable. Tabloid-style gossip and scandal was not a strong feature of broadcast networks, but came principally from cable. Older media (press and broadcast networks) were broader, more diverse and less likely to be dominated by mega-stories. Broadcast networks and newspapers focused least attention on the 'big' stories and carried the highest number of topic areas. Only newspapers included business, health and medical affairs among top priorities. Broadcast networks provided the most coverage of disasters and accidents. Cable devoted the most time to crime.

TRANSITIONING FROM NETWORK TO POST-NETWORK AGE

Geoffrey Baym (2010) identifies three phases in the evolution of television news. The 'network age', protected by FCC policies that generally privileged incumbent players against newcomers, endured from the 1940s through to the emergence of cable as a significant competitor in the 1980s and was characterized by an oligopoly of the 'Big 3': ABC, CBS and NBC. News in this period was conceived as a professional avocation distinct from entertainment and concentrated on 'rational-critical investigation of matters of assumed public interest'. On the downside, Baym cites James Carey to the effect that this approach to news 'reduced the role of citizens, speaking at and for them but allowing them no role in the conversation except as the audience' (Baym, 2010, p. 125). This era was succeeded by the period of market-driven multi-channel transformations coincident with the emergence of cable and satellite, and the removal of 'regulations that mandated news and public affairs programming', as in the 1996 Telecommunications Act. This in turn has been succeeded by the 'post-network' age. Baym enlists the help of Henry Jenkins in establishing the dominant features of this age (see below). It is the product not just of technology developments (principally VCR, CDs, DVD, internet, DVR, personal computers, MP3 players, 3G mobile phones, internet-enabled television), but also of shifts in political economy and culture, including the shift from national to global cultures, the corporatization of culture, and cultural hybridization.

1. A shift of content from medium-specific to flows across multiple media channels;
2. Increased interdependence of communications systems;
3. Multiple ways of accessing media content (anywhere, anytime convenience);
4. More complex relations between top-down corporate media and bottom-up participatory experimentation;
5. Experimentation;
6. Greater genre instability, including infotainment ('aestheticiztion of the political-normative sphere' and 'politicization of the aesthetic-expressive sphere' as exemplified by the popularity of satirists who come to be regarded as significant sources of political news and comments, including Stephen Colbert (*The Colbert Report* started in 2005), Bill Maher and Jon Stewart (*The Daily Show* started in 2004);
7. Discursive integration of different and sometimes contradictory 'styles, standards and assumptions'.

The emergence of a post-network age in the 2000s has radically destabilized the business model of network television. The distinction between 'terrestrial' and cable/satellite television news has grown progressively blurred as a result of the mandatory shift from analogue to digital television in 2009 and the absorption of most commercial television suppliers within the major media conglomerates, notably Disney, General Electric, Microsoft, Time Warner and Viacom/CBS. These link together national/local terrestrial, cable and satellite networks with the Hollywood studios and distributors. Drawing on the analysis of Baym and Jenkins (above), I identify the principal roots of these trends:

1. Relentless decades-long decline in audience ratings for the 'old' networks and achievement of only modest ratings for the 'new' networks;
2. An aging audience;
3. Diffusion of total advertising revenue across more and more outlets, including online sites and social network media, leaving less to spend on traditional television. Advertisers showing greater independence of media content altogether as they explore new methods of direct access to their preferred customers;
4. Increasing opportunities for audiences to access news content without exposure to advertising reduces advertiser interest in supporting traditional television;
5. Increasing audience resistance, especially among younger audiences, to being tied down to a particular time to watch news;
6. More competition for audience leisure time, including the internet and video-games;
7. Collapse in the perceived 'authority' of news, coincident with (a) the rise of opinionated journalism, (b) infotainment, (c) the inability of television news programming to hold the powerful to convincing account for abuses that have included the launch of wars on false pretexts, and (d) susceptibility to gross manipulations of information and logic by powerful lobbyists 'turning public information into marketing the purpose of manufacturing public perception' (Baym, 2010, 22).

In line with Baym, Lotz (2009) chronicles a move from the 'network' era through the 'multi-channel' transition of the 1980s–1990s, to the 'post-network' era. The network era was characterized by minimal viewer choice and control under conditions of high concentration or monopoly of supply (justified, as Uricchio (2009) notes, by the logic of 'scarcity' – a logic that had a lot less to do with the limitations of spectrum, as commonly claimed, and much more to do with the politics of allocation), predictable schedules, clustered advertising packages (not sponsored programs as in the early days of television), family viewing, non-offensive homogeneous content, and expensive news operations that bought the sympathy of policymakers. The model was also characterized by commitment to public service. All these features are transitioning in the post-network era towards heightened audience control and choice from multiple sources of anytime-anywhere television, financed by sponsorship and subscription as well as clustered advertising, with more heterogeneous, segmented or customized content, supported by weaker, less-profitable news operations whose news sometimes migrates to other categories of programming, including comedy (Blondheim and Liebes, 2009). Although market logic appears to have vanquished that of the public sphere as determinant of content, Lunt (2009) finds evidence of a new kind of public service by which popular television genres 'reflect a normative social order oriented to self-regulation and development in which expertise is constituted as an aid to everyday living … a site of social control, mobilizing individuals in processes of rationalization (the adoption of psychological discourse), self-surveillance, and confession' (Lunt, 2009, pp. 140–141).

CONCLUSION

US Network television is not in imminent threat of annihilation, but is in danger of gradual evaporation from a thousand cuts, many of which I have reviewed in the preceding pages:

- The loss of an audience willing to watch at fixed hours of day or night;
- Competition from others sources of information and entertainment, including electronic books and video-games;
- An overwhelming loss of younger viewers with sufficient knowledge and news skills to

appreciate anything more than extremely simplified, selective and misleading content;

- Growing consumer preference for 'anywhere-anytime' viewing which opens the door to secondary providers and/or spreads the available advertising expenditure across a broader range of outlets, including online news services of both traditional print and traditional broadcast organizations, mobile outlets, YouTube and Hulu;
- Greater desperation for advertising revenue, scratching down Chinese walls that once separated editorial from advertising;
- The continuing loss of news 'quality' in the sense of services that offer news that is authentic, independent, and truly critical, of power, being both trustworthy and of relevance to people's real needs as opposed to the manufactured and inauthentic wants imposed on them by media conglomerates in their bid to sell larger numbers of viewers to advertisers.

In this chapter I have reviewed the development of the original 'network' structure of US television news, noting the considerable degree to which this has been contested by the arrival of cable and satellite providers. These in turn helped to transform the field towards one of channel abundance, in news as in most other genres, within a context of industry concentration and conglomeration. Since this connects to issues of power, I have considered the extent to which the Republican charge of 'liberal media' holds up in the case of the three original 'terrestrial' channels and the three newer 'cable' channels. I have found little evidence to support that charge in the context of what is perhaps the gravest controversy of all, US overseas military engagements, and quite a lot of evidence that suggests that mainstream US television is principally a conduit for authoritative pronouncements from agents of the state and the corporatocracy. The transition from the network era to a 'post-network' era of poorly-resourced news abundance and anytime-everywhere television, has not eluded the deadening and orthodox hand of US conglomerate capitalism, but it does contain such potential for rupture and fragmentation that may make it increasingly difficult for the state to successfully and comprehensively impose on media its hegemonic accounts of world affairs through the lenses of neo-liberal capitalism and imperial interest.

REFERENCES

Ackerman, S. (2001). Al-Aqsa intifada and the U.S. media. *Journal of Palestine Studies* 30(2), 61–74.

Alexander, A., Owers, J., Carveth, R. and Hollifield, C. (2003). *Media economics: Theory and practice*. London: Routledge.

Allan, S. and Zelizer, B. (2004). *Reporting war: Journalism in wartime*. London: Routledge.

Auletta, K. (1992). *Three Blind Mice: How the TV Networks Lost Their Way*. New York: Vintage.

Barkin, S.M. (2002). *American television news: The media marketplace and the public interest*. New York: M.E. Sharpe.

Barnouw, E. (1968). *A history of broadcasting in the United States*. New York: Oxford University Press.

Barstow, D. (2008). Behind TV analysts: Pentagon's hidden hand, *New York Times*, 20 April.

Baym, G. (2010). *From Cronkite to Colbert: The evolution of broadcast news*. Boulder: Paradigm Publishers

Blondheim, M. and Liebes, T. (2009). Television news and the nation: The end? In E. Katz and P. Scannell, *The end of television? Its impact on the world (so far)*. The Annals of the American Academy of Political and Social Science, Vol. 625, September. Thousand Oaks, CA: Sage, pp. 182–191.

Blum, W. (2008). *Killing hope: U.S. military and C.I.A. interventions since World War II*. Monroe, ME: Common Courage Press.

Chossudovsky, M. (2005). *America's 'war on terrorism'*. Montreal: Global Research.

Cottle, S. (2008). *Global crisis reporting*. Milton Keynes: Open University Press.

Croteau, D. and Hoynes, W. (2005). *The business of media: Corporate media and the public interest*. Thousand Oaks, CA: Pine Forge Press.

Curtis, A. (2004). *The power of nightmares*. British Broadcasting Corporation.

Eke, C. (2008). Darfur: Coverage of genocide by three major U.S. networks on their evening news. *International Journal of Media and Cultural Politics*, (4)3, 277–292.

Engelman, R. (1996). *Public radio and television in America: A political history*. Thousand Oaks, CA: Sage Publications.

Epstein, J. (2000). *News from nowhere: Television and the news*. New York: Ivan R. Dee

Franken, A. (2003). *Lies and the lying liars who tell them*. New York: Dutton.

Goldenson, L. and Wolf, M. (1991). *Beating the odds: The untold story behind the rise of ABC*. New York: Scribners.

Greenwald, G. (2004). *Outfoxed: Rupert Murdoch's War on Journalism*. WantMedia Inc.

Grey, B. (1999). Pentagon pressure behind CNN firing of Peter Arnett. *World Socialist Web Site*, available at http://www.wsws.org/en/articles/1999/04/cnn-a22.html (last accessed 1 Aug 2014)

Griffin, D. (2008). *The new Pearl Harbor revisited: 9/11, the cover-up, and the exposé*. Northampton, MA: Olive Branch Press.

Griffin, D. (2009). *Osama Bin Laden: Dead or Alive*. Northampton, MA: Olive Branch Press.

Hallin, D. (1986). *The 'uncensored war': The media and Vietnam*. New York: Oxford University Press.

Herman, E. and Chomsky, N. (1988). *Manufacturing consent: The political economy of the mass media*. New York: Pantheon.

Hoskins, A. (2004). *Televising war: From Vietnam to Iraq*. London: Continuum.

Jaramillo, D.L. (2009). *Ugly war, pretty package: How CNN and Fox News made the invasion of Iraq high concept*. Bloomington and Indianapolis: Indiana University Press.

Jones, S. (2010). The media-lobbying complex, *The Nation*, 11 March, www.thenation.com/doc/20100301/jones/single (accessed March 3, 2010).

Katz, E. and Scannell, P. (2009). *The end of television? Its impact on the world (so far)*. The Annals of the American Academy of Political and Social Science, Vol. 625, September. Thousand Oaks, CA: Sage.

Kellner, D. (1992). *The Persian Gulf TV war*. Boulder, CO: Westview Press.

Kellner, D. (2003). *Media spectacle*. New York: Routledge.

Kovach, B. and Rosenstiel, T. (1999). *Warp speed: America in the age of mixed media*. New York: The Century Foundation Press.

Lotz, A. (2009). What is U.S. Television now? In E. Katz and P. Scannell, *The end of television? Its impact on the world (so far)*. The Annals of the American Academy of Political and Social Science, Vol. 625, September. Thousand Oaks, CA: Sage, pp. 49–59.

Lunt, P. (2009). Television, public participation, and public service: From value consensus to the politics of identity. In E. Katz and P. Scannell, *The end of television? Its impact on the world (so far)*. The Annals of the American Academy of Political and Social Science, Vol. 625, September. Thousand Oaks, CA: Sage, pp. 128–138.

Melber, A. (2009). The Nation, October 11, available at http://www.thenation.com/blog/new-white-house-line-against-fox-its-war (last accessed 1 Aut 2014).

Meyrowitz, J. (2009). We liked to watch: Television as progenitor of the surveillance society. In E. Katz and P. Scannell, *The end of television? Its impact on the world (so far)*. The Annals of the American Academy of Political and Social Science, Vol. 625, September. Thousand Oaks, CA: Sage, pp. 32–48.

Moyers, B. (2007). Buying the war. Bill Moyers Journal (April 25), available at https://www.youtube.com/watch?v=_XJ5dmkkFno (last accessed 1 August, 2014)

Paper, L.J. (1987). *Empire: William S. Paley and the making of CBS*. New York: St Martin's Press.

Reid, R. (2009). Al-Qaida showing smaller presence in Afghanistan. Associated Press, October 6.

Scott, P.D. (2008). *The road to 9/11: Wealth, empire and the future of America*. Berkeley and Los Angeles: University of California Press.

Solomon, N. (2007). *Media love, got war: Close encounters with America's warfare state*. Sausalito, CA: Polipoint Press. http://www.nytimes.com/2008/05/12/business/media/12ratings.html

Stelter, B. (2008). In the age of TiVo and web video, what is prime time? *New York Times*, 12 May (accessed July 31, 2014).

Stelter, B. and Carter, B. (2010). Network news at a crossroads, *New York Times*, 28 February, www.nytimes.com/2010/03/01/business/media/01network.html?scp=1&sq=Stelter%20and%20Carter&st=cse (accessed May 1, 2010).

Thussu, D.K. (2000). Legitimizing humanitarian intervention? CNN, NATO and the Kosovo Crisis. *European Journal of Communication*, (15)3, 345–362.

Thussu, D.K. (2007). *News as entertainment: The rise of global infotainment*. London: Sage.

Thussu, D.K. and Freedman, D. (2003). *War and the media: Reporting conflict 24/7*. London: Sage.

Uricchio, W. (2009). Contextualizing the broadcast era: Nation, commerce, and constraint. In E. Katz and P. Scannell, *The end of television? Its impact on the world (so far)*. The Annals of the American Academy of Political and Social Science, Vol. 625, September. Thousand Oaks, CA: Sage, pp. 60–73.

Volkmer, I. (2005). *News in the global sphere: A study of CNN and its impact on global communications*. London: John Libbey.

Voorhees, C.W., Vick, J. and Perkins, D.D. (2007). 'Came hell and high water': The intersection of Hurricane Katrina, the news media, race and poverty. *Journal of Community and Applied Social Psychology*, 17, 415–429.

Weaver, D., Beam, R., Brownlee, B., Voakes, P. and Wilhoit, G.C. (2007). *The American journalist in the 21st century: U.S. news people at the dawn of new millennium*. Lawrence Erlbaum Associates, Mahwah: New Jersey.

Whittemore, H. (1990). *CNN: The inside story: How a band of mavericks changed the face of television news*. New York: Little Brown and Co.

12

Hollywood Story: Diversity, Writing and the End of Television as We Know It

Darnell Hunt

Storytelling is a fundamental element of the human condition. Not long after our species first acquired speech, we began to play with words and gestures to spin tales that both tickled the imagination and offered new ways to reflect on the truths of the moment (Deacon, 1998). Storytelling constitutes 'complex communicative events', culturally grounded exchanges that may work to either reinforce or challenge social structures (Georges, 1969, p. 317). To be sure, how we make sense of our day-to-day experiences is continuously shaped by the stories we consume. We live vicariously through the pleasures and pains of characters and their predicaments, trying these adventures on for size as we reflect on who we are, who we are not, and who we hope to be. In a media-saturated society, stories are available to us everywhere and anytime, and we rely on them to help us cope with the particulars of our mundane existence (Gitlin, 2002). Stories have become such an integral part of our lives that we are not always conscious of the degree to which they hold us in their grip:

> We are, as a species, addicted to story. Even when the body goes to sleep, the mind stays up all night, telling itself stories. (Gottschall, 2012, p. xiv)

This chapter focuses on a mode of storytelling that undoubtedly has a profound impact on our lives: the screenwriting behind the myriad situation comedies and dramas that populate network television. By 'network television', I am referring to the 'cultural forum' (Newcomb and Hirsch, 1994) in the United States, comprised, in the second decade of the 21st century, of scores of broadcast and cable networks owned by just a handful of multinational corporations (Bagdikian, 2004). Though the written script constitutes the base of the network television 'text',[1] actors, directors and network executives also have a hand in assembling what ultimately makes it to the screen. But this is not the end of the story. Audiences bring to their encounters with television the stories of their lives, which necessarily color the way (and degree to which) they relate to what is on the screen (Fiske, 1987; Hall, 1973; Morley, 1992). Against the backdrop of the diversity of lived experiences in contemporary society, this chapter examines the process by which stories are sold and told in network television. Whose stories are being told and what does this portend for the future of television?

SELLING AND TELLING STORIES

In network television, telling stories has everything to do with selling stories (Einstein, 2002b; Gitlin, 1983; Stempel, 1996; Wild, 1999). Writer-producers[2] sell story ideas to networks for development. Networks then sell developed stories not only to audiences but also to advertisers, who place within the program breaks mini-stories about their products – commercials – in hopes of selling these products to captive audiences. And each week members of a television show's writing staff sell stories to each other (and to the network) as they work together to produce new episodes of the show. In short, telling stories on network television is a profoundly collaborative and contingent process. It is collaborative in the sense that television screenwriting is invariably shaped by multiple writers, network executives and consultants (Phalen & Osellame, 2012; Stempel, 1996). It is contingent because the stories that finally make it to screen must first pass through the filter of network favor and purchase (Bielby & Bielby, 1994, 2003; Gitlin, 1983; Wild, 1999).

The storytelling process commences the moment a writer-producer pitches an idea for a television show to a network in the hopes that the show will be picked up for inclusion on the network's program schedule (Einstein, 2002b; Stempel, 1996; Wild, 1999). Networks are critical gatekeepers in the storytelling process; they are uniquely positioned, because of their current programming and resulting brand visibility, to provide screenwriters with access to large audiences. As Wild (1999) puts it, 'A network is a circus tent' (p. 107). Moreover, as the Hollywood entertainment industry has become increasingly consolidated and vertically integrated, the networks themselves have become more directly involved in the television production process (Bielby & Bielby, 2003; Einstein, 2002a),[3] which means that in many cases a writer-producer making a pitch for a show is, in effect, making a pitch for employment with a production company associated with the network. All of these factors combine to give networks a tremendous amount of power over what actually makes it to the screen (Bielby & Bielby, 1994; Gitlin, 1983; Stempel, 1996).

Regardless of who ultimately 'owns' a show, the initial pitch typically involves a succinct and provocative outline of the show's basic concept to network executives, who, if the pitch is successful, will keep the project in play by authorizing its movement to subsequent steps of consideration. These steps typically follow a standard sequence, moving from commissioning of a pilot script, to green-lighting production of the pilot, to the actual pick-up of the series for a specified number of episodes (Einstein, 2002b). Financial considerations come into play during each of these subsequent steps. That is, the network must pay for the creation of the pilot script,[4] fund the production of the actual pilot, and pay a license fee for airing episodes of the green-lit show throughout the season. In any given season, there are hundreds of pilot projects pitched but only a small fraction of them advance through the pipeline all of the way to series status.[5]

Once a successful pitch garners a series commitment from the network, the 'showrunner'[6] associated with the project must quickly assemble a writing team. Seasoned writer-producers may themselves hold the title of showrunner, the executive producers responsible for hiring writers and other creative talent and keeping production on track from week to week. Alternatively, the network may handpick a showrunner or showrunners to manage a show it green lights.[7] In any event, a showrunner seeks to hire writers with whom he or she feels comfortable – which means that the writers finally selected to work on staff tend to have similar or at least relatable sensibilities when it comes to the show's subject matter (Phalen & Osellame, 2012). Network executives often reserve the right to veto a specific hiring decision of a showrunner if they do not feel comfortable with the choice or if they have other writing prospects they are keen on adding to the team (Gitlin, 1983). The general rule: the less experienced the showrunner, the more interference by the network (Einstein, 2002b). But there is always some interference,[8] such as the network script notes that circulate throughout the writers' room each week (Stempel, 1996; Wild, 1999).

The writers' room is where stories come to life. It is a celebrated space where a show's writing team meets throughout the project's production, spending 12 to 14 hours a day fleshing out the concept originally sold to the network (Phalen & Osellame, 2012). It is here where character and story arcs are developed, where members of the writing team attempt to sell ideas to one another for predicaments and jokes. It is here where story ideas, shaped by the conventions and formulae associated with a show's genre, [9] are finally fashioned into scripts for each episode. The object of both reverence and derision,[10] the writers' room has become almost mythic in industry lore. Depictions of what happens here oscillate between images of magic and banality, between those of creative minds at work and workaday yes-men and yes-women.

On a 2013 panel entitled 'How Not to Get Fired from the Writers' Room', a young Latina screenwriter echoed other panelists when she offered the following advice:

> It always comes down to personality. It always comes down to, 'Do I want to be with this person for 14 hours a day?' And you don't want to be with the nag.
>
> You don't want to be with the person who isn't a team player. ... Just be nice.[11]

Being a team player, the first one in and the last one out, going beyond the call of duty to convince showrunners they can count on you –these are the qualities that appear to keep those fortunate enough to land a job in the writers' room *in.* Though celebrated with fervor in writing circles and thus a valuable asset, talent may not be, in fact, the determining factor in one's success. 'You don't have to have a talent to write for television', noted one veteran comedy writer.[12] 'Thought it was writing, but it's not. It's a craft. It's like a tailor. You want cuffs? You've got cuffs.'

In many respects, the path by which an idea for a network television show ultimately finds its way to the screen resembles a manufacturing process comprised of several constraining 'filters' (cf. Herman & Chomsky, 1988): concentration of ownership and profit motives in the Hollywood industry generate incentives for television networks to produce their own programming, thereby freezing independent producers (and voices) out of the production process (Bielby & Bielby, 2003; Einstein, 2002a); network reliance on advertising revenue narrows the domain of representational possibilities as networks scramble to exploit tried and true formulae (what they think advertisers will buy) in hopes of locking in billions in advertiser commitments following 'upfronts' each year (Lotz, 2007a); networks rely on a revolving door of producers for show ideas because these producers' track records of producing hits in the past provide comfort for executives in an industry where most new shows fail (Bielby & Bielby, 1994, 1999; Gitlin, 1983; Stempel, 1996). This process is further underpinned by the conventional wisdom that successful programming must, first and foremost, comfort white audiences – the largest (but shrinking) racial bloc of viewers throughout most of commercial television's history – which is like a religious doctrine whose implications for storytelling are accepted as matters of faith (Hunt, 2000, 2005). As a result, the primetime television product that typically emerges out of the production pipeline appears almost Fordist in terms of process, structure and perspective.

'WHOSE STORIES ARE WE TELLING?'[13]

Hollywood has always been a 'bastion of whiteness' (Quinn, 2012, p. 469), a highly lucrative and insular industry in which white men dominate the power positions. In the arena of television screenwriting, this truism has been consistently supported over the years by numerous studies (Bielby & Bielby, 1998, 2002; Hunt, 2005, 2007, 2009; U.S. Commission on Civil Rights, 1977). Consider, for example, a television staffing report released in 2013 by the Writers Guild of America, West (WGAW), the Hollywood-based union representing television and film screenwriters (Hunt, 2013). The report showed that of the 566 pilots in the pipeline at one point during the 2010–11 television season, only 51 (9%) had at least one writer of color associated with them – despite the fact that minorities collectively accounted for 36.3% of the nation's population in 2010.[14] Moreover, just 7.8% of the executive producers working on the 190 current broadcast and cable network shows examined for the 2011–12 season were people of color. In other words, the report found that diverse writers were underrepresented by a factor of about 4 to 1 among writer-producers with the most decision-making authority, both in the development of original network show concepts *and* in the day-to-day management of the storytelling process.

The figure was a little better for the minority share of staff writers across all ranks working on the shows examined that season, 15.6%. But even here people of color were still underrepresented by a factor of more than 2 to 1. Meanwhile, though women fared better than minorities on measures of inclusion in the writers' room – for example, they accounted for 30.5% of all staff writers and 18.6% of executive producers – at slightly more than half of the population, they were still far from reaching proportionate representation on any of the measures. White men, as they had done since the earliest days of network television, controlled the lion's share of staff writing positions, particularly the more powerful ones.

It should be noted that amid familiar statistics like these, interethnic conflicts have emerged among media activists around the relative status of different ethnic and racial groups in the industry. The notion that African Americans were akin to a 'model minority' in network television fueled much of this tension.[15] While the misconception can be partially attributed to the fact that blacks had been overrepresented in front of the camera since at least the turn of the last century (Hunt, 2000), it also was enabled by a few exceptional moments in which a critical mass of black storytelling managed to make it through the network production pipeline. The first moment is marked by the rise of two defunct television networks, United Paramount Network (UPN) and Warner Brothers Network (WB). Both of these networks established footholds in the broadcast television market in the late 1990s and first few years of the 2000s by marketing a slate of black-oriented and black showrunner-led sitcoms, which were 'ghettoized' over just a few nights of the week in order to attract large African American audiences neglected by the major networks (Hunt, 2000). [16] When the UPN and WB networks folded in late 2006, the number of employed black television writers and showrunners plummeted (Hunt, 2009).

Another moment was ushered in by a $200 million deal the black director, writer and actor Tyler Perry struck with the Turner Broadcasting System (TBS) cable network in 2006 for Perry to produce 100 episodes of an original situation comedy series for the network.[17] Capitalizing on the success of his gospel stage plays, which were particularly popular among black church-goers, and his independently produced films based on this work, Perry established a television brand characterized by raucous black comedy laced with religious moralizing. In 2013, Perry ventured into the genre of primetime soap opera with a series that was picked up by the struggling OWN network, owned by the black billionaire talk show host Oprah Winfrey. Though successful in attracting black audiences starving for television that features black self-representation, Perry's shows generated considerable controversy within the black community over the stereotypical quality of the self-representations they circulated.[18]

A final exceptional moment worth considering is Black Entertainment Television's (BET) foray into original, scripted programming in early 2011, when it picked up a popular black-oriented and black-run situation comedy that the CW Network[19] cancelled in 2009. Soon after, the cable network announced the development of a handful of other black-oriented, scripted series headed by black showrunners. Sold by founder Robert L. Johnson in 2000 to Viacom for $2.34 billion in stock, BET had been the nation's only black-owned television network with the potential to reach large audiences since its 1980 founding.[20] But, prior to the sale, BET primarily ran music videos and recycled black situation comedies and movies, offering few opportunities for the telling of new stories. It remains to be seen whether the conglomerate that bought BET will accommodate, in a sustained way, departures from a network production process that had stymied diverse storytelling throughout most of television's history.

The idea that there is a necessary trade-off between diversity and quality – one of Hollywood's most persistent and pernicious canards – has enabled this industry status quo (Hunt, 2005). When confronted with abysmal diversity numbers, industry decision makers have usually resorted to the 'small pool' argument as a justification for the situation. There are so few talented and motivated minority writers in the pool, the argument goes, that it is no wonder the employment numbers for minorities are so low. Then, in an effort to placate those who demand more minority inclusion, industry decision makers have routinely promoted a number of programs ostensibly designed to develop minority talent.

In television, for example, there have been a handful of high-profile writing programs run by the studios and networks that, starting in the late 1990s, provided some writers from underrepresented groups with a leg up in the competition to land a seat in the writers' room.[21] But these public relations programs effectively constituted a lottery system that provided just one or two real opportunities for the hundreds of talented screenwriters competing for them. The major broadcast networks (with the exception of CBS) also offered programs that subsidized the placement of a 'diverse' writer on each of their current primetime shows. But whether this foot in the door actually led to long-term career opportunities for these writers is unclear. Anecdotal evidence suggests that the writers who filled the 'diverse' writer slots were sometimes marginalized in the writers' room, pegged as 'freebies' by other writers on staff, and frequently not asked to return after the initial season of employment.[22] Meanwhile, other black writers, like those with track records established on black situation comedies during UPN's heyday, were often typecast as only being able to write 'black' shows (cf. Bielby & Bielby, 2002).

Since at least the late 1960s, the Hollywood industry has spoken from both sides of its mouth on the issue of diversity (Quinn, 2012). Industry management has attempted to claim a 'reasonable middle ground' on the issue by playing labor unions like the WGAW – whose collective bargaining has been most successful in securing gains for the (white and male) writers already dominating employment – against activists demanding increased access for underrepresented groups (Quinn, 2012, p. 480). Meanwhile, power brokers at the networks, studios and talent agencies continued to reproduce themselves with comfortable replacements, decision makers who more often than not looked and thought like them. Rather than change organizational cultures, the few token women and people of color who made it to these prized positions were usually themselves changed by the organization as they managed tremendous pressures to be seen as team players, as promotion-worthy.

This business-as-usual environment motivated network decision makers to green light 'safe' projects, projects that resembled other successful shows in terms of genre and formula (Bielby & Bielby, 1994). In the name of self-preservation, network executives also were motivated to favor showrunners who had a track record of producing successful shows (Gitlin, 1983; Wild, 1999), writer-producers who were represented by the most powerful talent agencies (Bielby & Bielby, 1999). And because showrunners confronted incredible pressures to bring in a successful show on schedule and within budget, they were motivated to assemble writing teams with which they felt comfortable (Phalen & Osellame, 2012), people who tended to think like them. All of these motivations combined to favor those already entrenched in the industry, white males. There was little motivation to tackle the lack of industry diversity head on, a truth that was commonly reflected in the statements of network heads, middle managers and showrunners alike. When activists periodically challenged the Hollywood status quo, industry decision makers routinely responded by invoking a discourse of color-blind neo-conservatism that proclaimed the industry's openness to those who work hard and possess talent (Quinn, 2012). 'I don't care what they are as long as they can write. Color, sex means nothing to me. Just gimme a writer.'[23]

But 'a writer' is not just *any* writer. Writers tend to write from the reservoir of their personal experiences, to imagine stories that resonate with the challenges faced and overcome in their lives and in the lives of the people they know. Writing is inflected and nuanced by an array of cultural attributes, such as dialect, humor, and folk wisdom. When the men who dominate network television write, they more often than not write from the vantage point of their whiteness – from particular raced 'ways of seeing' (Berger, 1972; Hunt, 1997) that color their selection and treatment of settings, the attention they give to different characters in the narrative, and the situations these characters encounter. This is not to say that white writers cannot write stories that involve people of color; they obviously do, but rarely is the story the character of color's story (Bielby & Bielby, 1998; Hunt, 2005). In short, when white writers write about people of color, the stories they tell are likely to be different from the ones people of color choose to tell.[24] And this reality lies at the core of a prominent disconnect between the stories that routinely emerge from network television's production pipeline and the lived experiences of increasingly diverse audiences.

THE END OF TELEVISION AS WE KNOW IT: A CASE STUDY

Anacostia is a section of southeast Washington DC that became synonymous with crime and drugs in the 1980s after a rash of shootings and homicides caught the attention of local and national news media. It is also the backdrop for a successful web series of the same name that was independently produced by actor and writer Anthony Anderson, a resident and native of the largely black and working-class area. *Anacostia: The Web Series* is a provocative soap opera that centers around the lives and loves of three close friends who grew up together in the area: Sean (played by Anderson), a lonely, 30-something black gay male whose life partner is presumed dead during most of the first season; Salina, a good-hearted sage whose alcoholism has strained her otherwise ideal marriage; and Mia, an attractive housewife who endures daily insults from an emotionally and physically abusive husband. The 10 webisodes comprising Season 2 of the web series (which were uploaded to the show's YouTube channel in late 2010 and early 2011) range from just over 14 minutes to 34 minutes in running time, were shot on a shoe-string budget, [25] and attracted an average of about 10,000 viewers. The latter figure is modest by web series standards,[26] despite the numerous awards the web series has won at festivals throughout the country and the decision by veteran soap opera actress Martha

Byrne (CBS's *As the World Turns*) to join the cast in Season 3.

Anderson, who said he is a member of the Writers Guild of America, East, wrote all of the scripts for each season of *Anacostia* in advance, which he distributed one at a time for actors to memorize on the Thursday prior to each weekend shoot (there was no ad-libbing by actors). Although the web series appropriates formulae common to the soap opera genre (e.g., interrelated storylines, a focus on betrayal and revenge, the return of 'dead' characters and episodic cliff hangers), Anderson nonetheless managed to infuse into its stories a freshness that is in no small part a reflection of it being rooted in a specific place. *Anacostia* captures the flavor of a community not typically represented in network productions shot on carefully constructed studio sets or at iconic locations. At the expense of acting that may be a little uneven at times, it provides its audiences with regular peeks inside the homes and offices of regular Anacostia folk. It features in its establishing shots area landmarks, such as the 'Big Chair', that only locals would know. And most importantly, Anderson took plotlines one might see on network television and converted them into altogether different stories by placing black characters, a black community and contemporary black issues at the very center of the web series' narratives. All of this lends *Anacostia* an air of authenticity; it taps into a certain rawness that the better so-called 'reality' shows exploit to hook their audiences. Consider a sampling of viewer comments posted on the show's YouTube channel for the Season 2 finale:

> THE BEST URBAN SOAP OPERA YET!!!!!!!!!!!!!!!!!!
>
> omg..this is the best webseries ever!!! im hooked.. im gonna start season 3 now..is there anyway I can see season 1?
>
> this was really addictive. I couldnt take my eyes off the screen. I cant believe that fool survived a near death experience again. lol
>
> I love this series... U have to watch 2 seasons in one day...On to season 3. Thanks for showing the good side of Anacostia. Best wishes...

According to Anderson, one of his prime motivations for creating the web series was to cast the Anacostia area as a character, a move he hoped would challenge media stereotypes and more accurately reflect recent efforts by residents to redeem a historically significant community that was once home to ex-slave, abolitionist and author Frederick Douglass. Producing the independent web series locally, he added, also provided a relatively rare opportunity for local talent to be seen and develop profiles that may lead to creative opportunities outside of the DC area. This consideration, the free reign Anderson enjoyed to tell the stories he wanted to tell, and his growing acclaim in the community, all combined to explain why he had little interest in pursuing a network development deal:

> Money doesn't motivate me. The sheer reward I get from walking on the street and having people walk up to me – screaming about the last episode, and quoting lines from me. As long as the cast is good, I'm good.[27]

The creative impulse behind local productions like *Anacostia,* of course, is the same force that for generations has animated community theater, public access TV, and amateur home movies. What was new in the early years of the 21st century was the widespread adoption of technologies, such as low-cost, high-quality digital video cameras and 'free'[28] internet hosting sites like YouTube and Vimeo (introduced in 2004 and 2005, respectively) that fundamentally altered storytelling possibilities (Lotz, 2007b; Turner & Tay, 2009). The former dramatically lowered the barriers to entry for independent web productions that matched, in some cases, the production values of network projects with multimillion-dollar budgets. The latter gave independent writer-producers access to audiences that had the potential at least to rival those that network programming typically attracted,[29] particularly given the decades-long decline in network ratings.

Indeed, between 1980 and 2011, the share of household audiences claimed by the major broadcast networks dropped from about

50% to less than 20%.[30] Some of this slide can be attributed to the explosion of other network options during the period (e.g., the rise of scores of new cable networks) and to the emergence of alternative modes of watching television content (e.g., 'time-shifting devices' like TiVo, Hulu and Netflix). These developments were emblematic of an era of 'narrowcasting' in which what once had been large national audiences were splintered into ever smaller, specialized (and marginalized) audience segments (Smith-Shomade, 2004; Spigel & Olsson, 2004). The multinational conglomerates that owned the networks adapted to this contemporary environment by vertically integrating (controlling both the distribution and means of production) and diversifying (expanding their portfolios of specialized networks) in efforts to increase, or at least hold on to, overall market share.

But this new Hollywood oligarchy, which aggressively marketed the illusion of an ever-increasing diversity of choices, actually churned out a menu of stories through its carefully managed production pipeline that was much more uniform than it had to be. Diverse perspectives were absent at each stage of this process, from the writer-producers who made an initial pitch to a network, to the writers who produced commissioned pilot scripts, to the showrunners who managed the writers' room for newly green-lit and continuing series. It was an insular process that resisted the infiltration of new ways of seeing, except, on occasion, around the margins.

It is no wonder that there was an explosion of thousands of eclectically written and produced web series like *Anacostia* during this period, some 'going viral' and attracting sizable and devoted audiences, expanding the boundaries of what we think of as 'television' (Lotz, 2007b; Pierce & Tang, 2012). Accessible anytime and anywhere – on laptops, tablets, and smartphones – a seemingly endless menu of webisodes constituted a new way of telling (and engaging with) the stories on which our lives depend. These locally rooted and independent productions filled an important cultural gap; free from institutional and market pressures, they endeavored to tell stories that network television is not structured to tell.

NOTES

1 While scholars have noted that it is difficult to objectively define the boundaries of the television 'text' (cf. Fiske, 1987), I am referring here simply to the audio and visual form in which stories are packaged.
2 In network television, the producers who manage the storytelling process are also writers (Hunt, 2007, 2009, 2013).
3 The Federal Communication Commission (FCC) repealed the Financial Interest and Syndication Rules ('Fin-Syn') in 1995, thus permitting networks to participate more aggressively in the ownership of the programming they air. In the absence of regulation curtailing this type of vertical integration, corporate pressure is placed on networks to acquire programs from production companies owned by their parent companies, due to anticipated increases in synergy and profits (Einstein, 2002b, p. 2). By 2002, the top six suppliers of primetime television programming, which included the six broadcast networks and Universal, accounted for 82% of programming hours (Einstein, 2002a, pp. 27–8).
4 According to the 2011 Writers Guild of America, West (WGAW) basic industry agreement, the minimum compensation for the story plus teleplay for a primetime show of 60 minutes or less was $35,568 in 2013. But the *actual* cost of a pilot script was likely to be much higher, as the typical pilot script cost between $75,000 and $100,000 as early as the first few years of the 2000s (Einstein, 2002b).
5 For example, Hunt (2013) shows that of the 566 pilots in the pipeline for the 2010–2011 season, only about 20%t had been picked up for series as of July 2012.
6 In this chapter, I use the common industry term 'showrunner' to refer to the writer-producers specifically charged with overseeing all operations related to a show's production. While all showrunners are 'executive producers,' not all executive producers are showrunners (Wild, 1999).
7 This was the case for a show I consulted on with the Nickelodeon cable network.
8 Even during the so-called 'Golden Age' of live television drama in the 1950s, Stempel (1996) notes, there was network interference in the screenwriting process, albeit less intrusive.

9 As Gottschall (2012, p. 186) notes, the basic model of fiction is 'character + predicament + attempted extrication'. Within network television, this basic model is adapted to a variety of genres, such as the situation comedy (cf. Smith, 1999), which generate stories and scripts that follow certain taken-for-granted formulae.

10 See the web series *TV Writers' Room* (first uploaded to YouTube in 2011) for a derisive treatment of what goes on inside the writers' room.

11 Silvia Cardenas Olivas, 'How Not to Get Fired from the Writers' Room', WonderCon panel, uploaded to YouTube on April 29, 2013.

12 Quoted in Gitlin (1983, p. 71).

13 This subheading was taken from the main title of the Writers Guild of America, West's *2007 Hollywood Writers Report* (Hunt, 2007).

14 Source: 2010 Census of the United States.

15 See, 'NBC-NAACP Diversity Plan Irks Coalition', http://articles.latimes.com/2000/jan/07/news/mn-51628

16 The Fox network pioneered this strategy, albeit on a smaller scale, after it launched in late 1986 (Zook, 1999).

17 See www.broadcastingcable.com/article/105512-Debmar_Mercury_Secures_200_Million_Distribution_Deal_for_Tyler_Perry_s_House_of_Payne.php.

18 For example, see 'Tyler Perry Hates Black Women: 5 Thoughts on the Haves and Have Nots', www.racialicious.com/2013/06/03/tyler-perry-hates-black-women-5-thoughts-on-the-haves-and-have-nots/

19 UPN, which originally developed the show, *The Game*, and the WB were merged into the CW network in 2006.

20 The smaller, black-owned cable network TVONE did not launch until 2004 and has of this writing not been able to consistently produce original, scripted programming.

21 Examples include the ABC/Disney's Writing Program, NBCUniversal's 'Writers on the Verge' Program, and the Fox Diverse Writers Program.

22 Interview with Hollywood industry insider, July 9, 2013.

23 E. Jack Neuman quoted in Stempel (1996, p. 249).

24 For a case study of how raced ways of seeing can affect storytelling in network television, see Coleman and Cavalcante (2012).

25 According to Anderson, his Southeast Boy Productions delivered the typical webisode of *Anacostia* for about $1,100, which covered incidentals such as camera batteries, make-up, and food. These minimal costs were mostly covered by local sponsors who subsidized the costs for a given webisode or block of webisodes. To offset costs further, Anderson also relied upon product placement, in which the products or services of local businesses were featured within the narrative of a given webisode in exchange for financial or in-kind support. Meanwhile, the cast members all had '9–5 jobs' and volunteered their talents for the series, in some cases driving to Washington DC from as far away as New York City for five consecutive weekends to shoot a season's 10 webisodes. A goal for a forthcoming Season 4, Anderson said, was to quadruple the production budget, which would allow him to continue to increase the web series's production values and grow its audience.

26 Anderson noted the web series had also attracted about 300,000 viewers in the Washington DC area on public access television DCTV and on other websites, combining to make up more than 1 million viewers over the three seasons by late 2013. By contrast, the premiere webisode of *Diary of an Awkward Black Girl* boasted more than 1.5 million views since it was uploaded to creator Issa Rae's YouTube channel in early 2011. A figure like this, which rivals ratings for lower-tier broadcast and network offerings, is in no small measure why Issa Rae was offered in late 2012 a development deal with Shonda Rhimes, the black female showrunner (a rare breed in Hollywood) who is responsible for the ABC hits *Grey's Anatomy* and *Scandal*.

27 Personal Interview with Anthony Anderson, March 2013.

28 While the typical account holder does not have to pay a user fee, YouTube, like most 'free' commercial internet sites, relies upon advertising and the marketing of user data for revenues.

29 The future of independent internet-based storytelling – as a contender to network television – will likely hinge on how questions of 'net neutrality' (Peha, 2006), audience manufacturing, and audience measurement (Bermejo, 2009) are resolved.

30 www.businessinsider.com/brutal-50-decline-in-tv-viewership-shows-why-your-cable-bill-is-so-high-2013-1. Networks include ABC, CBS and NBC.

REFERENCES

Bagdikian, B. (2004). The New Media Monopoly. Boston: Beacon Press.

Berger, J. (1972). *Ways of Seeing: Based on the Television Series with John Berger*. New York, NY: Viking Press.

Bermejo, F. (2009). Audience Manufacture in Historical Perspective: From Broadcasting to Google. *New Media & Society*, 11, 133–154.

Bielby, D.D. and Bielby, W.T. (2002). Hollywood Dreams, Harsh Realities: Writing For Film and Television. *Contexts*, Fall/Winter, 21–27.

Bielby, W.T. and Bielby, D.D. (1994). All Hits Are Fluke: Institutionalized Decision Making and the Rhetoric of Network Prime-Time Program Development. *American Journal of Sociology*, 99, 1287–1313.

Bielby, W.T. and Bielby, D.D. (1998). *Telling ALL Our Stories: The 1998 Hollywood Writers Report*. Los Angeles, CA: Writers Guild of America, West.

Bielby, W.T. and Bielby, D.D. (1999). Organizational Mediation of Project-Based Labor Markets: Talent Agencies and the Careers of Screenwriters. *American Sociological Review*, 64, 64–85.

Bielby, W.T. and Bielby, D.D. (2003). Controlling Prime-Time: Organizational Concentration and Network Television Programming Strategies. *Journal of Broadcasting & Electronic Media*, 47, 573–596.

Coleman, R. M. and Cavalcante, A. M. (2012). Two Different Worlds: Television as a Producer's Medium. *In Watching While Black: Centering the Television of Black Audiences*. B. E. Smith-Shomade, ed. 33–48. New Brunswick, NJ: Rutgers University Press.

Deacon, T.W. (1998). *The Symbolic Species: The Co-Evolution of Language and the Brain*. New York, NY: W.W. Norton & Company.

Einstein, M. (2002a). A Historical Perspective on Program Diversity. *Program Diversity and the Program Selection Process on Broadcast Network Television*. Washington, DC: Federal Communications Commission, Media Ownership Working Group.

Einstein, M. (2002b). The Program Selection Process. *Program Diversity and the Program Selection Process on Broadcast Network Television*. Washington, DC: Federal Communications Commission, Media Ownership Working Group.

Fiske, J. (1987). *Television Culture*. New York, NY: Routledge.

Georges, R.A. (1969). Toward an Understanding of Storytelling Events. *The Journal of American Folklore*, 82, 313–328.

Gitlin, T. (1983). *Inside Prime Time: How the Networks Decide About the Shows that Rise and Fall in the Real World Behind the TV Screen*. New York, NY: Pantheon.

Gitlin, T. (2002). *Media Unlimited: How the Torrent of Images and Sounds Overwhelms Our Lives*. New York, NY: Henry Holt.

Gottschall, J. (2012). *The Storytelling Animal: How Stories Make Us Human*. New York, NY: Houghton Mifflin Harcourt Publishing.

Hall, S. (1973). *Encoding and Decoding in the Television Discourse*. Birmingham: Centre for Contemporary Cultural Studies.

Herman, E.S. and Chomsky, N. (1988). *Manufacturing Consent: The Political Economy of the Mass Media*. New York, NY: Pantheon.

Hunt, D.M. (1997). *Screening the Los Angeles 'Riots': Race, Seeing, and Resistance*. Cambridge: Cambridge University Press.

Hunt, D.M. (2000). *The African American Television Report: Progress and Retreat*. Los Angeles, CA: Screen Actors Guild.

Hunt, D.M. (2005). Black Content, White Control. In *Channeling Blackness: Studies on Television and Race in America*. D.M. Hunt, ed., 267–302. New York, NY: Oxford University Press.

Hunt, D.M. (2007). *Whose Stories Are We Telling? The 2007 Hollywood Writers Report*. Los Angeles, CA: Writers Guild of America, West.

Hunt, D.M. (2009). *Re-Writing an All-Too-Familiar Story? The 2009 Hollywood Writers Report*. Los Angeles: Writers Guild of America, West.

Hunt, D.M. (2013). WGAW 2013 TV Staffing Brief. Los Angeles, CA: Writers Guild of America, West.

Lotz, A.D. (2007a). The Promotional Role of the Network Upfront Presentations in the Production of Culture. *Television and New Media*, 8, 3–24.

Lotz, A.D. (2007b). *The Television Will Be Revolutionized*. New York, NY: New York University Press.

Morley, D. (1992). *Television, Audiences and Cultural Studies*. New York, NY: Routledge.

Newcomb, H. and Hirsch, P.M. (1994). Television as a Cultural Forum. In *Television: the Critical View*. H. Newcomb, ed., 503–515. New York, NY: Oxford University Press.

Peha, J. (2006). The Benefits and Risks of Mandating Network Neutrality, and The Quest for a Balanced Policy. Paper presented at the 34th Telecommunications Policy Research Conference.

Phalen, P. and Osellame, J. (2012). Writing Hollywood: Rooms With a Point of View. *Journal of Broadcasting & Electronic Media*, 56, 3–20.

Pierce, L.M. and Tang, T. (2012). Refashioning Television: Business Opportunities and Challenges of Webisodes. *International Journal of Business and Social Science*, 3, 163–171.

Quinn, E. (2012). Closing Doors: Hollywood, Affirmative Action, and the Revitalization of Conservative Racial Politics. *The Journal of History*, September, 466–491.

Smith, E.S. (1999). *Writing Television Sitcoms*. New York, NY: Perigee.

Smith-Shomade, B. (2004). Narrowcasting in the New World Information Order: A Space for the Audience? *Television & New Media*, 5, 69–81.

Spigel, L. and Olsson, J. (2004). *Television After TV: Essays on a Medium in Transition*. Durham, NC: Duke University Press.

Stempel, T. (1996). *Storytellers to the Nation: A History of American Television Writing*. Syracuse, NY: Syracuse University Press.

Turner, G. and Tay, J. (2009). *Television Studies After TV: Understanding Television in the Post-Broadcast Era*. New York, NY: Routledge.

U.S. Commission on Civil Rights. (1977). *Window Dressing on the Set: Women and Minorities in Television*. Washington, DC.

Wild, D. (1999). *The Showrunners: A Season Inside the Billion-Dollar, Death-Defying, Madcap World of Television's Real Stars*. New York, NY: HarperCollins.

Zook, K.B. (1999). *Color by Fox: The Fox Network and the Revolution in Black Television*. New York, NY: Oxford University Press.

13

Television Cinematography

Deborah Tudor

Television cinematography is a set of highly complex technological, social, aesthetic and industrial-economic practices. Understanding these relationships and relating them to practices in other media requires a multi-dimensional approach that historically contextualizes cinematography industrially, sociologically, technologically and aesthetically. This chapter outlines a number of fields of research that help shape the television image using a combinative approach that allows a fuller understanding of the forces that shape the cinematographic styles of television. Although the chapter presents areas of research as discrete sections, it will become apparent that they are heavily interdependent, and that these fields of study address questions across boundaries.

THE STRUCTURES OF PRODUCTION

The field of production culture offers a paradigm that examines the impact of cultures of laborers on the work of film production, and the ways in which its practitioners understand it. This culture is revealed through a proliferating number of channels: a wide array of production personnel interviews available to the public through trade journals, popular magazines and blogs, as well as DVD commentaries. These texts constitute an apparent 'behind the scenes' look at production culture whose transparency demands critical interrogation. John Caldwell (2008) emphasizes the contemporary Hollywood industry management of publicity 'to market its production practices as a spin on the "authentic" production process and as on screen entertainment', and claims that these publicity items allow media producers to control the cultural narratives about production, as well as to transform them into profit-generating mechanisms.

The structure of media production culture has a large impact on style, since television and film crews draw from same labor pool in Southern California and other large production centers. Industry convergence also promotes technological crossover among productions for different platforms: theatrical, network, cable network, internet and mobile media. Recent analyses of style tend to incorporate production context; Jeremy Butler (2010) describes his study of television style as a merger of 'problem solving' production practices with close textual readings.

Political economy studies have demonstrated repeatedly the importance of film and television holdings as conglomerates, as well as the interdependent nature of film and television (Holt 2010; Winston 1996), in addition to showing the importance of understanding the television industry's systems to understand the final product that shows up on television screens (Meehan 2010). This industrial convergence produces effects that influence all aspects of television production, not just cinematography. Caldwell (2008) identifies the following trends that are shaping current practices of television production: 'migratory labor and churn, outsourcing's bid culture, speed shooting and hyperproduction, the digital sweatshop, the director/producer as bible, masculinized tools and worker masochism, gendered production space, industrial contact zones, studio tracking boards and counter-tracking boards, criticism as stealth marketing, branding as industrial viral practice and the collapsed workflow caused by the DI (digital internegative).

Feminist scholars have added further nuances to the study of production culture through an examination of ways that media production genders labor. Studies of the production of texts by women producers and directors expand production culture to include issues of gendered labor and production by examining the work and conditions of labor for 'below the line' workers, many of whom are female (D'Acci 1994; Levine 2001). Miranda Banks (2009) points out the critical differences that emerge in such studies of above and below the line workers. She finds the distinction 'crucial to understanding the nature of production … to seeing different possibilities for intervention by feminist production studies scholars, and she puts special emphasis on the definition of above the line as 'creative' and below the line as 'craft'. Such markers can be unpacked to find economic and cultural values that accrue to work defined as either Creative (above the line) or Technical (below the line) (Banks 2009; Caldwell 2008). The field of feminist production studies is developing paradigms that unpack the descriptions of masculinized labor and production practices, which allows a more detailed study of the practice of cinematography.

TELEVISION PRODUCTION TECHNOLOGIES

Both analog and digital production technologies and practices co-exist in the early 21st century. These practices are often combined in production. Analog capture material is digitized and processed through a digital workflow. These negotiations among technologies occur at the level of production, and debate flourishes in the professional trade journals and their online bulletin boards.

While some directors and cinematographers conduct extensive field tests of HD cameras and 35 mm cameras to make a shooting choice, others make the choice on less closely reasoned grounds (Bosley 2009). Ongoing discussions of the advantages and disadvantages of film versus HD capture and extensive critiques of various cameras dominate threads on cinematography discussion boards. For many high profile HBO and cable network series, capture occurs on either 35mm film or HD. Such high-end series may also use film dailies for the pilots, but generally switch to digital dailies for the rest of the episodes (Thomson 2010).

A survey of trade journals reveals that capture formats, though clustered around HD or 35mm, also include other non-standard formats, such as Super 16 film, the capture medium used for AMC's *Walking Dead*

(Sweeney 2007), and Super 8 film, used for filming a small portion of an episode of *Life on Mars* (Holben et al. 2009). Production interviews contain numerous references to choice of format being one that cinematographers make by considering 'whatever serves the show'. This statement can be viewed as an exemplar of Caldwell's industrial self-publicizing (Holben et al. 2009).

Many network series shoot on HD. The series *Lie to Me* (Fox) switched from film to digital capture for its second season (Oppenheimer 2010), although the Director of Photography reports that they chose the Arri D-21 camera because of its ability to deliver 'film like' images. Fox's new series *Chicago Code* shoots on HD, primarily with the MX Red One (Mullen 2010). DSLRs are also being used in some cases for television commercial shoots and for some series episodes, like the season finale of *House* in May 2010 (Crow 2010).

For postproduction, distributed workflow may be added to the labor conditions and factors of industry structure affecting the practice of cinematography. The distribution of files on a server enables several editors to work simultaneously, even from different geographic locations. While there is currently no systematic data to support the claim that this practice is as yet an industry standard, it is used frequently on television series with identified 'high production values' – ones that follow the stylistic model of theatrical film production, i.e. single camera filming. For HBO's *True Blood*, shot on 35mm, Technicolor develops and telecines the footage. The practice uses server-based editing of files that allows multiple edit units to work on the same core files. Except for dailies and delivery processes, the output is file-based. *True Blood* cinematographer Matthew Jensen claims that the post process benefits him, as increased access to grayscale allows for 'much subtler contrast changes and color combinations' (Goldman 2010, p. 77).This process is still rare enough to be described as a 'major test case for an all file-based workflow for an episodic TV series originating on film' (p. 77).

TELEVISION STYLES AND CONVERGENCE: CINEMATIC, TELEVISUAL AND VIDEOGRAPHIC

As indicated, media convergence and the commonality of the production labor pool help distribute stylistic elements across television, cinema, internet and mobile media. The wide range of televised material – sports, soap operas, game shows, shopping shows, cable movies, mini-series, dramas, comedies, news and reality shows – also complicates the study of televisual style. Caren Deming (2010) points out that the television industry has undergone several periods of convergence. She examines the early post-World-War-II years of the industry as a 'period of experimentation and innovation characteristic of a period when new technologies' begin the process of interpenetration with an established technology, and she concludes that since its earliest days, television has resisted any easy dichotomization between media, and that 'convergence between cinematic and televisual styles is evolutionary' (Deming 2010, p. 139), and needs to be understood as such. This conclusion implies that the realms of style in television are not static, but are in constant states of flux, as a brief synopsis of the debates over these categories indicates.

In their most basic forms, the categories *cinematic* and *televisual* signify two different broadly conceptualized categories of program style, with a range of variations in each. These categories generally differ at the levels of camerawork and lighting, set design and sound. Shows using a cinematic style tend to shoot with one camera, using precise framing and movement within sets constructed in depth. This allows camera movements to occur laterally and in depth, or on both the x and y-axis. These factors distinguish cinematic style on television from a style more commonly perceived as 'televisual', which is identified by little camera movement on the y-axis, use of shallow set construction, and three-camera production (Butler 2010). Further stylistic differentiation occurs between shows that utilize Steadicam shots versus those that use handheld, shakier camerawork. This is one of the characteristics of a visual style that refers to documentary cinematography: shaky image, whip pans and generally unsteady framing.

However, the concept of the televisual expands from this narrow definition in the work of other theorists, including John Caldwell (1995) and Caren Deming (2010). Caldwell's televisuality describes a complex series of determinations that produce varying visual styles on television. While Caldwell uses the same definition of 'cinematic' as does Butler, he inserts another category standing against the cinematic, instead of relying on Butler's narrow use of televisual. The category of the 'videographic', defined as a hyperkinetic style that uses a high number of effects, opposes the cinematic in this model (Caldwell 1995). This redefinition expands the televisual to contain a number of different presentational styles; indeed it is its stylization that distinguishes it, along with an inversion of the importance of style and substance. The televisual cannot be analyzed apart from the industrial circumstances producing it, so he firmly embeds the televisual within the cultural and economic circumstances of the television industry in the 1980s (Caldwell 1995). Deming's (2010) approach branches out further, defining televisuality as a 'synthesis of stylistic, technological, and ideological characteristics' (p. 126). This is a fluid definition of the term. She explores the constitutive elements of televisuality: spatiality, aurality, femininity and hybridism while locating this formation historically. She finds elements of televisuality present in her analysis of *The Goldbergs* (1948–1954), a series from television's 'golden age', indicating a long historical timeline for this concept (Deming 2010).

Shows like *24* have used multiple frames onscreen as an alternative to editing. Critics have classified this stylistic technique as an artifact of technological convergence and the social usage of internet and mobile media (Friedberg 2009; Manovich 2001; Tudor 2008). The usage of multiple frames enables audiences to experience several simultaneous points of view within one scene. This acts as an alternative to continuity editing. Multiple frames can also provide points of view into simultaneous scenes occurring in separate locations, creating a stylistic alternative to traditional crosscutting. Shifts in framing and editing attributable to convergence, and earlier, the importance of television channels as an ancillary market for theatrical releases (Bordwell and Thompson 2001), have produced a mutant version of continuity editing called 'intensified continuity'. This system of editing is marked by the absence of establishing shots, increased use of close ups and shots that include multiple points of narrative focus, and shorter average shot length, leading to faster editing pace (Bordwell and Thompson 2001).

Convergence also has created a production environment in which advertising and show content are blurred and combined in different ways (Jenkins 2002). In addition to product placement, the framing of shots on television now is sometimes partially obliterated under a pop up advertisement for another show on the same network, usually one following the current show or one aimed at the same audience. These pop up ads block part of the screen, rendering precise framing somewhat incoherent and splitting the audience's attention between the narrative and the moving pop up ad. This practice seems limited as yet to episodic network television, yet, significantly, it crosses show categories, appearing on everything from network dramas like *House* (Fox) to reality-style DIY shows like *The Barefoot Contessa* (The Food Network), and complicates categories of style. Following Herbert Zettl (1989), it would seem that this reconfigures television's onscreen space as graphical space, where such overlays call attention to a perceived lack of depth (as cited in Deming 2010, p. 133). The use of these pop ups creates a layered onscreen image that resembles a computer screen, thus activating some of the spectatorial relations explored in the research cited below.

MOBILITY AND AESTHETICS

The ability to watch television on mobile media generates a discussion of technological, industrial, economic and aesthetic ramifications. Convergence has been discussed in terms of its ability to 'unbundle' shows into segments for mobile media (Dawson 2007). Theoretical discussions of segmentation and flow and remediation give some context to contemporary unbundling of television shows for migration to other platforms (mobility). These discussions can be traced to the work of Raymond Williams (1974) on television's stream of programming, advertising and previews of new shows as a 'seamless, irreversible flow that elided the differences between its constitutive elements' (as cited in Dawson 2007, p. 239). John Ellis' work on the segment as the foundational unit of television answers Williams by arguing that flow does not erase the distinctions or 'seams' between short segments (Ellis 1992 [1982]). Jane Feuer assimilated both Williams and Ellis' points of view through proposing a 'dialectic of segmentation and flow ... segmentation without closure' (Feuer 1983). The debate on flow has also been given a long historical timeline through the claim that the medium of television was designed with movement in mind, a claim that links the development of portable television to an increasingly mobile culture beginning in the 1960s (Spigel 2004). Erik Dawson (2007) provides the preceding summary of the flow argument in context with the notions of 'remediation' (Bolter and Grusin 2000). This term refers to the 'representation of one medium in another' and Bolter and Grusin embed this argument, derived from Marshall McLuhan's (1964) statement that the content of a medium is always another medium (as cited in Bolter and Grusin 2000, p. 45). They also argue that a medium is itself something which remediates; that is, 'that which appropriates the techniques, forms and social significance of other media and attempts to refashion them in the name of the real' (Bolter and Grusin 2000, p. 65). This constructs an inherent relationship between media, as no one medium can ever operate without reference to other media.

It is clear that 'unbundling' a television program can produce another revenue stream for the owners of the material, and the proprietary 'locking' of some unbundled television shows, like *24: Conspiracy*, discussed below, can limit the migration of content to multiple platforms. Dawson (2007) argues that this locking also can produce a hardware-specific aesthetic that prevents media texts from being able to move across platforms.

Producers also use mobile media to present auxiliary material that is not televised but is related to the episode. Many shows, such as *30 Rock* (NBC) or *Battlestar Galactica* (SyFy) have associated web series that run on the network website, while a number of series episodes are available online through either independent or network affiliated sites, usually the day after airing on television. Workflow and production issues for web and mobile media form a large category of discussion at media producer websites and bulletin boards such as www.cinematography.com or www.digitalcontentproducer.com. The former is primarily a bulletin board and digital content producer, run by *Millimeter* magazine and publishes online articles on production in addition to hosting extensive forums on all aspects of cinematography. A typical article recounts a specific, detailed production experience, such as the process of capture, editing and publishing to mobile media (Ozer 2008).

The adaption or reorganization of television aesthetics for the small screens of mobile devices has taken several approaches. These distinctive approaches are not simply modifications of television production to suit technological requirements of mobile media, but rather are discursive formations formed by the same complex determinations that have impacted television aesthetics (Dawson 2007). *24: Conspiracy*, a mobile series derived from *24*, provides a case study of how a very stylized television aesthetic adapts into mobile screens. This series was developed for delivery over the Vodaphone system in the UK. Holson (2005) describes the look of the mobile version as one that used an overwhelmingly high proportion of tight close-ups, flat, high-key lighting, center-weighted character blocking, very shallow depth of field, and exaggerated size for elements of *mise-en-scène* deemed too small to be seen easily on a small screen such as gunshot wounds, which 'became the size of grapefruit for the mobile version'. When the web series was included on a season boxed set of *24* DVDs, the hardware specific mobile aesthetic of the Vodaphone series did not translate well to the television screen. However, episodes of the regular TV series *24* remain consistently one of iTunes most heavily downloaded items, indicating that its television aesthetic moves well among types of screens, indicating that its style represents a flexible aesthetic that derives from television production's historical concern with the variable quality of the range of television receivers available at any one time (Dawson 2007).

NEWER DEVELOPMENTS IN RESEARCH

Auteur theory is another area of study that has shifted to accommodate newer models of production in the industry. The designation and role of an auteur is part of the underpinnings of studies that examine the organization of images on television. The recruitment of emerging and established cinematic auteurs in the 1980s into television commercial advertising and series production forms an influential strand in the emergence of Caldwell's (1995) concept of televisuality. However, the traditional positioning of the director as auteur has been displaced by the series showrunner (executive producer) who controls the look of show, holds the 'bible' and hires directors (Caldwell 2008). The showrunner may also be or become a cinema auteur: for example, J.J. Abrams, whose series *Alias* and *Lost* garnered him a cinema career (*Cloverfield* 2008, *Star Trek* 2009) and have promoted him as an auteur on the basis of his oeuvre. This provides an example of a case in which a television showrunner working in a videographic (Caldwell) style appropriates a foundational element that allows recognition of style, the status of auteur. The creation of auteurs in television, a mass industrial production, also raises a question about the perceived 'need' for auteurs. Theorists view the valorization of auteurs as particularly problematic in US television, due to its mode of industrial production (Shattuc 2010). Why are some show creators dubbed auteurs, with 'vision', while others remain completely anonymous? These arguments trace the lineage of auteurism from the Western Renaissance through literary studies of authorship, art house cinema, authorship studies of US studio directors, to contemporary production culture studies that view the 'author' as one more element of a controlled industrial discourse. In these public interviews and 'behind the scenes articles' (discussed above) cinematographic choices can be attributed to terms like 'inspiration', or 'individual style', without situating choices in current ideological, industrial practices.

Research in the area of culture and technology situates television cinematography in broader, historicized studies of the social functions of technologies. In this field, technology is widely viewed as more than hardware; it is a type of matrix of contexts that permits a range of types of image productions (Caldwell 1995), and a set of instrumentations that carry cultural fears, aspirations and other concerns related to the role of technology in our lives. Studies in this field are careful to avoid questions of vulgar technological determinism, pointing out those ideological forces at work on the invention and diffusion of technologies. The introduction of new technologies bases itself on older, more widely accepted ones, while the appeal of the new technology is presented as a dream or vision of progress (Sobchack 2004).

Localized studies in this area analyze specific problems, such as the ways that the rhetoric of filmmakers reveals how they rationalize the introduction of digital technologies, including 3D, into already existing analog processes, and how these technological shifts can enable different choices (Turkle 2004), which are then rhetorically transformed into individual creative aesthetic choices.

Research on technological, aesthetic and industrial convergence is not monolithic; it is inflected by studies of a simultaneous splintered field of narrower local production cultures (Caldwell 2008), and also by ways in which one of the basic metaphoric and actual elements of media – the screen – bears multiple meanings and uses in a culture heavily penetrated by media devices that are in many cases only a screen, like a tablet computer or a smart phone. The large variety and number of screens that a person living in a developed nation encounters ranges across technologies and formats: analog projection, television screens, older cathode ray tubes, plasma and LED screens in many sizes and aspect ratios. Passive spectatorship oscillates with interactive uses of a screen as device. This development is being studied in more localized ways, which means that classical film theory's universal approach to cinematography, editing, sound, performance and reception studies is undergoing adaptation into multiple, fragmented perspectives. Work on screen studies for the so-called 'new' media produces models of spectatorship that rest upon an assumption that a computer window is not coterminous with the screen, since in fact, multiple windows may be open on a computer screen at the same time. These multiple windows can give access to local or global data, thereby creating a disunified spectator position very different from an image organized around a Renaissance style single perspective image. In this situation, classical theories of spectator and screen do not explain the multiple perspectives and positions that viewers may take, alternating spectatorship with interactivity or immersion (Friedberg 2009; Manovich 2001; Tudor 2008). Since the window, a term from computer usage, is not congruent with either television or cinema screen (Friedberg 2009), the unpacking of the processes of working with screens (spectating and using), and the ideological meanings of screens, begins to diverge from a strict Lacanian model. In this model, the 'screen' acts as a mirror, and is heavily implicated in the formation of subjectivity through lack, or presence and absence (Grosz 1989). This lack then anchors spectatorship. An alternative view to spectatorship is emerging through the consideration of interactivity and the use of multiple windows, one that is tentatively predicated on pattern and randomness rather than absence or presence (Hailes 1999). Such discussions of multiple windows and/or screens unpack the metaphors and ideology of the window, the screen and the frame, and, through such analysis of the fractured perspectives as they offer, may point us toward aesthetic forms in which the 'multiple and simultaneous' supersede sequentiality (Friedberg 2009, p. 243). This approach will become crucial in 21st-century media research, as the image undergoes further repositioning with relationship to any type of screen. We currently see a renewed interest in 3D movies, in which the image can escape the boundaries of the frame and seem to float between the spectator and the physical screen of the theatre. Holographic projection systems that produce images seeming to occupy three-dimensional space, and which can be interacted with, will further challenge theories of the relationships of human viewers and users to images.

REFERENCES

Banks, Miranda J. (2009). 'Gender Below-the-Line: Defining Feminist Production Studies'. In Vicki Mayer, Miranda J. Banks and John T. Caldwell (eds), *Production Studies* (Kindle version). New York: Routledge.

Bolter, Jay and Richard Grusin (2000). *Remediation: Understanding New Media*. Cambridge MA: MIT Press.

Bordwell, David and Kristin Thompson (2001). *Film Art* (7th Edn). New York: McGraw-Hill.

Bosley, Rachael K. (2009). Pitch Perfect. *American Cinematographer*, 90(10), 30–43.

Butler, Jeremy (2010) *Television Style*. New York: Routledge.

Caldwell, John (1995). *Televisuality: Style, Crisis and Authority in American Television* (Kindle version). New Brunswick, NJ: Rutgers University Press. Retrieved from www.amazon.com

Caldwell, John (2008). *Production Culture* (Kindle version). Durham, NC: Duke University Press. Retrieved from Amazon.com

Crow, Brian (2010, April 14). *House* Season Finale Shot Entirely w/ Canon 5DMark Two. (Msg 1) Message posted to: www.cinematography.wonderhowto.com/forum/house-season-finale-shot-entirely-w-canon-5d-mark-ii-85/

D'Acci, Julie (1994). *Defining Women: Television and the Case of Cagney & Lacey*. Chapel Hill, NC: University of North Carolina Press.

Dawson, Max (2007). Little Players, Big Shows. *The International Journal of Research into New Media Technologies*, 13(3), 231–250.

Deming, Caren (2010). Locating the Televisual in Golden Age Television. In Janet Wasko (ed.), *A Companion to Television*. London: Wiley Blackwell, pp. 126–141.

Ellis, John (1992[1982]). *Visible Fictions: Cinema, Television, Video*. London: Routledge.

Feuer, Jane (1983). The Concept of Live Television: Ontology as Ideology. In E. Ann Kaplan (ed.), *Regarding Television: Critical Approaches*. Frederick MD: University Publications of America, pp. 12–22.

Friedberg, Anne (2009). The Virtual Window: From Alberti to Microsoft. Cambridge, MA: MIT Press.

Goldman, Michael (2010). 'Post-Focus'. *American Cinematographer*, 91(8), 76–78.

Grosz, Elizabeth (1989). *Sexual Subversions*. Sydney: Allen and Unwin.

Hailes, N. Katherine (1999). *How We Became Posthuman*. Chicago: University of Chicago Press.

Holben, Jay, Jean Oppenheimer, Iain Stasukevich and Patricia Thomson (2009). Cutting Edge Cinematography. *American Cinematographer*, 90(3), 20–45.

Holson, L.M. (2005, October 17). Is That a One-Minute Soap Opera, Or Is It Mom Calling? *New York Times*, p. 1C.

Holt, Jennifer (2010). It's Not Film, It's TV: Rethinking Industrial Identity. *Jump Cut #52*. http://ejumpcut.org/currentissue/Holt/NotFilmTV/index.html

Jenkins, Henry (2002). *Convergence Culture: Where Old and New Media Collide*. New York: New York University Press.

Levine, Elana (2001). 'Toward a Paradigm for Media Production Research: Behind the Scenes at General Hospital. *Critical Studies in Media Communication*, 18, 66–82.

Manovich, Lev (2001). *The Language of New Media*. Cambridge MA: MIT Press.

McLuhan, Marshall (1964). *Understanding Media: The Extensions of Man*. New York: New American Library; Times Mirror.

Meehan, Eileen (2010). Watching Television: A Political Economic Approach. In Janet Wasko (ed.), *A Companion To Television*. London: Wiley-Blackwell, pp. 238–255.

Mullen, David (2010, December 3). Chicago Code (Msg 1). Message posted to www.cinematography.com/index.php?showtopic=50270

Oppenheimer, Jean (2010). Telltale Tics. *American Cinematographer*, 91(7),16–20.

Ozer, Jan (2008, February 1). Onsite Workflow. *Digital Content Producer*. Retrieved from: www.digitalcontentproducer.com/dcc/revfeat/onsite_workflow/index.html

Shattuc, Jane (2010). Who Makes American TV? In Janet Wasko (ed.), *A Companion To Television*. London: Wiley-Blackwell, pp. 142–154.

Sobchack, Vivian (2004). Science Fiction Films and the Technological Imagination. In Marita Sturken, Douglas Thomas and Sandra Ball-Rokeach (eds) *Technological Visions: The Hopes and Fears That Shape New Technologies*. Philadelphia: Temple University Press, pp. 145–158.

Spigel, Lynn. (2004). Portable TV: Studies in Domestic Space Travels. In Marita Sturken, Douglas Thomas and Sandra Ball-Rokeach (eds), *Technological Visions: The Hopes and Fears That Shape New Technologies*. Philadelphia: Temple University Press, pp. 110–144.

Sweeney, Vincent (2007, August 7). The Walking Dead on ABC (Msg 1). Message posted to www.cinematography.com/index.php?showtopic=47584

Thomson, Patricia (2010). Mob Money. *American Cinematographer*, 91(9), 34–49.

Tudor, Deborah (2008). Eye of the Frog: Reconceptualizing Space in Cinema. *Cinema Journal*, 48(1), 90–110.

Turkle, Sherry (2004). Spinning Technology: What We Are Not Thinking About When We Are Thinking About Computers. In Marita Sturken, Douglas Thomas and Sandra Ball-Rokeach (eds), *Technological Visions: The Hopes and Fears That Shape New Technologies*. Philadelphia: Temple University Press, pp. 110–144.

Williams, Raymond (1974). *Television: Technology and Cultural Form*. New York: Schocken Books.

Winston, Brian (1996). *Technologies of Seeing*. London: BFI Publishing.

Zettl, Herbert (1989). *Graphication in Television Studies: Textual Analysis*. New York: Praeger, pp. 137–163.

14

Options and Exclusivity: Economic Pressures on TV Writers' Compensation and the Effects on Writers' Room Culture

Felicia D. Henderson

> If this doesn't change, we are entering a new culture, a culture of indentured servitude. A culture where the studios are being allowed to change the minimum wage without the necessity of negotiating with the Union that's there to protect the writers. [Billy Ray, Co-Chair WGA Negotiating Committee, Writer, *Captain Phillips*, *The Hunger Games*] (B. Ray, 2014)

Studios and networks argue that the rising cost of production is among the most pressing reasons for seeking ways to lower production costs. Shorter production seasons and controlling above-the-line costs have often been primary targets for cost-cutting measures. Where a short-order season once was a money-losing proposition for a studio, money from international and digital rights can sometimes now justify such a move (Schneider, 2012). Also, shorter production seasons can potentially attract high-profile talent in front of the camera who would otherwise shrink from the commitment of traditional television schedules. Cable and cable-equivalent outlets have taken advantage of the lure of shorter productions seasons to entice feature film talent for some time. Networks have begun to follow the lead of cable outlets to attract the same level of talent. The popularity of serialized dramas, which garner low ratings when repeated, is another studio and network justification for shorter seasons.

Some high-profile writers have been vocal about the benefits of shorter seasons. The co-creator and executive producer of *Lost* (2004–2010), Damon Lindelof, stated:

> To any broadcast network willing to reduce an episodic order I say simply, 'Bravo!' ... I do wish that the [networks] would make these decisions based on the show itself – not just to lure top talent – as some series would benefit significantly by having less episodes. But this is certainly a hell of a start. (Schneider, 2012)

Neal Baer, who executive produced *ER* (1994–2009) and *Law and Order: Special Victims Unit* (1999–) is also an advocate. 'Certainly there's more pressure doing 22 episodes. The train leaves in July and never stops 'til the end of April or mid-May' (Schneider, 2012).[1]

Baer subscribes to a popular argument among those in favor of shorter seasons – the opportunity to eliminate the pressure to come up with an exhaustive number of episodic stories (usually 22).

What has long been ignored by studios, cable outlets and networks is how short-order seasons affect television writers' compensation. For, while the manner in which shows are produced is transitioning to shorter schedules, writers contracts continue to contain clauses that were not problematic under long production seasons, but compromise their earning capacity in an increasingly short-order production work world. What is not taken into consideration in the argument for shorter seasons is how the antiquated contract *options* and *exclusivity* clauses create downward economic pressure on writers' salaries and consequently affects the cultural political environment and therefore the production culture in the writers' room. Therefore, while the talent pool interested in working in television but unwilling to forego film careers to do so, and top-tier, well-compensated showrunners may consider shorter seasons preferable, what was once hailed as the major appeal of cable as a superior creative environment – shorter series orders – has become a practical problem for writers working on such shows.

PRODUCTION OF CULTURE: HISTORICAL VS CONTEMPORARY TELEVISION SEASONS

In television's formative years, writers on variety shows such as Sid Caesar's *Your Show of Shows* (1950–1954), half-hour comedies like *I Love Lucy* (1951–1954), *The Twilight Zone* (1959–1964), and other dramas, worked under exclusive employment contracts for approximately 10 months of the year to produce approximately 28–39 episodes of prime-time television. Then, the writers experienced a hiatus – an unpaid interruption in the continuity of production – for approximately two months before returning to produce another season of episodes of a show. During hiatus periods, writers were still under contract, but did not earn a salary. Employment in US-based prime-time television operated on the relative certainties of this production calendar for over fifty years.

A critical part of the production process has been the creation of narrative content by television writers. Many have discussed the collaborative or collective authorship that occurs in the writers' conference room in order for what I term *situational authorship* to thrive. The culture in the writers' rooms – which often mimics a dysfunctional family unit[2] – and the manner in which television has been authored since the famed and much celebrated writers' (conference) room was first popularized behind the scenes of Sid Caesar's *Your Show of Shows*, continues to be the core narrative content creation in contemporary television. Often considered the most talented *dream team* of comedy writers in television history, the media focus on the behind-the-scenes maneuverings of *Your Show of Shows* – specifically, the creative process that revolved around Caesar, Imogene Coca and the writing staff – is responsible for much of the early myth-making surrounding the television comedy writing process. The much honored writers included head writer, Mel Tolkin, Lucille Kallen (the only woman on the writing staff), Howard Morris, and those who would go on to become comedy luminaries: Carl Reiner, Neil Simon and Mel Brooks. The idea for the series belonged to Sylvester 'Pat' Weaver, the President of NBC from 1949 to 1956, and Max Liebman, a Broadway theater writer, producer, director and composer. Yet, Weaver and Liebman are rarely mentioned as being integral to the success of the groundbreaking comedy-variety show. Instead, the mythic history around comic output was created in the writers' conference room – a cigar smoke-filled small office where comic 'geniuses', boisterous men and one woman, created comedy by committee and invoked humor that spoke to a broad audience. The actions in the television writers' room have been of popular and critical interest ever since.

In contemporary television, whether in dramas, comedies or talk shows, on-air narratives continue to attempt to capture and creatively depict this space for viewers. Witness, Aaron Sorkin's ill-fated NBC drama *Studio 60 on the Sunset Strip* (2006–2007), Tina Fey's popular sitcom, 30 Rock (2006–2013), is a comical and critical exploration of the same space, and most recently, the Sundance Channel's *The Writers' Room* (2013), a talk show in which the host is allowed behind-the-scenes access to the writers' rooms of successful comedies and dramas and one-on-one interviews with the writers and producers. Interest in depicting this space on TV began with the *Dick van Dyke Show* (1961–1966), and the cultural activities, social relations and institutional constraints operative in the writers' room remain relevant to defining the space as one that is central to production culture in prime-time television series production. Given the significance of the space, the increasingly prickly relationship between the Writers Guild of America (WGA) and the Alliance of Motion Picture and Television Producers (AMPTP) over what the WGA argues has become antiquated contractual controls, presents an opportunity to examine the shift in cultural politics in the writers' room as a result of the rift between labor and management.

In contemporary network television, a typical season consists of a more manageable 20–24 episodes over approximately nine months of production, followed by a three month period of unpaid employment if you are employed on a show that will be returning to production for another season. Historically, such periods of down time have been welcome reprieves from intense, albeit creatively rewarding, work in the business of writing television shows. Over the past decade, the business model in cable television and cable-equivalent internet outlets – 6–12 episode series orders – has led to longer periods of unpaid employment. As explained to Guild members in a letter from WGA President Chris Keyser and Negotiating Committee Co-Chairs Chip Johannessen and Billy Ray, during the 2014 contract negotiations with producers, 'Writers on short-order shows now find themselves working for half a year or less, then stuck on unpaid hiatus for open-ended periods while waiting to learn if their shows – and their contract – will be renewed' (Patten, 2014).

Although networks are also experimenting with short-orders, the majority of network television still operates on a fall through spring schedule with some exceptions and experimental programming efforts. With cable channels, long lag times between seasons and extended options on writers' services that contractually obligate them for up to six months after the previous season finale airs or up to nine months after the season premiere, hampers a writer's ability to seek employment on other shows. As further explained in the Keyser, Johannessen and Ray letter:

> During this period they are virtually unemployed because studios demand 'exclusivity' and 'first position,' preventing writers from seeking other work, their ability to make a living cut off. At a time of unprecedented prosperity for the companies, writers' economic circumstances should not be deteriorating. These contract terms are clearly unfair and require a collective solution. (Patten, 2014)

As mentioned earlier, in recent years, network television has been testing the short-order seasons, as well. Perhaps most notably, ABC's *Scandal* (2012–) debuted in its premiere season with an eight-episode order (including the pilot episode). Fox's *Sleepy Hollow* (2013–), CBS's *Under the Dome* (2013–), and ABC's *Once Upon A Time in Wonderland* (2013–), similarly, launched with abbreviated, by network standards, 13 episode seasons, instead of the standard 22 episodes,[3] compared to the 8–12 on cable and cable equivalent outlets.[4] Unfortunately for writers contractually bound to series, the reduced orders, whether in network or cable, have not been accompanied by corresponding changes in their contractual obligations to the studios that employ them. These constraints discussed above are problematic for writers because although they are working less – 6–13 episodes instead of 22 – yet they remain contractually exclusive to the studio for the entire year, just as they did when the studio produced twice as many episodes. The model for a US television season is changing, yet the studios contractual hold on writers has remained the same.

CULTURE OF PRODUCTION: ECONOMIC PRESSURES AND CULTURAL RUPTURES IN THE WRITERS' ROOM

The economic and creative pressures placed on writers as a result of working under employment contracts that do not reflect contemporary trends in television production creates fissures in a system that is dependent on the relationship between management (producers) and labor (writers) as business partners. When this relationship is fractured, bitter union contract disputes, work stoppages and ruptures in the cultural politics in the writers' workspace may potentially erupt. When analyzing this disconnect from a broad cultural production perspective not unlike that offered by Bernard Miège (1989) in his seminal work, *The Capitalization of Cultural Production*, further explored by David Hesmondhalgh in *The Cultural Industries*, and expanded upon by John Thornton Caldwell (2008) in *Production Culture*, the contested space between content creators and studio is clear. The downward economic pressures on television writers' compensation as a result of the trend toward shorter series orders leads to the potential for political, creative and cultural ruptures in the writers' workspace. In other words, if the model of a television season is changing in US prime-time television for the first time in more than fifty years, yet the studios' contractual hold on writers is antiquated, the very foundation of what allows writers' rooms to be effective workspaces is threatened.

Regarding the exclusivity language in writers' contracts, once a writer has been engaged by a studio, generally, the studio demands language in the contract that gives them exclusive rights to a writer's services. Such exclusivity can be sweeping – no allowance for any outside work during the production period – or somewhat accommodating – no outside work on other television series, but writers may seek employment on movies of the week (MOWs), for example. Often, in an attempt to continue to build their careers, writers will work on projects *in the background* because they are contractually bound to the show on which they are currently employed. They are allowed to work on other projects during hiatus periods, but they are obligated to the studio to which they are contracted and the studio has the right to recall them for services at any time. This makes those with available employment opportunities very reluctant to consider such writers because those writers are encumbered by a contract that is entitled to the said writers' services in *first position*. Complicating matters further is the option clause in a writer's contract. This clause allows the studio to delay a decision about whether or not to engage the writer's services on the show or pilot after the writer has signed a contract. During the option period the studio has the right to preclude, or at least interfere with, the writer's ability to consider other work (Stankevich, 2014). The result is that there may be prolonged periods of time when the writer may not be able to gain employment and generate other revenue.

WRITERS' COMPENSATION AND UNION INTERVENTION

Founded in 1933 to represent screenwriters, the Screen Writers Guild (the precursor to the WGA) began representing television writers in 1948. Since, the early days of network television writers have been represented by a union that utilizes collective bargaining to negotiate, set and enforce minimum salaries for its members. Every three years, the WGA negotiates a collective bargaining agreement with the AMPTP. The contract covers screen, television and new media writers (Minimum Basic Agreement, 2014). The Guild's early strikes of the 1950s and 1960s focused on defining a residuals format whereby the writers who had written episodes could be compensated when their work was repeated on television.[5] The 1960s strikes expanded focus on residuals to include screenwriters' compensation when their feature films aired on television and health benefits. The focus in the 1970s was on improving the definitions of residuals. As recently as the 1980s, management and labor struggled to agree on fair terms and apportionments for compensation beyond initial salary obtained from creating television content – residuals. On this front, 'At least five strikes ensued in the 1980s, three of them lengthy. Writers' participation in the home video market (VHS and later DVDs) was lost in the 1985 contract negotiations, setting the stage for the 1988 writers' strike, which lasted more than five months' (Handel, 2011). After a relativity long period of calm, in 2007, it was residuals – how to define and compensate writers for the use of creative content utilized on burgeoning digital media platforms – that led to yet another strike.

For the Minimum Basic Agreement (MBA) contract ending in 2014, residuals are not at the core of contract negotiations. The points of contention are unique in that the central issue is restrictive contracts clauses that encumber writers' ability to seek employment or primary compensation. Previous labor disputes that led to work stoppages were not based on primary compensation for content creation. At its core, primary compensation is a more simple calculation than residual compensation, (which is always a fraction of primary compensation). The minimum compensation is outlined in the WGA's MBA and usually increases – usually by 3% – per negotiated contract period (Minimum Basic Agreement, 2014). As the 2014 negotiations between the WGA and the AMPTP demonstrate, the battle over the options and exclusivity language in writers' contracts – as cable and cable-equivalents are offering nearly as many staff writing positions as network television – is one that affects writers' primary compensation now and will do so in the future.

> During calendar year 2013, broadcast networks introduced 23 new series, while cable/digital debuted almost 40, not counting kids fare. That means that soon there may be more writers working in cable and digital than in broadcast, all of them facing the underemployment problem that is at the heart of the current WGA-AMPTP stand-off. (Andreeva, 2014)

With the increase in short-order employment in cable and more and more network examples following suit, a larger number of writers – a critical mass of writers – are working on shows under the restrictive contracts. Union intervention as the current contract was expiring was the next logical step. Over the last 19 years, I have held every position on a writing staff – from writer trainee to executive producer on ten half-hour comedies and one-hour dramas.[6] It has been my experience that one of the discussions that binds the writers on a staff is light-hearted complaining about their Union, the cost of private school for the kids, the network that should run more promos for the show, or the studio that should stand up to the network for the creative integrity of the show. Still, at the end of most gripe sessions the writing staff generally agreed that what they were grousing about are *first world* or *high class*, problems – an acknowledgement that such conversations amounted to little more than good-natured venting with individuals with whom they shared a unique and protected space. There is an implicit understanding of their socio-economic status relative to the average American citizen. The life of a steadily employed television writer is one of privilege, and expressing such gripes outside of this space would be ill-mannered. In the writers' room, sharing in this manner is simply a ritual common in this workspace that only those in the workspace would understand.

Yet in 2014, in the months leading up to the new contract negotiations between the WGA and the AMPTP, writers began speaking to outsiders.[7] From their private and protected workspace, in a manner inconsistent with the cultural politics of the writers' room, they complained about unlimited holds and exclusivity being contractually forced on them. In the past, it was not uncommon for writers' room conversations to turn to talk of which talent agency had the better literary department, which agency was known for its lazy agents, and whether or not television writers need agents and managers in today's transitional industry. Equally common has been the writer who pipes up to say that he/she feels lucky to be paid to do something he/she would do for free or to comment on the satisfaction of getting the opportunity to do exactly what he/she had hoped to do. My goal is not to paint a picture of an idyllic world of television writers that doesn't exist. There have been plenty of stories – many made public through interviews, memoirs, biographies, trades publications, history books, and news reporting – that demonstrate how creatively challenging, dysfunctional and emotionally and mentally unhealthy this environment can be. I am simply noting that, in my experience, writers very reluctantly chatter about their *first world problems* to anyone other than other writers, anywhere other than in the writers' room. In the 2014 MBA contract negotiations writers showed extreme concern for the erosion of their ability to earn primary compensation and this concern was manifested in a shift in the cultural politics in the writers' room.

WRITERS' FRUSTRATIONS AND A POLITICAL CULTURAL SHIFT IN THE WRITERS' ROOM

Evidence of writers' discontent and the shift in how they exhibited their frustration with the outdated clauses in their contracts can be found in both the letters written to union members by the WGA President and Negotiating Committee Co-chairs and in my interview with Billy Ray, one of the Co-Chairs. In a letter dated January 30, 2014, before negotiations with the AMPTP began, the Co-Chairs Johannessen and Ray mentioned that they would be standing by the goal of 'limiting unpaid exclusivity and holding arrangements that increasingly prevent television writers from making a living' (C.J. Ray, 2014). More telling is their second letter, addressed to Guild members on March 13, 2014:

> Last week we concluded our second round of contract talks with the AMPTP. Every aspect of our contract has been negotiated and agreed upon with two exceptions – options and exclusivity – which remain points of contention between us ... But with the advent of basic cable, pay TV and now Netflix-type Internet shows, that changed. Short-orders of 13, 10, or even 8 episodes are now the norm, and the predictability of the old broadcast season has vanished. As you have told us in showrunner meetings and scores of show visits, these two developments – short-orders and uncertain schedules – have combined to create a serious problem for many writers. (Patten, 2014)

This communication to WGA members is an AMPTP negotiations update. The tone regarding options and exclusivity is intentionally firm and speaks of the Negotiating Committee Co-Chairs getting their directive on the need to act regarding 'short-orders and uncertain schedules' directly from writers, in 'showrunner meetings and scores of show visits'. Showrunner meetings typically take place in hotel ballrooms, restaurants, or at the WGA headquarters. *Show visits* are official visits from a WGA representative (paid administrators who are not WGA members) to individual shows to speak with writers about the various services the Guild offers and to listen to writers' concerns. Such visits are largely ceremonial, with writers, who are usually anxious to get back to work, asking a minimal amount of questions. It is only during contentious contract negotiations and threats of work stoppages that such visits become important from a writing staff's point of view. What is clear in the above missive from the

Guild is that writers were behaving out of character. The politics of their protected space calls for them to quickly dismiss any interlopers, even those affiliated with the show. Actors, studio and network executives and production integration representatives alike may all be welcomed into the writers' room, but the goal is always to escort them out as expeditiously as possible. What is demonstrated here, however, is that in large enough numbers to get the negotiating committee's attention, writers shared their dissatisfaction with compensation constraints – also something writers rarely do with anyone outside of their protected community.

The fact that writers engaged WGA representatives on many show visits about their compensation, speaks to how seriously writers felt about the inability to seek other employment while waiting through unpaid hiatuses that sometimes last as long as a year. In speaking specifically about dissatisfaction with their compensation, not only did they act in a manner not consistent with writers' room production culture, they did so with an outsider in a ritual outside of the established cultural politics of their private space. Doing so delivered a clear message to the WGA Negotiating Committee about what television writers considered the most important issue of the current contract negotiations with the AMPTP. Co-Chair Ray provides further proof of the level of writer discontent that led to a significant rupture in writers' room production culture – made public through the use of the media – declaring options and exclusivity issues for which the WGA would consider a work stoppage:

> The reason that we made an issue of this, is that we've gone to writers' meetings and writers' rooms, over and over again and the issue of options and exclusivity has come up over and over again. What we're seeing is a subset of [a] larger issue – the race to the bottom. The downward pressure on writers' compensation that is directly related to contracts that block them from competing for jobs. It's a corporate mentality of cost-cutting. We believe that [the] language of the studios is the language of indentured servitude. Not unlike the sharecropper system in [the] reconstruction south. We have heard about it over and over from writers because now, writing as a proposition, feels shakier. (B. Ray, 2014)

Instead of lively discussions about story, character development, episode, season and story arcs in a workspace that normally supports collective authorship, writers are now wondering if they can make a middle-class living as a writer, yielding many conversations about personal finances. Under these conditions, a shift in culture takes place whereby writers do ask other writers to discuss their contracts and whether, individually, they have been able to negotiate away unlimited exclusivity clauses. Hostility is bred between more powerful writers who are more highly sought after and, therefore, have more negotiating clout, and lower-level writers who have little or no negotiating leverage. The latter often must witness the former finish his or her 6–13 episode order and then go off to work on another show for another 6–13 episodes, in an effort to cobble together a full season of compensation because he/she has the negotiating leverage to do so. This breeds resentment in the writer who must stand by while the cable or network show he/she is contracted to decides on airdates, whether or not to split one season of episodes into two season, whether or not they will air new episodes in a few months or in a year. Initially, being staffed on a critically acclaimed, award winning series may be enough to sustain a writer for a season or two. It is also true that all shows are not critical successes, and earning roughly one-third to one-half of what traditional network series writers earn because your show is producing one-third or one-half the number of shows becomes economically and culturally problematic. According to Ray:

> Studios and networks don't take into consideration that writers have to start thinking about economics and that division happens between them when they do. On top of that, because they're squeezing wages, they're forcing writers to look for other sources of income whether that be writing spec scripts or selling real estate, it makes them less effective in the writers room, doing their jobs. (B. Ray, 2014)

PROGRESS TOWARD AN IMPERFECT FUTURE

On April 2, 2014 a letter from the Co-Chairs to WGA members announced that a tentative new MBA agreement for the 2014–2017 contract had been reached and:

> Importantly, we have now placed limits on the options and exclusivity requirements often imposed on episodic television writers. Our negotiations on these issues were complicated and protracted, but the companies worked with us to find solutions. As a result, the endless unpaid holds that have become more and more commonplace in television have now been addressed in the MBA for the first time ever. (C.J. Ray, 2014)

On April 30, 2014, 'The new film and television contract was ratified by an overwhelming 98.5% of WGA members who voted. Of the 1,193 valid votes cast there were 1,175 "yes" votes and 18 "no" votes (1.5 %). There were 8,218 eligible voters' (Verrier, 2014). Although the WGA President Keyser and the 20-member Negotiating Committee that includes Co-Chairs Johannessen and Ray lauded the precedent setting victory in relations to holds and unlimited exclusivity on writers that severely compromised their ability to earn a competitive wage, the victory deserves further scrutiny. In the actual language from the announcement on the WGA website, the specifics of the deal are somewhat problematic.

> (1)The agreement may not require that the writer be exclusive to the Company except during periods when the writer is being paid for his or her writing services; and (2) the Company may not hold a writer for more than 90 days under a negotiated option agreement without paying a holding fee of at least 1/3 of the MBA minimum for the writer's services. If the Company chooses not to pay the holding fee, the Company must allow the writer to accept an offer of other series employment, or, if the series has been renewed, it may exercise the option and put the writer back to work. On January 1, 2016 the threshold will increase by 5% to $210,000 per season. (C.J. Ray, 2014)

The WGA, in general, and the MBA specifically, fights for policies and compensation patterns that cover minimum compensation for all writers. Under the new agreement writers earning $200,000 or less will no longer be forced into what Ray called 'Indentured servitude'. However, writers and writer-producers who earn more than $200,000 are not covered by this agreement. This means that most upper-level writers who do not have the influence to negotiate out of the limiting clauses will continue to be forced into agreements that hamper or outright preclude them from working on more than one 6–13 episode ordered series per contract period. The reward for experience and promotion is a contractual obligation to downward pressure on economic opportunity. This new deal also does not allow the culture in the writers' room to return to culture as usual. In fact, it may shift the power dynamics a bit in the direction of lower-level writers, who now have more career opportunities when measured by contractual freedom than upper-level writers.

The primary purpose of collective bargaining is to come together to accomplish what no one voice can alone. The WGA – film, television, radio and digital media's collective bargaining arm – negotiates a morally acceptable floor and morally acceptable working conditions. This is what unions are charged with protecting. It certainly makes sense to adamantly challenge exclusivity language because such clauses limit a writer's ability to work. Rather than attempting to legally change the MBA – a covered writer's minimum wage – the AMPTP has attempted to change it in effect by controlling options and exclusivity. Shorter seasons without a contractual means of foregoing the unlimited waiting periods is, in effect, union-busting – a method of manipulating minimum compensation without negotiating with the WGA.

The 2014 WGA/AMPTP contract negotiations were not the first contentious negotiations over primary compensation showing evidence of how far the AMPTP is willing to go to exert downward economic pressure on writers' salaries. In fact, I would argue that the request with which the AMPTP's opened the negotiations – $60 million in rollbacks in health, pension and screenwriter minimums – stems from residual hostility from the 2007–2008 WGA strike. As writers experienced after that strike, producers randomly decreased writers' salaries and discontinued many lucrative production deals with writer-producers, claiming that due to the strike they could no longer afford to pay writers pre-strike compensation (Henderson, 2010). As Hesmondhalgh argues, cultural industries and the texts they produce are complex, ambivalent and contested (Hesmondhalgh, 2003). Television production is an example of this complex relationship. The studios' long-standing stance on the need for holds and exclusivities despite critical changes in the length of production seasons and writers' strong reaction to that stance is a specific example of the site of contestation. The producers who fund creative output focus a great deal of attention on justifying their position regarding holds and exclusivity as corporate strategies and fiscal responsibility without much concern for how the content creators on whom they are dependent will react to such corporate thinking about what they see as creative endeavors. Often content creators have difficulty viewing their art as commerce. Even when they are able to do so, experiencing their contributions being treated as relatively inconsequential to a business's overall success is difficult to process.

In the 2014 MBA contract negotiations between the WGA and the AMPTP, management's focus on contractual strangleholds relied on historical arguments about the way business has always been conducted. It is unreasonable to negotiate contracts based on not wanting change because how business is being carried out is how it has always been carried out. It would not have been unreasonable for producers to claim that it is with their funding that writers and producers on their shows are trained and become better writers and producers with a unique skill set learned while earning a salary on a set paid for by the producer. The fact is that writers are unique commodities with special skills that studios should not want to allow competitors to access. During contract negotiations, the position in which studios and writers found them is not unlike the relationship between professional sports team owners and the athletes that are contracted to play for them. Players cannot play for one team and then another in the same season unless they are traded to that team for all future play. Because the Los Angeles Lakers season is over does not mean the players should now be allowed to go play with the Los Angeles Clippers during their playoff run, and then return to the Lakers when the new season begins. It is all a gamble. You play for the team that drafts you. If that team has a season of eight episodes, you produce eight episodes and wait until next season. In this argument, team owners or studios who spend millions of dollars compensating athletes and writers to play in their system, and who give them the opportunity to learn their trade secrets and gain valuable experience that will make them more employable, are not unreasonable for not wanting to share their investments with the competition in the off season.

NOTES

1 It is worth noting that both Lindelof and Baer are among the television showrunner/executive producer elite. They are highly compensated and sought after. Currently, Lindelof is under an exclusive multi-year, multi-million dollar contract to create shows for Warner Bros. Television and Baer has a similarly exclusive deal with CBS Productions. Their opinions about whether the length of a season should be shorter should not be considered the views of the average television writer.

2 I believe writers' rooms can be compared to dysfunctional families in that the relationships between the writers are familial in nature. Each writer, including the showrunner, naturally falls into his or her familial roll, i.e., showrunner/executive producers as paternal/maternal figures, who are the head of household; upper-level writers such as the *know it all* big sister, the *crazy uncle*, and the *eternally adolescent* big brother are the other archetypes. Inexperienced young writers and lower-level writers playing the roles of the *too anxious to please* little brother, the *mysteriously aloof* middle sister, and the *screw up but*

charismatic beautiful young aunt who, at least once per month, gets her act together just before she's fired. Together, their individual strengths make them a strong *family* that can accomplish the task at hand.

3 Although a standard network season is considered 22 episodes, series are ordered for 12 episodes plus the pilot and a *back nine* episodes that are ordered if the ratings substantiate it.

4 The Writers Guild of America represents film, television, radio and digital media writers. Therefore, whether working in cable or network television, television writers are represented under the same bargaining agreement.

5 Residuals or 'Payments that studios and producers make to actors, directors, writers, and others when movies and television programs are rerun or are reused in another media' (Handel, 2011).

6 In addition to my cinema and media studies research as a doctoral student at UCLA, I am a working television comedy and drama writer and producer, who has worked on the writing staffs of several sitcoms and dramas including *The Fresh Prince of Bel Air*, *Moesha*, *Everybody Hates Chris*, *Soul Food*, *Gossip Girl* and *Fringe*.

7 According the Alliance of Motion Picture and Television Producers (AMPTP) website, 'Since 1982, The Alliance of Motion Picture and Television Producers (AMPTP) has been the trade association responsible for negotiating virtually all the industry-wide guild and union contracts, including the American Federation of Musicians (AFM); American Federation of Television and Radio Artists (AFTRA); Directors Guild of America (DGA); International Alliance of Theatrical Stage Employees (IATSE); International Brotherhood of Electrical Workers (IBEW); Laborers Local 724; Screen Actors Guild (SAG); Teamsters Local 399; and Writers Guild of America (WGA)'. It negotiates 80 industry-wide collective bargaining agreements on behalf of over 350 motion picture and television producers (www.amptp.org/).

REFERENCES

Andreeva, N. (2014, March 14). *Breaking News*. Retrieved from www.deadline.com: www.deadline.com/2014/03/wga-amptp-close-to-new-contract-more-talks-march-31/

Caldwell, J.T. (2008). Production Culture: Industrial Reflexivity and Cricial Practice in Film and Television. Durham: Duke University Press.

Handel, J. (2011). *Hollywood on Strike! An Industry at War in the Internet Age*. Los Angeles, CA: Hollywood Analytics.

Henderson, F.D. (2010). It's Your Own Fault: How Post-Strike Hollywood Continues to Punish Writers for Striking. *Popular Communication: The International Journal of Media and Culture*, 8(3), 232–239.

Hesmondhalgh, D. (2003). *The Cultural Industries*. London: Sage Publications.

Minimum Basic Agreement. (2014, May 5). Retrieved from www.wgaeast.org: www.wgaeast.org/index.php?id=162 (accessed 03 11, 2014).

Miège, B. (1989). *The Capitalization of Cultural Production*. Bagnolet: International General.

Patten, D. (2014, March 13). No WGA Deal Yet, Talks Set to Resume on March 31. *Breaking News*. Retrieved from www.deadline.com/2014/: www.deadline.com/2014/03/wga-amptp-close-to-new-contract-more-talks-march-31/

Ray, B. (2014, March 16). (F.D. Henderson, Interviewer)

Ray, C.J. (2014, April 30). *Communication to Members*. Retrieved from www.my.wgaw.org: www.my.wgaw.org/content/subpage_secure.aspx?id=5396

Schneider, M. (2012, March 9). Networks Explore More Cable-Sized Short Order Series. TV Guide. *Today's News: Our Take*. Retrieved from www.tvguide.com: www.tvguide.com/news/networks-short-order-series-1044601.aspx

Stankevich, M. (2014, March 21). Partner, Stankevich-Gochman (F.D. Henderson, Interviewer).

Verrier, R. (2014, April 30). Writers Guild of America Members Ratify Contract. Retrieved from www.latimes.com: www.latimes.com/entertainment/envelope/cotown/la-et-ct-wga-vote-20140430-story.html

15

A Greener Screening Future: Manufacturing and Recycling as the Subjects of Television Studies

Vicki Mayer and Clare Cannon

> The technology of broadcasting was introduced as a marginal element in very complex social structures. (Raymond Williams 1974, p. 26)

An activist of the New Left, Williams probably had little notion that his book *Television: Technology and cultural form* would become a canonical text for television studies scholars. Yet the founder of such key concepts as *flow* and *mobile privatization* would be remembered largely by the words addressing the latter half of the subtitle. Technology, its manufacture and its disposal, would be largely forgotten, left in the same landfill as the sets themselves. Television studies moved on to consider production as a relatively closed system that begins with a program concept and ends with a program viewer.

Nevertheless, television set manufacturing and recycling are key concepts to the study of television in all its varied technological forms. In this chapter we review this slow trajectory in recognizing the material production of television as a technology, from the late-1960s to the early 2010s. From there, we examine three lenses through which television and media scholars have looked at the creation, distribution and eventual discard of the technological product. While these ways of seeing television position content as but 'a marginal element' to these 'very complex social factors', the political and economic implications of their studies have redirected television studies to consider television beyond its texts.

FROM TECHNOLOGY OF FREEDOM TO TECHNOLOGIES OF TOXICITY: THREE HISTORICAL AWAKENINGS

Although the critique of electronics production and its incumbent political effects might be drawn back to V.I. Lenin's (1916/2011) withering assessment of American and German electrical cartels in the early 1900s, the critical study of television manufacturing per se can be traced back to parallel strains of scholarship that addressed the relationship between national sovereignty, geopolitics and economic domination. Williams was not alone in recognizing that the development of radio and then television broadcasting was guided, first, by the political interests of the nation-state, and, second, by the economic interests of the manufacturers of the sets. Pointing to the uncompromising defense of private US communications companies in the negotiations with other countries after World War II, Herbert Schiller described some of the ways that 'economic might and communications know-how could complement each other to effectively promote the creation of an American century' (1969, p. 49). First, the know-how and capital investment of manufacturers would help assist in the national economy by creating a Fordist cycle for the production and consumption of manufactured goods. US-based factories would produce the sets that the working masses could buy themselves. Financed by commercial advertising, the set manufacturers then became the first producers of television contents, promoting further consumption and production. From there, television set manufacturers would be key to US foreign policy, opening new markets for technological transfer, programming know-how, and a host of US imports. Through the marriage of electronics and political economy, Schiller envisioned television as both tool and technique of US empire. Schiller, along with Williams, Armand Mattelart (1983), and a host of others, first recognized the ways television set manufacturing could grow national gross domestic product, while dominating new communications markets abroad. Cloaked in the political rhetoric that television was a 'technology of freedom' (qua De Sola Pool, 1983), the US by the early 1980s successfully shifted the focus of international debates in the postcolonial era from the national control of communications manufacturing to the freedom of speech and the free flow of information.

As this tight alliance between a Fordist-led cycle of capital accumulation in the US and Western Europe and a Western-led dominance of television set manufacturing abroad slowly unraveled over the next forty years, scholars once again recognized the importance of set manufacturing, this time in terms of global liberalization policies, and their impacts. The ebbing of nationalist import-substitution policies around the world paved the way for the establishment of free trade zones in which electronics manufacturers could find cheaper labor and relaxed environmental standards. Envisioned as a 'race to the bottom' for the cheapest labor and resources, critical scholars witnessed the transfer of manufacturing companies from the US and Western Europe to East and Southeast Asia, Mexico and other locations throughout Latin America. In North America, this movement was most punctuated by the passage of the North American Free Trade Agreement in 1994, which reduced trade and tariff barriers between the US, Mexico and Canada (Casteñeda Paredes, 2003; Mosco 1990). In Brazil and India, the liberalization of the imports market, including tariffs reductions, allowed domestic companies to import technology 'kits' of parts that could then be cheaply assembled in-house (Mayer 2011, p. 37; Pendakur 1991, p. 245). Those companies, often subsidiaries of the largest multinationals, then extended their reach back into the territories of the home offices as well as into neighboring countries (Reygadas 2002). Jefferson Cowie's (2001) history of RCA's television factories shows that the corporation moved operations, first within the United States and then to Mexico in search of greater labor control. New destinations for manufacturing and assembly experienced an accelerated degree of class disparities between the new managerial class fostered by the electronics boom and the class of factory workers, many of whom emigrated for temporary contracts in often dangerous working conditions (cf. Barrera 1996, pp. 193–194). Hazardous industrial waste poisoned local communities (cf. Casteñeda Paredes, 2003; Smith and Raphael 2003). Conversely, communities that had once depended on manufacturing went into economic decline (Cowie 2001, pp. 38–40). Communities with landfills began to recognize the economic and environmental burdens of handling and recycling television sets as the digital revolution progressed, producing mountains of e-waste (Raphael 2003).

The digital transformation of television set the final stage for television scholars to recognize the importance of television manufacturing and the ways that the machinery itself contributed to ongoing trends in complex social structures. The televisions themselves of the last decade resemble computers as television content has merged with other digital streams on flat screens and handheld electronic devices. Their manufacture largely mimics that of other high-tech industries in that production is no longer governed by single corporations. Rather, production is modularized between highly coordinated corporate units and their subcontractors, from highly-specialized firms of designers, developers and marketers, to the mass manufacturing plants where chips are made, circuit boards wired, plastics molded, and components assembled for a wide range of electronic goods. In the new television economy, manufacturing is no longer a 'core competency for market control' as profit margins have shifted to product innovation and intellectual property rights (Lüthje 2006, p. 22). Following the policies that precipitated this sea-change (cf. Drake 1993; Schiller 1999), media and communication scholars asserted that a global wave of government-led mandates for digital broadcasting and the deregulation of digital technology trade follow an old geopolitics of television's marketization and empire building (Berland 2009; Castañeda 2013; Galperin 2004). In the US and Western Europe, digital television would be a way to continue monopolizing the most profitable sectors over the manufacturing of new electronic goods; while in other countries, governments have worked in conjunction with corporate interests to create a new market for digital television and stimulate foreign investment. Meanwhile, electronics producers can be assured a windfall from the guaranteed obsolescence of analogue television sets along with the planned obsolescence of new television-viewing gadgets (Parks 2007). Critics have noted that these trends not only reflect but also accelerate the environmental despoiling of the Earth (Boyce and Lewis 2009; Maxwell and Miller 2012) and the development of a new international division of labor, one in which outsourcing and contracting has placed the greatest economic risks on the backs of those workers in the most precarious sectors of television production (cf. Mayer 2009; McKercher and Mosco 2007; Qiu 2009).

The periodic revival of the study of television manufacturing and recycling demonstrates the importance of these processes to international politics and economic globalization, as well as the social contexts of people's lives and their landscapes. This history also marks the growing awareness among media scholars that the study of television should include a longer history of the medium and its life cycle as a material object. Despite these awakenings, the twin topics of television manufacturing and recycling could be identified still as the slumbering giants in the house that television studies has built. Those who have tried to stir the giants have taken three theoretical approaches in making these issues heard.

'UNTRASHING' TELEVISION; THEREBY TRASHING TELEVISION STUDIES

Even with its claims to interdisciplinarity, media studies in general and television studies in particular, have been apt to fetishize their own object. In other words, the importance of television as a set of representations passing through a medium has replaced any sense of actual medium and how it became the tool of choice for what these scholars consider the most important cultural forum in the late 20th century. The pre-eminence of television then as a particular kind of box made of vacuum tubes, hand-soldered wires, a wooden case and concave glass was simply assumed, even if in fact, there were many other means to broadcast an image. The selection of the television set in its particular material form had to do with the availability of materials and a particular labor process in national and industrial contexts.

To unearth these contexts, media scholars have taken various critical frameworks and applied them to television manufacturing and recycling. While no scholar could be said to be rigid in their theories or methods, we offer them in three strands that describe various approaches towards the topics: media archeology, historical ecological materialism and grounded ethnography. Each approach commonly starts with a corrective to the underlying assumptions about media's value in television studies today.

Media Archeology

Media archeology is an approach that considers the material processes that led to the cultural construction of particular objects as television over all others and the subsequent replacement of television objects as these material processes change. Referencing Walter Benjamin's consideration of the 'aura' of obsolete objects as the product of industrial transformations and Michel Foucault's insights into the disciplinary power of discourses, media archeology is a way to ground discourses about television and new media in material processes, while observing the ways the discourses govern how objects are created, used and discarded (cf. Parikka and Huhtamo 2011). Through this lens, discarded media are still communicative of 'a relationship between our past and our future' (Sterne 2007, p. 16). Addressing the past, the junked television set is a fossil, writes Jennifer Gabrys (2011). Like a fossil, the set contains a remainder or residue of past decisions made across its lifespan. Projecting those decisions to the future, Jonathan Sterne (2007) writes that new media are already marked by their future decomposition, a result of a consumer society that defines obsolescence by its association with the latest functionality. Over time, the sedimentation of televisions as fossils, both physically in the dump and rhetorically in the discourse, produce a record of what counts as media. In other words, the newness of digital television relied not only on innovations on the production line, but also on a discursive shift away from 'newness' as a completely different communication means, and towards a fetish of modern consumerism. Media archeology demonstrates that the line between the television set and trash is arbitrary. As Gabrys (2011) illustrates, there is no hard and fast rule between when a television functions perfectly well and when the same television becomes e-waste. At the same time, the arbitrary line is not accidental; it is grounded instead in contingent relationships between social networks. From the making of chips and circuits to salvaging the copper from cathode tubes, each stage in the life of the television involves valuing only some materials that generate profit and discarding the rest as waste. Media archeology thus employs historical and organizational methods to question teleological assumptions in television studies that new television technologies are always better and inherently progressive in improving societies.

Ecological Materialism

Bringing together ecological history together with a critical political economy, ecological materialism seeks to expose the power relations embedded in the making and unmaking of television sets, screens and other digital media technologies. Following structural patterns of oppression and injustice across the 'nexus of management, empire, labor, and the media', ecological materialism lays waste to the twin ideologies in media studies of technological empowerment and the autonomous media producer/consumer (Maxwell and Miller 2012, p. 13). To the former ideology, Richard Maxwell and Toby Miller chart the involvement of the military-industrial complex in dominating the production of film stock, the laying of international cables, and a dangerous regimen of radiation and chemical experiments, which have consequently poisoned workers and their environs. To the latter ideology, they provide an alternative narrative of the colonial and postcolonial labor conditions in the Global South that support others' creative production in the Global North. Ecological materialism thus attempts to show that 'the digital era has not eradicated social inequalities, but in fact has exacerbated the negative dynamics that maintain the status quo' in terms of race, gender, nation and region (Castañeda 2013, p. 118). These dynamics seem most obscured through television's programming, which frequently promotes eco-celebrity causes, 'green' lifestyles and middle-class ingenuity in salvaging and repurposing garbage (cf. Parks 2007). Greenwashing as a way of promoting consumer goods through faulty advertising claims is threaded throughout the branded discourses of television networks, news programs and entertainment, making corporations seem like 'socially responsible citizens' and middle-class citizens seem like the sources for environmental heroism (cf. Miller 2006). Meanwhile, the profit imperatives built into the television spur new ways to extract surplus labor through new corporate processes of recycling and e-waste picking. Lisa Parks (2007) explains the chain of managers that bring her monitor from a recycling plant 30 miles from her hometown in Southern California to Australian-owned facilities in Malaysia, Vietnam and the Philippines. At the ends of those chains, Ravi Agarwal and Kishore Wankhade (2006) document the trade, labor and environmental laws that conspire to keep India's recycling workforce sick and destitute. In these ways, ecological materialism considers television's extractive technologies in the current capitalist system, but the point of raising consciousness about this system is to change it.

Cultural Geographies of Manufacturing and Recycling Labor

Whereas ecological materialism charts the impacts of political and economic structures on a global geography of the television's life cycle, cultural geographies of manufacturing and recycling labor are most concerned with how those structural processes are experienced in specific locations. Questioning what television means to the subjects who work in the electronics factories, live in the free-trade zones, and operate on the margins of the global economy, these studies show how transnational systems of production help create the cultural and class identities of their workers. As Luis Reygadas' (2002) comparative study of transnational manufacturing companies shows, each factory operates as its own ecosystem, hybridizing the cultures of managers and laborers, all of whom often migrate from other parts of the globe in search of work. In each factory, then, television workers are made to fit the criteria of the corporation. Leslie Salzinger (2003) asserts multinational electronics corporations fantasize a docile and nimble-fingered labor force, which each corporation then has to produce through working bodies. Her study of assembly work at various electronics factories in Cuidad Juarez on the US-Mexican border reveals that different hiring and disciplinary control mechanisms feminized and, conversely, masculinized workers in very different ways. As a result, Lisa McLaughlin (2007) found that in a Malaysian electronics factory women complied as docile working-class subjects who would not demand either higher wages or better working conditions. The notion of co-implication means that women may use their feminine identities to their advantage at times, while reinforcing stereotypes and hiding other forms of racial, religious or age oppression they may suffer. In these situations, notions of cultural or class resistance are highly contextualized by geographic circumstances. Pun Ngai's (2005) ethnography of women in a Chinese chip and circuit factory found that the high levels of surveillance meant that pain, as in bodily or emotional collapse, was the only way to find a release from the toils and poisons of the daily production processes. Vicki Mayer's (2011) ethnography of women in circuit and assembly factories in Brazil found more variation between stricter and more permissive factories, though any form of organized resistance would be immediate grounds for dismissal and blacklisting from employment in the entire industrial district. By highlighting the importance of gender, race, class and national identity, cultural geographies of manufacturing and recycling labor illustrate how power dynamics in factories infiltrate the communities where they concentrate, empowering some and marginalizing others. At the same time, ethnographies tend to show that even at the bottom of social hierarchies, people are not technological 'have-nots', but have their own social networks and cultural registers for using technologies (cf. Qiu 2009). An ethnographic study of the e-waste recycling in Dhaka, Bangladesh found that even term 'e-waste' may have little meaning to those who see used electronics as sources of income and potential social mobility, as Josh Lepawsky and Mostaem Billah (2011, pp. 40–41) wrote: 'We expected we would end up in dumpsites, in piles of waste. Instead we wound up in production sites. We hadn't followed things ejected from the global economy – we were right in the middle of it'. These grounded studies of what televisions mean to the person making them provide a corrective to television studies that ascribe notions of creativity, knowledge production, pleasure and resistance only to the most privileged classes of media producers and consumers.

TOWARDS A PROACTIVE STUDY OF TELEVISION MANUFACTURING AND RECYCLING

Media scholarship on television manufacturing and recycling has followed on the heels of countless scores of nongovernmental agencies, labor and environmental rights movements and activists in talking about the role of the physical object in hindering the well-being of the Earth and its peoples. Numerous independent documentaries have provided counter-narratives to those that tout television, its attendant technologies and mass media consumerism as the *fait accompli* of modernity.[1] In this sense, this stream of critical writings within television studies could be considered generally reactive rather than proactive. Within the discipline, however, these studies are proactive both in pushing media scholars to recognize television within wider networks of power and in relation to a deeper ethics of what it means to study television.

For more than a decade now, environmental and labor groups have put forth recommendations towards a greener screening future. Since individual consumers cannot affect the necessary systemic changes, activists have come together in order to expose human rights abuses and environmental degradation and to change policy in order to better protect labor and the environment from the detrimental effects of the production, consumption and disposal of our media technologies (Maxwell and Miller, 2012). Among the notable activist groups working on these issues are the Silicon Valley Toxics Coalition (SVTC) in the US, the Basel Action Network International (BAN), Centro de Reflexión y Acción Laboral (CEREAL) in Mexico, and Scholars Against Corporate Misbehavior (SACOM) in Hong Kong. SVTC, founded in 1982, has helped pass legislation that educates, protects and monitors chemical pollution of groundwater systems. They pressured the EPA to name 29 sites around the Santa Clara area as Superfund sites, the largest concentration of such sites in the US. Clean up continues today. SVTC continues their work in pressuring policymakers to address the e-waste crisis and prevent further human and environmental rights abuses. BAN has worked since 1997 to end the global trade of toxic waste (Silicon Valley Toxics Coalition, 2013). BAN serves as a major source for information: statistics, photo-documentation and field research on the trade in toxic waste across the USA, Europe and Asia. BAN, working closely with the United Nations Environment Program, also develops international policy on the movement of toxic waste (Basel Action Network, 2013). CEREAL was founded in 1997 by workers in the *maquiladora* industry as a means of providing greater environmental and labor protection for workers in the electronics manufacturing industry. They investigate and publicize the abuses and conditions of workers within global factories in Mexico (CEREAL, 2013). Founded in 2005, SACOM brings together students, scholars, labor activists, and consumers in order to promote labor rights through critiques of corporate behavior and policy intervention. They work to document and publicize the labor conditions at Apple-producer Foxconn and other electronics manufacturers in China. Their work has stimulated consumer outrage, pushing Apple Inc. to negotiate with Foxconn in order to improve some working conditions and to increase wages (SACOM, 2013).

And what of television scholars' outrage? As researchers, media studies writ large could take a cue from activist organizations to work collaboratively on these global issues that are hidden from their immediate surroundings. By pooling research and resources to push for green citizenship and green governance, the International Campaign for Responsible Technology (ICRT), for example, connects a network of activist groups across global regions of dispersed factory towns and waste sites. Their 2012 summit in South Korea created a global action plan to make Samsung and Apple more accountable for making sustainable electronics. They also developed a research agenda for investigating legal frameworks for compensation of those poisoned workers in the Asian electronics industry. Some governments have responded in kind to these coordinated actions. In 2003, the European Union passed the Waste Electrical and Electronic Equipment Directive that promoted a strategy of 'Extended Producer Responsibility'. Rather than a strategy of traditional regulation that governs how companies had to produce, recycle and dispose of their products, the EU law creates incentives for companies to reduce the environmental impact of their products altogether. Furthermore, this directive made the disposal and recycling of electronics the responsibility of the manufacturer, and no longer the responsibility of the public (Smith and Raphael, 2003). The directive was updated in 2012 with more stringent target disposal and recycling goals for member countries, as well as stronger legislation against the shipment of hazardous waste between member countries (WEEE, 2012). This form of green governance connects media objects to the global environmental costs of producing, transporting and disposing of the commodities that contribute to their gross domestic product, while legislating in a way that is fairer to workers and consumers.

Beyond joining the picket, however, Maxwell and Miller (2012, 31–36) suggest that those who study media technologies should adopt and promote an eco-centric ethics, one that values nature, affords it rights, and thus makes judgments on its behalf. They argue an eco-centric ethics are the most disruptive in that they challenge us to re-orient our relationship to nature from personal hubris to deep regard. This turn would go beyond an anthropocentric ethics that values only workers' rights without attending to the environmental pollution that endangers the planet. Ultimately, television scholars need be concerned with both workers and nature through the interlocking political economic systems of production, distribution, consumption and disposal. In this sense, activists have not only illuminated the environmental and human abuses of media technology production and disposal systems to television scholars, they also have charted a future for television scholars to be more ethical in their own claims about the newest media and their shiny technologies.

NOTE

1 They include: *Maquilopolis, Deconstructing Foxconn, eDump, Terra Blight, The Story of Electronics, Digital Dumping Ground.*

REFERENCES

Agarwal, R. and Wankhade, K. (2006). Hi-tech Heaps, Forsaken Lives: E-waste in Delhi. In Ted Smith, David Sonnenfeld and David Naguib Pellow (Eds), *Challenging the Chip: Labor Rights and Environmental Justice in the Global Electronics Industry* (pp. 234–246). Philadelphia: Temple University Press.

Barrera, E. (1996). The US–Mexico Border as post-NAFTA Mexico. In Emile McAnany and Kenton Wilkinson (Eds), *Mass Media and Free Trade: NAFTA and the Cultural Industries*. Austin, TX: University of Texas Press.

Basel Action Network (2013). www.ban.org/

Berland, J. (2009). *North of Empire: Essays on the Cultural Technologies of Space*. Durham, NC: Duke University Press.

Boyce, T. and Lewis, J. (2009). *Climate Change and the Media: Global Crises and the Media*. New York: Peter Lang Publishing.

Castañeda Paredes, M. (2003). Television at the Border: Trade Policies and High-tech Manufacturing for the Global Market. In J. Lewis and T. Miller (Eds.), *Critical Cultural Policy: A Reader* (pp. 272–281). Malden, MA: Blackwell.

Castañeda Paredes, M. (2013). Television-set Production in the Era of Digital TV. In Angharad N. Valdivia and Vicki Mayer (Eds), *The International Encyclopedia of Media Studies Volume II: Media Production* (pp. 105–125). Massachusetts: Blackwell.

Centro de Reflexion y Accion Laboral (CEREAL) (2013). www.cerealgdl.org/index.php/en/

Cowie, J. (2001). *Capital Moves: RCA's Seventy-year Quest for Cheap Labor*. New York, NY: The New Press.

De Sola Pool, I. (1983). *Technologies without Boundaries: On Telecommunications in a Global Age*. Cambridge, MA: Harvard University Press.

Drake, W.J. (1993). Territoriality and Intangibility: Transborder Data Flows and National Sovereignty. In K. Nordenstreng and H. Schiller (Eds), *Beyond National Sovereignty: International Communications in the 1990s* (pp. 259–313). Norwood, NJ: Ablex.

Gabrys, J. (2011). *Digital Rubbish: A Natural History of Electronics*. Ann Arbor, MI: University of Michigan Press.

Galperin, H. (2004). *New Television, Old Politics: The Transition to Digital TV in the United States and Britain*. Cambridge: Cambridge University Press.

Lenin, V. I. (2011). *Imperialism the Highest Stage of Capitalism*. Eastford, CT: Martino Fine Books. (Original work published in 1916)

Lepawsky, J. and Billah, M. (2011) Centricity, Periphery, Boundary, and Edge: Assembling Urban Orders from Rubbish Electronics. *Inter-Asia Roundtable 2011: Recycling Cities*. National University of Singapore Asia Research Institute, pp. 35–67.

Lüthje, B. (2006). The Changing Map of Global Electronics: Networks of Mass Production in the New Economy. In Ted Smith, David Sonnenfeld and David Naguib Pellow (Eds), *Challenging the Chip: Labor Rights and Environmental Justice in the Global Electronics Industry* (pp.17–30). Philadelphia: Temple University Press.

Mattelart, A. (1983). *Transnationals and the Third World: The Struggle for Culture*. New York, NY: Bergin and Gravey Publishers.

Maxwell, R. and Miller, T. (2012). *Greening the Media*. New York, NY: Oxford University Press.

Mayer, V. (2009). Digitally not Yours: Spatial Discourses and Discursive Spaces for Brazilian Digital Television Policy in Manaus. *The Communication Review*, 12, 1–19.

Mayer, V. (2011). *Below the Line: Producers and Production Studies in the New Television Economy*. Durham, NC: Duke University Press.

McLaughlin, L. and Johnson, H. (2007). Women and Knowledge Work in the Asia-Pacific: Complicating Technological Empowerment. In Catherine McKercher and Vincent Mosco (Eds), *Knowledge Workers in the Information Society* (pp. 249–266). Maryland, MD: Lexington Books.

McKercher, C. and Mosco, V (eds) (2007): Knowledge Workers in the Information Economy. Lanham, MD: Lexington Books.

Miller, T. (2006). Cultural Citizenship: Cosmopolitanism, Consumerism, and Television in a Neoliberal Age. Philadelphia: Temple University Press.

Mosco, V. (1990). Toward a Transnational World Information Order: The U.S.-Canada free Trade Agreement. *Canadian Journal of Communications*, 15(2), 46–64.

Ngai, P. (2005). *Made in China: Women Factory Workers in a Global Workplace*. Durham, NC: Duke University Press.

Paredes, M. Castaneda (2003). Television Set Production at the US-Mexico Border: Trade Policy and Advanced Electronics for the Global Market. In Justin Lewis and Toby Miller (Eds), *Critical Cultural Policy Studies: A Reader* (pp. 272–281). Massachusetts: Blackwell.

Parikka, J. and Huhtamo E. (Eds) (2011). *Media Archaeology: Approaches, Applications, and Implications*. Berkeley: University of California Press.

Parks, L. (2007). Falling Apart: Electronics Salvaging and the Global Media Economy. In Charles Acland (Ed), *Residual Media* (pp. 32–47). Minneapolis: University of Minnesota Press.

Pendakur, M. (1991). *Canadian Dreams and American Control: The Political Economy of the Canadian Film Industry*. Michigan: Wayne State Press.

Qiu, J. (2009). *Working Class Network Society: Communication Technology and Information have-less in Urban China*. Cambridge, MA: MIT Press.

Raphael, C. (2003). E-waste and the Greening of the Information Age. *STS NEXUS*, 3(2), 23–28.

Reygadas, L. (2002). *Ensamblando culturas: Diversidad y conflicto en la globalizacion de la industria*. Barcelona: Gedisa.

Salzinger, L. (2003). *Genders in Production: Making Workers in Mexico's Global Factories*. Berkeley: University of California Press.

Schiller, D. (1999). *Digital Capitalism: Networking the Global Market Systems*. Cambridge, MA: MIT Press.

Schiller, H. (1969). *Mass Communications and American Empire*. Boulder, CO: Westview Press.

Silicon Valley Toxics Coalition (2013). www.svtc.org/

Students and Scholars Against Corporate Misbehavior (SACOM) (2013). www.sacom.hk/

Smith, T. and Raphael, C. (2003). High Tech goes Green. *YES! A Journal of Positive Futures*, 25, 28–30.

Sterne, Jonathan (2007). Out with the Trash: On the Future of New Media. In Charles Acland (Ed.), *Residual Media* (pp. 16–31). Minneapolis: University of Minnesota Press.

Waste Electrical and Electronic Equipment (WEEE) (2012). *Official Journal of the European Union*, 55, 38.

Williams, R. (1974). *Television: Technology and Cultural Form*. New York: Schocken Books.

Films

Maquilopolis: City of Factories, dir. Vicki Funari and Sergio de la Torre, 2006.

Deconstructing Foxconn, dir. Jack Qiu, 2010.

eDump, dir. Michael Zhao, 2007.

Terra Blight, dir. Isaac Brown, 2012.

The Story of Electronics, dir. Annie Leonard, 2010.

Ghana: Digital Dumping Ground, PBS Frontline, 1999.

PART III

Cultural Forms

16

Television Program Formats: Their Making and Meaning

Albert Moran

INTRODUCTION

It is over a dozen years since the *Big Brother* TV program format was devised in the Netherlands by producer John de Mol for his production company Endemol. As Bazalgette (2005) has shown, the program broke new ground in terms of its premise of bringing together a group of young people and having them live in a confined space where they were filmed by a battery of hidden cameras. Beginning in the Netherlands, and soon spawning a string of franchised productions across the world, *Big Brother* has been arguably the most watched program in world television history. Its accumulated audience around the globe for its different national versions regularly produced up to 2005 was estimated at 740 million viewers. It has also proved to be an extremely valuable franchise worldwide, generating more than US$10 billion in profits for its owner in the same period to 2005 (Bazalgette, 2005, pp. 316–329).

The phenomenal success of the program also heralded the maturation of TV formats as industry craft and cultural enigma. This way of doing television is now a highly significant component of business and social practice in modern television, at both the national and international levels. Whether in the form of reality, quiz show, infotainment programming, makeover, talent show, sitcom or drama, the advent of the television program format seems to signal the triumph of media globalization even while asserting the continued importance of local or domestic programming (Moran, 2008; Moran & Malbon, 2006; Tunstall, 2008). What does the paradox imply? How can the TV program format best be understood in the rapidly changing mediascape of present-day international television?

This chapter explores the practice and significance of TV format programming. The discussion is divided into six sections. The first offers a short overview of TV program formats, identifying some of their obvious and not so obvious features. The routine of adapting broadcast programs from other places and times is far from new, although still relatively little understood and researched; thus the second section consists of an extended three-part history outlining the development of the practice of remaking radio and television programs, with particular attention given to cross-national borrowings. The TV format has only become a widespread business routine in the past 20 years despite the longevity of practices of cultural replication and modification in the creative industries. Therefore, in the third section of the article, I ask why this has come about at the present time. The answer lies in an understanding of formats as industrial and cultural commodities. TV formats are helping to reorganize the institution of television along international lines. Accordingly, the next two sections of the discussion examine how production and labor relations, circulation and viewing have been affected by TV formats. In the final section of the chapter, I pay attention to format scholarship. A small but growing band of media researchers has been delving into the phenomenon and meaning of TV formats for over 20 years. The chapter concludes by considering the context and results of the phenomenon of TV formats offered by this body of investigation.

OVERVIEW

What exactly is a television (program) format? In its everyday use, the term 'format' relates to a way of presenting textual materials in a formula or blueprint. The label is also now used to mark a particular way of organizing business and cultural knowledge within the international television industry. Briefly, a television program format constitutes a set of experiences and skills that make it easier to imitate or remake a television program from one time to another and from one place to another (Moran, 1998, pp. 6–12). Of course, any human artifact can variously be copied, imitated, counterfeited or mimicked. The results of such activity may or may not be especially accurate, persuasive, credible or convincing remakes of a predecessor, but that is beside the point. What matters with a television format remake is the fact that it is possible to amplify or facilitate such adaptation. Making available such supports as handy resources, relevant information, background detail and skilled assistance can speed up and smooth out the path of imitation. The copyist is helped to a more faithful likeness or facsimile without, perhaps, making the mistakes, suffering the pitfalls and engendering the same unnecessary expense encountered in the original operation. The TV format is simply the application of this principle to the area of television program remaking.

A TV program format can thus be understood as the deliberate enhancement of the adaptability of a program. It is that complex and coherent body of knowledge assembled by a devisor or owner that permits and facilitates the imitation of a TV program by another. The format builds business and cultural versatility. Preferring the notion of knowledge flow to that of information exchange, geographer Michael Storper (2002) describes this tendency in a more general way that is congruent with present-day global forces in business and culture. He defines this knowledge franchising as 'the transfer of economically useful practices, routines and conventions, which are complex and coherent assemblages of different kinds of information' (p. 16).

Sometimes – especially in the international TV industry – more down-to-earth explanations of television format adaptation hold sway. Two domestic metaphors to do with cooking are popular with media professionals who are providing their own insight into these processes. From one point of view, the emphasis falls on the idea of a core or a structuring center that helps to organize industry practice, including the on-air program. This view is at work in accounts that see a TV format as equivalent to a cooking recipe. In turn, the television format is understood as that set of invariable elements in a program out of which the variable elements of an individual episode can be produced (Moran, 1998, pp. 22–29; Moran & Malbon, 2006, pp. 7–9; van Manen, 1994, pp. 1–8). Equally, though, the stress may fall on the idea of the format providing an outer shell or organizing framework that permits and facilitates the making of a program. According to this second view, a format can be understood as a means of ordering individual episodes of a program. As a television producer once put it, 'The "crust" (of the pie) is the same from week to week but the filling changes' (Moran, 1998, p. 13). Using either metaphor, the idea is essentially the same. A television format is the arrangement of elements that assist in the remaking of a television program in another place at another time.

One other distinction is in order at this point. Although international television industries talk confidently of the format as a single object, it is in fact a complex multiple entity that typically manifests in a series of overlapping but separate forms. At the point of programming and distribution, a format takes the cultural form of different episodes of the same program. Meanwhile, at the production end of the commodity chain, these different industrial manifestations include up to a dozen different resource and knowledge components. The varying number and diversity of these elements brought together for purposes of program adaptation underline the point that a format is not a single or a simple entity. Philosophically, one might assert that a format is a kind of abstraction known only through the various shadows that it throws, although these are collectively deemed sufficiently tangible to persuade many of its actual existence. The industry ignores this conceptual fuzziness for good pragmatic reasons, for the television format has become one of the most important means of functioning industrially in the contemporary environment of international television remaking. As an economic and cultural technology of exchange inside the television institution, formats facilitate cultural and business operations. The format, then, has meaning not so much because of a principle but because of a function or effect. The crucial point about the format is not what it is, but rather what it permits or facilitates (Moran, 1998, p. 9).

HISTORY

The first historical stage happened between 1935 and 1955. These years saw the worldwide establishment of radio broadcasting and a fledgling innovation of television. Program ideas were transferred recurrently within and across borders, although this usually occurred with little immediate cultural or economic reward accruing to the source of the content.

The flow had Anglo-American origins, with the United States in particular displaying a proliferation of internal markets that secured the need for constant program reuse. The markets were based on different time zones/audiences (prime-time, daytime, and even children's versions), on geography (East Coast, West Coast as well as regional and local broadcast) and on different media platforms, with both radio and television versions of programs hitting the airwaves. Collectively, these outlets drove the practice of program copying (Moran, 2013a, p. 5).

The overseas expansion of large US manufacturing industries led several domestic advertising agencies to follow their home-grown clients and set up local offices in centers as widely dispersed as London, Montreal, Sao Paolo and Sydney. These not only prepared radio commercials for clients, but also produced the programming vehicles that would carry these advertisements. And if overseas branch plant clients reused and adapted advertising and promotional copy, along with designs developed for the home market, then the logic of operations dictated that many original programming vehicles be adapted and reused as broadcasting environments for commercial messages elsewhere in the world (Moran, 2013a, pp. 5–6).

Transnational moves towards program content borrowing were underway by the early 1940s. Preferred genres included audience-participation programs, comedy and quiz shows and drama series, serials and anthologies. Program scripts were exported in large quantities, turning up in centers from Australia to Brazil, Cuba and Quebec.

This early flow of program know-how was not confined to commercial radio broadcasters in the United States and elsewhere. After all, the BBC served as a source of program inspiration to public service broadcasters in various parts of the British Empire – a tutor function that expanded after 1945. But it too was ready to look to the United States. Brunt (1985) had earlier noted the BBC's adaptation of the US Mark Goodson radio quiz format *What's My Line?* in the early 1940s, and this historical connection has been explored and documented more fully by recent investigation (Camparesi, 2000; Chalaby, 2012; Hilmes, 2012). In particular, the point has been made that this adaptation involved the first licensing or goodwill payment to an owner/devisor for the reuse of the program idea – first for a UK radio remake and then for a television version (Chalaby, 2012, p. 43).

The second stage, from 1955 to 1980, was accelerated by the steady national expansion of television services in the West. Broadcasters sourced content through local production and the import of canned and format programs (Waisbord, 2004), although the import process tended to be casual, one-off and frequently unrecognized. Early European instances usually involved US quiz show formats, such as a highly successful Italian remake of the US show *The $64,000 Question* by the national broadcaster RAI (Ferrari, 2012). Imitation also involved other program forms, such as the Finnish remake of the popular BBC comedy series *Hancock's Half Hour* (Keinonen, 2009)

This period also saw the emergence of two cross-national entrepreneurs who would subsequently turn out to be highly significant format gatekeepers. The American Paul Talbot established Fremantle International as a global distributor in 1952, and from 1960 remade the children's series *Romper Room* in more than 30 different territories under license from its US owner/devisor (Moran, unpublished). Meanwhile, Reg Grundy systematically remade US quiz shows in Australia for over 20 years before internationalizing his program distribution and production from 1980 onwards (Moran, 2013b).

US television was still the richest source of know-how, although the directions of global traffic were beginning to multiply. For example, UK sit-coms crossed the Atlantic, with *Till Death Us Do Part* (1965–68) remade in the United States as *All in the Family* (1971–78), and *Steptoe and Son* (1962–67) adapted as *Sanford and Son* (1972–77). An even more complicated movement concerned a European-derived comedy sporting program, which began as the French *Intervilles* in the 1950s and ran from 1965 to 1999 as the European Broadcasting Union-sponsored *Jeux Sans Frontieres* (Games Without Borders). A UK remake, *It's a Knockout*, ran from 1966 to 2001, while a US clone, *Almost Anything Goes*, was subsequently aired in national, children's and syndicated celebrity versions (Moran, 2013a, p. 8).

Television industries in national settings mostly paid little attention to movements in program format exchange. In fact, producers were sometimes unaware that adaptations were happening elsewhere. US program owners were especially ill-equipped to pursue any legal action against the unauthorized use of their program ideas by overseas producers. The United States was not a signatory to the Berne Convention on Copyright, and legal action concerning copyright infringement had to be pursued in the overseas jurisdictions where the alleged violation took place. One owner who did undertake exhaustive legal action was UK host and producer Hughie Green. In 1978, Green learned that an unauthorized version of his British quiz show, *Opportunity Knocks*, was being aired by the New Zealand Broadcasting Corporation. He took legal action but lost, with the courts failing to find that there was a copyright in the formula of the program (Moran, 2013a, p. 9). Even so, the case signaled that program devisors and producers might use legal measures in the future to attempt to police the circulation of their program ideas.

Another sign of the same widespread ignorance of program remaking can be detected on the part of professional media watchers. There is no mention of the subject of television program remaking in the 1977 *New York Times Encyclopedia of Television* (Brown, 1977), while the UNESCO 'Television Traffic' worldwide investigation concentrated only on 'canned' programs (Nordenstreng & Varis, 1974).

A further stage of know-how circulation belongs to the last two decades of the 20th century, when content exchange became format business. Market expansion was a key determinant of this era. It resulted in an increased program demand, triggering the search for low-priced content. 'Live' entertainment filler forms of television increasingly were preferred to more expensive content that included drama series, serials and mini-series, situation comedy and current affairs. Game show formats – suitably 'localized' to fit homegrown tastes – became one of the staples of the new trend.

The era also witnessed an increasing formalization of program content exchange. A large-scale business with a visible set of arrangements, practitioners and dominant institutional players was coming into existence. This included annual or biannual world television trade fairs, where format programs as well as canned programs were increasingly being licensed. Indeed, 1988 saw the formation of Action Group, the first network of companies interested in swapping program formats for the national territories in which they operated (Moran, 2013a, p. 13).

Between 1978 and 1980, the United States had ratified the Berne Convention on Copyright. This move seemed to extend the possibility of copyright protection over what was deemed to be intellectual property – which, potentially, included television program formats. However, the outcome of the *Opportunity Knocks* case did not encourage legal action, although the threat of the latter was often sufficient to cause a perceived violation to cease. There were plenty of exceptions to this trend in the form of piracies and rip-offs, but licensings and fee payment became more regular, normalized and customary in the cross-border exchange of television program know-how.

Bourdon (2012, p. 122) usefully sees this era in terms of the entry of European television into the world of formats. The three major format production and distribution organizations that made their presence felt towards the end of this era support his claim. Endemol was formed by the merger of two homegrown Dutch production companies, and benefited enormously from both EU broadcasting policy and preferential treatment from Europe-wide broadcasters. The BBC formalized and commercialized its connections in other Anglophone territories, involving itself first in program format licensing, and eventually moving to offshore format production. In The UK company, Pearson Television (shortly to become FremantleMedia), went fishing for format libraries in the 1990s, with a catch that included Grundy Worldwide and Paul Talbot's Fremantle International (Moran, 2013a, p. 13).

Reflexivity grew, and from 1990 onwards, the term 'format' became common within the trade. The label was used more consistently in the television business press as program adaptation and remaking became more newsworthy and widely acknowledged across the international television industry. Coincident with this came extended discussion about matters of legal protection and legislative change. Finally, the 1990s also witnessed early signs of critical interest on the part of media scholars and researchers.

Finally, a contemporary phase of the format trade is still with us. This stage commenced in 1998 with a string of mega-formats, beginning with *Big Brother*. The format and its companions received near-simultaneous licensings, resulting in saturation distribution as well as great profitability and widespread local and cross-border visibility. The United Kingdom and the Netherlands became national frontrunners in the worldwide format business, although triumph in the US market was still deemed a vital step for a successful round of global licensings (Chalaby, 2012).

This practice became an industry shibboleth, although *Big Brother* – the most successful, talked about and written about television program of all time – received less than jubilant ratings in the US market. However, other format trajectories – such as the highly successful re-launch of the German-derived reality crime-magazine format *Aktenzeichen XY* (File XY: Unsolved) as *America's Most Wanted* – did confirm US primacy while also highlighting the ever increasing, multidirectional flow of global formats (Moran, 2013a, p. 14).

The present stage has seen more and more television markets undergoing integration into the global format network. Among the newcomers, we might include hitherto-marginal television markets such as those of Columbia, Outer Mongolia and even the People's Republic of China (PRC). This kind of commercial and cultural integration has larger political implications of course, a fact highlighted in 2013 when the PRC moved to protect its television producers with a 'One Format' policy that drastically curtailed imports (Keane, 2013).

Meanwhile, the industry trade fairs remain important mechanisms of global television format integration, with the 2000 establishment of the Format Recognition and Protection Association (FRAPA) constituting one of many moves towards greater formalization. Training courses in education, practical manuals and handbooks, and the publication of a weekly trade magazine are further signs of this professionalization.

Copyright law has mostly remained helpless to protect formats. Instead, company owners have resorted to various pragmatic measures to defeat poachers and copycats. One practice involves the simultaneous licensed release of formats into the global marketplace, a shift away from an earlier circulation system that involved delayed, one market at a time, 'trickle-down' operation in favor of a blanket-saturation, near-simultaneous format release pattern. Another has involved legal bullying, where only the wealthiest can afford to embark on costly and uncertain legal action (Moran, 2013a, p. 14).

A final change involved the degree of public attention given to television formats. Remakes have become highly visible, with overseas connections highlighted and flaunted in promotion and marketing. Guests on live programs are also often briefly interchanged between different national versions of mega-formats. The global reach of a format is now part of its publicity and promotion.

THE REMAKE AS INDUSTRIAL AND CULTURAL COMMODITY

Previous sections have already touched on the fact that the remake is both an industrial and a social technology. This idea needs further exploration, so in this section I focus on both the commercial and cultural dimensions of the TV program format in turn.

Commercial Entity

As an industrial commodity, the TV program format warrants such names as 'franchise' or 'brand name', a useful reminder that, from a business perspective, the product in question is the total package of commercial rights together with the supporting television vehicle that carries such names as *Big Brother* and *Who Wants to Be a Millionaire?* (Bellamy & Trott, 2000; Dicke, 1992; Todreas, 1999). The television vehicle is available not only through broadcast TV but also through various other platforms, including internet, mobile phone, iPod, radio, advertising, telephone, billboards, merchandising and personal management. The franchising associated with a format involves the renting of a bundle of services that attempts to ensure successful profitability for the broadcasting network licensing the format set of services. Managing a package of rights rather than an individual right ensures the generation of incomes from a series of ancillary operations that are not normally licensed, but rather exist in the area of publicity (Moran & Malbon, 2006, pp. 1–19). At the same time, in exploiting these rights a company like Endemol, owner of the *Big Brother* format, also needed to exercise some commercial prudence. The balancing act is to secure the most profitable returns, together with ensuring the longest period of licensed rental. Like any franchising parent, a format company such as Endemol was disposed to build its product, to maintain and strengthen the overall international commercial reputation of *Big Brother*, and to secure local licensing fees in different territories that were compatible with this objective.

The guises under which the format might have appeared are clear, given this general disposition (Hill, 2002; Johnson-Woods, 2002; Roscoe, 2001). Take, for example, the Australian remake of *Big Brother*. Merchandizing included such standard elements as videos, DVDs, CDs, magazines, t-shirts, bunny ears and masks. The format also conferred celebrityhood (both temporary and more durable) on former housemates; in turn, these 'celebrities' made 'star' appearances not only in subsequent series of the program but on other shows, including talk shows, game shows, and even drama serials. Even the *Big Brother* house basked in this glory, being opened to the public as a commercial attraction at its *Dream World* theme park site on the Gold Coast (Johnson-Woods, 2002; Roscoe, 2001). While on air, the Australian *Big Brother* also briefly pervaded the world of local commerce, where it was coupled with brand names for other commercial products and services. This kind of linkage was consolidated and extended by other marketing strategies, such as onscreen product placement and phone line voting facilities.

Remaking already successful materials and content offers a good chance of duplicating past and existing successes. Media producers, including those operating in the field of television, attempt to take out financial and cultural guarantees by using material that is in some way familiar to audiences (Fiddy, 1997; Gitlin, 1983; Moran, 1998). Having invested in the brand, it makes good business sense to derive further value from it in these different ways (Bellamy & Trott, 2000; Todreas, 1999). In turn, this tendency of recycling is further facilitated by owning the copyright on the property in the first place. In the age of television remaking, there is a clearly identified need to derive as much financial mileage from an ownership as possible – hence the idea of intellectual property. This move to safeguard and control content-related ideas formalizes ownership under the protection of intellectual property laws such as those relating to trademarks, brand names and registered designs, as well as those of copyright law (Freedman, 2003; Lane, 1992; Lane & Bridge, 1990a, 1990b; Moran, 1998; van Manen, 1994). Indeed, the present era of television may come to be characterized as one of a heightened awareness of and emphasis on program rights.

The interests in rights held by television companies are not defined abstractly, but change with commercial circumstances. For example, the income generated from the licensing of a TV program into public usage has to be measured against its use as a means of promotion. As Frith (1987) has pointed out, copyright is generally used to make money rather than to control use. Nevertheless, this emphasis on rights helps to secure the general conditions for the process of selling the same content over and over again across a series of different media platforms as a key feature of the present epoch. In the particular case of multichannel international television, the process of worldwide geographic dispersal and recycling of existing content goes under the specific name of TV format adaptation. Existing TV program formats are capable of producing new elements that expand the range of possible applications or adaptations. The spinoff program is both new and original, but its seeds and potential are seen to lie in the pre-existing program format (Gitlin, 1983, pp. 63–85; Thompson, 2003, pp. 74–105).

Cultural Dimensions

The television formatted program is equally ambiguous as a cultural commodity. One form of remake, such as that involving *Who Wants to Be a Millionaire?* and *The Weakest Link*, has involved a minimum of change and variation from one place to another. Thus, as will be discussed later in the chapter, formats also raise issues concerning cultural standardization, social representation, national identity and globalization. Some television program format adaptations aim at translation equivalence, where cultural variations between remakes are kept to a minimum. In such instances, remaking is part of a process of intended cultural standardization. The aim, in effect, is to reproduce the equivalent local version of a format program to the point where a Swedish version of, say, *Who Wants to Be a Millionaire?* mostly looks and feels much the same as a Singaporean remake of the program, even allowing for differences of language, social outlook, domestic cultures, history, religion and ethnicity. This cultural trajectory of formats has the ambition to distribute a standardized social product in much the same way as the Coca-Cola corporation distributes a universal soft drink product across the world.

For the most part, however, format programs are permitted – and even encouraged – to engage in bounded variation from versions produced elsewhere. The perception is that cultures can vary considerably from one place on the planet to the next, and that the program format must take account of this difference. History, geography, ethnicity, custom, language and religion can have significant effects on populations, so producers of television program formats should make necessary changes to the form and style of programs. These differences are seen to be necessary in order that any particular format program has local popular appeal with home audiences. One can speak here of an effect that might variously be described as nationalizing, domesticating, or even indigenizing the format to achieve this popular appeal. Still, particular qualifications need to be borne in mind.

Various researchers who have studied this process of cultural decision-making in a particular market setting have noted a process of culturally selective gatekeeping at work (see Jaflin & Waisbord, 2009; Moran, 2008). The screening process in operation on the part of media organizations is a highly personal one, even if it is being carried out by a handful of media professionals on behalf of a national television broadcaster in a particular television market. The process is in no way democratic. Instead, it imposes the cultural will of the few on the many. This professional decision-making becomes even more apparent when a viewing population may be culturally diverse, with different languages, religions and ethnicities spread across the viewing population. This was the case in Singapore with the making of *Who Wants to Be a Millionaire?*, as reported by Lim and Keane (2004). Nevertheless, the gatekeeping decision was taken to record the program in Mandarin Chinese, thereby privileging those in the viewing population who were of Chinese origin and discriminating against viewers of Indian and Malaysian extraction (Lim & Keane, 2004, p. 114).

It hardly comes as a surprise to realize that TV producers are selective where cultural decision-making in program production is concerned. Nor is it news that viewing populations are constantly addressed as socially homogeneous when that may be far from the case. All the same, even if this kind of gatekeeping is at work in format program remaking, this is more a condition of broadcast television generally rather than a particular problem associated with format program adaptation. Nevertheless, this social determination does serve to highlight conceptual difficulties surrounding terms that might be used to describe the kind of flavoring or coloring that program formats undergo to fit them for a particular viewing population. Various terms come to mind, including 'nationalize', 'localize' and 'indigenize', and these labels are useful provided the above caveat is borne in mind. One might even appropriate this kind of customizing as representative of that every day, banal nationalism that Michael Billig (1995, pp. 1–14) has identified.

Mention should also be made concerning one other aspect of the cultural contextualization of the TV program format. This concerns the engines and effects of globalization. The advent of formalized international trade in TV program formats might seem to confirm Ritzer's (1998) McDonaldization thesis. After all, the business is now organized to bring about multiple versions of hit formats such as the talent contest *Pop Idol* and the comedy-soap opera *Ugly Betty* in one television market after another across the globe. However, as I have pointed out elsewhere, the 'global' trade in program formats never reaches half the number of countries across the planet, and much less of the world's population, that industry publicity tends to suggest (Moran, 2009b). Despite the fact that television program format templates are widespread, bounded variation within cultural repetition is the order of most program franchising. This makes the McDonaldization thesis difficult to sustain. In fact, television program format franchising might be seen as a force pulling in a different direction. Pioneer media researcher Jeremy Tunstall (2008) has recently returned to a topic that he first raised in the 1970s: the extent to which there is a worldwide media culture. In 1977, Tunstall published a large-scale analysis asserting that such a culture did exist and its center was the United States. In 2008, his view was that the global media pattern, including television, was progressively fragmented and decentered, a view that is becoming more accepted among researchers (cf. Sinclair, Jacka, & Cunningham, 1996; Straubhaar, 2007; Tunstall, 2008). Media territories increasingly are emphasizing their own broadcasting identities, and television program formats are significant in driving this trend (Moran, 2009b).

TV FORMATS REORGANIZE DISTRIBUTION AND PRODUCTION

The maturation of the TV format business heralds a significant change in television program production and distribution. This reorganization is every bit as significant as the farming out of production that began in US network television in the early 1950s (Gitlin, 1983). Earlier, television production had occurred in-house at television stations using production staff who were permanently employed. The new system witnessed a farming out of production responsibility and budgeting to independent producers. In the main, these subcontractors guaranteed to deliver so many units of programming to the commissioning station or network, on schedule and within budget. The primary relationship under such a system was between the commissioning broadcaster – whether a television station or a network – and an independent producer, who in turn hired in the creative and other labor necessary to produce these units of content. Under this system of post-Fordist production, the independent producer was usually responsible for originating and devising the program in the first place so that the outside production company, very often in collaboration with the original broadcaster, acted to license the program for rebroadcast elsewhere.

This earlier development is mentioned because it acts as a precursor to changes associated with formats. Under the regime of television program remaking, the post-Fordist relationships have been reconfigured further so that a new system for generating and circulating programming comes into existence. A new commodity form emerges in the shape of a kind of universal, abstract television program. It is assumed to be culturally flexible, capable of being made over to suit different industrial and social conditions that it will meet in the different television territories of the world into which it is licensed. The format is assumed to be a perfect vehicle for systematic distribution on a worldwide basis. The finished or canned program (Waisbord, 2004) had far less market fit, imposing itself culturally as well as commercially in the overseas television territories in which it played, with only subtitling or dubbing signaling accommodation to local audiences (Kilborn, 1993). With the format program, a set of industrial knowledges are circulated for purposes of program remaking in various territories across the globe. The primary relationship under this contemporary regime is that between an owner/devisor and a broadcaster in another territory who is keen to transmit a program based on that body of a successful package of industrial knowledge. Initially, though, a transmittable program must come into existence. This entails a secondary relationship under which the format owner and devisor must first license a domestic producer who, working with either or both the international distributor and the domestic broadcaster, produces a local, 'original' version of the program (Moran & Malbon, 2006).

This system of television remaking is a highly centralized international configuration. What was once local and domestic – namely the origination and development of a program concept – has now been geographically recentered, sometimes in another territory. Television program production under this system involves a franchising operation in which local production refers only to the manufacturing or fabricating of the program at the point of consumption, rather than to its creation. Television program production has always had both a white-collar and a blue-collar component, a devising and development stage on the one hand and a manufacturing, constructivist stage on the other. These stages were joined together as part of the same process of bringing a program into existence.

Under the newer system of relations entailed by TV program formats, the phases of the cycle are split off from each other. They now tend to occur at different times and in different places. Distribution is the key activity ordering this agenda. Where this process once functioned simply to circulate already existing programs, now it acts to orchestrate production, separating off the manufacturing or making of programs from their creation, and initially concentrating on the latter. Under this regime, program manufacturing is both delayed and exported, as discussed below.

We should understand the era of television remaking in terms of the specific connotations that program production has now acquired. On a world scale, there may or may not have been a significant reduction in program creation. Now, instead of many programs being originated locally or nationally to serve local tastes and interests, there has emerged a second system whereby fewer program formats are created. However, local tastes and interests are still being met. Because program origination now occurs on a centralized basis, there remains a transnational capacity to meet this need, although this happens thanks to the particular nationalizations of program manufacturing that occur in specific territories. Production is no longer the single, unified process of construction whereby what begins as a television idea finally emerges fully formed in the shape of a television program. Rather, this activity is split into two different routines. The first comprises creation and origination, while the second involves remaking.

To the extent that 'remaking' suggests the manufacturing of television programming units in different places that are mostly the same (although they do display certain sanctioned differences), a consideration of the global system of television program production under the 'remaking' label is apposite in the current era. Indeed, it is worth taking this concept of manufacturing a little further. Two poles of flexibility concerning remaking should also be mentioned.

One can, for instance, pinpoint many format adaptations where there is a good deal of flexibility in a particular local adaptation. Variations may arise because of technical, budgetary and other immediate considerations, and may also happen for broader cultural and social reasons. For instance, the adaptation of the format of *Pop Idol* falls into this category, with particular national producers being permitted a degree of latitude so far as any adaptation is concerned. In other words, this kind of fabricating of a particular program from its format does not preclude a degree of creativity that is exercised in the specific details of the adaptation. Poetic license often allows a national producer to vary elements from those laid down in the format documentation blueprint.

On the other hand, as already noted, other format owners and licensors do not permit this degree of flexibility. Instead, they insist that the elements specified in the format be rigidly adhered to in the assembling process. This seems to be the attitude adopted by the BBC with regard to the remaking of several program formats that it owns and licenses, such as *The Weakest Link* and *Dancing with the Stars*. Such cultural standardization is not restricted to television formats, however.

It is, for example, the attitude reported by Burston (2000, 2009) regarding the remaking of such stage musicals as Disney's *The Lion King* and *Beauty and the Beast* in major centers such as London, Toronto and Sydney. The attitude of the owner or licensor seems to be that the format has already given rise to commercial and popular success, and that there is therefore no need to tamper with it. In instances such as this, where the format specifications are strictly adhered to, the term 'manufacturing' is a particularly apt one. There is little or no aesthetic choice or flexibility in the mounting of a production based on a pre-existing format. Most – if not all – creative work has taken place at the front end of the process. All that is required at the back end is adherence to the existing design.

Lack of flexibility can also be observed in the system of choice, whereby some program format companies come to dominate the market. As Gitlin (1983) notes in his classic study of the operation of US network television in Hollywood, the tendency is for those working on the inside track to remain there and for those on the outside to occasionally break in, but for the most part to remain on the margins (1983, pp. 67–93). Hence, although small broadcasting bodies and private companies will devise formats, most of these are probably unlikely to achieve wide distribution or industry clout. Over the longer term, this tendency seems unlikely to lead to a significant buildup on the margins of the world industry of any particular creative expertise in format origination and development. The organization of the format programming industry tends to promote the development of creative under-development at the margins as part of an ongoing re-division of cultural labor occurring in global media.

Another effect of format programming lies more directly in the domain of finance. As noted elsewhere, the advent of privatized multichannel television has been accompanied by a search for ways to lower production costs (Moran, 2008). Three overlapping factors are at work. First, there is the need by private operators to show greater profitability (Turner and Tay, 2009). New services and narrowcasting combine to reduce likely audience sizes and advertiser revenues so that other areas of cost saving and financial income have to be brought online. Reality TV and format programming have been two means among others of achieving lower-cost programming. Despite frequent claims to the contrary, reality TV – with its use of unpaid participants and its elimination of the need for professional scripts – has been very successful in pruning production expenses at a time when television operators desperately need to contain costs. Similarly, format programming achieves further considerable savings by obviating the need for expensive research and development (R&D) at the site of subsequent productions of a format. Magdar (2004), for example, points out that in US network television at least 80% of new productions in canned programming will result in commercial failure, whereas the rate of failure of format programming remakes is less than 20% (pp. 147–149). Hence, although reality TV can also be distributed in the form of canned programming, and although format programming can include other genres such as drama serials and current affairs, it makes good economic sense to distribute new reality programs in the format form.

The upshot of such a conjunction is that there is likely to be long-term underemployment of particular types of creative labor in television industries, most especially those at the margins of the system. Researchers in various territories have already noted the drop in output in more expensive forms of content, including drama series and current affairs television (Brunsdon, Johnson, Moseley & Wheatley, 2001). This trend seems likely to persist. Over the longer term, the development must lead to the erosion of particular creative capacities outside the main centers of television production. Actors, writers, journalists, and others move elsewhere or find other forms of employment in the face of this new reality.

Format programming necessitates a different way of thinking about local television production. Content franchising is a highly centralized system coincident with the ongoing reorganization of national television structures into a larger international pattern. Under its production regime, that which was once local and domestic – namely, the origination and development of a program concept – has now been relocated elsewhere. This reconfigures what is meant by production. Television program making under this system involves a franchising operation. Local production refers only to the manufacturing or fabricating of the program at the point of local circulation or broadcast, but not to its creation or devising (Moran & Malbon, 2006, pp. 18–26). This is a novel and significant development.

Television program production has always had both a white-collar element and a blue-collar component – a devising and development stage on the one hand and a manufacturing phase on the other. Under the system of canned program manufacture, these two activities were seamlessly joined together as part of the process of bringing a program into existence. Under the system associated with format programming, the phases are split off from each other. They usually occur at different times and in different places. Devising and development are centralized and promoted while production is dispersed, decentralized and postponed. And it is distribution that orders this agenda. Where the latter activity once functioned simply to circulate already existing canned programming, now it intervenes to orchestrate production, separating the making or manufacturing of programs from their creation and origination, and initially concentrating on origination. Under the new regime, program manufacturing is both delayed because it is not immediately required and exported because simultaneous manufacturing is required at multiple local television sites around the world. This process serves to again underline the blue-collarization that is occurring in secondary industries in the era of format programming and a more global television system.

Two other effects of this localization of format production should also be mentioned. The first has to do with the manufacturing of local programming based on a format package. This again returns us to the 'McDonaldization of TV' label and points to the standardization and routinization involved in the local fabrication of a format program (cf. Ritzer, 1998). The term 'manufacturing' is appropriate because of the minimal involvement – or sometimes non-involvement – of 'creative' personnel in the local making of a format program. Under the format programming regime, this kind of 'white-collar' work has already occurred elsewhere, and the remaining 'blue-collar' work has mostly been regularized and routinized. Licensed owners often insist that the form and style of each local production be faithful and exact copies of the original. Local production staff can often be given little opportunity to improvise and introduce variations that might be thought to be necessary to suit local audiences (Moran & Malbon, 2006, pp. 30–39). Even where this requirement is not the case, and the local production team is allowed more flexibility and input in its adaptation of a format, caution frequently intervenes. There is a reluctance to vary the production elements as specified in the format package in case this results in the new version of the program failing to reproduce the success of other versions, including the original. The general industry maxim is 'If it's not broken, then don't fix it!' In fast-food franchising, the notion of 'hamburgerology' appears to allow little variation in how hamburgers are to be made in individual outlets, and the same idea prevails in at least some franchised program making.

In any case, there is another means at hand to ensure general conformity to the pattern to be followed in the production process. This comes in the shape of the production service that the licensor will provide to the licensee on the occasion of the first local production of a format. The main element of this arrangement is the presence of a 'travelling producer' made available by the licensor to oversee the new production (Moran, 2009a; Moran & Malbon, 2006, pp. 25ff.). Although the industry frequently refers to this process as one of 'hand holding' with the local producer, clearly the 'travelling producer' is involved in a gatekeeping process as part of the facilitation exercise (see Jaflin & Waisbord, 2009). The aim is to ensure that the pattern is followed, with variations introduced only when these are necessitated by production exigencies, and by local cultural conditions and considerations. Otherwise, the purpose is to ensure that there is a high degree of conformity with the articulated model of production for the format. This process is clearly a complex matter involving bounded innovation as well as routine. At one level, there is an important element of training and skilling at work. Even so, the overall situation of adaptive remaking means that the prospects for innovative, creative labor are systematically lessened.

A final element motivating this outcome has to do with the fact that format program remaking is most valuable in the area of production fees rather than in the domain of license fees. A company benefits more from practical, on-the-ground involvement in an adaptation than from any rent derived from ownership of its intellectual property (Moran & Malbon, 2006, p. 33). There is, then, a tendency for international format giants to set up branch-plants for national production of the formats they own. Highlighting this situation, Endemol, FremantleMedia and the BBC have all established local production facilities in Australia over the past decade. This internationalizing of national production sectors of television tends to further restrict local opportunities for original creative labor on the part of national media professionals and organizations.

REORGANIZING TELEVISION AUDIENCES

If the practice of TV program format circulation rearranges distribution and production, the craft also helps in reorganizing the consumption of programming by viewers. Recent years have seen significant changes in central genres and programs of television (Wasko, 2005). Part of this new regime includes the deployment of the internet, telephony and other new media technologies in order to achieve greater viewer involvement in 'live' events on screen. At this level, the new programming propelled by television program franchising invites audiences to be active, participatory and engaged with the course of what is occurring in a particular program. This new environment makes talk and fandom compulsive, even irresistible, as on-screen figures attempt to negotiate their television fate and on-screen longevity.

But whether 'live' reality programs, documentaries, quiz shows and other participatory genres, or fictional forms such as situation comedy, telenovela drama, and serialized fiction, franchised program making has a general audience effect involving the creation of dispersed, decentered communities of viewing. Each adaptation is given a national setting, whether the locality of viewing is limited to a province, is nationwide, or even serves a group of adjoining television markets such as those occurring in the Nordic countries or the Middle East. The format program functions to unite nations of viewers. It represents an arrival of the global TV event, which commands audiences not only on a 'local' basis but also at a global level. Dayan and Katz (1992) have already drawn attention to the phenomenon of public event television – the transmission and broadcast of a happening that is occurring in the outside world, and in which the public in many different parts of the nation-state and increasingly across the planet as a whole is led to take interest. This kind of television is variously resistant to being made over and customized for 'local' viewing. Although different national mediations can occur, when it comes to international public event television, there is likely to be a high degree of similarity between what audiences in particular national settings are watching.

The TV program format heralds a new stage in international public event television, and in global television viewing as a whole, for the event itself is not simultaneous so far as its different local manifestations are concerned. It is proximate rather than exact. The national adaptations of such popular formats as *Pop Idol* and *Dancing with the Stars* vary because of domestic industry considerations. Typically, these have to do with timing, duration, program form and style, as well as the context of surrounding programs. The result is that each national audience witnesses a version of these formats that is the same but different, common but localized. They occur within the same broad timespan, but their individual timing is quite distinct. A TV program format is viewed as a series of parallel audience watchings, complementary rather than equivalent, like each other but not the same, similar but not identical.

TV program format adaptation and remaking brought about through the circulation of TV formats are ushering in the phenomenon of diasporic public-event television. Media public events such as *Pop Idol* and *Dancing with the Stars* occur not as single massive TV phenomena so far as audience involvement is concerned. Instead, there are many different but parallel audiences, which view a series of approximately similar programs over a near-common stretch of time. Each has a distinct sense of the program so its members can cumulatively be said to be part of its worldwide audience, yet the knowledge of each of these audiences is in fact subtly different and distinct. In other words, while the audience for a TV program format can be massive, the fact is that this audience is not present as a single mass.

RESEARCHING FORMAT FRANCHISING

If the activity of TV program remaking is not especially new, the same cannot be said about its critical investigation. The television studies sub-field of critical format inquiry has really only come into existence over the past 20 years. Nevertheless, there has been incidental observation and comment that stretches back much further. In Australia, for example, the routine of program remaking had long been noticed by the general public, newspaper critics, and other observers. Cultural commentators in the 1960s and 1970s were fond of publicly mocking producer Reg Grundy for his routine of making Australian television versions of well-known US network quiz shows (Moran, 2013b). Ironically, this derision was accompanied by a singular failure to notice the same tendency on the part of the public broadcaster, the Australian Broadcasting Corporation (ABC), which regularly borrowed and reworked program ideas taken from the BBC and British commercial television (Moran, 1998). In any case, such commentary tended to be inconsequential, and was only offered in passing. The practice of format remaking was understood merely as an indicator of how banal and unoriginal commercial television and its practitioners could be.

Format franchising myopia also extended to more professional media watchers.

As already noted, there was no entry concerning the subject in TV critic Les Brown's (1977) *New York Times encyclopedia of television*. It was not until the early 1990s that more careful attention began to be given to the matter. Three signs of this new interest can be cited. First, as television companies started to formalize their interest in format program remaking, media journalists and professional television watchers started to systematically feature format trade matters in the pages of magazines such as *TV World* and *Broadcast and Television Business International*, alongside shorter, more incidental notices in such publications as *Variety* and *Billboard*. Feature-length attention began to be given to TV program formats and franchising. Features took the form of industry surveys and roundups identifying the major companies involved in the trade, and the various current formats on offer.

There were also discussions of format company mergers and market ambitions, new formats in particular genres, and even more imponderable topics such as what constituted a format and how formats might be legally protected. This latter interest signaled a second focus of critical attention. In 1994, the first book on TV formats appeared (van Manen, 1994). Its scope was limited to matters concerning the commodity's legal status. Unfortunately, the book's appeal was even more circumscribed by the fact that it was written in Dutch by an Amsterdam-based attorney, and was not subsequently made available in other languages. Even so, as van Manen's bibliography makes clear, the handbook's publication helped focus a growing interest in legal matters regarding format copyright and intellectual property law.

The last decade of the millennium was also an era of emerging scholarly attention in program franchising. The phenomenon of international proliferation and remaking of formats in such television genres as the quiz show, the news program, and the soap opera was the focus of interest from the early 1990s (see Cooper-Chen, 1994; Lee, 1991; Sinclair, Jacka & Cunningham, 1996; Skovmand, 1992). Newcomb's (1997) *Museum of Broadcasting Communication encyclopedia of television* marked this development with a lucid entry on the subject (Fiddy, 1997). I published the first English-language book-length study of formats, *Copycat TV*, in 1998 (Moran, 1998). For the most part, these *fin de siècle* format studies were concerned with the analysis of textual features of national program adaptations in what was still a developing trade.

The new century has seen a second and even a third wave of scholarly inquiry. This has coincided with the advent of TV program mega-formats that began with *Do You Want to Be a Millionaire?*, *Big Brother* and *Survivor*, and continues to the present with formats such as *Master Chef*, *Pop Idol* and *Ugly Betty*. Mega-program franchises have been spread across various format genres, including reality TV, live entertainment, and to a lesser extent filmed series, including sit-coms and telenovelas. Their popular international success has prompted an impressive output of shorter and longer critical studies, with two trends evident. One group of scholarly investigators has been fascinated by the advent of reality television, and especially with the global success of programs such as *Big Brother* (Frau-Meigs, 2006; Hill, 2005; Holmes & Jermyn, 2004; Mathijs & Jones, 2004; Murray & Ouellette, 2004). In turn, teams of international researchers have responded to the challenge of pinpointing the cultural meaning of these format remakes, so that successive book collections and shorter articles have addressed the meaning of such popular format programs as *Survivor*, *Do You Want to Be a Millionaire?* and *Pop Idol*, the makeover genre generally, and the drama serial *Ugly Betty* (Cooper-Chen, 2005; Heller, 2007; Lewis 2009a, 2009b; McCabe & Akass, 2013; Smith & Wood, 2003; Zwan & de Bruin, 2012).

A second group of studies has been more concerned with the underlying cultural and commercial practice of TV program format franchising (Bielby & Harrington, 2008; Chalaby, 2010; Esser, 2010; Havens, 2006; Magdar, 2008; Moran & Keane, 2004; Moran & Malbon, 2006; Waisbord 2004).

I followed *Copycat TV* (1998) with several other studies, including an industry guide (Moran & Malbon, 2006) and investigations of format franchising in Asia (Keane, Fung & Moran, 2007; Moran & Keane, 2004). This industry focus has been amplified by other important studies, including an insider media-professional account of the behind-the-scenes development of key mega-formats (Bazalgette, 2005) and two world-regional survey investigations of format trade on behalf of the world franchising association, FRAPA (Jager & Behrens, 2009; Schmitt, Bisson, & Fey, 2005). Two more recent collections by international scholars have further widened this research circle (Moran, 2009b; Oren & Shahaf, 2012).

Finally, some thoughts about future critical research are warranted. Investigating TV format remakes in national settings offers various challenges to both experienced and emerging scholars. Format adaptation is now a key component of contemporary television across much of the world, so this kind of study has immediate purpose and interest. Such inquiry is implicitly comparative and international, as highlighted in the 2009 and 2012 edited collections of format scholarship mentioned in the last paragraph (Moran, 2009b; Oren & Shahaf, 2012), where chapters that address national variations of the same cultural phenomenon are juxtaposed. Frequently, researchers are part of particular national academic diasporas, so they have the linguistic and cultural competencies to analyze format remakes that may not necessarily be available to some of their colleagues – for example, the phenomenon of an African-born researcher based in a North American academic institution writing about an African adaptation of the format of the British-originated *Pop Idol* (Ndela, 2012). TV format study has the potential to force researchers, teachers and students to pay more attention to the particularity and uniqueness of television industries and audiences in different parts of the world as part of the process of understanding TV program format dispersal.

Another area of possible inquiry involves developing a more robust theoretical account of format adaptation and remaking. From this viewpoint, format remaking may be new, but the principle is not. Semiotics and structuralism are two areas of reflection that hold out promise for increasing overall understanding of this kind of adaptation (Lotman, 1990; Taussig, 1993). Additionally, the phenomenon of franchised remaking is very common across many sectors of the knowledge industries, including those to do with culture, education, health and science (Moran & Keane, 2010). More broad-ranging research that links developments across some of these sectors has the capacity to throw greater light on the meaning of formats.

A second type of investigation of TV formats might concentrate on more historical inquiry, studying some of the many instances of broadcast remaking inside particular national boundaries, thereby revealing more complex patterns of international contact than is often supposed (Ferrari, 2012; Keinonen, 2009). More micro-studies are also needed. Analysis of particular national program remakes has tended to concentrate on textual comparison, so empirical process studies of adaptation involving media organizations could make a valuable contribution to the field. Finally, there is the classic area of inquiry focusing on audience negotiation of formats (Hill, 2005). What understandings do audiences have of formatted programs? How do viewers respond to their universality? Why do formatted programs seem to fit well into the current social environment of television? In short, there are many worthwhile research issues to pursue.

SUMMARY

This chapter has outlined the practices and meanings of TV program formats. The possibility of remaking a television program either in another time or in another place has existed since the start of television broadcasting in the

1930s and 1940s. Nevertheless, the industrial and cultural activity of program franchising has only matured in the contemporary environment of television. I have sought to understand this paradox by a multipronged investigation of the notion of the format. The subject has been examined in terms of definition and history, in terms of how it reorganizes business and culture, and in terms of how it has been understood by audiences. The latter category includes not only the viewing public but also critical researchers and scholars. By ending with an account of how researchers and scholars have grappled with the complex phenomenon of the format, I hope to have offered an original and stimulating contribution, posing new questions and opening new horizons to this fascinating area of study and practice.

FURTHER READING

Allen, R.C. (Ed.) (1995). *To be continued: Soap opera around the world*. New York: Routledge.

Banerjee, I. (2002). The locals strike back? Media globalization and localization in the new Asian television landscape. *Gazette: International Journal for Communication Studies*, 64(6), 515–535.

Bellamy, R.V., McDonald, D.G. & Walker, J.R. (1990). The spin off as television program: Form and strategy. *Journal of Broadcasting and Electronic Media*, 34, 283–297.

Bielby, W.T. & Bielby, D. (1994). All hits are flukes: Institutionalized decision making and the rhetoric of network primetime program development. *American Journal of Sociology*, 99(5), 1287–1313.

Brenton, S. & Cohen, R. (2003). *Shooting people: Adventures in reality TV*. London: Verso.

Clausen, L. (2004). Localizing the global: 'Domestication' processes in international news production. *Media, Culture & Society*, 26(1), 25–44.

Crisell, A. (1996). *Understanding radio*. New York: Routledge.

Dell, C.E. (2003). The history of 'travellers': Recycling in prime time American network programming. *Journal of Broadcasting and Electronic Media*, 47(2), 260–279.

Ek, R. (2006). Media studies, geographical imaginations and relational space. In J. Folkheimer & A. Jamson (Eds), *Geographies of communication: The spatial turn in media studies* (pp. 10–29). Goteberg: Nordicom.

Escoffery, D.S. (Ed.) (2006). *How real is reality TV? Essays on representation and truth*. Jefferson, NC: McFarland & Co.

Fine, F.L. (1988). A case for the federal protection of television formats: Teasing the limits of 'expression'. *Pacific Law Journal*, 17, 49–75.

Franko, E. (2006). Marketing 'reality' to the world: Survivor, post-Fordism, and reality television. In D.S. Escoffery (Ed.), *How real is reality TV? Essays on representation and truth*. Jefferson, NC: McFarland & Co.

Ganguly, L. (2010). Global television formats and the political economy of cultural adaptation: *Who Wants to Be a Millionaire?* in India. In T. Oran & S. Shahaf (Eds), *Global television formats: Understanding television across borders* (n.p.). New York: Routledge.

Gillespie, M. (1995). *Television, ethnicity and cultural change*. New York: Routledge.

Heller, D. (2003). Russian 'sitcom' adaptation: The Pushkin effect. *Journal of Popular Film and Television*, 31(2), 60–86.

Hetsroni, A. (2005). Rule Britannia! Britannia rules the waves: A cross-cultural study of five English-speaking versions of a British quiz show format. *Communications*, 30(2), 129–153.

Hilmes, M. (2003a). British quality, American chaos: Historical dualisms and what they leave out. *The Radio Journal*, 1(1), 42–65.

Hilmes, M. (2003b). Radio nations: The importance of transnational media study. In *Atlantic Communications: Political, social and cultural perspectives on media and media technology in American and German history in the 20th century*. London: Berg.

Hyatt, W. (1997). *Encyclopedia of daytime television*. New York: Billboard.

Kean, C. (1991, October). Ideas in need of protection. *Television*, 20–21.

Keane, M. (2004) A revolution in television and a great leap forward for innovation? China in the global television format business. In A. Moran & M. Keane, *Television across Asia* (pp. 88–104). London: Curzon/Routledge.

Komplan, D. (2005). *Rerun America*. New York: Routledge.

McDermott, M. (2004). Mark Goodson and Bill Todman. In H. Newcomb, *Encyclopedia of television* (pp. 1183–1184). London: Routledge.
Miller, J. (2000). *Something completely different: British television and American culture*. Minneapolis: University of Minnesota Press.
Moran, A. (2005). Configurations of the new television landscape. In J. Wasko (Ed.), *A companion to television* (pp. 291–307). Blackwell Companions to Cultural Studies. Malden, MA: Blackwell.
Moran, A. (2007). Localizing global television. *Media International Australia*, 124, 145–155.
Moran, A. & Keating, C. (2007). *A historical dictionary of Australian radio and television*. Lanham, MD: Scarecrow.
Morris, N. & Waisbord, S. (Eds) (2001). *Media and globalization: Why the state matters*. Lanham, MD: Rowman and Littlefield.

REFERENCES

Bazalgette, P. (2005). *Billion dollar game: How three men risked it all and changed the face of television*. London: Time Warner.
Bellamy, R.V. & Trott, P.L. (2000). Television branding as promotion. In S. Eastman (Ed.), *Research in media promotion* (pp. 127–159). Mahwah, NJ: Lawrence Erlbaum.
Bielby, D. D. & Harrington, C. L. (2008). Global TV: Exporting television and culture in the world market. New York: New York University Press.
Billig, M. (1995). *Banal nationalism*. London: Sage.
Bourdon, J. (2012). From discrete adaptation to hard copies: The rise of television formats in European television. In T. Oren and S. Shahaf (Eds), *Global television formats: Understanding television across borders* (pp. 111–127). New York: Routledge.
Brown, L. (1977). *The New York Times encyclopedia of television*. New York: Times Press.
Brunsdon, C., Johnson, C., Moseley, R. & Wheatley, H. (2001). Factual entertainment on British television: The Midlands Research Groups 89 project. *European Journal of Cultural Studies*, 4(1), 29–63.
Brunt, R. (1985). *What's My Line?* In L. Masterman (Ed.), *Television mythologies* (pp. 21–28). London: Comedia.
Burston, J. (2000). Spectacle, synergy and megamusicals: The global industrialisation of the live entertainment economy. In J. Curran (Ed.), *Media organization and society* (pp. 44–63). New York: Oxford University Press.
Burston, J. (2009). Recombinant Broadway. *Continuum*, 23(2), 159–169.
Camparesi, V. (2000). *Mass culture and national traditions: The BBC and American Broadcasting*, 1922–1954. Fucecchо, Italy: European Press Academic.
Chalaby, J.K. (2010). Broadcasting in a postnational environment: The rise of transnational TV groups. *Critical Studies in Television*, 4(1), 39–64.
Chalaby, J. (2012). 'At the origin of a global industry: The TV format trade as an Anglo-American invention', *Media, Culture & Society*, 34(1), 36–52.
Cooper-Chen, A. (1994). *Games in the global village: A 50 nation study*. Bowling Green, OH: Bowling Green University Popular Press.
Cooper-Chen, A. (2005). A world of 'millionaires': Global, local and 'glocal' TV game shows. In A. Cooper-Chen (Ed.), *Global entertainment media: Content, audiences, issues* (pp. 237–251). Mahwah, NJ: Lawrence Erlbaum.
Dayan, D. & Katz, E. (1992). *Media events: The live broadcasting of history*. Cambridge, MA: Harvard University Press.
Dicke, T.S. (1992). *Franchising in America*, 1840–1980. Chapel Hill, NC: University of North Carolina Press.
Esser, A. (2010). Television formats: Primetime staple, global market. *Popular Communication*, 8, 273–292.
Ferrari, C. (2012). 'National Mike': Global host, global format in early Italian television. In T. Oren and S. Shahaf (Eds), *Global television formats* (pp. 128–147). New York: Routledge.
Fiddy, D. (1997). Format sales, international. In H. Newcomb (Ed.), *Museum of Broadcasting Communication encyclopedia of television* (pp. 623–624). Chicago: Fitzroy Dearborn.
Frau-Meigs, D. (2006). *Big Brother* and reality TV in Europe. *European Journal of Communication*, 21(1), 33–56.
Freedman, D. (2003). *Who Wants to Be a Millionaire?* The politics of television exports. *Information, Communication & Society*, 6(1): 24–41.
Frith, S. (1987). Copyright and the music industry. *Popular Music*, 71, 57–75.

Gitlin, T. (1983). *Inside prime time*. New York: Pantheon.

Havens, T. (2006). *Global television marketplace*. London: BFI.

Heller, D. (2007). *Makeover television: Realities remodelled*. New York: Taris.

Hill, A. (2002). *Big Brother*: The real audience. *Television and New Media*, 3(3), 323–340.

Hill, A. (2005). *Reality TV: Audiences and popular factual television*. London: Routledge.

Hilmes, M. (2012). *Network nation: A transnational history of British and American broadcasting*. New York: Routledge.

Holmes, S. & Jermyn, D. (Eds) (2004). *Understanding reality television*. London: Routledge.

Jaflin, S. & Waisbord, S. (2009). Imagining the national: Television gatekeepers and the adaptation of global franchises in Argentina. In A. Moran (Ed.), *TV formats worldwide: Localising global programs*. Bristol: Intellect.

Jager, E. & Behrens, S. (2009). *The FRAPA report 2009: TV formats to the world*. Cologne: FRAPA.

Johnson-Woods, T. (2002). *Big bother: Why did that reality TV show become such a phenomenon?* Brisbane: University of Queensland Press.

Keane, M. (2013). Personal interview with Albert Moran, Brisbane.

Keane, M., Fung, A. & Moran, A. (2007). *Out of nowhere? New television formats and the East Asian cultural imagination*. Hong Kong: Hong Kong University Press.

Keinonen, H. (2009). Constructing a sense of cultural proximity in a Finnish adaptation of *Hancock's Half Hour*. *Critical Studies in Television: Scholarly Studies in Small Screen Fictions*, 4(2), 37–52.

Kilborn, R. (1993). Read my lips: Re-evaluating subtitling and dubbing in Europe. *Media Culture and Society*, 15, 641–650.

Lane, S. (1992). Format rights in television shows: Law and the legislative process. *Statute Law Review*, 13(1), 24–49.

Lane, S. & Bridge, R.M. (1990a). The protection of formats under English law, part I. *Entertainment Law Review*, 96–102.

Lane, S. & Bridge, R.M. (1990b). The protection of formats under English law, part II. *Entertainment Law Review*, 131–142.

Lee, P. (1991). The absorption and indigenization of foreign media cultures: A study on a cultural meeting point of the East and West: Hong Kong. *Asian Journal of Communication*, 1(2), 52–72.

Lewis, T. (2009a). From global to glocal: Australianizing the makeover format. In A. Moran (Ed.), *TV formats worldwide: Localizing global programs* (pp. 291–305). Bristol: Intellect.

Lewis, T. (Ed.) (2009b). *TV transformations: Revealing the makeover show*. London: Routledge.

Lim, T. & Keane, M. (2004). Let the contests begin! 'Singapore slings' into action: Singapore in the global television format business. In A. Moran & M. Keane (Eds), *Television across Asia* (pp. 105–121). London York: Curzon/Routledge.

Lotman, Y. (1990). *Universe of the mind: A semiotic theory of culture*. Trans. A. Shukman, Introduction U. Eco. London: I.B. Tauris.

Magdar, T. (2004). The end of TV 101: Reality programs, formats and the real business of television. In S. Murray & L. Ouellette (Eds), *Reality TV: Remaking television culture* (pp. 137–152). New York: New York University Press.

Magder, T. (2008). 'Television 2.0 The business of American television in transition', In S Murray & L Ouellette (Eds.), Reality TV remaking television culture, 2nd ed. (pp 141–164). New York: NY New York University Press.

Mathijs, E. & Jones, J. (Eds) (2004). *Big Brother international: Formats, critics, and publics*. New York: Wallflower.

McCabe, J. & Akass, K. (Eds) (2013). *Reading Ugly Betty: TV Betty goes global*. New York: I.B. Taurus.

Moran, A. (1998). *Copycat TV: Globalization, program formats and cultural identity*. Luton: University of Luton Press.

Moran, A. (2008). Cultural power in international television markets. In J. Wasko & M. Erickson (Eds), *Cross-border cultural production: Electronic runaway or globalization?* (pp. 333–359). New York: Cambria Press.

Moran, A. (2009a). *New flows in global TV*. Bristol: Intellect.

Moran, A. (Ed.) (2009b). TV formats worldwide: Localizing global programs. Bristol: Intellect.

Moran, A. (2009c). Practising localization: Regimes of television production in the era of migrating media. In G. Elmer (Ed.), *Migrating media: Space, technology and global film and television*. Lanham, MD: Rowman and Littlefield.

Moran, A. (2013a). Global television formats: Genesis and growth. *Critical Studies in Television*, 1(1), 3–12.

Moran, A. (2013b). *TV format mogul: Reg Grundy's transnational career*. Bristol: Intellect.

Moran, A. (unpublished). Transnational television trader: Paul Talbot and Australian broadcasting 1949–71.

Moran, A. & Keane, M. (Eds). (2004). *Television across Asia: Globalization, industry and formats*. London: Curzon/Routledge.

Moran, A. & Keane, M. (Eds). (2010). *Cultural adaptation*. London: Routledge.

Moran, A. & Malbon, J. (2006). *Understanding the global TV format*. Bristol: Intellect.

Murray, S. & Ouellette, L. (Eds) (2004). *Reality TV: Remaking television culture*. New York: New York University Press.

Ndela, N. (2012). Global television formats in Africa: Localizing *Idols*. In T. Oran & S. Shahaf (Eds), *Global television formats: Understanding television across borders* (pp. 357–381). New York: Routledge.

Newcomb, H. (1997). *Museum of Broadcast Communications encyclopedia of television*. London: Fitzroy Dearborn.

Nordenstreng, K. & Varis, T. (1974). *Television traffic: A one way street?* New York: UNESCO.

Oren, T. & S. Shahaf eds. (2012) Global television formats: Understanding television across borders. New York and London: Routledge.

Ritzer, G. (1998). *The McDonaldization thesis: Explorations and extensions*. London: Sage.

Roscoe, J. (2001). *Big Brother* Australia: Performing the 'real' twenty-four-seven. *International Journal of Cultural Studies*, 4(1), 473–488.

Schmitt, D., Bisson, G. & Fey, C. (2005). *The global trade in television formats*. London: Screen Digest.

Sinclair, J., Jacka, E. & Cunningham, S. (Eds) (1996). *New patterns in global television: Peripheral vision*. London: Oxford University Press.

Skovmand, M. (1992). Barbarous TV international: Syndicated *Wheels of Fortune*. In M. Skovmand & K.C. Schroeder (Eds), *Media cultures: Reappraising transnational media* (pp. 84–103). London: Routledge.

Smith, M. & Wood, A. (Eds) (2003). Survivor *lessons: Essays on communication and reality television*. Jefferson, NC: McFarland.

Storper, J. (2002). *Globalization and knowledge flows: An industrial geographer's perspective in regions, geography and the knowledge-based economy*. New York: Verso.

Straubhaar, J.D. (2003). Choosing national TV: Cultural capital, language, and cultural proximity in Brazil (pp. 77–112). In M. G. Elasmar (Ed.), *The impact of international television: A paradigm shift*. Mahwah, NJ: Lawrence Erlbaum.

Straubhaar, J.D. (2007). *World television: From global to local*. Austin, TX: University of Texas Press.

Taussig, M. (1993). *Mimesis and alterity*. San Francisco: Ignatius Press.

Thompson, K. (2003). *Story telling in film and television*. Cambridge, MA: Harvard University Press.

Todreas, T. (1999). *Value creation and branding in television's digital age*. London: Quorum.

Tunstall, J. (2008). *The media were American: US mass media in decline*. New York: Oxford University Press.

Turner, G. & Tay, J. (Eds) (2009). *Television studies after TV*. London: Routledge.

van Manen, J.M. (1994). *Televisiefor mats: en iden nar Nederlands recht Otto Cranwinckle*. Amsterdam: Uitgever.

Wasko, J. ed. (2005). A Companion to television. Malden: Blackwell.

Waisbord, S. (2004). McTV: Understanding the global popularity of television formats. *Television & New Media*, 5, 359–383.

Zwan, K. & de Bruin, J. (2012). *Adapting* Idols: *Authenticity, identity and performance in a global television format*. Farnham: Ashgate.

17

Cultural Forms of Television: Sport

David Rowe

INTRODUCTION: TELEVISION, SPORT AND CULTURAL TEAM FORMATION

Of all the cultural forms of television, sport stands out as the most conspicuous case where television found an external partner that produced something immeasurably more significant than either party could have managed if operating in isolation. Television creates its texts through many different arrangements. For example, it can take a cultural form like film that is designed for another context, shrink it, punctuate it with advertisements (as indirect 'payment' for viewing), and display it. It has taken the established formats of newspapers and radio stations, modified them considerably, and produced the most important mode of news delivery that has been seen (literally) over the last century. It did something analogous with magazine content and created television current affairs, similarly working with staged drama and its radio renditions to produce television soap opera, drama and the mini-series (see various contributions to Miller, 2003).

Television can take the output of another cultural industry, like popular music, and give it a widely circulated visual presence beyond the concert venue and the radio broadcast by providing space for the TV pop video, while also diffusing and enhancing the unique concert or 'gig' by covering it from many different viewing positions and incorporating coverage of setting up and 'bumping out', interviews with the performers, and so on. Television can create its own genres, like 'reality TV', and foster new links between broadcast, online media and telecommunications. However, of all the hybridic genres and synergistic relationships that television has formed in the constitution of contemporary popular culture, it is difficult to conceive of one that has had a more thoroughgoing cultural significance than its association with sport.

In bringing two social institutions together, and fashioning such a complex interdependence as to render any disentanglement inconceivable to the point of mutually assured destruction, TV sport represents the 'killer application' so much desired, and so often elusive, as the medium has developed from modest analogue origins to digital omnipresence (Hutchins & Rowe, 2009, 2012). Importantly, this is not a nexus that can be confined to the Anglo-US axis (see, for example, Mehta, 2008; Miller, Lawrence, McKay & Rowe, 2001). While there is no doubt that television sport first emerged in this domain – albeit producing two rather different models of presentation, economics and relationships with the state (Chandler, 1988; Goldlust, 1987) – sport and television have been intimately involved everywhere that they can be found in the same physical or technical space. Television sport may have been, and still remains in significant ways, an export from the Anglo world to the rest of the globe, but each national television system has appropriated and incorporated it – and sometimes sent it back to its historical point of origin (see, for example, Gupta, 2009) – in the making of a global 'media sports cultural complex' (Rowe, 2004, 2011a).

In this chapter,[1] I will set out to probe the development and significance of sports television. In the process, I suggest that this 'marriage made in heaven' has been marked by deep passions of both a positive and negative kind, evidencing a level of co-dependency that has continually raised issues of power, both over who controls sport and how much sport can 'extract' from television. These structural issues, though, must be considered alongside those pertaining to the uses and abuses of sports television in seeking to understand its massive appeal across the globe by operating as a giant engine of 'social texts' that trades on high emotion, and routinely divides that world into a myriad of competing camps.

SPORT AND TELEVISION: CONTEST AND NEXUS

Scholars of sport and television – as well as of sports organizations, fans, viewers, governments and media companies – have been concerned about the mutual power of the institutions since they first came together. Most of the anxiety has been on the side of sport. This is because while the institutions of sport and media, as the products of modernity, are more or less contemporaries, professional sport is well over a century older than television and, though commercial in some respects from the outset and increasingly industrialized, sport is also deeply wedded to nostalgic, pre-modern notions of heroic athletic culture as exemplified by the Olympics, and to a communitarian ethos typified by folk games (Guttmann, 1978). Thus, while there is no doubt that television can be credited with the cultural pervasiveness and economic wealth that sport first achieved in the second half of the 20th century, it has always produced an ambivalence among sports *aficionados* (Cashman & McKernan, 1981; Klatell & Marcus, 1988). These misgivings about television have a long history and wide socio-cultural resonance that go well beyond its relationship with sport. Just as the 'box in the corner' occupied more key domestic space and leisure time as people willingly acquired this new consumer durable and sat before its screen, mass cultural theorists, behavioural psychologists and cultural entrepreneurs told them that it was very bad for them (see, for example, Winn, 1977). Television would, the populace was widely and consistently told, degrade social relations, stimulate anti-social activity, and have negative psychological and physiological consequences, especially for the young – indeed, critics of television have accused its producers of infantilizing their audiences (Hartley, 1999). Thus, as with other forms of morally questioned leisure, such as drinking alcohol and gambling, watching television for the purposes of pleasure has been widely practised but also subject to external and self-surveillance and disciplining. Such discourses are heavily scored by social class (Clarke, Critcher & Johnson, 1987), as well as by gender and age, and are represented in the tension between the educational, self-improving functions of television and its entertainment dimensions.[2] The place of sport, a major component of television programming, along this continuum has been similarly contested because, as another form of popular culture, it has also faced class-inflected disputes about its cultural worthiness and social implications (Whannel, 1992, 2008).

Thus, sports programming and large viewing audiences can come to represent one of the most deplorable indications that the medium has been given over to 'mindless' mass culture. When wedded to socially approved violence, aggressive expressions of masculinity, and various forms of in-group chauvinism, the passivity felt by its critics to be encouraged by the distracting sports televisual text (*viz* the familiar figure of the 'couch potato') becomes actively malign (Bryant, Zillmann & Raney, 1998). There is, then, a prior dispute about the cultural status and social implications of sport before it was ever manifest on the small screen (Elias & Dunning, 1986). As noted above, sports culture has been broadly split between popular, pre-modern expressions of folk culture – the amateur, edifying refinement as exemplified by the Olympian ideal – and the professional, commercially acquisitive aspects of the sports industry. The forms of popular sport that first emerged out of the physical culture of (especially, in the first instance, British) village life tended to be carnivalesque, unruly and often dangerous. The social disorder that they represented, including their negative impact on both fitness for agricultural and then industrial labour, and also military combat, demanded the introduction of binding rules, determinate time spans, and regulated levels of violence. But as this disciplinary shaping of folk physical play took place through state sanction, and by means of the compulsory integration of approved sports and exercise routines into the growing institution of formal education (Hargreaves, 1986), sport became better suited to commercial exploitation because it was more organized, practised, known and understood. The enclosure and regulation of entry into the sites of sports contests, and the 'spin-off' activities of gambling on results and consuming beverages, food and other products, became more prominent and systematic under industrial capitalism. It is this unashamedly pleasurable side of working-class sport and leisure (Clarke & Critcher, 1985) that has attracted most elite criticism, and which has carried over into class-based disapproval of sport on television.

But sport is not an entirely working-class activity, and some modes of sport, such as rugby union in the elite schools of England and France and the ivy league universities of the USA, have been championed for their capacity to provide socially valuable forms of training in leadership, cooperation, competition and goal-orientation, both to educate and reinforce the status of a dominant class (Bourdieu, 1978). Olympism, as envisaged by its successful modern architect-cum-revivalist Baron Pierre de Coubertin, saw competitive sport as a valuable medium of interpersonal, international communication that was untainted by the profit motive (Hill, 1992). Although, Olympism (later, as it quasi-professionalized, also referred to as 'Prolympism' – Donnelly, 1996) became in the late 20th century deeply enmeshed with corporate capitalism, and sports such as rugby union and cricket have long since left behind a primarily amateur orientation, proponents of sport could still suggest that its representation on television does much more than agreeably pass the time or attract viewers with displays of approved forms of violence. Sport can also be presented as a virtuous activity that its viewers might imitate with positive results. Such favourable repercussions of showing the 'right' sports in the 'right' spirit could be extended and deepened by a strategic articulation with a third crucial manifestation of modernity – the nation – even in very different societies from the Western countries that revived the Olympics and championed sport, such as China (Xu, 2008).

The 'fit' between sport and nation was advanced through a combination of cultural, technological, ideological and economic processes. In cultural terms, sport developed both as popular leisure and state-sanctioned (indeed, mandated) physical activity. So, familiarity with it as cultural practice was both voluntary in civil society and compulsory within school institutions. Integral to the codes of sport was the uncertainty of outcome and propensity to encourage identification with one of the participants arising from structured competition and score-keeping – an essential attribute for its emergent appeal to television. The technology of television made it possible for spatially-bounded events not only to be carried as moving image and aural actuality far beyond the fixed locations in which they took place, but to create in symbolic form a new phenomenon – the mediated sports audience. It is easy in the current television and sport saturated moment to underestimate the shocking power of a nation rendered to itself through a televised sport event. Wheen (1985, p. 223) reminds us of the 'astonished' reaction to John Logie Baird's seminal broadcast of the Epsom Derby horse race (a hierarchical combination of ruling-class display and working-class involvement) in Britain in 1931 – and serially repeated in virtually all of the world's nations in their specific cultural contexts since.

As Whannel (1992, p. 20) notes, the BBC's pioneering of sports television was intimately connected to a nation-building project:

> The BBC, in short, had become a primary definer of national identity, a forger of national unity. In contributing to the production of the imaginary coherence of national identity, it was articulating two elements of national culture most decisively de-politicised – sport and the monarchy. It was not merely accident or coincidence that found the BBC at the focus of this powerful ideological articulation.

Whannel (1992, p. 21) quotes BBC Television Controller Norman Collins describing the national role of television sport in the crucial post-Second-World-War period:

> And sportsmen who play cricket on the green or follow a local football team or belong to a church tennis club will have [the major sports venues] Lords, Wembley and Wimbledon all paraded before them for their pleasure. In short this country will at last discover what goes on there. And the discovery will be well worth making.

This happy self 'discovery' of the nation through television's illumination of great sports events is compounded by the construction of the 'other' in televising, especially in real time, their national sporting representatives in competition. The clear cultural and ideological role of national sports television systems (not unprecedented in media terms, given the previous importance of the press and radio, but unquestionably more effective because of its enhanced capacity to give a sense of 'being there') can be repeatedly demonstrated in the televising of sports events such as the Olympics and the World Cup of association football. For example, the Opening Ceremonies of the summer Olympic Games – which produce persistently the world's single largest television audience – can through television announce to the world that a nation claims superpower dominance (Russia 1980, Los Angeles 1984, Atlanta 2006), rising power status (Seoul 1988, Barcelona, 1992, Beijing 2008), or can function as a reminder of its historical influence (Athens 2004, London 2012) and its reintegration into the community of nations (Rome 1960, Tokyo 1964, Munich 1972).

However, this foregrounding of the nation state is at the same time intimately linked to the economic importance of sports television. Not only does it advertise the nation, but it is a key advertising space in which national and multinational goods and services can be displayed. Since, in particular, the so-called 'Hamburger Olympics' of 1984 in Los Angeles, the Games have used television's audience reach for brand sponsors and advertisers to broker substantial broadcast rights contracts. The value of television to the Olympics is thus both direct and indirect. The elite Olympic Partner (TOP) programme, which affords exclusive worldwide marketing rights both to the Summer and Winter Olympic Games, involves a very substantial financial contribution in recent Olympiads from the prominent brands Coca Cola, General Electric, McDonalds and Visa (USA); Acer (Taiwan); Atos Origin (France); Omega (Switzerland); Panasonic (Japan); and Samsung (Korea). There is a range of other sponsorship, marketing, supplier, ticketing and licensing arrangements, as well as broadcast rights revenue. Although the value of television rights cannot be neatly disaggregated from other sources of revenue (which are themselves dependent on television coverage, such as sponsorship) and, indeed, from other broadcast and media platforms (radio, mobile and internet), according to the IOC (2012a) broadcasting provides 47% of revenue, followed by 45% for corporate sponsorship, 5% for ticketing and 3% for licensing.[3] The historical shift in economic significance from place-based to mediated sport spectatorship described above is starkly revealed here, with the latter directly generating approaching ten times the revenue of the former.

REVENUE FOR SPORT, REGARD FOR TELEVISION

According to the *Olympic Marketing Fact File, 2008 Edition* (IOC, 2012b, p. 26):

> Olympic broadcast partnerships have provided the Olympic Movement with an unprecedented financial base and helped to ensure the future viability of the Olympic Games.

The global broadcast revenue figure for the 2004 Olympic Games represents a fivefold increase on the 1984 Los Angeles broadcast revenue two decades earlier, while the 2006 Olympic Winter Games saw an eightfold increase on the 1984 Sarajevo broadcast revenue a little over two decades earlier. These Olympic broadcast partnerships have been the single greatest source of revenue for the Olympic Movement for more than three decades.

The main source of broadcast revenue for the Olympics is clearly television (in conventional broadcast form), with sponsorship (as noted, highly dependent on broadcast exposure) not far behind. As will be discussed later, technological convergence and the development of networked digital media is challenging the dominance of broadcast sports television, but the value of television-dominated broadcast rights for the summer Olympics over the last half-century is striking. It has risen, as Table 17.1 reveals, from US$1.2 million for the Rome 1960 Games to US$1,739 million for Beijing 2008.

There is no sign of Olympic media rights (still dominated by broadcast television) being in decline – their combined estimated value for the 2010 Vancouver Winter Games and 2012 London Summer Games is US$3.91 billion (*Sportcal*, 2012, p.7)

The extent to which television companies can earn back, and profit from, these major investments is contested, but the web of interests is such that, as I have argued previously, it is possible to 'make money while losing it in sports television' (Rowe, 2004, pp.74–79) if the value of sports rights is viewed holistically as a pivotal marker of broadcaster status and legitimacy. However, Folkenflik (2006) notes with regard to the long-time premier Olympic broadcaster, NBC, that billing for advertisements and on-selling broadcast

Table 17.1. Olympic Summer Games broadcast revenue history, 1960–2008* (US$ million)

1960	Rome	1.2
1964	Tokyo	1.6
1968	Mexico City	9.8
1972	Munich	17.8
1976	Montreal	34.9
1980	Moscow	88
1984	Los Angeles	286.9
1988	Seoul	402.6
1992	Barcelona	636.1
1996	Atlanta	898.3
2000	Sydney	1,331.6
2004	Athens	1,494
2008	Beijing	1,739

*Figures not corrected for inflation.

Source: International Olympic Committee (2012b, p. 26).

rights may be insufficient to offset broadcast rights payments:

> No union of sport and media is closer than that between NBC and the Olympics. NBC has the exclusive broadcast rights for the powerful American market and a degree of influence over how the Olympic Games are organized. But that control comes with a steep price – and an increasingly uncertain payoff.

Similarly, Paul Swangard, Managing Director of the Warsaw Sports Marketing Center, University of Oregon, in response to NBC Sports Chairman Dick Ebersol's prediction of a profit from the 2006 Winter Games broadcasting of [US]$50 million, argued that after the Games if audiences watch other NBC programming it is economically worthwhile, but 'If it's another two years until they win a Nielsen week [US television viewing as measured by that company] again, I think you begin to question why you spend that much money to win the Nielsen ratings once every two years' (quoted in Folkenflik, 2006).

If premium sports broadcasting produces only expensive, intermittent television 'events', then it may be regarded as a liability. Yet, as Folkenflik (2006) also recognizes, NBC has historically long, multi-Games rights contracts, while 'NBC's parent company, General Electric, is a global sponsor. And a senior executive from NBC Sports sits on the governing board of the International Olympic Committee'. Thus, broadcast rights buy sporting influence, broadcaster longevity and parent company profile (although this, too, must be paid for through the TOP programme). For Singer (1998), then, possession of significant sport 'properties', even extending to ownership of sports clubs and stadia, is a non-negotiable requirement of operating as a major media company.

That sport is able to demand and receive such bounty (or, perhaps, a dowry) from television is striking evidence of their mutual appeal (Rowe, 1996). This apparently smooth synergy invites inspection given that, in the early era of television, there was considerable suspicion in sport towards television, which television returned commensurately with scepticism (Stoddart, 1986). Professional sport's initial reliance on charging for admission provoked the not-unreasonable fear that paying spectators would be discouraged from attending a sports event that they could watch from home and without charge on television. Extracting substantial rights fees from television, though, was more than handsome compensation to sport for any loss of gate takings. Television, in bringing sport (ultimately) into every household of affluent societies and (increasingly) into the homes and other spaces of all other nations, would provide the incalculable additional benefit of astonishing cultural reach and a clear incentive for television to advertise and promote the sports industry as integral to the interests of television itself. This arrangement has set the terms for something of an institutional struggle over who controls sport – sports organizations themselves or the broadcasters who, as noted, directly fund around half their revenue (and often more), and transport them in moving image form beyond the stadium. But the economic relationship between sport and media also raises questions about who controls those institutions – their economic elites, workers (such as sportspeople and journalists) or appreciators (fans, spectators, viewers, listeners, readers).

For this reason, a strong theoretical approach running through research and scholarship on sports television has been political economy (Boyle & Haynes, 2000; Lawrence & Rowe, 1986). The '*cui bono*?' question has tended to be submerged by self interest within the sports television industry, or answered naively and superficially as 'everybody', with no apparently substantial transaction costs arising from the exchanges of cultural content, visibility and financial compensation. However, several theorists have been critical of the role of television in accelerating and extending the commodification of sport and, less commonly, in advancing commodity logic within television itself. Sut Jhally's (1984) influential idea of the 'sports-media complex', for example, developed out of a general critique of the capitalist exploitation of culture to embrace sport in particular (for example, Hoch, 1972), as well as preliminary explanations of sports television's wider cultural-ideological role (Buscombe, 1975). For Jhally, the key moment in the development of sports television was the realization in the United States context in the late 1930s (subsequently to be highly influential around the world) that the insertion of advertisements in and around major televised sports contests could have a dramatic effect on product sales (in this case Gillette razors, thereby reinforcing the key relationship between sport and masculinity).

This development, though, is not merely a case of capital accumulation in and through media sport. It is about the transformation of the practice of sport and, in turn, its associated meanings. Regarding the first, Jhally notes that:

> There has been a great deal written about the effect that [this] commodity structure has had on the organization and nature of professional sports. It has led to sports leagues changing the rules of the game to provide a better television package; clubs moving from one city to another based not upon stadium support, but upon the television audience; the flow and momentum of the game being interrupted as the game is stopped for time-outs that are called so that television can show commercials; the creation and destruction of entire sports leagues based upon whether or not television support could be found; and the ability (or inability) of teams to sign players, depending on the size of the television market a team controls. (Jhally, 2006, p.139)

There is an implication here that television has in some way 'corrupted' sport. As noted above, this is a historically dubious argument – as Jhally (2006, p.138) himself notes, 'sports have always been tied into a commodity sphere of one kind or another, their shape and organization always dependent on their level of profitability'. If, though, sport was never entirely 'pure' (regarding commerce) in modernity, television has taken it unquestionably beyond selling 'bites' of athlete time and stadium space involving a limited set of ancillary industries such as betting, transport and hospitality, into a much more abstract commodity space. Commercial television captures and sells audiences for other purposes – including inducements to buy a range of advertised products, raising aggregate viewer ratings so as to charge more for advertising across whole networks, and, increasingly, to buy other television and communication services (including 'bytes' of data).

However, as with other cultural products, there is a tension within televised sport between standardization and diversification. In any given context, sport attracts different viewers. For example, 'snow' sports are much more significant in northern continental Europe than in sub-Saharan Africa, or even in the UK, where different combinations of climate, opportunity and cultural history produce different levels of sporting participation and interest. Sport must, then, have audience appeal – indeed, 'made for television' sports programming which cannot summon a substantial co-present crowd constitutes a greatly diminished television spectacle that only serves to reinforce the key connection between in-stadium and at-home audiences. As Jhally and others (for example, Horne 2006) have noted, the 'massification' of audiences is necessary in order to turn localized sports events into widely diffused sports visual texts. The development of individual sports as sets of codified rules that enable coherent and comparable competitive contests already requires a 'disciplinary' governmental logic (Miller et al., 2001), but television imposes its own restrictions in order to fit sports events into programme schedules and to be able carry them across distinct television markets.

A GLOBAL SPORTS TELEVISION APPARATUS

Over the last two decades, a global television market has emerged (Rowe, 2011a) which constitutes a very large, multi-channel space that can be 'populated' with live sports events of extensive interest around the world, as well as provide niche programming for minority sports interests in various countries (including servicing the interests of expatriates living in 'enemy' sports territories, such as Indian cricket fans resident in the USA, Scots association football fans living in North America, and Australian rules football fans located anywhere other than Australia). Both of these forms of sports television can operate as 'subscription drivers' of sufficient appeal to persuade viewers to pay for access to them. But much sports television is also 'filler', consisting of multiple repeats of sports contests and programming of events with at-best marginal interest to viewers in various territories. Nonetheless, the marginal cost of such programming is negligible where the broadcast structure already exists, and it has, for specific sports, the additional utility of being able to declare that they are broadcast (and, increasingly, streamed) on a worldwide basis. For example, the sport of rugby league, played mainly in parts of northern Europe and Australasia, can claim to be a 'global sport' on account of its audio-visual reach. In 2009, the third match in its annual domestic State of Origin series between New South Wales and Queensland was made accessible in many countries, with the Australian Rugby League announcing:

> Millions of fans around the world will be able to tune into Game 3 of the Harvey Norman State of Origin Series thanks to NRL.com and Telstra Bigpond.
>
> For the first time NRL.com will provide live streaming of the match into Europe and South America – and access is free, simply by logging onto NRL.com and following the links.
>
> The live streaming service will complement the game's international broadcasting arrangements that will see the match televised around the globe via the Australia Network, Sky NZ, Setanta and Fox Sports International.
>
> 'State of Origin is a world-class event and with the assistance of our broadcast partners, NRL.com and Telstra Bigpond we are taking it to an even bigger world-wide audience,' ARL Chief Executive Mr Geoff Carr said today. (Australian Rugby League, 2009)

International broadcasting arrangements were listed as involving scores of countries, including the USA (incorporating Puerto Rico) and Canada, Afghanistan, Chad, Cote D'lvoire, Iraq, Kyrgyzstan, Macau, North Korea, Nuie, Oman, Palestine, Slovakia, Vietnam and Yemen. Live streaming was also available to countries including Bolivia, Bosnia/Herzegovina, Gibraltar, Paraguay, Suriname and the Vatican City. The intensity and scope of interest in rugby league in Kyrgyzstan, Chad and the Vatican City is difficult to ascertain, but can be safely assumed to be slight. Nonetheless, broadcast sports television and the internet can confer on sport a *de facto* global status simply on the basis of such widespread availability.

At the same time, television companies can cite their global circulation of sports texts to demonstrate their power and influence in the world of sport. All of the so-called 'Big Five' transnational (though US-based) media corporations – Disney, Viacom, Time Warner, NBC Universal and News Corporation (conventionally called News Corp and split in 2013 into two publicly listed companies, News Corp and 21st Century Fox) – are involved in sports television, but it is the last that is the most influential and most closely associated with sport on a global basis. News Corp/21st Century Fox has aggressively pursued the acquisition of sports broadcast rights as, under the leadership of Rupert Murdoch, it developed from a minor Australia-based newspaper company into an international media conglomerate. Murdoch's most famous declaration of intent in sport was that it would be used as a 'battering ram' to open up the subscription television and television services market (Andrews, 2006), but he also deployed sport to signal the intent of 'demoralizing' other networks by depriving them of their most cherished sport properties. In the USA, News Corp's FOX Sports broke into the television sports market by outbidding the established networks – ABC, NBC and CBS – in the early 1990s. According to its 2008 Annual Report, News Corp (2009) retained broadcast network leadership in the USA, the world's most lucrative television sports market:

> FOX Sports remained in first place among all broadcast networks for the 12th consecutive year on the strength of major broadcasts, including the Major League Baseball All-Star Game, the World Series, the BCS National Championship Game, the NFC Championship Game and the 50th Daytona 500.
>
> Our Super Bowl broadcast last year achieved the largest audience in Super Bowl history. The broadcast was the most-watched program in television history, with almost 150 million viewers.
>
> FOX Sports also led all broadcast networks with five Sports Emmy Awards® – including outstanding sports series for *NASCAR on FOX,* bringing the total number of Emmy Awards® won by FOX Sports to 76 since its inception in 1994. (News Corp, 2009, p. 35)

The company's 'commitments, borrowings and contractual obligations' regarding sports programming rights were, it said, US$16,866,000 (p. 70). Although its 2012 Report was a little less 'upbeat', it stated that:

> In sports, the NFL, World Cup and Sky Sports are examples of the exceptional content we deliver to our customers. We are also investing in high-growth markets, and Fox Sports Brazil is a good example of this dynamism. (News Corp, 2012, p. 6)

Across the Atlantic, 21st Century Fox has a very significant sports presence, the most important of which is English Premier League (EPL) football, which was a crucial television property that saved the entire BSkyB (of which it is the largest shareholder) subscription service when it was acquired in 1992 (Horsman, 1997). Having consistently outbid and outmanoeuvred broadcast rivals, including the BBC and ITV (although sometimes forging strategic partnerships with competitors in which it has been the dominant party), 'seeing off' both the collapsed ITV Digital and the British operation of Setanta, and resisting with some success attempts by the European Commission to break up its subscription sports oligopoly, 21st Century Fox/News Corp has been the major force in European television sport for several years, as it reveals below in its 2009 Annual Report:

> Sky Sports showed more than 40,000 hours of sports across five core channels in 2008 and secured a new three-year agreement for live coverage of the UEFA Champions League.
>
> Last year, we acquired a 25% equity interest in Germany's leading pay-TV operator, Premiere AG [now Sky Deutschland]. The German television market is the largest in Europe, with approximately 38 million television households, and offers News Corporation tremendous opportunities. (2009, p. 39)

However, in 2011, in the light of the now-infamous phone hacking scandal, (the then) a full takeover bid for the highly lucrative BSkyB was withdrawn and the large telecommunications company BT entered the British football television market a year later (Rowe, 2013).

But 21st Century Fox/News Corp's sports television operations extend far beyond the USA and Europe, connecting different TV sports products and territories, and making possible the circulation of the same media content across time and space. Table 17.2 reveals (as a 'snap shot' of a constantly shifting field) this global sweep of its sports television operation.

The combination of technological infrastructure, financial resources and transportable broadcast rights enables media corporations to span the globe, and confers on them a distinct advantage in the provision and exploitation of sports television. However, this is not

Table 17.2. A sample of 21st Century Fox/News Corp's global sports television platforms, regions and rights

News Corp TV platform	*Region/country*	*Key sports rights*
Fox	USA	NFL
BSkyB	UK & Ireland	EPL
Sky TV	New Zealand	Rugby Union [SANZAR]
Star TV	Asia	EPL
Star TV	India	Cricket
Sky Italia	Italy	Serie A
Fox Sports	Australia	NRL, A League
Sky Deutschland	Germany	Bundesliga
Fox Pan American Sports	Central/South America and the Caribbean	Copa Bridgestone Libertadores/ Sudamericana

necessarily a seamless process of gaining control. Nationally protective broadcast policies may favour local media companies (Scherer & Rowe, 2014), and there might also be other political sensitivities that provide a barrier to market entry. For example, Rupert Murdoch's widely reported 1993 speech declaring that advances in communications technology were a threat to 'totalitarian regimes everywhere' provoked a reaction from the Chinese Communist Party leadership that blocked the progress of News Corp for well over a decade in that country (Dover, 2008).

These obstacles apply, of course, not only to sports television and entertainment-related activities, but to critical journalism, including broadcast and print sports journalism. As McChesney (2008, p. 404) has noted:

> The media giants are not interested in pursuing dangerous stories that cost a lot of time and money to pursue, promise little financial payoff, and can antagonize governmental authorities with whom the media barons desperately want to stay on good terms. Most indicative of this trend has been the manner in which four of the five largest media firms in the world have fallen over themselves attempting to please the government of China. Disney's and News Corp's campaign to please Chinese rulers by watering down their journalism and operations has been chronicled elsewhere. Time Warner and Viacom entered the fray in the fall of 1999. What these episodes make clear is that no viable system of journalism can be expected from a system under the thumb of massive self-interested commercial organizations.

Dover's (2008) book is one of the works that has 'chronicled elsewhere' the pandering to the Chinese government for commercial purposes that McChesney mentions, including removing the BBC from Star TV and cancelling a News Corp commissioned book by former Hong Kong Governor Chris Patten (Page, 2003). While there is, in any case, a common practice of seeking to de-politicize sport and sport-related media and journalism from within (Rowe, 2007), it can be even more difficult to achieve, when, for example, in 2008 China hosted a global mega media sports event like the Olympics, and garnered very bad publicity through the news media when sport was used as an avenue of political protest, as occurred with regard to the Torch Relay and the campaign to highlight the treatment of Tibetans (Qing and Richeri, 2012). The dynamic nature of television sport ownership was later revealed when, in January 2014, 21st Century Fox sold its 47% stake in the non-sport oriented China Star TV in switching investment to India (especially in TV sport), having previously acquired full control of ESPN Star Sports in China (*Financial Review*, 2014).

It is also possible for the balance of power to shift in particular sports, with specific, negative consequences for Western-dominated sports television regimes. For example, cricket, which is played predominantly in countries that were once part of the British Empire, has in the last two decades seen a pivotal shift towards domination by the Indian sub-continent, having formerly been under the control of the London-based Marylebone Cricket Club (MCC),[4] which has responsibility for the game's 'laws', and the formerly white-dominated governing body the International Cricket Council (ICC). The 2005 move of the ICC to Dubai in the United Arab Emirates was not only a result of effective, and historically unusual, political cooperation among India, Pakistan, Sri Lanka and Bangladesh. It was also a reflection of the emergent economic power of over a billion people in one of the world's most rapidly developing nations (India), and the consequent market dominance, through a combination of television and (often illegal) sport betting, that has resulted in approximately 80% of all financial turnover in world cricket occurring in that country.

The three main forms of the contemporary game offer very different temporal structures: Test match cricket lasts up to five days with about 30 hours of play, not including breaks for lunch and tea; one-day cricket runs either from mid-morning to early evening, or from after lunch to late evening, meaning a total spectator time commitment of about ten hours, including a between-innings break; and the Twenty20 format involves only about three hours actual play, and a break between innings. These cricket formats now compete in various ways as sports event, industrial format, television spectacle and in-stadium experience. The five-day game, in taking up a much large part of the television schedule, can be useful programming 'fodder' but also a disruption to year-round television viewing habits and programmes (such as watching news, current affairs and drama series). If, though, there is little broadcast atmosphere created by the visual and aural contribution of the crowd, then the 'live' event loses much of its compelling appeal given that the in-stadium audience is a crucial part of the televisual text itself (Rowe, 2004). Furthermore, the reliance of all sport on unpredictable actions and outcomes means that, if the contest is unequal, the justification for viewing is reduced, especially for casual observers, and the negative impact on ratings- or even subscription-dependent television is exacerbated by the increased length of the broadcast.

Thus, the viewing (from home and in-stadium) preferences of the dominant market has a clear impact on the sport itself, meaning that, for example, even if five-day Test matches were the preferred game form of Western broadcasters and spectators, they would still be subject to the cricketing tastes of the Indian sub-continent. When, therefore, the Marylebone Cricket Club, as part of its brief to be 'committed to the good of the game', surveyed cricket fans in three countries (a sample size of 500 in each of India, South Africa and New Zealand) where the most historically prestigious form of the game, the Test match, is in decline, the findings were that 'only seven per cent of cricket followers in the sport's modern heartland, India, nominate Tests as their preferred form of the game' (Saltau, 2009). The short Twenty20 format, invented in England but extensively developed in India through the establishment of the franchise-based Indian Premier League (IPL) in 2007, had now emerged as the most important cricket form in its most important market:

> [But] in India, Twenty20 internationals rule, with 58 per cent of the respondents declaring these as their favourite compared with 31 per cent for ODIs [One-Day Internationals], 7 per cent for Tests and 4 per cent for the billion-dollar Indian Premier League. Overall, 13 per cent said they preferred Tests to the limited-overs contests ...
>
> ... The study also challenges the assumption that slow over-rates and dull pitches are most instrumental in keeping people away from Tests. Instead, it suggests ticket prices and timeslots are key factors. The study also reaffirms the power of television – 89 per cent of people surveyed watch cricket live on TV. Sachin Tendulkar last October broke the world run-scoring record in a Test against Australia in Mohali in front of only a few thousand people at the ground. (Saltau, 2009)

Here the pivotal place of sports television is evident, as well as the importance of in-stadium spectators to the television text. Despite attempts by television cameras to avoid showing large, empty areas of the stadium and to concentrate on the sporting action, the televisual spectacle is depleted where the significance of a major sporting achievement cannot be signified by a large co-present watching crowd being itself watched by a dispersed television audience cross-checking the 'must-see' importance of the moment. These questions surrounding the televising of cricket are familiar and can be traced, as previously argued, to the origins of sport on television, where the latter has been feared as a force that would remake sport in its own image. Television-friendly forms of sport, endowed with the economic dominance of broadcasting within the sports industry, have been favoured and further modified by being sped up, subjected to editorial choices that alter ways in which they are and can be seen, interpreted and amplified by commentary, arranged to fit advertising breaks and programme schedules, and so on. This ability to set the terms of sporting practice and spectatorship is still largely the prerogative of Western media companies, which have successfully exported sport in general and specific brands such as the EPL, the NBA and MLB (US domestic basketball and baseball, respectively) to new markets (Rowe & Gilmour, 2009). But, as noted in the case of cricket, the balance of power in sport, sponsorship and television can change, especially in favour of the ascendant Asia-Pacific region.

However, concentrating predominantly on the global political economy of television sport can lead the cultural object itself to become a little obscure. The television sports texts shaped by their desirability as economic agencies are also the products of other forces, from the received cultural practices whose pleasures made them available for their rationalization and governmentalization as sport in the first place, to the socio-cultural relations in which they are embedded. Decoding the social, cultural and political significance of television sport has, then, been as central to its research and scholarship as its ownership, control and distribution, although these dimensions cannot be neatly separated into economic, cultural and ideological compartments. Indeed, Jhally (2006, pp.141–2) and other critical media sport analysts (discussed below) explicitly link mediated sport to issues of militarism, nationalism, social class, gender and race as required 'readings [that] are very important follow-ups to the focus on production and encoding – a shift from process to product'. Rather than representing 'important follow-ups', critical interrogations of sports texts and their attendant ideologies are integral to the analysis of mediated sport.

TELEVISION SPORT TEXTS: FORMS, MEANINGS, IDEOLOGIES

Sport, irrespective of any connection to television, can never be cut free from the web of social connections of which it is a creation. Sports television in the first instance, then, had to deal with a socio-cultural formation that pre-dated it and was marked by its histories, structures and power dynamics. For example, as already observed, sport has historically been dominated by men. Indeed, it has been argued forcefully by various scholars of sport and gender that it has been an institution and field of culture that has been critical to the very constitution of dominant forms of heterosexual masculinity (while simultaneously functioning as a homosocial environment where men can engage with each other in heteronormatively approved ways, including openly expressing affection, admiring each other's bodies and touching). Toby Miller describes this phenomenon of sport-licensed-and-unusual gender behaviour in an instructive anecdote of watching sportsmen perform on television in a space set aside for (mainly) men to view them from a distance and to connect with other men in close proximity:

> My friend Rob Nixon and I watched the 1998 men's World Cup soccer match between Colombia and England in a lower-west-side Manhattan nightclub that had opened its daytime doors to soccer fans. The crowd was English. The man next to me was wearing a T-shirt that identified him as someone from a working-class area of London. His voice went with it. We never spoke, but he clutched at me whenever England did well and embraced me with each goal and the ultimate victory. Initially uncomfortable, I found myself looking forward to these shows of emotion. That encounter made me think about the masculinities on display in the competition, the strange gamut of passions and passionate exchanges – players blowing kisses, driving one another into the ground with projectile cuddles, and adopting strange poses for the crowd, for all the world like fey, queer-acting models. (Miller, 2001, pp.7–8)

Here, sportsmen can be seen as performers of a kind of masculinity that could often incur negative (even violent) homophobic sanctions (Tomsen, 2003), but who choose to act in such a way in a 'live' mediated cultural space occupied by some of the largest audiences ever assembled,[5] and which seemingly 'licenses' other men to set aside the usual social protocols in environments that have not been designated explicitly as homosexual. If television sport is, as is to be expected, implicated in wider gender relations, these may in various ways reproduce, resist, reinforce or innovate in constructions of gendered identities and power. The accumulated evidence has not been favourable to sport or television in these respects given that, separately and in combination, both institutions can confer very considerable social prestige, cultural profile and pecuniary reward on those that it 'anoints' – and deny them to others. If television can be regarded as simultaneously reflecting sporting status for (selected) men and enhancing that elevated social position, then it can generally be regarded as complicit in an unequal gender order (Creedon, 1994; Fuller, 2009). Successive content analyses over several decades have consistently revealed (with occasional anomalies, such as the Olympics), an enormous disparity in the televisual coverage of sportsmen and women, usually reflecting a ratio of at least 20 to 1 across all televised sport (Rowe, 2004, p.89–90; Messner & Cooky, 2010). This is not only a question of quantity – the quality of the coverage of women's sport on television has frequently privileged sportsmen (Duncan & Messner, 2005; Eastman & Billings, 2000), thereby leading some sportswomen to seek greater visibility and reward through the projection of sexualized images commonly at odds with recognition of their sporting prowess and achievement.

This sexualized mode of presentation has been something of a trap (if a lucrative one) for sportswomen and has been exacerbated by television, with its propensity for using sexuality, especially female sexuality, as a lure for viewers. In the case of sport, sex appeal can both enhance interest in particular sport stars among fans, and stimulate it (perhaps literally) among those who may care little for sport *per se*, but can still be counted among its audience and so have considerable exchange value. Tennis, for example, demonstrates conspicuously television's capacity for sexual objectification and *voyeurism* in accentuating spectatorial hierarchies that already exist within and outside the spatialized sports event itself. For example, during the 2009 Wimbledon tennis tournament – which is regarded as the most conservative of all the so-called 'Grand Slam' events in being run by the All-England club and televised by the BBC – it was admitted that calculations of 'box office appeal' were applied, especially to female tennis players, in determining hierarchies of co-present and mediated visibility:

> When it comes to choosing which women play on Centre Court, good looks count for more than big shots.
>
> While a succession of easy-on-the-eye unknowns have appeared in Wimbledon's prime arena, the top women's seeds have been relegated to lesser courts.
>
> And last night, the All England Club admitted that physical attractiveness is taken into consideration.
>
> Spokesman Johnny Perkins said: 'Good looks are a factor'. (Andrews & Fernandez, 2009)

The manner in which sport and television work together here to privilege the appearance of female tennis players, especially in the absence of a national focus, is revealed in the comments of the event broadcaster:

> ... A BBC source said: 'It's the Wimbledon play committee, not us who decides on the order of play.
>
> 'But obviously it's advantageous to us if there are good-looking women players on Centre Court...
>
> ... Our preference would always be a Brit or a babe as this always delivers high viewing figures.'... (Andrews & Fernandez, 2009)

If this is an accurate representation of the BBC's position on sports television, then it is apparent that, even where the broadcaster is not commercial and their right to broadcast the event is guaranteed by state regulation, the logic of audience maximization through sexual objectification – and, as noted, nationalism – is nonetheless pivotal. The 'logic of television', then, can be said to overdetermine sporting excellence, and has been internalized by the sport organization that is dependent on it.

This hyper-sexualization of female athletes in which television is centrally, though not exclusively, implicated (given similar examples in newspapers and magazines; see Bruce, Hovden & Markula, 2010), is now accompanied, though not matched, by a new-found emphasis on the appearance of male athletes.

The reasons for this development are fairly obvious – as Miller (1998) has noted, sport's and sport media's monosexual concentration on the male spectatorial subject and performative object have limited the possibilities of its commercial exploitation across the whole population. Women were historically treated by commercial television in particular as 'accompanying' viewers of heterosexual men or as a market segment that was difficult to attract and, as a consequence, long neglected. With the saturation and even, in some instances (such as association football in Britain in the 1980s), decline of the male media sport market, heterosexual women and gay men (although lesbians largely remain outside the televisual marketing frame) have been increasingly recruited to watch sportsmen in sexually *voyeuristic* terms, while the marketing-induced expansion of clothing, cosmetic, exercise and other presentational industries in the previously underexploited male (especially heterosexual) sector has seen celebrity sportsmen enlisted as style models (Smart, 2005; Whannel, 2001).

Most notably, the now-retired English footballer David Beckham's trajectory from suburban working-class London, to Manchester and then Madrid, Los Angeles, Milan, Paris, central London and Miami has signified a stylistic re-positioning – culturally and spatially – of the celebrity sportsman (Cashmore, 2002; Coad, 2008). Television and other media have, therefore, been instrumental in transporting sport not only from the stadium to the home, but also from the restricted realm of sporting performance to the worlds of the gossip column, the fashion pages and the investment bulletin. Thus, the media sport text can pass freely across cultural zones, especially when sexual scandal and prurience are given full rein – as occurred with two of the most important male sport icons, Beckham and Tiger Woods, over their respective marital infidelity scandals (Rowe, 2009). Such events raise questions about what actually constitutes sports television – the primary sports contest itself in live, recorded or highlight form, or the much larger apparatus of coverage, comment and critique that surrounds it in a manner by which the link to the practice of sport is often tenuous and can seem to be almost incidental. Under these circumstances, it is hardly surprising that sex – a staple of television programming and advertising – should be closely connected to sport.

It should be noted that this is not an argument against any connection between sport, television and sexuality. Sport above all involves dispositions of the body, and it would be improbable – indeed, undesirable – if it were compulsorily de-sexualized (Guttmann, 1996; Miller, 2001). Thus, the ideological question is not so much whether sport can or might be 'sexy', but how it treats sexuality and gender power. Television and other media still operate within a largely heteronormative framework that is itself hierarchical in its treatment of men and women. It continues to have difficulty with managing and presenting queer and homosexual sexualities and their relationship to sport (Caudwell, 2006; Hemphill & Symonds, 2002), tending to interpret departures from the heteronormal as scandalous *per se*. Sports television, broadly defined, has been both a vehicle for the carriage and reproduction of ideologies of dominance, and a volatile site where highly charged political issues come to the fore.

Thus, while the politics of gender have, in the domain of sport, tended to be conventionally patriarchal or hegemonically masculine, several cases of the ill treatment of women by sportsmen (Benedict, 1997) have prompted important public debates and required state, media and sports organizational responses to the subordinate position of women within sport and in the society of which it is a part. Paradoxically, then, television has been a party to the erosion in certain respects of male-dominated sport's capacity to seal itself off from scrutiny of appropriate norms of conduct and attitude. It has also, both deliberately and inadvertently, led to critical reconsiderations of the gender order of whole societies in which sport is a celebrated institution. For example, in Australia in 2009 a current affairs programme (*Four Corners*) from a public broadcaster entitled 'Code of Silence' (Australian Broadcasting Corporation, 2009) addressed several incidents involving the treatment of women and the largely inadequate response to it within the sport of rugby league. Another television programme, the *Footy Show* (on the commercial 9 Network), sought to minimize the reputational damage caused by allegations (of a non-criminal nature) made in the programme against one of its key presenters, Matthew Johns. In this task the *Footy Show* was manifestly unsuccessful, with a public outcry leading to Johns being suspended, and the programme holding, quite uncharacteristically, a serious discussion of attitudes to women in the code. Although Johns was subsequently 'rehabilitated', this example indicates that television does not operate monolithically, and that it can also be a progressive force within sport and society, in this case bringing significant pressure to bear on authorities and clubs in their handling of such matters, and producing changes in their practices.

The *Four Corners* case deliberately raised an important social issue, but the unpredictability of televised sport can also engender significant social debates. For example, when Zinedine Zidane, the captain of France (and of north African descent), was sent off for violent conduct in the FIFA World Cup Final of 2006, enormous coverage ensued as to what prompted him to head-butt his Italian opponent, Marco Materazzi (Jiwani, 2008; Rowe, 2010). Although allegations of racism, especially that Materazzi called Zidane 'the son of a terrorist whore' (although the former admitted only to offensive comments about the latter's sister), were never proven, the phenomenon of racial vilification in sport was given considerable attention, as it was in the case of the alleged on-field description by Indian cricketer Harbhajan Singh of the black Australian cricketer Andrew Symonds as a 'monkey' (Vivek, 2008; Rowe, 2011b). Broadcast television, with its instantly available but generally inconclusive audio-visual record, has for several decades been a powerful vehicle for communicating structural inequalities arising in and through sport, not least among which is the politics of 'race' (Carrington, 2010). Such representations of the use and abuse of power are now given enormous additional reach and multiple, detailed (but often bizarre, crude and cruel) treatments on the internet and in mobile media (Hutchins & Rowe, 2012, 2013). These rapidly emergent 'new apps' for media sport question television's primacy, and so are prompting not only changes to the production and style of sports television, but also to its critical academic analysis.

CONCLUSION: MEDIA SPORT AFTER TELEVISION?

Sports television has had a remarkable rise, becoming, within a few decades in the middle of the last century, the most important medium, and sport is its most important partner institution. But the age of broadcast television is undergoing transformation as convergent digital media bring together broadcasting, telephony and computing, and mobile media technologies create mobile sports viewing experiences (Boyle & Whannel, 2010; Hutchins & Rowe, 2012, 2013). Television sport researchers, who have tended to concentrate on the ownership, control and global expansion of television, the embedded ideologies and cultural politics of media sport texts and, to a lesser extent, production and labour processes in the making of TV sport, must now confront the coming dethronement of sport 'television as we knew it'. Tay and Turner (2010, p. 32) note 'the current difficulty in finding a simple and comprehensive means of measuring and mapping – and therefore of properly understanding – what is becoming of what we still call "television"'. Watching sport on television is, as with other viewing genres, a far cry from a singular, scheduled appointment with one domestic technology, except in one key respect – the predominant temporality and sociality of the live sport event.

Most of what we call sports television is not live, but it is watching sport 'in the moment' and 'in company' that is the most prized and pleasurable experience. Live streamed sport through computers, tablets, game consoles and on miniature mobile telephone screens may carry key sports events in real time, but they are far inferior spectatorial opportunities than, say, watching a high definition, wide screen television with good sound among family and friends. The 'time shifting' that so preoccupies those who announce the death of television can never match the 'live broadcasting of history' (Dayan & Katz, 1992) that is a live sports event. This socio-cultural status is recognized in many countries across the world, with free-to-air broadcasts of major (inter)national sports events often protected by the state in the interests of cultural citizenship (Scherer & Rowe, 2014). In the broadest sense, the premium sports event will still be televised, and will remain its most powerful economic, social and cultural force. Sport will still be viewed from a distance via audiovisual technology irrespective of the erosion of the conventional broadcast model in favour of other methods of delivery, such as Internet Protocol Television (IPTV). Understanding the control, production and deep structural meaning of these texts will still, then, be crucial for TV sport scholars.

However, with regard to media sports culture *in toto*, there is urgent new work to be done on what now 'counts as sports television' and its networked, digital relationships with other media, forms and genres, often in evanescent, hybridic forms (see Leonard, 2009). For example, the contemporary sport fan can keep one eye on a live sports event while playing a *wii* or sport management simulation of the game, pausing to contribute to an online debate with fellow fans in another hemisphere about preferred tactics and selections, or watching another match streamed live, catching up on some YouTube favourite sporting moments, or listening to a podcast about the class oppression of appreciators of sport by acquisitive media entrepreneurs (Ruddock, Hutchins & Rowe, 2010). Under these dynamic circumstances sport and television are still a 'match made in heaven' – but one that in the 21st century tends to be polyandrous rather than monogamous, and with rather more intricate pre-nuptial agreements and post-nuptial affiliations. Television's place in the media sports cultural complex is for now assured, as signified by its audience popularity, cultural visibility and the persistently high cost of premium broadcast rights. However, the once hazy challenge from competing technologies, audience expectations and media sport conglomerates from outside the hegemonic Western sphere is acquiring greater clarity and is restlessly in motion.

NOTES

1 The research on which this chapter is based has been supported by two Australian Research Council Discovery Grants. *Handling the 'Battering Ram': Rupert Murdoch, News Corporation and the Global Contest for Dominance in Sports Television* (DP0556973) concentrated on the development of global sports television, especially subscription-based, over the last decade. *Struggling for Possession: The Control and Use of Online Media Sport* (DP0877777, with Brett Hutchins) addressed the transition from broadcast to converged network media sport. I thank Callum Gilmour, in particular, for his research assistant contribution to both projects and Janine Mikosza, Genna Burrows and Vibha Bhattarai Upadhyay for their research work on the latter.

2 As the world's longest established and probably most prestigious national broadcaster, the British Broadcasting Corporation's (BBC's) (2009) mission to 'To enrich people's lives with programmes and services that inform, educate and entertain' has provoked constant debate, pitting 'serious programming' against 'light entertainment' (Born, 2005).

3 It is worth noting that in the period 2009 to 2012 there was a pronounced shift towards sponsorship (rising by 11% from 34%) and a small rise in licensing (1% from 2%), with the proportion (though not the quantum) of broadcasting revenue falling 6% from 53% and a sharp proportional decline in ticketing revenue (also by 6% from 11%).

The 2015 Marketing Report will reveal if this is a trend or is limited to that triennium.

4 On its website the MCC (2013) states: 'MCC is the world's most active cricket club, the owner of Lord's Ground and the guardian of the Laws of the game. Founded in 1787, it issued the *first code of Laws* in 1788 and moved to its current home at Lord's in 1814. It remains, in the words of a former President, Plum Warner, "a private club with a public function"'. The problem, of course, is that 'the world's most active cricket club' is also a symbol of the imperial power that is overtly challenged in the post-imperial, post-colonial context in which India openly asserts its recently established hegemony.

5 For example, as Cone (2006) notes 'Italy's soccer World Cup final win over France in July was television's most-watched sports event in 2006, drawing more than twice as many viewers as any other program, a report by Initiative Worldwide said', with an 'average of 260 million people tuned in … more than 600 million viewers watched some part of the Berlin match'. Even in the USA, where association football is not the dominant sport, the World Cup Final 'relegated the Super Bowl to second spot in this year's ratings', drawing 'a global average audience of 98 million'. Mediated sport events, then, play across a range of national and transnational contexts, but can only intermittently lay claim to *bona fide* global status.

REFERENCES

Andrews, D.L. (2006). *Sport, commerce, culture: Essays on sport in late-capitalist America*. New York: Peter Lang.

Andrews, E. & Fernandez, C. (2009, 29 June). Babe, set and match: Why looks count for more than talent when Wimbledon decides which girls will play on Centre Court. *Mail Online*. Retrieved from www.dailymail.co.uk/news/article-1196155/Babe-set-match-How-looks-count-talent-Wimbledon-decides-girls-play-Centre-Court.html

Australian Broadcasting Corporation (2009). Code of silence. Retrieved from www.abc.net.au/4corners/content/2009/s2565007.htm (last accessed 26 May, 2013).

Australian Rugby League (2009). Origin beamed and streamed around the world. Retrieved from http://web.archive.org/web/20090913140951/www.australianrugbyleague.com.au/origin/article.php?id=1317

BBC (2009). *About the BBC*. Retrieved from www.bbc.co.uk/info/purpose/

Benedict, J. (1997). *Public heroes, private felons: Athletes and crimes against women*. Boston: Northeastern University Press.

Born, G. (2005). *Uncertain vision: Birt, Dyke and the reinvention of the BBC*. London: Vintage.

Bourdieu, P. (1978). Sport and social class. *Social Science Information*, 17, 819–840.

Boyle, R. & Haynes, R. (2000). *Power play: Sport, the media & popular culture*. Harlow, UK: Pearson Education.

Boyle, R. & Whannel, G. (Eds) (2010). Sport in new media cultures [Special Issue]. *Convergence: The International Journal of Research into New Media Technologies*, 16.

Bruce, T. Hovden, J. & Markula, P. (Eds) (2010). *Sportswomen at the Olympics: A global content analysis of newspaper coverage*. Rotterdam: Sense.

Bryant, J., Zillmann, D. & Raney, A.A. (1998). Violence and the enjoyment of media sports. In L. Wenner (Ed.), *MediaSport* (pp. 252–265). London and New York: Routledge.

Buscombe, E. (Ed.) (1975). *Football on television*. London: British Film Institute.

Carrington, B. (2010). *Race, sport and politics: The sporting black diaspora*. London: Sage.

Cashman, R. & McKernan, M. (Eds) (1981). *Sport, money, morality and the media*. Sydney: New South Wales University Press.

Cashmore, E. (2002). *Beckham*. Cambridge: Polity.

Caudwell, J. (Ed.) (2006). *Sport,sexualities and queer/theory*. London and New York: Routledge.

Chandler, J.M. (1988). *Television and national sport: The United States and Britain*. Urbana, IL: University of Illinois Press.

Clarke, J. & Critcher, C. (1985). *The devil makes work: Leisure in capitalist Britain*. London: Macmillan.

Clarke, J., Critcher, C. & Johnson, R. (Eds) (1987). *Working-class culture: Studies in history and theory*. London: Hutchinson.

Coad, D. (2008). *The metrosexual: Gender, sexuality, and sport*. Albany, NY: State University of New York Press.

Cone, J. (2006, 19 December). Soccer World Cup final had sport's largest TV audience of 2006. *Bloomberg.com*. Retrieved from www.bloomberg.com/apps/news?pid=20601077&sid=aOYdMOmxoV4c&refer=intsports

Creedon, P.J. (Ed.) (1994). *Women, media and sport: Challenging gender values*. Thousand Oaks, CA: SAGE.

Dayan, D. & Katz, E. (1992). *Media events: The live broadcasting of history*. Cambridge, MA: Harvard University Press.

Donnelly, P. (1996). Prolympism: Sport monoculture as crisis and opportunity. *Quest*, 48, 25–42.

Dover, B. (2008). *Rupert Murdoch's China adventures: How the world's most powerful media mogul lost a fortune and found a wife*. Melbourne: Penguin Australia.

Duncan, M.C. & Messner, M.A. (2005). *Gender in televised sports: News and highlights shows, 1989–2004*. Los Angeles: Amateur Athletic Foundation of Los Angeles.

Eastman, S.T. & Billings, A.C. (2000). Sportscasting and sports reporting: The power of gender bias. *Journal of Sport and Social Issues*, 24, 192–213.

Elias, N. & Dunning, E. (1986). *Quest for excitement: Sport and leisure in the civilising process*. Oxford: Basil Blackwell.

Financial Review (2014, 6 January). Murdoch's 21st Century Fox sells out of China. Retrieved from www.afr.com/p/business/marketing_media/murdoch_st_century_fox_sells_out_LipcnhO0mCDY86I1mwVTSJ

Folkenflik, D. (2006, 24 February). NBC evaluates investment in Olympic broadcast rights. NPR. Retrieved from www.npr.org/templates/story/story.php?storyId=5232275

Fuller, L.K. (Ed). (2009). *Sexual sports rhetoric: Global and universal contexts*. New York: Peter Lang.

Goldlust, J. (1987). *Playing for keeps: Sport, the media and society*. Melbourne: Longman Cheshire.

Gupta, A. (2009). India and the IPL: Cricket's globalized empire. *The Round Table*, 98, 201–211.

Guttmann, A. (1978). *From ritual to record: The nature of modern sports*. New York: Columbia University Press.

Guttmann, A. (1996). *The erotic in sports*. New York: Columbia University Press.

Hargreaves, J. (1986). *Sport, power and culture: A social and historical analysis of popular sports in Britain*. Cambridge: Polity.

Hartley, J. (1999). *Uses of television*. London and New York: Routledge.

Hemphill, D. & Symons, C. (Eds) (2002). *Gender, Sexuality and Sport: A Dangerous Mix*. Petersham, NSW: Walla Walla Press.

Hill, C. (1992). *Olympic politics*. Manchester: Manchester University Press.

Hoch, P. (1972). *Rip off the big game: The exploitation of sports by the power elite*. New York: Anchor.

Horne, J. (2006). *Sport in consumer culture*. Basingstoke, UK. Palgrave Macmillan.

Horsman, M. (1997). *Sky high: The inside story of BSkyB*. London: Orion Books.

Hutchins, B. & Rowe, D. (2009). From broadcast rationing to digital plenitude: The changing dynamics of the media sport content economy. *Television & New Media*, 10, 354–370.

Hutchins, B. & Rowe, D. (2012). *Sport beyond television: The internet, digital media and the rise of networked media sport*. New York: Routledge.

Hutchins, B. & Rowe, D. (Eds) (2013). *Digital media sport: Technology, power and culture in the network society*. New York: Routledge.

International Olympic Committee (IOC) (2012a). *Revenue Sources and Distribution*. Retrieved from www.olympic.org/ioc-financing-revenue-sources-distribution

International Olympic Committee (IOC) (2012b). *Revenue Olympic marketing fact file 2008 edition*. Retrieved from www.olympic.org/Documents/marketing_fact_file_en.pdf

Jhally, S. (1984). The spectacle of accumulation: Material and cultural factors in the evolution of the sports/media complex. *The Insurgent Sociologist*, 12, 41–57.

Jhally, S. (2006). *The spectacle of accumulation: Essays in culture, media, and politics*. New York: Peter Lang.

Jiwani, Y. (2008). Sports as a civilizing mission: Zinedine Zidane and the infamous head-butt. *Topia: Canadian Journal of Cultural Studies*, 19, 11–33.

Klatell, D. & Marcus, N. (1988). *Sports for sale: Television, money and the fans*. New York: Oxford University Press.

Lawrence, G. & Rowe, D. (Eds) (1986). *Power play: Essays in the sociology of Australian sport*. Sydney: Hale and Iremonger.

Leonard, D.J. (Ed.) (2009). New media and global sporting cultures [Special Issue]. *Sociology of Sport Journal*, 26.

Marylebone Cricket Club (MCC) (2013). What is MCC? Retrieved from www.lords.org/mcc/the-club/what-is-mcc/

McChesney, R.W. (2008). *The political economy of media: Enduring issues, emerging dilemmas*. New York: Monthly Review Press.
Mehta, N. (2008). *Television in India: Satellites, politics and cultural change*. London and New York: Routledge.
Messner, M.A. & Cooky, C. (2010). *Gender in television sports: News and highlights shows, 1989–2009*. Center for Feminist Research: University of Southern California.
Miller, T. (1998). *Technologies of truth: Cultural citizenship and the popular media*. Minneapolis, MN: University of Minnesota Press.
Miller, T. (2001). *Sportsex*. Philadelphia, PA: Temple University Press.
Miller, T. (Ed.) (2003). Television: Critical concepts in cultural and media studies. London and New York: Routledge.
Miller, T., Lawrence, G., McKay, J. & Rowe, D. (2001). *Globalization and sport: playing the world.* London: SAGE.
NewsCorp Annual Report 2008 (published 2009). New York, NY: 21st Century Fox/ News Corp. http://newscorp.com/AR2008Flash/NC08.html
NewsCorp Annual Report (2012). Retrieved from http://reports.21cf.com/Report2012/2012AR.pdf
Page, B. (2003). *The Murdoch archipelago*. London: Simon & Schuster.
Qing, L. & Richeri, G. (Eds) (2012). *Encoding the Olympics: Comparative analysis on international reporting of Beijing 2008: A communication perspective*. London: Routledge.
Rowe, D. (1996). The global love-match: Sport and television. *Media, Culture & Society*, 18, 565–582.
Rowe, D. (2004). *Sport, culture and the media: The unruly trinity* (second edition). Maidenhead, UK: Open University Press.
Rowe, D. (2007). Sports journalism: Still the 'toy department' of the news media? *Journalism*, 8, 385–405.
Rowe, D. (2009). Attention la femme! Intimate relationships and male sports performance. In L.K. Fuller (Ed.), *Sexual sports rhetoric: Global and universal contexts* (pp. 69–81). New York: Peter Lang.
Rowe, D. (2010). Stages of the global: Media, sport, racialization and the last temptation of Zinedine Zidane. *International Review for the Sociology of Sport*, 45, 355–371.
Rowe, D. (2011a). *Global media sport: Flows, forms and futures.* London and New York: Bloomsbury Academic.
Rowe, D. (2011b). The televised sport 'monkey' trial: 'Race' and the politics of postcolonial cricket. *Sport in Society*, 14, 786–798.
Rowe, D. (2013). The sport/media complex: Formation, flowering and future. In D.L. Andrews & B. Carrington (Eds), *A companion to sport* (pp. 61–77). Oxford: Blackwell.
Rowe, D. & Gilmour, C. (2009). Global sport: Where Wembley way meets Bollywood boulevard. *Continuum*, 23,171–182.
Ruddock, A., Hutchins, B. & Rowe, D. (2010). Contradictions in media sport culture: 'MyFootballClub' and the reinscription of football supporter traditions through online media. *European Journal of Cultural Studies*, 13, 323–339.
Saltau, C. (2009, 18 November). Blow for test match cricket. *The Melbourne Age.* Retrieved from www.theage.com.au/articles/2009/11/17/1258219839902.html?page=fullpage#contentSwap1
Scherer, J. & Rowe, D. (Eds) (2014). *Sport, public broadcasting, and cultural citizenship: Signal lost?* New York: Routledge.
Singer, T. (1998). Not so-remote-control. *Sport*, March: 3–6.
Smart, B. (2005). *The Sport star: Modern sport and the cultural economy of sporting celebrity*. London: SAGE.
Sportcal (2012, July). Special feature: London 2012 Olympic Games, 26. Retrieved from www.sportcal.com/pdf/gsi/Sportcal_Issue26_6-9.pdf
Stoddart, B. (1986). *Saturday afternoon fever: Sport in the Australian culture*. North Ryde, NSW: Angus and Robertson.
Tay, J. & Turner, G. (2010). Not the apocalypse: Television futures in the digital age. *International Journal of Digital Television*, 1, 31–50.
Tomsen, S. (2003). 'A gross overreaction': Violence, honour and the sanctified heterosexual male body. In S. Tomsen & M. Donaldson (Eds), *Male trouble: looking at Australian masculinities* (pp. 91–107). Melbourne: Pluto Press.
Vivek, V.V. (2008, 5 January). Harbhajan in trouble over verbal doosra to Symonds. *Indian Express*. Retrieved from www.indianexpress.com/news/harbhajan-in-trouble-over-verbal-doosra-to-symonds/257882/

Whannel, G. (1992). *Fields in vision: Television sport and cultural transformation*. London: Routledge.

Whannel, G. (2001). *Media sport stars: Masculinities and moralities*. London: Routledge.

Whannel, G. (2008). *Culture, politics and sport: Blowing the whistle, revisited*. London: Routledge.

Wheen, F. (1985). *Television: A history*. London: Century.

Winn, M. (1977). *The plug-in drug: Television, children and the family*. New York: Viking.

Xu, G. (2008). *Olympic dreams: China and sports, 1895–2008*. Cambridge, MA and London, UK: Harvard University Press.

18

Latin American Telenovelas: Affect, Citizenship and Interculturality

André Dorcé

The telenovela[1] is a Latin American melodramatic fictional narrative about romantic love organized in a serial televisual form. Originally introduced to TV audiences more than 50 years ago in countries that first adopted TV technology, Telenovelas are famous for representing all sorts of stereotypical Latin American characters in complex and passionate situations causing tears, laughter and pleasure to millions of people. These dramas have also been key vehicles for the vast telenovela star-system which has catapulted many actresses/actors to global fame. Depending on its commercial success, a given title could span from 70 to 200 or more episodes, broadcast on a daily basis, from noon until evening, including primetime on most national TV networks across the continent. Fifty-three per cent of people aged 12–64 in Latin America identify as regular telenovelas viewers (Miller, 2010: 200). The genre has become the most popular fictional TV form produced and consumed in the region and the single most important audiovisual product exported from Latin America to the global cultural market. However, recent studies are documenting the steadily rise in popularity of other locally produced fictional genres and formats. Fifty-two per cent of all fictional TV produced throughout 2008 in Obitel[2] countries were telenovelas while the other 47% were series (Orozco & Vassallo, 2009: 34). Telenovelas' success has played a significant role in the expansion of powerful corporate media in Mexico, Brazil, Venezuela, Chile, Argentina and Colombia as well as in these countries' efforts to counteract American and European dominance in the televisual industry. Recent titles such as the Argentinian *Rebelde Way* and *Montecristo*, or the Colombian *Café con aroma de mujer* and *Betty la fea*, have established new benchmarks for the genre in the international format and canned programming business, having been broadcasted and adapted in more than 77 and 84 countries respectively (Medina & Barrón, 2010: 10).

The genre's social relevance and its articulation with wider economic, cultural and political processes in the history of various national contexts has been the object of rich and heterogeneous analyses. Scholars from diverse intellectual and disciplinary traditions, mostly from the humanities and social sciences, have produced insightful knowledge about such a complex subject matter. However, telenovelas have not always been an obvious and legitimate object of academic research (Chernitsky, 1996; Marques de Melo, 2001). What can now be characterized as an established and growing subfield of media and cultural studies in the region, specializing in televisual melodrama, is the result of intense academic debate that has provided a fertile ground for questioning conventional ideas about the constitutive role that popular culture, the industrial media and the nation-state have played in shaping contemporary subjectivity. In this chapter I will sketch a panoramic account of key theoretical inflexions in this field of study and enumerate relevant empirical work carried out to date in this area. In doing so I aim to describe what I consider the core ideas of this vast body of research, while at the same time suggesting some possible ways to think further about fictional television, melodramatic or otherwise, and its relations with broader socio-political processes. The exercise of a complete and exhaustive survey of all pertinent material produced in the region is a task that goes well beyond the possibilities of this chapter.

INITIAL CONSIDERATIONS

Academic research on media and culture in Latin America, produced from a polycentric circuit of metropolitan universities, share some common theoretical backgrounds and similar socio-historical processes of institutional development. In fact, it could be argued that media research, as a regional project – mobilized by strong imaginaries of *latinoamericanidad* – has a common critical starting point, which, as we will show, catalysed and devised the emergence of television, particularly of telenovelas, as a strategic object of analysis.

Latin America in the 1960s was a period of profound socio-political transformations,[3] which included the establishment of increasingly powerful private media monopolies and pro-government public broadcasting systems in several countries of the region. Understandings about the role media played in shaping social dynamics in those contexts were mostly informed by positivist models of communication ('Who says what in which channel, to whom and with what effect'). In tune with such instrumentalist conception of communication, mass media were conceived of as fundamental tools for development and modernization (Fox, 1997). Thus, a substantial amount of mass media research in Latin America was narrowly oriented towards investigating the technical and 'concrete' functions of the media in order to promote the development of modernized consumer societies (Martín-Barbero, 1987; Saintout, 1998; Tufte, 2000). In that theoretical horizon fictional TV genres, including serials and telenovelas, were not considered as relevant objects of analysis, as was factual news.

The New World Information and Communication Order (NWICO) debate promoted by the UN in the early 1970s became a significant development that had a key role in the implementation of developmental communication research, and in the subsequent rise of critical anti-imperialist communication studies. Such debate heralded by UNESCO had multiple ramifications and useful outcomes for media research as Ecuador's government together with its national university supported, in 1971, the establishment of the Regional Centre for Advanced Studies in Journalism (CIESPAL in Spanish) in Quito, Ecuador. In 1973 this institution organized the first Latin American Media Conference in Costa Rica, where a significant number of Ibero-American media researchers met in an institutional forum for the first time, in order to debate the theoretical and methodological contours of their field. In the meeting one area of major consensus was the critique of

North American empirical studies, which assumed a functionalist model of 'objective' and 'neutral' research, and which failed to recognize its teleological ethos as ideologically problematic. Unlike functionalists, the emergent critical approach conceived the social phenomenon of mass media as being strongly determined by political and economic forces, highlighting the fact that most media were privately owned and problematically linked to authoritarian regimes. To fail to recognize these determinations – it was argued – was to ignore the power these media have had as instruments for maintaining the *status quo* (Tufte, 2000).

Against the growing optimism surrounding discourses of development fomented by the UN, media research advanced by these scholars was articulated with dependency theory, a strand of critical theory that emerged in the late 1960s and early 70s as a sceptical counter-hegemonic discourse. It was argued that international capitalism, imposed from the metropolitan centres of the North, favoured unequal material and symbolic exchanges (domination) that were the source of many problematic socio-political contradictions in peripheral/developing countries. Most of those contradictions were thought to destabilize national cultures by installing consumerist ideology as a legitimate and natural way of life, to the detriment of traditional and sovereign local symbolic practices (Fox, 1997; García Canclini 1988; Marques de Melo, 1988).

Within that context, popular culture – understood not as expression of contemporary Western modernity, but as a stronghold of conflicting pre- and post-colonial imaginaries – began to be constituted as a key site for epistemological debate and political intervention. In that sense, mass media's genres and formats, particularly those from television and cinema imports, were generally seen either as culturally invasive, or politically numbing. By incorporating theories developed by French structuralist and European semiotic traditions, together with behaviourist psychology models (stimulus-response) into their frameworks, media researchers in Latin America were inspired to 'unmask' the underlying 'negative' ideological nature of mass media messages, and to criticize the ways in which 'corporate media promoted cultural imperialism'. Such an 'ideologized' investigative approach was characteristic of most research produced at that time to explain the media's modes of social operation, including, of course, the development of reductionist accounts of the telenovela's role in everyday life (Marques de Melo, 1971; Mattelart & Dorfman, 1972; Pasquali, 1963).

By the mid-1970s, though, the development of significant theoretical shifts in the international field of critical social sciences and humanities allowed for the reconsideration of some of the main assumptions underlying the then leading theoretical perspectives in media and culture studies. Researchers in Brazil and Venezuela working on semiotics were among the first to analyse the formal and discursive specificity of telenovelas as a genre within their national settings (Rector, Tilburg, Coccato & Morana in Escudero, 1996: 10). These scholars inaugurated one fundamental and very persistent line of research on audiovisual melodramatic fiction in Latin America. That body of work gave a unique analytical visibility to televisual fiction and to the genre's textual devices that went beyond established theoretical models. Thus, they introduced into the wider debate a heuristic model that explained the telenovela's semantics in relation to an ideal reader/audience. That theoretical construct –the ideal reader – was susceptible to being contrasted, challenged or complemented by emerging empirical socio-demographic data produced by theoretical frameworks concerned with the relationship between subjectivity, culture and power. Yet at that time there were few 'empirical' observations that could support the idea that actual media consumption worked differently to what established analytical models predicted.

For example, in 1974 Argentine and Belgian researchers Mabel Piccini and Armand Mattelart, carried out empirical research on television viewership in Chile that contributed to changing in significant ways the understandings of television and its imagined audience. They documented how a group of working-class TV viewers re-signified televisual content in unexpected ways, partly because of their socio-cultural position. Media power, particularly that of television, began to be conceptualized in a more complex manner, as far as its audiences were thought of not only as receivers of univocal messages, but also as 'producers of meaning' through the reception process (Mattelart & Piccini, 1974, in Grimson & Varela, 1999).

Interestingly, in this conception of working-class audience there was an important displacement in the observed process of signification; from an 'all mighty ideological message' to a 'proletarian army of meaning liberation', which in many ways established the basis for the debate about culture during the early 1980s in Latin America.[4] More importantly, by reframing key issues about mass media's reception process, and its constitutive and contradictory relation to popular culture, future research on telenovelas would take reception analysis as another of its founding pillars. Such innovative ways of thinking about media and its audiences were also in dialogue with two other key regional intellectual practices that were re-conceptualizing subjectivity and agency in that politically convoluted time in Latin America: Liberation Theology [5] and Paulo Freire's contributions to critical thought, through his literacy and consciousness-raising work amongst adult peasants in Brazil. Freire's method was intended to contribute to the extermination of the 'culture of silence', that is, the 'condition of oppression wherein people are deprived of the ability to know, to express and to transform their lives' (Rowe & Schelling, 1991: 176). In fact, such a pedagogical perspective has informed different educational approaches in the field of audience research in Latin America, particularly in relation to telenovelas. Back then, Argentine researcher Carlos A. Douhourq took part in a small educational project for popular audiences that sought to teach people the 'correct' critical decoding, with the assistance of the 'senior decoder' and other pedagogical tools (Grimson and Varela, 1999). In these conceptualizations, the imagined audience is theoretically empowered, yet it was thought that this power had to be catalysed by means of a political intervention. Popular audiences in this new model are then only potentially rebellious (or immune) against the structuring nature of cultural industries (in the Adornian sense) if they receive the 'correct' ideological resources for such a critical decoding. From this tradition emerged an important line of research that during the 1980s conceptualized telenovelas as powerful vehicles for positive social change through the incorporation of educational content (Singhal et al., 1994).

We should bear in mind that, in the 1960s and 70s, the already unequal and contested process of democratization in Latin America suffered a major setback with the rise of dictatorships and authoritarian regimes throughout the continent (for instance in Haiti, Brazil Argentina, Uruguay, Paraguay, Chile, Venezuela and Nicaragua). This also explains the marginalization – and in some instances radicalization – of critical thought both inside Latin American academia and in the broader field of intellectual practices. Military regimes forced many social sciences scholars to go into exile to other countries in the region and beyond, where they were reincorporated into academia. From there, some of them combined their academic work with their active intervention and participation in diverse social and political movements emerging in their home countries as well as in their new spaces of exile.

This – many times involuntary – mobility of scholars throughout the continent also nourished important theoretical and epistemological fluctuations already taking place: culture, media and its consumption became crucial subjects of analysis from different disciplinary locations, and, in some cases, from perspectives conceived beyond disciplinary frontiers. Furthermore, the physical dislocation of several researchers working in

Latin American media studies – anthropologists, sociologists, semiologists, economists, philosophers and psychologists – allowed the intensification of transcontinental dialogue. Thus a certain degree of consensus allowed the implementation of common research themes in the regional field of critical social sciences. At that point, the already normalized appropriation of continental theory and its critical assimilation also encouraged an increasing dialogue with British cultural studies, particularly with the reading of key scholars such as Edward P. Thompson, Raymond Williams and Stuart Hall. The growing interest of these authors (particularly about Hall and his collaborators at Birmingham) 'to explore the ideological significance of the structure of media products outside the news categories' (Morley, 1992) went down well with a group of Latin American scholars whose work – as we have shown – had emphasized the importance of understanding popular culture and popular classes, not merely as carriers of dominant ideology, but as active/resistant producers of meaning, culture, and thus of hegemony (Borelli, 2001). The appropriation of those theoretical models helped to re-articulate, from a regional perspective, a substantial shift in the terms in which culture and power were understood in relation to differentiated and hierarchized scales/analytical categories of social processes (popular, folklore, elite, mass, national) (Zubieta, 2000).

Grimson and Varela point out that the meeting of the Communication Council of CLACSO[6] held in Buenos Aires in 1983 was a point of condensation of key deliberations about the 'culturalization' of media studies in Latin America,[7] and, as we shall see in the subsequent sections, this peculiar 'cultural turn' would relocate melodrama and telenovelas in a completely different theoretical position. According to these authors Jesús Martín-Barbero's and Néstor García Canclini's interventions in that discussion, helped establish productive lines of research across diverse disciplines.[8] If it is true that Barbero's and Canclini's theoretical contribution did not necessarily have a foundational character for telenovela studies, it has certainly played a substantial role in placing previous and ongoing research on media reception and textual analysis in relation to complex historical processes of cultural hybridization. The main topic of that meeting was 'Communication and popular cultures in Latin America'. For Martín-Barbero the incorporation of the cultural dimension of social processes of communication did not mean a dreaded withdrawal from the political, as many of his detractors would suggest, but a reconsideration of unseen dimensions of social life, such as the formation of new subjects and forms of resistance. The real 'withdrawal' was, rather, from the (then dominant) informational paradigm of communication whose perspective obscured the ability of popular audiences to mobilize different meanings to those rallied by hegemonic culture.

Martín-Barbero, then, sets out the theoretical basis that will frame his very influential *From Media to Mediations* (1987), by asserting the need to study communication from the standpoint of the popular, a task that requires making different epistemological displacements. Firstly by recovering and analysing history it is possible, he asserts, to understand how *the popular* amalgamates with *mass formations*; secondly, by tracing social memory it is possible to discern what and how mass formations are constitutively related to the popular; finally, by identifying the new functions that popular sectors assign to mass culture it becomes possible to recognise differential uses of mass culture[9] Hegemony for him (as for Gramsci) is never 'completed', as the popular classes have a strong nucleus of memory, which is activated in the presence of conflict. Martín-Barbero goes on to stress the importance of the development of reception studies, yet not reception as a discernable moment in the process of communication (production-message-circulation-reception), but as a site from which communication should be thought about. Hence for this author to speak of reception is to speak about anachronism, different temporalities between both audiences and producers, between the messages and the media and between texts and contexts.[10]

For his part García Canclini suggested that discussions about 'the popular' must take into consideration the investigation of the cultural dimension of consumption, where the hegemonic messages and symbolic goods socially available are interpreted in conjunction with the perceptual codes and quotidian habits of subaltern classes. However, he suggests that if the hegemonic culture determines, to some extent, the options available to popular classes, then '[the popular classes] select and combine the materials received – in their perception, memory and in the function they give to those [symbolic] goods – and build with them bricolages, other [meaningful] systems that are never the precise echo of the hegemonic "offer"' (in Grimson and Varela, 1999:70). Garcia Canclini's interest in art and popular culture, tradition and modernization overlapped with Martín-Barbero's perspective on the epistemological relevance of popular classes in the configuration of culture, while emphasizing the combinatory modalities of *hybrid* symbolic practices. For García Canclini, hybrid cultures are fundamentally liminal, and have porous borders where struggles and negotiations are confronted. In such sites, power is not defined by 'confrontations and vertical actions', but following Foucault, by 'interwoven relations' whose cultural and political effectiveness is not explained by the imposition of power, but by the play of differences in the fabric of social life: 'what gives [hybrid cultures] their efficacy is the obliqueness that is established in their fabric. How can we discern where ethnic power ends and where family power begins, or the borders between political and economic power?' (Yúdice, 2001: xii).

Melodrama under such theoretical assumptions came to stand as a fine instance of the complex workings of modernization and globalization in post-colonial Latin America; particularly as a residual and creative cultural force (as a key cultural matrix) compelling millions of people to watch telenovelas everyday all over the world.

Tradition and Modernity: Melodrama Takes Over Television

Latin American telenovelas have inherited key features from older European melodramatic traditions, deeply rooted in the intersection of popular culture and 'highbrow' culture in 19th-century Latin America. As a cultural product, telenovelas can be thought to be the point of convergence of different cultural sensibilities, a 'compendium' of different 'structures of feeling' (Williams, 1961) articulated to create a powerful and emotional representation of life. The telenovela 'embodies' and enacts a substantive portion of human tragedy as lived in societies experiencing multiple and often contradictory processes of modernization, transnationalization and globalization. In turn, melodrama, as a *technology of feeling* which governs the narrative and aesthetic output of telenovelas, has given a very distinctive sort of 'emotional coherence' to the complex cultural configurations that have taken place throughout the second half of the 20th century in Latin America.

In that sense, research on telenovelas has considered symbolic media forms in relation to their contexts of production and consumption, taking into account the historical backgrounds of such contexts and of the text itself. For Martín-Barbero, television, through its genres, ' … stimulates cultural competence and the recognition of cultural differences. The genres, articulating the narration of serials, constitute a fundamental mediation between the logic of the system of production and that of consumption, between the logic of the format and how that format is read and used' (1993: 220–221). Consequently, Martín-Barbero suggests that we move into a cultural conception of the genre as a 'communication strategy', following the work of Italian researchers such as Mauro Wolf. The genre then gains visibility in a dialogical sense, as it appeals to 'communicability', which in turn allows us to analyse it as a text. So the failure to acknowledge the genre beyond a literary perspective will not allow us to recognize its social function and its 'methodological significance'. This 'methodological role is of key importance in the analysis of texts of the mass media and especially television' (Martín-Barbero, 1993:241).

The genre, then, is one of the socio-symbolic mediations that need analytical attention when studying the relationship that binds telenovelas with residual components of previous cultural formations deeply rooted in social memory, as is the case with melodrama. Melodrama – as a hybrid genre – has been constituted as a dense narrative structure that allegorically represents a complex and unequal historical process of incorporation of the popular classes to key spaces of a heterogeneous modernity. It condenses in its history the moment of the social emergence and visibility of the masses, the transformation and conversion of the popular into masses and vice versa. Thus, melodrama interpellates and mobilizes anachronistic structures of feeling in those audiences who experience pleasure and affecting engagement from this type of storytelling. As Hermann Herlinghaus puts it, melodrama has a fundamentally intermedial character:

> [as its] versatility allows it to move across diverse genres and communication media generating interstices and new conceptual bridges. Such versatility responds in almost all cases to a re-articulation of non-discursive 'excesses' that, nevertheless, belong to a part of a narrative or a narrative matrix ... the melodramatic shows that beneath institutionalized rationality (discourse) lurk the imaginaries of life and action from which, in turn, a symbolic access to the spheres of 'ordered' imagination is negotiated. (Herlinghaus, 2002: 40–41)

In that sense, melodrama's historical vitality allows Martín-Barbero to sketch a history of the matrices of popular forms, from melo-theatre, circus and carnival to serialized literature, from the latter to film and radio (*tangos*, *corridos* and *boleros*), and finally to television. The peculiar stylization of popular narratives shaped by melodrama wherein binary oppositions (between good/bad, rich/poor, ugly/beautiful, and so on) represent a very Manichean interplay of moral values, takes on a whole new meaning when read from this epistemological position.

This perspective not only entails the consideration of the long historicity of melodrama but also of other key mediations (daily family life, social temporality and cultural competence) that constitute dynamic areas of social reproduction and creativity. As Martín-Barbero puts it, mediations are the 'sites of emanation from where the constrictions that delineate and configure the social materiality and cultural expressivity of television emerge' (Martín-Barbero, 1987: 232).

Research on telenovelas draws its robustness partially from the multiple theoretical and intellectual traditions that inform it. It is this chapter's contention that the analytic framework so far described here is the result of convergent epistemological guidelines that certainly supersede Martín-Barbero's and García Canclini's work. Nevertheless their work has been constantly invoked in this field[11] as presenting a grand narrative about the economic, social and cultural repercussions of modernization/globalization many times lacking, sometimes explicit, in much of the case studies about telenovelas to be explained in the next section. Of course it will soon be clear that there are cases where specific instances actually do go beyond these canonical authors and set out original emergent frameworks.

Telenovela Studies

By the 1980s the pioneer research previously made by semiotic studies on telenovelas was substantially complemented by an interdisciplinary interest in broader aspects of the subject. Academics from Brazil, Mexico, Argentina, Peru, Venezuela and Colombia established a regional network to study telenovelas at a local level while sharing some common theoretical premises. According to Ana Uribe there are three recognizable lines of research that have produced rich insights about televisual melodramatic serials: production processes; textual analyses; and reception studies (Uribe, 2009). All work encompassed in these areas presents both sophisticated theoretical and methodological modulations developed accordingly to suit specific instances of study. The next sections will provide a general perspective on the most relevant work carried out in the field.

THE COMPLEX PROCESSES OF TELENOVELA PRODUCTION AND COMMERCIALIZATION

The first line of research is concerned with telenovela's processes of production, distribution and broadcasting. These are fundamental questions that need to be addressed in order to better understand how 'a genre like the novella [telenovela] refers not only to a social imagination, but also to collective representations, to a historical memory as well as to the genealogy of media institutions in which the "specificities" of serialized production are anchored' (Mattelart & Mattelart, 1990: 103). Questions about the way in which specific patterns of industrial organization and public regulation determine the storylines, aesthetics, the cast and market cycle of a melodramatic product are put under scrutiny. Ethnography as well as other qualitative and quantitative methods have been used to describe the contexts and modalities of production/distribution on a local and transnational scale (Chávez, 1992; Gómez Gutiérrez, 2005; González, 1994; Gutiérrez Espíndola, 1988; López, 1998; Miller, 2010).

There have also been some attempts to develop historical accounts of national broadcasting systems in relation to other social and political developments, by closely looking at the role that politicians, businessmen, intellectuals, artists and formats/genres (like telenovela) have played in setting up peculiar socio-cultural regimes (Dorcé, 2005; Fernández & Paxman, 2000; Martín-Barbero & Rey, 1999; Ortiz et al., 1989; Tufte, 2000; Vassallo, 2004). Unfortunately these valuable efforts remain marginal, because, among other things, media historiography has not yet been fully incorporated as a strategic line of enquiry; also because private media corporations tend to have a rather tight 'closed doors' policy that many times restricts archive, observational or documental research. There has also been interest in studying the politics in the many times unequal forms of international commercial exchange of audiovisual products and the ways in which telenovelas have changed the terms of the game for most major Latin American broadcasters (Biltereyst & Meers, 2000; Rogers & Antola, 1985; Straubhaar, 1982). In that sense, Miami[12] has become one of the decisive nodes for Latin American and international media conglomerates in the struggle for the global markets (Mato, 2003). Some of the central findings produced from this approach can be summarized as follows:

- The production of a telenovela is a complex and contested process within an organization and structure of differential power positions subordinated to corporate criteria of capitalist production (González, 1998).
- The industry has granted relative autonomy and higher status to producers and writers, as they have developed valuable authorial styles and prestige (Dorcé, 2005; Leal & Oliven, 1988; Tufte, 2000; Vassallo, 2004). Authorship has become a marker of specific production values followed by niche audiences within the genre. However, recent analyses suggest that emergent trends are shifting attention from writers as central authorial figures, to the entire production companies that employ them (Orozco and Vassallo, 2009).
- For major broadcasting networks and TV producers in Latin America, commercialization of telenovelas has been the source of the majority of their profits (Televisa and TV Azteca in Mexico, Rede Globo in Brazil, RCN and Caracol Televisión in Colombia, RCTV in Venezuela, Telemundo in the USA (Marques de Melo in Biltereyst & Meers, 2000). The industry is increasingly transnational as its exports within Latin America are now consolidated, and demand from Eastern Europe, Africa, Asia, the Middle East and United States keeps growing (López-Pumajero, 2007; Mato, 2003). The centrality of telenovelas in the

production line of major media corporations has been consolidated by merchandising paraphernalia of their most popular titles (music, clothing, posters, comic books, etc.) and commodifying telenovelas stars' private lives (in both popular press as well as in TV formats such as talk shows, gossip shows, reality shows and the news).

- All telenovelas are cultural hybrids, yet each telenovela production company develops distinguishable trademarks (formal configuration of their titles: cast, aesthetics, narrative style, etc.) within the confines of its national mosaic of cultures. As Daniel Mato points out, many recent telenovelas are significantly investing in developing a 'strong local flavour' by accentuating local references in response to formidable success in the exportation of titles that 'include *Café con Aroma de Mujer* (exported to seventy-seven countries and related to the coffee industry); *Sueños y Espejos* (related to the kidnapping industry); the telenovelas made by Carlos Vives about Escalona, the well-known vallenato author and singer; and *La Caponera* and *Las Juanas*, both of which display strong local flavour' (Mato, 2005: 427). The mobilization of dominant stereotypes working in local imaginaries embedded in 'universal love stories' allows these telenovelas to be symbolically and economically proficient with differentiated domestic audiences/markets, and, when pertinent, effective as 'normalized exotic' imports in regional and foreign markets, thus performing as 'glocalized' products (Mato, 2002; Paxman, 2003). In that sense, the successful commercialization of 'formats'[13] in the international audiovisual trade circuit as a 'packaging' strategy, accounts for the perceived effectiveness of culturally 'neutral' formulas (*Betty la fea* exported to over 70 countries) in reducing what Colin Hoskins and Rolf Mirus once called the *cultural discount*[14] (1988), thus reducing economic uncertainty in the process of importing symbolic goods. By developing locally adaptable formats, the industry sought to habilitate local audiovisual production companies to eventually increase total value (use, exchange and sign value)[15] of the original investment by enhancing what Straubhaar originally termed *cultural proximity*[16] (1991).
- According to the Ibero-American Observatory of Television (OBITEL), Brazil, Argentina and Mexico are the biggest telenovela producers in the region. Telenovelas in these countries, as well as in Uruguay and Portugal, have the highest ratings. Throughout 2008 it was the televisual genre with the most titles produced in all cases analysed,[17] except for Chile and Spain where local series are far more successful than telenovelas. Interestingly, the popularity of nationally produced series is on the rise in all other countries, now being the second most watched televisual form (Orozco & Vassallo, 2009). This indicates a tendency for telenovelas to be slowly but significantly losing ground to other locally produced fictional genres. During 2008, of all televisual fiction broadcast in Brazil of Ibero American origin,[18] 11% was imported, while the remaining 89% was Brazilian. The same correlation for Mexico reveals that 33% was of Ibero-American origin, and the other 67% was Mexican made. Fifty per cent of all Ibero-American fiction broadcasted in Argentina was produced outside that country and the other 50% was made in Argentina (Aprea & Kirchheimer, 2009: 70; Orozco et al., 2009: 196; Vassallo, 2009: 118).

Text and Sensibility

The second line of enquiry has been focused on exploring the intertextual complexities that constitute the genre as a paradigm, based on recurrent and dynamic semiotic features identified in specific titles (Escudero, 1996; Gómez, 1996; Mazziotti, 1996; Petris, 1996). Much of what we currently know about the multiple narrative devices working in telenovelas has derived from this line of work. Given the detailed analytical work carried out, it has been possible to identify thematic and stylistic differences between texts produced both in different Latin American countries and at different periods of time. In order to broadly illustrate how variability in the genre has been characterized by scholars working on telenovelas' textuality, I will use Oscar Steimberg's model of classification of Argentinean telenovelas (1997: 17), as well as Thomas Tufte's model for Brazilian *novellas* (2000: 98). They identify three historically contextualized modalities of textual configuration: the foundational model; the social-realist model; and the post-realist model. These are not clear-cut categories devised to designate a linear and progressive development of the genre but are designed to indicate how dominant formal and thematic tendencies change over time and, in certain instances, overlap with emergent and residual melodramatic forms.

The Foundational Model

The first group of telenovelas Tufte's and Steimberg's models identify were produced and broadcast – in different national settings – throughout the second half of the 1950s and most of the 1960s. For both authors this is the *foundational* moment of telenovelas where the aesthetic, narrative and thematic specificities of the genre were set up as a televisual form. In a similar vein, the periodization made by Nora Mazziotti, following work made by Mabel Coccato (1979), takes the advent of videotape in the early 1960s as the closure point of what she calls the initial phase or the 'prehistory' of telenovelas and the starting point for the *handcrafted phase* (*etapa artesanal*) (Mazziotti, 1996: 29). In either case there is clear academic agreement that in this moment the highly melodramatic canon was introduced as the guiding principle of these romantic narratives, together with a peculiar stylization derived from the historical, technical and aesthetic resources available to the production teams at that point.

The first telenovelas were inextricably linked to the imaginaries produced by film and radionovelas since the 1940s. The first radio melodrama to be acknowledged as having a vast impact in Latin American media was the Cuban radionovela[19] *El Derecho de Nacer* (*The Right to be Born*) (1941). Written by Cuban Felix B. Caignet, this story has been described by different authors as the epitome of classic Latin American melodrama (Martín-Barbero & Rey, Mazziotti). Reynaldo González observes that B. Caignet made a unique cultural adaptation of the American soap opera into the radionovela:

> Caignet knew how to adapt the soap operas to the Latin American taste and achieved an unprecedented success with it. He knew how to retain the daily dramatic tension and suspense in each episode [the 'cliff-hanger'] by using the sad sobbing of actors and the effective story telling of the narrator. He displaced the action from the kitchen and the living room – characteristic of the American Soap – to exotic locations as the coffee plantations of Palma Soriano. And while the American soap operas tended to avoid controversies around sex, Caignet opened up a space for passion [carnal desire], abortion and illegitimacy [in kinship ties]. (Quoted in Oroz, 1995: 25)

El Derecho de Nacer proved to be one of the first successful multimedia narratives in the continent as it was produced on radio,[20] comic books,[21] cinema and television, always with overwhelming success.

When adapted to television, many of these stories originally written in Cuba and Argentina as *radionovelas* (Mazziotti, 1996), would maintain a classical composition: the narrative starts by setting off a fictional state of affairs, a *mise-en-scène*, the leading characters, a dramatic conflict which confronts characters and situations, and, finally, a resolution of the conflict. The plot in this group of programmes is divided into different story lines, although it tends to be exclusively focused on countless misfortunes that impede the 'legitimate' romantic union of the protagonist couple. A vital component for dramatic tension in this fictional form is intra-diegetic secrecy, which, when dispersed by substantial revelations (true identity of key characters), generates reifying moral restorations (Escudero, 1996). The narrative is progressed in a linear way, with slow pace and reiterative style (in every sequence, a character comments repeatedly on situations shown in precedent sequences). It also relies heavily on the use of ellipses. Most of the *mise-en-scène* is placed in domestic spaces, and the foundational TV camera gaze (three cameras alternating shots) relies heavily on close-ups in order to amplify the facial expressions (Traversa, 1997). This type of choreographic preparation gives a peculiar aesthetic style to this group of telenovelas.

In Mexico the telenovelas *Senda Prohibida* (1957), *Teresa* (1959), *Pecado mortal* (1960) and *El derecho de nacer* (1966) set out most of the conventions of what is known in Latin America as the Mexican model:

> based on the Cuban radio soap [this model] has given form to a serious genre, in which heart-rending, tragic suffering predominates. This format depicts exclusively primordial feelings and passions, excluding all ambiguity or complexity from the dramatic space. That is to say, [explicit] references to places and times are blurred or neutralized. (Martín-Barbero, 1995b: 279)

The evasiveness with regard to social complexity and its highly moralized perspective on private life, characteristic of the telenovelas produced within this foundational model, has given to the classic Mexican telenovela an identity that has been characterized as 'notorious for [its] "weepiness" [and its] extraordinarily Manichean vision of the world and lack of specific historical referents' (López, 1995: 261). The majority of Mexican telenovelas 'tend to privilege primal emotions over socio-historical context, but they substitute dialogue and utterly Spartan sets for the signifying baroqueness of their *mise-en-scene*', as Ana López puts it (1995: 262).

At this early stage of television's first appearance in the continent, many emblematic telenovelas produced in Brazil (*25499 Ocupado*, 1963), Colombia (*En nombre del amor*, 1963) and Argentina (*Dos a querese*, 1963) were also developed under this model. Colgate Palmolive, a major sponsor of telenovelas in the region, owned the rights of many scripts that circulated fluidly across national borders, thus giving a certain thematic and formal homogeneity to most telenovelas made at that time (Mazziotti, 1996).

There is plenty of media historiography and archival research to be done yet about such a definitive period of time in all different national contexts, but what is clear for the most part is that the foundational model has remained as a strong generic influence over other telenovela texts up to now. In Mexico where recent research has shown a steadily decline in narrative and aesthetic innovation (Orozco, 2006), Televisa has opted to produce remakes in contemporary styles of many of the foundational titles, such as *Corazón Salvaje*, *El Derecho de Nacer*, *Teresa*, etc. According to Medina and Barrón, 60% of Televisa's telenovelas are remakes (2010: 9).

The Social Realist Model

The second moment in the history of telenovelas, which begins in the last half of the 1960s, is recognized as the 'modern' stage of the genre, where 'social realism' is used to represent a more complex melodramatic interpretation of life. The topics remain centred on intimacy and romance, yet the set of causal factors that customarily originate the dramatic conflict in the foundational telenovelas is diversified. As Steimberg points out:

> The telenovela of this period remains focused on familiar and romantic relationships but it relates them to social problems and conflicts from the [local] region and epoch that transcend the kinship and friendship dynamics. The main characters' virtues and defects are based on individual psychopathology conditioned [and determined] by its socio-historical position. (1997: 21)

At this point of the late 1960s, the popularity and profitability of the genre became evident; TV corporations took full control over the production of telenovelas, so sponsors were removed from that venture. In order to maximize profits the production process had to be efficient and professional, and new technologies were incorporated to assist in this rationalization (videotape, colour TV, new production studios, the prompter, mobile TV units, exclusivity contracts with TV stars, etc.). In terms of the narratives, the *mise-en-scène* started to take place outside the domestic realm, in recognizable streets and neighbourhoods from big cities. The visual rhetoric began to be more diverse and rich. It was in 1968 when in Brazilian television emerged the discourse that embraced a socio-political realist aesthetic with high quality values and a narrative configuration that became a singular tradition now recognized by many scholars as a powerful and persistent trend in Brazil (Hamburger, 2000; Martín-Barbero, 1995b: 280; Mattelart

& Mattelart, 1989; Mazziotti, 2006). A perfect example of this cultural trend is represented by the Brazilian telenovela *Beto Rockefeller*[22](1968) and the Argentinean *Rolando Rivas Taxista* (1972) and *Rosa de lejos* (1967/1980), among others. The Mexican telenovela *María Isabel* (1966) and the Peruvian *Simplemente María* (1969) have also been associated with this significant change in the genre (Dorcé, 2005; Singhal et al., 1994). In many ways, the process of massive migration from rural villages to new urban cities taking place in Latin America at that time, with all the socio-cultural traumas that involved, emerged sharply represented in fictional melodramatic narratives (Martín-Barbero & Rey, 1999).

The inclusion of many topics in telenovelas that were socially perceived as pertinent not only responded to the logic of a creative genius. As the role of many writers and producers of televisual melodrama became increasingly recognized as trademarks of different narrative styles and peculiar production values, this social notoriety and certain disposition to interact with the audience endorsed a novel way to incorporate the audiences' opinions and suggestions into the on-going telenovelas' plot; thus many times significantly altering the story according to the changing perceptions of the public. As Brazilian author Aguinaldo Silva puts it: 'My cleaning woman returns home every evening to the favela. Every day, she tells me the reactions of people from her neighbourhood and on the bus to the previous evening's episode. Yesterday, for example, they told her to tell me not to banish Roque Santeiro from the town'…' (quoted in Mattelart & Mattelart, 1989: 43). Apart from this less systematic mode of audience surveying, big production companies such as Rede Globo and Televisa were among the first to commission opinion polls and focus groups where viewers 'are regularly invited to express their opinion on scripts and characters. These measurements of public reaction, some of which are extremely sophisticated, are accompanied by an analysis of the large volume of letters addressed to the channel and the author' (Mattelart & Mattelart, 1989: 42).

Another relevant trend identified within the social realist model was the introduction in Mexico of the historical telenovela, a new subtype of televisual melodrama (*Carlota y Maximiliano*, *El carruaje*, *Senda de gloria*, *La antorcha encendida*). Its appearance is significant in a country where until 1996 all explicit references to parliamentary politics in televisual fiction were systematically avoided or banned. If there is something quite peculiar about these historical telenovelas, in contrast to the foundational ones, it is the fact that 'history' – with its implied aura of being grounded in 'reality'– worked as a pedagogical discourse meant to educate viewers on seminal national historic facts. In trying to educate this citizen, the telenovela genre experienced a vital transformation: conflicts no longer occurred exclusively inside the home, the bedroom or the kitchen, but also in presidential palaces, in public squares, in battlefields. Personal problems and misery were no longer only due to familial disputes, but also because of social injustices. Similarly, romantic quarrels and break-ups were no longer exclusively caused by evil and 'pathologized' mothers, sisters and envious lovers, but also by disputing ideologies. Ultimately, the interest and investment in articulating televisual melodramatic accounts about the processes of nation building entailed the transformation of the domestic melodrama. In response to this sort of historical representation, even one of the most respectable Mexican historians of that time, Daniel Cosío Villegas, approved *La Tormenta*, as the programme represented a kind of 'democratic television that will make the people (*el pueblo*) conscious of themselves, through historical evolution' (quoted in Fernández & Paxman, 2000: 132; Charlois, 2010; Dorcé, 2005).

The Post-Realist Model

Finally, the third period in the history of telenovelas is identified by Steimberg as the post-modern phase and by Tufte as the post-realist phase of telenovelas. This is the period in which the vast majority of academic analyses have been carried out from the mid-1980s onwards. Steimberg characterizes this period in the development of Latin American TV in aesthetic and narrative terms using Argentine telenovelas as examples for this model: there are multiple story lines with more or less the same hierarchy, that is, instead of there being a single protagonist romantic couple, there are other characters whose storylines are also central to the whole narrative. The style used to frame these stories tends to be overtly *neo-baroque*,[23] in the sense that it makes multiple references to other aesthetic and narrative resources from popular genres: video-clips, action movies, advertising, comedy, musicals, documentaries, and so on (Steimberg, 1997: 22). In some specific cases, the intertextual construction of these programmes is structured as a parody aimed to make fun of many generic conventions of telenovela or other genres, thus pointing – see the case of *Antonella* – to the 'constructiveness' of telenovelas[24] (Monfazani, 1996; Petris, 1996; Soto & Kirchheimer, 1996). Another important shift in these titles is the broadening of their thematic horizon: sexuality, religion, ecology, race and gender issues are incorporated into the narratives from different ideological positions, sometimes critical, but predominantly conservative. Colombian, Mexican and Venezuelan telenovelas have significantly broadened the local socio-cultural diversity represented in their melodramatic fictional worlds with variable degrees of formal innovation and explicit politicization (Acosta-Alzuru, 2003; Beard, 2003; Cervantes, 2005; Dorcé, 2005; Morrissey, 2002; Rincón & Percy, 2007).

In the case of Brazilian telenovelas, which followed a similar path to their Argentine equivalents, Tufte observes that the democratization of Brazilian politics during the mid-1980s and early 1990s allowed telenovelas[25] to address current political issues in an explicit and sometimes playful way 'thereby sparking often very large and profound public debates about those issues. Telenovelas thus become media-based public spheres articulating and promoting debates on issues of common concern' (2000: 100).[26] In that sense, Brazilian telenovelas have played a paradigmatic role in Latin America as examples of the rich and diversified possibilities of the genre. However, the incorporation of many conjunctural factual topics in their diegesis has been a rather ambivalent process with significant tensions between transgressive thematic inclusion and conservative exclusions.[27] The extent of the social relevance of telenovelas in Brazil is such that when fictional characters that are socially perceived as representative of given social groups or communities are given unfair and abusive dramatic treatment in a telenovela, very visible public demonstrations ensue. In that sense Brazilian telenovelas have become strategic sites where the nation is also imagined (Vassallo, 2004).

The Situated Audience: Telenovelas, Contexts and Readings

The third analytical approach has been working on the social uses and appropriation of telenovelas. This investigative practice results from a broader concern with cultural consumption. As pointed out earlier, one fundamental point of agreement in the CLACSO meeting was about the need to move beyond speculative readings of ideological effects in mass culture's goods and texts, in order to investigate how audiences elaborate meaning in reception and consumption practices within the wider context of culture (Grimson & Varela, 1999; Jacks, 1996; Saintout 1998).

As we have seen in the initial part of this chapter, the emergence of Latin American studies into media and culture provided the general framework within which most of telenovela research has been conducted over the past forty years. It has been argued that the development of this area of research has been marginal and often methodologically deficient, thus considerably restricting the potential of these fundamental investigations (Dorcé; 2005; Lozano & Frankenberg, 2008; McAnany & La Pastina, 1994). In spite of these limitations, we now have a better understanding of the cultural, technological and economic dynamics that constitute the telenovela, as well as a clearer picture of the process of the cultural consumption of the genre.

In 1974, when Stuart Hall published his influential encoding/decoding model in Birmingham, which laid the foundations for the 'reception paradigm' in the Anglo-Saxon world, Mabel Piccini and Armand Mattelart published in Chile the results of their research project about the relationship of television and the 'proletariat' based on similar theoretical assumptions as those in Hall's model. This paradigmatic shift –from informational models of communication to models of 'active' communication – was characterized by Perti Alasuutari as the 'first generation of reception studies' (1999), and took place in Latin America at more or less the same time as in the UK. In the early 1980s Brazilian researchers working on telenovelas (Fuenzalida, 1992; Leal & Oliven, 1988) were asking similar questions as their colleagues in Europe and America who were investigating soap opera's audiences (Ang, 1985; Brunsdon, 1981; Hobson, 1982; Katz & Liebes, 1985; Liebes & Katz, 1990). Following Alasuutari's schema, we see that the 'second generation of reception studies' finds, in its Latin American version, the same paradigmatic inflection towards audience ethnography as a prominent method of choice (Alfaro, 1988; Barrios, 1988; Covarrubias, 1998; Leal & Oliven, 1988; Tufte, 2000; Uribe, 1998).

These explicit moves into reception research and cultural consumption marked the starting point for a series of emergent paths of investigation into culture which have been, ever since, in permanent dialogue with each other, and whose perspectives share substantial theoretical and methodological intersections. Nevertheless, there are key differences between these perspectives, which give them relative autonomy from each other and a certain degree of specificity. Brazilian researcher Nilda Jacks suggests in this regard that such trends have had a rather influential role in the subsequent study of audiences in Latin America; thus she distinguishes five theoretically specific trends that have also framed research on telenovelas' audiences (Vassallo, 2004):

- *Cultural consumption*, developed by Garcia Canclini leading the Program for the Study of Urban Cultures at the Universidad Autónoma Metropolitana;
- *Cultural fronts*, which point to the symbolic struggle between different socio-cultural groups, developed by Jorge González and the group of the Culture Programme at the University of Colima, in Mexico;
- *The social use of media*, which focuses on the mediations between social movements and communicative practices, developed by Jesús Martín-Barbero;
- *Active reception*, developed by the Chilean, Valerio Fuenzalida, which stresses the cultural influence of television.
- *The multi-mediations* model elaborated by Guillermo Orozco Gómez in Mexico.

According to the research developed by these five trends, telenovelas are watched and *enjoyed* – as a pleasurable practice – by men and women from different social classes and age groups who everyday follow the romantic misfortunes of fictional characters in Latin America (González, 1998; Mattelart & Mattelart, 1989; Tufte, 2000). The televisual genre is used as mediation for intra-familiar socialization. Families gather around the television set to do all sorts of things while telenovelas are broadcast, including watching them with differential degrees of attention and engagement.[28] In many cases, television viewership in low-income neighbourhoods takes place in a symbolically constructed sphere – in the material interface between 'home' and its surroundings – that articulates private with public spaces in everyday ritualized and gendered interactions. This is a hybrid sphere of signification, an 'intermediary zone, neither private nor public. It implies a special organisation of time and space, linked with a special code of conduct, which together create a sphere that is central in the formation of self, of identity, underpinned by emotion, and with the telenovelas as central agents' (Tufte, 2000: 231).

Research has documented that audiences are active and construct a variety of meanings from telenovelas. Uribe in Mexico concludes her ethnographic analysis by posing some interesting hypotheses in this regard:

> Certainly television is the inter-structural flow with more presence among the families; the group and the individuals have strong exposure and contact with television. The viewers are active agents in their relationship with television. Such a relation is not direct or univocal; it is mediated by different situational, institutional and cultural elements within which receivers are subjects in situation [sic]. Through their social reading conformed by their cultural *habitus*, the receiver recognizes and legitimizes what he [sic] watches, it is he who ultimately defines and chooses his plausible or implausible relationship with the contents he receives. (Uribe, 1998: 271)

Research has also shown how audiences are clearly aware of the fictional nature of telenovelas and have a degree of expertise regarding genre conventions. Audiences also have a considerable repertoire of knowledge about telenovelas in their memories, and

> When watching one telenovela the family experiences different levels of involvement with the actions, situations and characters presented in those programmes. They share emotions, meanings and vicarious experiences that viewers live as real, though they know at a conscious level that the telenovela is a made up story, a fictional product. The situations and actions created and organized sequentially are recognized as implausible, yet they are experienced as 'effects of truth'. (Covarrubias, 1998: 309)

Telenovelas' hybridity, discussed above, operates in peculiar ways when constituted through rearticulated consumption contexts resulting from the international migration of the Latin American diaspora and the proliferation of broadcasting companies in the US eager to tap into a growing niche of avid media consumers by importing a number of canned telenovelas (Uribe, 2009). This is a key point, as Jade Miller suggests, for understanding how 'South-to-North flows of people via migration can be considered an engine of South-to-North flows of culture along with the converse: North-to-South flows of capital in payment for content' (Miller, 2010: 210).

In that sense:

> [T]elevisual melodrama is not only a site where the tensions among the national, the local and the global are articulated and made manifest, it is also a communicative bridge that links viewers across national, expanded regional and global realms of transmission and reception, working to shape new cultural and intercultural communities. (Benamou, 2009: 152)

Thus telenovela consumption is as much a modernizing national ritual as it is a practice that structures belonging for emergent transnational traditions and new intercultural patterns of communication (Piñon, 2011)

What about telenovela reception in socioeconomic differentiated groups though? Is class a significant marker in the construction of social difference mediated by TV? Research undertaken in Brazil in the early 1980s showed significant differences in inter-class readings of telenovelas. Working-class audiences were as proud to admit they watched telenovelas as they were keen to share their knowledge about details of actors' private lives, or enact representations of entire melodramatic dialogues. Contrastingly, upper-class viewers tended to critically distance themselves from telenovelas, at first by reluctantly admitting they watch those programmes: 'I watch telenovelas because I am very interested in news (broadcasted right before *novelas*)', 'I watch them in order to have something to talk about with my maid' or 'If we don't watch them, we won't know what is going on in terms of culture in this country' (Leal & Oliven, 1988: 87–88). Upper-class audiences drew a distance when talking about the narrative, referring to the actor's names and judging their performances, while using 'irony and slang to separate and distance themselves from what is recognized as a "popular commodity"':

> There is a profound contrast in the style of the narrative accounts between the working-class and upper-class discourses. These differences are found in the choice of words, use of grammar and the kind of relationship established with the ethnographers in the narrative situation. These distinctions, peculiar to two social domains, reflect difference in life trajectories, life perspectives, class situations and specific symbolic repertoires. (Leal & Oliven, 1988: 87–88)

In similar work carried out in Chile's rural communities, the telenovela was the preferred and most viewed televisual genre by adult women and men (peasants), even over American fiction:

> Unlike urban women, peasant women prefer Mexican and Venezuelan telenovelas instead of Chilean ones because, in spite of its 'weird' accents and different landscapes, they identify themselves with situations set out by the narrative: 'those things can happen to me', 'are based on real life, at least some of it', 'it shows how rich people depend on the poor', 'it represents us the poor and our poverty'. (Fuenzalida, 1992: 135)

Interestingly, from these types of findings a set of key questions were also formulated. If, as pointed above, according to Straubhaar, *cultural proximity* worked in this case to Chilean telenovelas advantage, why did these women prefer Mexican and Venezuelan telenovelas? La Pastina and Straubhaar carried out research in rural Brazil to further understand why people prefer telenovelas made outside their home countries, and, in doing so, tried shedding some light on the heuristic limitations of the *cultural proximity* model. The results of that study – where they analysed the contextual reception of the Mexican telenovelas *Marimar* and *María Mercedes* and the Venezuelan *Topacio* in Macambria, Brazil – revealed that a complex articulation of determinations are overlapped with cultural proximity to make possible the enjoyment of 'relatively alien' TV programmes:

> the concept of cultural proximity in this case could be stretched to imply … a text that seems to provide the local viewers with ideological content that does not challenge or question their own values and beliefs, but rather returns viewers to an idealized reception state in which telenovelas provide melodramatic, cathartic space that did not question these viewer's attitudes. (La Pastina & Straubhaar, 2005:283)

Such a logic operates across cultural boundaries within a national space as

> because of regional culture, urban rural divides or differences of class, [viewers] might feel more proximate to products originating in other cultures that might share some of the same values. Genre proximity, in the form of melodrama, also explains the complex relationship that existed between rural Brazilian viewers and these Mexican melodramatic texts. (La Pastina & Straubhaar, 2005: 283)

The preference for Mexican traditionalist telenovelas by these people over Rede Globo's, on the grounds of perceived differential degrees of realism portrayed in each programme, signals not only contrasting ideological and cultural constructions of the 'verisimilar' and its normative implications, but also the symbolic efficacy with which such difference is constitutive of identitarian frameworks that produce 'resistant' readings. Clearly, this is a socio-cultural dynamic that presents serious obstacles for the expected usefulness of telenovelas for both progressive politics and entertainment-education, as we shall see below.

In connection to this argument, Fuenzalida's research found that naturalism as descriptive aesthetics was not particularly interesting for rural Chilean audiences, and that formal experimentation in the genre was not desired or expected:

> Fictional characters are more interesting [and emotionally thrilling] for what they can do, for their behavioural patterns and actions [in the face of familiar conflicts, jealousy, love extra-matrimonial attraction], rather than by a static description of 'how they are' … Naturalism is not attractive to these audiences because it only repeats already known situations, it is not innovative as it eludes possible alternative transformations in the dramatic realm. (Fuenzalida, 1992: 152)

However, this proclivity to celebrate innovation in the dramatic structure of the genre is not extended to a generalized demand for formal and aesthetic innovation in telenovelas for this group of viewers. Thus, as Fuenzalida puts it, 'formal experimentation in telenovelas is hardly successful [for these viewers], and it would only be advisable for film and video' (1992: 152).

In a similar way, the findings made by Leal and Oliven's inter-class comparative reception research carried out in Brazil, suggests that even when both working-class and upper-class audiences recognize formal innovation in the same text, they interpret it differently:

> The working-class group indicates that they liked this telenovela [*Sol de Verão*] because is different, but at the same time they wish the characters would act according to traditional scripts. Solutions, conciliations and decisions are things they expect the telenovela to provide for them ... For the upper-class group the finale is not that important. It is important for the characters to find solutions, but this does not necessarily calls for ultimate dramatic situations in which death punishes the bad guys and marriage rewards the good ones. Rather the telenovela is valued for its dialogues, its photographic and technical quality, and aesthetic surface independently of the ending. (1988: 91–92)

In Venezuela, Acosta-Alzuru carried out research to understand how a socio-culturally diverse group of viewers made sense of feminist discourses in *El país de las mujeres* (*The country of women*). This televisual text had a clear feminist inclination that nevertheless stigmatized feminism as a political movement through negative representations of it. Acosta-Alzuru found that, according to audience members:

> *El país* [emphasized] the qualities, rights and struggles of Venezuelan women [through] realistic portrayals, which make for a balanced authentic representation of women ... however, they do not understand these themes as feminist – a term they reject. In other words, in the production and consumption of this telenovela, feminism is divorced from messages that seek to empower women and improve their living conditions. In this way, feminist ideas are co-opted by the text. As the audience is seduced by *El país*' stories and characters, the same elements that successfully deliver the telenovela's message (casting, plot and dialogue) also work successfully to discredit feminism. In the end, feminism is defused, robbed of its message and labelled as deviant. (2003: 280–288)

As in the other cases shown above, we see here how telenovelas' melodramatic pathos can play a central role in mobilizing and confronting imaginaries about intimacy in differentiated socio-cultural spaces, thus stimulating readings about the potentially political nature of private dramas in unexpected ways, not only in allegorized form, but in explicit ways. Hamburger characterizes this dynamic through the results of her research on the social appropriation of Brazilian telenovela *O rei do Gado*, a popular programme that used the agrarian reform debate in Brazil as a dramatic device to enhance its fiction's realist effect:

> This telenovela, recognized as a female program, also provided a shared repertoire through which viewers phrased private dramas in terms that were publicly legitimized, and they translated political issues into a private perspective. Telenovelas translate political topics to terms that are meaningful in everyday life. They also provide a dimension of shared repertoire to dramas that viewers deal with in their everyday lives ... In doing so, telenovelas might be thought of as ... constitutive of a space of virtual interaction [wherein] collective identities, such as national representations, are not shaped around shared conceptions of political organisation but rather are shaped around a common debate of ideal models of gender and generation relations and of family structure. In this virtual space, the relationship between politics and intimacy is redefined ... telenovelas constitute a virtual domain of everyday life in which viewers share and relate to each other, thus constituting political identities saturated with intimacy. (2000: 174)[29]

All this research shows the ways in which telenovelas play a substantial role in shaping multiple public sphericules (Gitlin, 1998) of various scales in Latin America and beyond. These studies put forward original knowledge about the relevance of entertainment in the formation of subjectivity and dense political cultures with different degrees of performativity, visibility and social legitimacy. They inaugurate innovative ways of thinking about how melodrama, affect and intimacy mediate and constitute broad intercultural spaces of social recognition, debate and conflict that challenge conventional ideas about citizenship in contemporary democracies; particularly in Latin America where socio-economic inequality and impunity remain strong. The telenovela industry has been an integral part in the constitution of complex global networks that mobilize the production and distribution of formats and programming that is now operating with other successful genres as well.

At the same time cultural industries in the region, especially the audiovisual sector, show worrying patterns of concentration that in concert with subordinated regulatory policies – and politicians – have precluded the expression of the rich local socio-cultural diversity that struggles to find an effective communicative platform to face globalized change (Fox & Waisbord, 2002; Mastrini & Becerra, 2007). Telenovelas have proved useful resources for this and other purposes, but future research should not overlook the possibility that fundamental struggles for recognition are also happening elsewhere, in audiovisual formats and genres articulated to melodrama lurking somewhere in the intermedial landscape facilitated by new symbolic practices and ICTs. There is great opportunity for this rich field of research to develop more integrated studies about how telenovelas are consumed and produced beyond Latin America and how different cultural adaptations of global formats may contribute to mobilize or repress resources for different and contradictory types of citizenships. Equally important is the understanding of how new patterns in the division of cultural work in transnational networks are affecting the material and symbolic modalities of production, distribution and capitalization of televisual melodrama.

NOTES

1 A term originally used to describe the stories broadcast on Mexican Channel 4.
2 Ibero-American Observatory on Television Fiction.
3 For instance, increasing industrialization, accelerated migration from rural to urban settings, the consolidation of the middle classes, the Cuban revolution, growing socio-political discontent with the instauration of authoritarian regimes, the consolidation of cultural industries (Mattelart and Mattelart, 1989).
4 In 1973, the Journal *Comunicación y Cultura* published an editorial which stated that 'it was necessary to go beyond the limits imposed by the massive ideological capitalist apparatus and empiricist science, in order to problematize mass communication using the socio-political alternatives produced by the dominated classes: "by pointing out the intimate articulation between the ideological struggle with other instances of the liberation process, the redefinition of mass communication must take into account the multitude of responses that the dominated sectors mobilise in the context of their daily practices by resisting the old system in order to construct a new order (1973: 4)"' (Vinelli, 2006: 3).
5 The progressive movement within the Catholic Church began with the Second Vatican Council, held in Rome in 1962–1965. Responding to the global decline in believers, not least in Latin America, the World Catholic Church approved a doctrine supporting renewed social commitment among many priests and laymen. Liberation Theology was a socially committed theology preaching heaven on earth and using Marxism to explain the dynamics and structures of society (Tufte, 2000). This doctrine had a fundamental role in the democratization process in Brazil, Colombia, Guatemala and Nicaragua. In Mexico this sector of the Catholic Church is rather marginal, yet it has been suggested that it has strong connections with the rise of the Zapatista Movement of National Liberation.
6 Latin American Council of Social Sciences.
7 This juncture has in turn been taken as a fundamental point for the articulation and emergence of what has been labelled – mostly from Anglo-American academia – as Latin American Cultural Studies. This denomination has been the subject of a crucial debate about the problematic terms in which British and American cultural studies have been institutionalized locally, the often unequal basis of intellectual exchange between 'north and south', and the way in which seminal intellectual work produced in the transdisciplinary field of media and culture research/practice in Latin America has been written off, thus reducing its rich genealogy and limiting its socio-historical specificity (Mato, 2002; Mattelart and Neveu, 2004).
8 In a recent survey of work published on television audiences in Latin America from 1992 to 2007

carried out by Lozano and Frankenberg, they found that García Canclini and Martín-Barbero were among the top five authors cited (Lozano and Frankenberg, 2008).

9 Here Martín-Barbero draws a line to differentiate his work from that of the 'uses and gratifications' model developed in the late 1950s until the late 1970s both in the United States and Britain by Blumler & Katz (1974) and McQuail & Windhal (1981). He establishes this difference by framing these uses within the logic of hegemony in cultural consumption.

10 As Saintout points out, this heterogeneity has to do with the division of gender, classes and generations, as what Williams calls the diversity of cultural formations in one society, that is, the archaic formations interacting with residual and emergent formations (Saintout, 1998).

11 See Tufte (2000), López (1995), Orozco (2006), Rincón and Percy (2007), La Pastina et al. (2003), Allen (2004) and Vassallo de Lopes (2004).

12 Many Latin American production companies are now opening offices in Spain, China and Russia to better promote format sales.

13 In a larger argument about the present tendencies in the transnational audiovisual market to develop both product differentiation and homogenization, Mato points out that 'these formats consist of the sale of the basic structure of the telenovela: the concept, the plot and the main characters. Subsequently, sub-plots and other characters are added to such a format using both local actors and modes of speech to "localize" the product and thus adapt to specific internal markets' (Mato, 2005: 427). Mato has documented that various telenovela producers are well aware of the counterproductive implications of homogenizing (erasing localisms as accents, places or even moral conventions) to culturally specific programmes.

14 According to the authors, cultural discount occurs when 'a particular programme rooted in one culture, and thus attractive in that environment, will have a diminished appeal elsewhere as viewers find it difficult to identify with the style, values, beliefs, institutions and behavioural patterns of the material in question ... As a result of the diminished appeal, fewer viewers will watch a foreign programme than a domestic one of the same type and quality' (Hoskins & Mirus, 1988: 500–501).

15 See Jean Baudrillard's different types of value and their relationship with culture in García Canclini (2004).

16 'The idea of cultural proximity tries to explain why television production is growing within Latin America and other regions of the world at both the national and regional levels. The argument, building on Pool (1977), is that all other things being equal, audiences will tend to prefer that programming which is closest or most proximate to their own culture: national programming if it can be supported by the local economy. This approach argues that popular imported genres will be adapted for local production, if the economy of the television market in question can support them. [Cultural proximity] argues that audiences will tend to choose to watch television programs that are closest, most proximate or most directly relevant to them in cultural and linguistic terms. Their first preference would be for material produced within their own language and local or national culture' (Straubhaar, 2005: 273).

17 The countries participating in that initiative producing data are Argentina, Brazil, Chile, Spain, USA, Mexico, Portugal and Uruguay.

18 That is, the total sum of all TV programmes made in Ibero-America, including Brazilian ones, is the 100%.

19 In 1950s Cuba, radionovelas (radio soaps) were extremely popular. In the tobacco factories, radionovelas were used by the owners to stimulate the workers – who were regular consumers of radio dramas – while they worked and to increase their daily attendance at the factory. In those days the workers could not afford to buy a radio and one way to follow up their favourite radionovelas was by going to work at the factory (Fernández and Paxman, 2001: 75).

20 According to Reynaldo González, in Cuba, radionovelas such as *El Placer de Sufrir* (The Pleasure of Suffering) were advertised as 'This radionovela will make you cry out of anguish and joy' (in Oroz, 1995: 27).

21 The famous collection of comic books which specialized in melodramatic stories was called *Lágrimas y Risas* (Tears and Laughter). From the 1950s to the 1980s it sold more than a million units each week (Reyes de la Maza, 1999: 23).

22 '*Beto Rockefeller* is considered the first realistic telenovela. It was written by Braulio Pedroso and was broadcast on TV Tupi in 1968 ... [it] introduced a quicker rhythm, colloquial language and a more relaxed style of acting. The telenovelas [in this period] became less theatrical and gradually came to resemble reality more and more ... [1968] was one of the darkest years in the political history of Brazil. The military coup was in 1964 but in 1968, with students, artists and intellectuals taking to the streets, and with an emerging urban guerrilla war in cities like Recife and Rio, the last constitutional civil rights were put out of function ... A new military constitution was installed in 1969. The years 1968–74 were the most repressive years of the military regime. The social routine of watching telenovelas every evening, established in these years, was thus rooted in a political reality: repression by the military

regime, which influenced everyday life, spreading fear and limiting options for social and cultural activity' (Tufte, 2000: 102–105).

23 Steimberg takes the term from Omar Calabrese and uses it in a rather cryptic sense. Yet the term seems to refer to a cluster of different styles which in a way saturates any given work of art, or popular culture, and which operates as a counter classicist discourse (1997: 22–25).

24 In the Argentine telenovela *Antonella* (1992/3), one of the characters shouts to another: 'you're talking like a bad telenovela actor' (Steimberg, 1997: 25).

25 Titles like *Corpo a corpo* (1984), *Barriga de Aluguel* (1990), *De corpo e alma* (1992), *Mujeres de arena* (1993), *Renaser* (1993), *Explode coração* (1995), *A indomada* (1996) *A próxima vítima* (1995), *Torre de Babel* (1998) and *O Clone* (2001).

26 These telenovelas have included diverse perspective on topics such as the agrarian reform, artificial insemination, *coronelismo* (the power of local oligarchies), ecological conflict, real state speculation, racism, homosexuality, political corruption, machismo, urban violence, interracial marriage, etc. (Vassallo, 2004).

27 For example, a couple of lesbian characters were removed from *Torre de Babel* because conservative media disapproved of such inclusion. Characters from Jorge Amado's novel *Porto dos milagres*, most of whom were originally black, were subjected to a 'whitening' in the telenovela adaptation produced in 2001 (Vassallo, 2004). In a different example, 'politicians who disagreed with the fictional representation of [Brazilian] National Parliament in the telenovela *O rei do gado* made their complaints public in the press and in the congressional tribune ... In an article, a federal representative criticizes *O rei do gado* for transmitting a bad image of politicians as lazy, corrupt and unscrupulous' (Hamburger, 2000: 166).

28 González (1998) suggest that according to the survey and the ethnographic case studies they conducted in Colima, Mexico, television is rarely watched attentively and uninterruptedly, because in the average Mexican middle-class household there is only one TV set, which customarily is located in the living room. Thus, there is a practical difficulty for any member of the family to watch TV alone in a focused way.

29 For a similar argument in relation to UK and American soap operas see Van Zoonen (2003).

FURTHER READING

Escosteguy, A.C. (2001). Cultural Studies: A Latin American Narrative. *Media, Culture & Society* 23(6): 861–873.

García Canclini, N. (1987). De los medios a las mediaciones. Comunicación, Cultura y Hegemonía. (De por qué el científico social es un cazador oculto.) *Diálogos* (19): 77–79.

García Canclini, N. (1989). *Culturas Híbridas: Estrategias Para Entrar Y Salir De La Modernidad*. México: Grijalbo.

García Canclini, N. (1997). Hybrid Cultures and Communicative Strategies. *Media Development* 44(1): 22–29.

García Canclini, N. (2001). *Consumers and Citizens: Globalization and Multicultural Conflicts*. Translated by Yúdice, G. Edited by Yúdice, G. Franco, Jean and Flores, J., Cultural Studies of the Americas. Minneapolis: University of Minnesota Press.

Hall, S. (1982). The Rediscovery of 'Ideology': Return of the Repressed in Media Studies. In Curran, J., Gurevitch, M., Bennet, T. and Woollacott, J. (Eds), *Culture, Society and the Media*. London: Routledge.

Modleski, T. (1981). The Art of Being Off-Center: Daytime Television and Women's Work. *Tabloid: A Review of Mass Culture and Everyday Life* (4): 18–24.

Morley, D. (2000). *Home Territories: Media, Mobility and Identity*. London: Routledge.

O'Connor, A. (2003). Consumers and Citizens: On Nestor Garcia Canclini. *Pretexts: Literary and Cultural Studies*, 12(1): 103–120.

Orozco Gómez, G., Hernández, F. & Huizar, A. (2009). Mexico: The Increasing Publicity of Fiction and its Stars. In Orozco Gómez, G & Vasallo de Lopes, M.I. (Eds), *Television Fiction in Ibero-America: Narratives, Formats and Advertising*. Sao Paulo: Globo.

Paxman A. and Fernández, C. (2001). *El Tigre: Emilio Azcarraga y su imperio Televisa*. México D.F.: Grijalbo.

Propp, V. (1975). *Morphology of the Folktale* (2nd edn). Austin: University of Texas Press.

Straubhaar, J.D. (1991). Beyond Media Imperialism: Assymetrical Interdependence and Cultural Proximity. *Critical Studies in Mass Communication* (8): 1–11.

Vassallo de Lopes, M.I. & Vilches, L. (Eds) (2008). *Global Markets, Local Stories: OBITEL Yearbook 2008*. Rio de Janeiro: Globo Universidade.

Vasallo de Lopes, M. I. (2009). Brazil: on the threshold of new paths. Television fiction in ibero-america: Narratives, formats and advertising. In G. Orozco Gómez and M. I. Vasallo de Lopes (pp. 98–147). Sao Paulo, Globo.

REFERENCES

Acosta-Alzuru, C. (2003). I'm Not a Feminist ... I Only Defend Women as Human Beings: The Production, Representation, and Consumption of Feminism in a Telenovela. *Critical Studies in Media Communication* 20(3): 269–294.

Alasuutari, P. (Ed.) (1999). *Rethinking the Media Audience*. London: SAGE.

Allen, R. (2004). Making Sense of Soaps. In Allen, R. & Hill, A. (Eds), *The Television Studies Reader*. London: Routledge.

Alfaro, R.M. (1988). Los Usos Sociales Populares de la Telenovela en el Mundo Urbano. *Culturas Contemporáneas* (4–5): 223–259.

Ang, Ien (1985). *Watching Dallas: Soap Opera and the Melodramatic Imagination*. London: Methuen.

Aprea, G. & Kirchheimer, M. (2009). Argentina: Crisis and Renovation from the Margins. In Orozco Gómez, G. & Vasallo de Lopes, M.I. (Eds), *Television Fiction in Ibero-America: Narratives, Formats and Advertising*. Sao Paulo: Globo.

Barrios, L. (1988). Television, Telenovelas and Family Life in Venezuela. In Lull, J. (Ed.), *World Families Watch Television*. Newbury Park: Sage.

Beard, L.J. (2003). Whose Life in the Mirror? Examining Three Mexican Telenovelas as Cultural and Commercial Products. *Studies in Latin American Popular Culture* 22: 73–88.

Benamou, C.L. (2009). Televisual Melodrama in an Era of Transnational Migration: Exporting the Folkloric Nation, Harvesting the Melancholic-Sublime. In Sadlier, D.J. (Ed.), *Passion, Pathos and Entertainment: Latin American Melodrama*. Chicago: University of Illinois Press.

Biltereyst, D. & Meers, P. (2000). The International Telenovela Debate and the Contra-Flow Argument: A Reappraisal. *Media Culture Society* 22(4): 393–413.

Blumler J. G. & E. Katz (1974). *The Uses of Mass Communication*. Newbury Park, CA: Sage.

Borelli, S.H.S. (2001). Telenovelas Brasileiras: Balanços e Perspectivas. *São Paulo em Perspectiva* 15(3).

Brunsdon, C. (1981). 'Crossroads' Notes on Soap Opera. *Screen* 22(4): 32–37.

Cervantes, A.C. (2005). La telenovela colombiana: un relato que reivindicó las identidades marginadas. *Investigación y desarrollo* 13(2).

Charlois, A. (2010). *Ficciones de la historia e historias en ficción: La historia en formato de telenovela. El caso de Senda de Gloria*. Guadalajara: Universidad de Guadalajara-CUCSH.

Chávez, G. (1992). El Análisis de Producción de la Telenovela Mexicana y su Forma Moderna y Racional de Operar. Unpublished Masters thesis, Universidad Iberoamericana, Ciudad de México.

Chernitsky, L. (1996). Telenovela: el placer del género. In Soto, M. (Ed.), *Telenovela/Telenovelas: los relatos de una historia de amor* (pp. 31–35). Buenos Aires: Atuel.

Coccato, M. (1979). 'Apuntes para una historia de la telenovela venezolana.' Video Forum (2 & 3).

Covarrubias Cuéllar, K.Y. (1998). ¡Ya sé lo que va a pasar! Anticipación y telenovela. In González, J.A. (Ed.), *La Cofradía de las Emociones (In)terminables: Miradas sobre Telenovelas en México*. Guadalajara: Universidad de Guadalajara.

Dorcé, A. (2005). The Politics of Melodrama: The Historical Development of the Mexican Telenovela, and the Representation of Politics in the Telenovela Nada Personal, in the Context of Transition to Democracy in Mexico. Unpublished PhD thesis, Goldsmiths College, University of London, London.

Escudero Chauvel, L. (1996). El Secreto Como Motor Narrativo. In Verón E. & Escudero Chauvel, L (Eds), *Telenovela: Ficción Popular Y Mutaciones Culturales*. Barcelona: Gedisa.

Fernández, C. & Paxman, A. (2000). El Tigre: *Emilio Azcárraga y su imperio Televisa*. Distrito Federal: Grijalbo.

Fox, E. (1997). Media and Culture in Latin America. In Corner, J., Schlesinger, P. & Silverstone, R. (Eds), *International Media Research: A Critical Survey*. London: Routledge.

Fox, E. & Waisbord, S. (Eds) (2002). *Latin Politics, Global Media*. Austin: University of Texas Press.

Fuenzalida, V. (1992). Qué Ven Los Campesinos Chilenos En La Telenovela? Del Reconocimiento a La Reivindicación Del Televidente. In Orozco, G. (Ed.), *Hablan Los Televidentes: Estudios De Recepción En Varios Países*. México D.F.: Universidad Iberoamericana.

García Canclini, N. (1988). Culture and Power: The State of Research. *Media Culture Society* 10(4): 467–497.

García Canclini, N. (2004). *Diferentes, Desiguales y Desconectados*. México: Gedisa.

Gitlin, T. (1998). Public Sphere or Public Sphericules? In Liebes, T and Curran, J. (Eds), *Media, Ritual and Identity*. London: Routledge.

Gómez, R. (1996). Temas articuladores del género telenovela. In Soto. M (Ed.), *Telenovela/Telenovelas: Los relatos de una historia de amor*. Buenos Aires: Atuel.

Gómez Gutierrez, C. (2005). *La construcción de la realidad en las telenovelas de TV Azteca. Un esquema de producción*. Instituto Tecnológico Y de Estudios Superiores de Occidente: Tlaquepaque, Jalisco.

González, J.A. (1994). Las estructuras del encanto: Ideas para analizar la producción social de las telenovelas. In Cervantes, C. & Sánchez Ruíz, E.E. (Eds), *Investigar la Comunicación. Propuestas Iberoamericanas*. Guadalajara: CEIC Universidad de Guadalajara/ALAIC.

González, J.A. (Ed.) (1998). La Cofradía De Las Emociones (in) *Terminables: Miradas Sobre Telenovelas en México*. Guadalajara: Universidad de Guadalajara.

Grimson, A. & Varela. M. (1999). *Audiencias, Cultura y Poder: Estudios Sobre Televisión*. Buenos Aires: Eudeba.

Gutiérrez Espíndola, J.L. (1988). La Industrialización Del Melodrama: Historia Y Estructura De La Telenovela Mexicana. In Trejo Delarbre, R. (Ed.) *Las Redes De Televisa*. Mexico City: Claves Latinoamericanas.

Hamburger, E.I. (2000). Politics and Intimacy: The Agrarian Reform in a Brazilian Telenovela. *Television & New Media* 1(2): 159–178.

Herlinghaus, H. (2002). La Imaginación Melodramática. Rasgos Intermediales y Heterogéneos De Una Categoría Precaria. In Herlinghaus, H. (Ed.), *Narraciones Anacrónicas De La Modernidad: Melodrama e Intermedialidad En América Latina*. Santiago, Chile: Cuarto Propio.

Hobson, D. (1982). *Crossroads – The Drama of a Soap*. London: Methuen.

Hoskins, C. and Mirus, R. (1988). Reasons for the US Dominance of the International Trade in Television Programmes. *Media, Culture and Society* 10(4): 499–504.

Jacks, N. (1996). Televisión, recepción e identidad: cuestiones e imbricaciones. *Miradas latinoamericanas a la TV*. In G. Orozco (ed.). México, PROIICOM/UIA.: 173–196.

Katz, E. & Liebes, T. (1985). Mutual Aid in the Decoding of Dallas: Preliminary Notes from a Cross-Cultural Study. In Drummond, P. & Patterson, R. (Eds), *Television in Transition*. London: British Film Institute.

La Pastina, A.C. & Straubhaar, J.D. (2005). Multiple Proximities between Television Genres and Audiences: The Schism between Telenovelas' Global Distribution and Local Consumption. *Gazette* 3(67): 271–288.

La Pastina, A.C., Rego, C.M. & Straubhaar, J.D. (2003). The Centrality of Telenovelas in Latin America's Everyday Life: Past Tendencies, Current Knowledge, and Future Research. *Global Media and Communication* 2(2).

Leal, O.F. & Oliven, R.G. (1988). Class Interpretations of a Soap Opera Narrative: The Case of the Brazilian Novela Summer Sun. *Theory, Culture and Society* 5(1): 81–89.

Liebes, T. and Katz, E. (1990). *The Export of Meaning: Cross-Cultural Readings of Dallas*. Oxford: Oxford University Press.

López, A.M. (1995). Our Welcomed Guests: Telenovelas in Latin America. In Allen R.C. (Ed.), *To Be Continued ... Soap Operas Around the World*. London: Routledge.

López, H. (1998). Estudio Base De Telenovelas En México: Nota Metodológica. In González, J.A. (Ed.), *La Cofradía De Las Emociones (in) Terminables: Miradas Sobre Telenovelas En México*, Universidad de Guadalajara.

López-Pumarejo, T. (2007). Telenovelas and the Israeli Television Market. *Television New Media* 8(3): 197–212.

Lozano, J.C. & Frankenberg, L. (2008). Enfoques teóricos y estrategias metodológicas en la investigación empírica de audiencias televisivas en América Latina: 1992–2007. *Comunicación y Sociedad* (10): 81–110.

McAnany, E.G. & La Pastina, A.C. (1994). Telenovela Audiences: A Review and Methodological Critique of Latin American Research. *Communication Research* (21): 828–849.

Marques de Melo, J. (1971). As telenovelas em São Paulo, estudo do público receptor. In Marques de Melo, J. (Ed.), *Comunicação Social: Teoria e Pesquisa*. Petrópolis: Vozes.

Marques de Melo, J. (1988). Communication Theory and Research in Latin America: A Preliminary Balance of the Past Twenty-Five Years. *Media, Culture and Society* 10(4): 405–418.

Marques de Melo, J. (2001). 'La Telenovela Brazileña: de gata Cenicienta a Reina Mediática.' *Estudios sobre las Culturas Contemporáneas* 7(13): 41–74.

Martín-Barbero, J. (1987). *De Los Medios a Las Mediaciones: Comunicación, Cultura y Hegemonía*. Mexico City: Gustavo Gili.

Martín-Barbero, J. (1993). *Communication Culture and Hegemony: From the Media to Mediations*. Translated by Fox, E. and White, R.A. London: SAGE.

Martín-Barbero, J. (1995a). Matrices Culturales De La Telenovela. In Peñamarin C. and López Díaz, P. (Eds), *Los Melodramas Televisivos y La Cultura Sentimental*. Madrid: Comunidad de Madrid-Instituto de Investigaciones Feministas de la Universidad Complutense de Madrid.

Martín-Barbero, J. (1995b). Memory and Form in the Latin American Soap Opera. In Allen, R.C. (Ed.) *To Be Continued ... Soap Operas Around the World*. London: Routledge.

Martín-Barbero, J. & Rey, G. (1999). *Los Ejercicios del Ver: Hegemonía Audiovisual y Ficción Televisiva*. Barcelona: Gedisa.

Mastrini, G. & Becerra, M. (2007). 'Presente y tendencias de la concentración de medios en América Latina.' *Zer: Revista de estudios de comunicación* (22): 15–40.

Mato, D. (Ed.) (2002). *Estudios y otras prácticas intelectuales latinoamericanas en cultura y poder*. Caracas: CLACSO.

Mato, D. (2003). The Telenovela Industry: Markets and Representations of Transnational Identities. *Media International Incorporating Culture and Policy* (106): 46–56.

Mato, D. (2005). The Transnationalization of the Telenovela Industry, Territorial References, and the Production of Markets and Representations of Transnational Identities. *Television New Media* 6(4): 423–444.

Mattelart, A. & Dorfman, A. (1972). *Para leer al Pato Donald. Comunicación de masa y colonialismo*. Buenos Aires: Siglo XXI.

Mattelart, A. & Mattelart, M. (1989). *Pensar Sobre Los Medios: Comunicación y Crítica Social*. México D.F.: Universidad Autónoma Metropolitana, Unidad Xochimilco.

Mattelart, A. & Mattelart, M. (1990). *The Carnival of Images: Brazilian Television Fiction*. Translated by Buxton, D. New York: Bergin & Garvey.

Mattelart, A. & Neveu, É. (2004). *Introducción a los estudios culturales*. México: Paidos.

Mazziotti, N. (1996). *La Industria De La Telenovela: La Producción De La Ficción En América Latina*. Buenos Aires: Paidos.

Mazziotti, N. (2006). *Telenovela: Industria y Prácticas Sociales*. Bogotá: Norma.

McQuail, D., & Windahl, S. (1981). *Communication models for the study of mass communications*. London: Longman.

Medina, M. & Barrón, L. (2010). La telenovela en el mundo. *Palabra Clave* 13(1): 77–97.

Miller, J.L. (2010). Ugly Betty Goes Global: Global Networks of Localized Content in the Telenovela Industry. *Global Media and Communication* 6(2): 198–217.

Monfazani, A. (1996). Antonella: una Eva y dos lecturas (al menos). In Soto, M. (Ed.), *Telenovela/ Telenovelas: Los relatos de una historia de amor* (pp. 73–90). Buenos Aires: Atuel.

Morley, D. (1992). *Television, Audiences and Cultural Studies*. London: Routledge.

Morrissey, M. (2002). Tres Mujeres: Reclaiming National Culture in the Post- Colonial Telenovela. *Studies in Latin American Popular Culture* 21: 221–232.

Oroz, S. (1995). *Melodrama: El Cine De Lágrimas De América Latina*. México D.F.: Filmoteca UNAM.

Orozco Gómez, G. (2006). La telenovela en México: ¿De una expresión cultural a un simple producto para la mercadotecnia? *Comunicación y Sociedad* (6): 11–35.

Orozco Gómez, G. & Vasallo de Lopes, M.I. (Eds) (2009). *Television Fiction in Ibero-America: Narratives, Formats and Advertising*. Sao Paulo: Globo.

Ortiz, R, Simões Borelli, S.H. & Ortiz Ramos, J.M. (Eds) (1989). *Telenovela: história e produção*. São Paulo: Editora Brasiliense.

Pasquali, A. (1963). *Comunicación y cultura de masas*. Caracas: Monte Avila Editores.

Paxman, A. (2003). Hybridized, Glocalized and Hecho En México: Foreign Influences on Mexican TV Programming since the 1950s. *Global Media Journal* 2(2).

Petris, J.L. (1996). Telenovela, paradoja de su metadiscurso teórico: despreciada por simple, respetada por compleja. In Soto, M. (Ed.), *Telenovela/Telenovelas: Los relatos de una historia de amor*. Buenos Aires: Atuel.

Piñon, J. (2011). Ugly Betty and the Emergence of the Latina/o Producers as Cultural Translators. *Communication Theory* 21(4): 392–412.

Reyes de la Maza, L. (1999). *Crónica de la telenovela: México sentimental*. México, D.F.: Clío.

Rincón, O. & Percy, C. (2007). Colombia: Cuando la ficción cuenta más que los informativos. In Vasallo de Lopes, M.I. & Vilches, L. (Eds), *Culturas y mercados de la ficción televisiva en Iberoamérica: Anuario OBITEL 2007* (pp. 133–154). Barcelona: Gedisa.

Rogers, E.M. & Antola, L. (1985). Telenovelas: A Latin American Success Story. *Journal of Communication* 35(4): 24–35.

Rostow, W.W. (1960). *The Stages of Economic Growth: A Non-Communistic Manifesto*. Cambridge: Cambridge University Press.

Rowe, W. & Schelling, V. (1991). *Memory and Modernity: Popular Culture in Latin America*. London: Verso.

Saintout, F. (1998). *Los Estudios De Recepción En América Latina*. Buenos Aires: Ediciones de Periodismo y Comunicación, Facultad de Periodismo y Comunicación Social de la Universidad Nacional de la Plata.

Singhal, A., Obregon, R. & Rogers, E.M. (1994). Reconstructing the Story of *Simplemente Maria*, the Most Popular Telenovela in Latin America of All Time. *Gazette*, no. 54: 1–15.

Soto, M. & Kirchheimer, M. (1996). Los cuerpos de Antonella. In Soto, M. (Ed.), *Telenovela/ Telenovelas: Los relatos de una historia de amor*. Buenos Aires: Atuel.

Steimberg, O. (1997). Estilo contemporáneo y desarticulación narrativa. Nuevos presentes, nuevos pasados de la telenovela. In Verón, E. & Escudero Chauvel, L. (Eds), *Telenovela: Ficción Popular y Mutaciones Culturales*. Barcelona: Gedisa.

Straubhaar, J.D. (1982). The Development of the Telenovela as the Paramount Form of Popular Culture in Brazil. *Studies in Latin American Popular Culture* (1): 138–150.

Traversa, O. (1997). La telenovela mira hacia el pasado. In Verón, E. & Escudero Chauvel, L. (Eds), *Telenovela: Ficción popular y mutaciones culturales* (pp. 63–72). Barcelona: Gedisa.

Tufte, T. (2000). *The Rubbish Queen: Telenovelas and Modernity in Brazil*. Luton: University of Luton Press.

Uribe Alvarado, A.B. (1998). La telenovela en la vida familiar cotidiana: Apuntes de investigación. In González, J.A. (Ed.), *La cofradía de las emociones (in)terminables: Miradas sobre telenovelas en México*. Guadalajara: Universidad de Guadalajara.

Uribe Alvarado, A.B. (2009). *Mi México imaginado*. Colima, México: Miguel Ángel Porrúa, Universidad de Colima, Colegio de la Frontera Norte.

Van Zoonen, L. (2003). After Dallas and Dynasty, we have .... Democracy. Articulating soap, gender and politics. Restyling politics. J. Corner and D. Pels (eds.). London: Sage.

Vassallo de Lopes, M.I. (2004). Narrativas televisivas y comunidades nacionales: el caso de la telenovela brasileña. *Comunicación y Sociedad* (2): 71–97.

Vinelli, N.A. (2006). Argentina: miradas sobre la recepción en los setenta. *Revista Question* 12(1).

Williams, R. (1961). *The Long Revolution*. London, Chatto & Windus.

Yúdice, G. (2001). From Hybridity to Policy: For a Purposeful Cultural Studies, Introduction to Néstor García Canclini's *Consumers and Citizens*. George Yúdice, Trans. In García Canclini, N., *Consumers and Citizens: Globalization and Multicultural Conflicts*. Mineapolis: University of Minnesota Press.

Zubieta, A.M. (Ed.) (2000). *Cultura popular y cultura de masas: conceptos, recorridos y polémicas*. Buenos Aires: Paidós.

19

Television News and Current Affairs

Kathleen M. Ryan, Lisa McLaughlin and David Sholle

INTRODUCTION

During the writing of this chapter, a major earthquake hit Haiti. The capital city of Port Au Prince was devastated. Tens of thousands of people were feared dead with at least three million others displaced from their homes. The international news media converged on the small, impoverished Caribbean nation to follow the evolving story. The resulting coverage was an illumination of the power, limitations and changing conditions and practices of contemporary television news.

Initially, images of the quake were scarce, largely because the country is one where few outside reporters are based full-time. Because of the extent of the damage to both property and infrastructure, few images were coming out of the country and reporters struggled to get flights into the capital city. Early reports mostly were based on satellite phone interviews with individuals on the scene. The first images came from social media such as Twitter and cell phones (photographs) or YouTube (video); the US-based cable network CNN made extensive use of its iReport platform, which relies on reports gathered by amateur, rather than professional, journalists. In the United States on the day after the earthquake, the three broadcast networks (ABC, CBS and NBC) each expanded their evening newscasts from one-half hour to a full hour but left the broadcast of the special programming at the discretion of the local affiliates. In at least one mid-sized market, two of the three stations cut to syndicated programming after the network newscast aired.

The networks sent a variety of talent to cover the Haiti story. NBC's team, for instance, included Brian Williams of NBC's *Nightly News* and long-time foreign correspondent Kerry Sanders as well as *Today* show veterans Ann Curry (news anchor) and Al Roker (weathercaster). The move seemed to signal that the network was treating this traditional 'hard news' story as a softer human interest piece, by using morning show talent such as a weathercaster known for guest appearances on NBC programs like *30 Rock* and a news person often associated with 'scoops' such as the story of the McCaughey septuplets or the birth of eight infants to the 'Octomom' Nadya Suleman. ABC and CBS had a similar mix of talent, including ABC's Robin Roberts, a former ESPN announcer, and CBS *60 Minutes* investigative reporter Byron Pitts.

Cable and network news also sent medical experts to cover the disaster, the most prominent of which were CNN's Dr Sanjay Gupta and MSNBC's Dr Nancy Snyderman. Howard Kurtz, the *Washington Post* columnist who hosts the current affairs 'pseudo-media-watchdog' program *Reliable Sources*, later criticized the two for being 'double duty doctors': 'The networks send their medical specialists to the tragedy in Haiti and they wind up doing triage. Are journalists crossing an ethnical line here?' (CNN, 2010). While on the show Snyderman staunchly defended the need to serve as a doctor in a context in which medical personnel were scarce, but neither the host nor the guests ventured the observation that the medical experts likely were sent to Haiti in order to promote the human interest quality of the disaster, to become part of the story as doctors helping to save the afflicted. Just days later, the networks pulled out the big guns, leaving just small teams, such as Sanders and his producer, to cover events following the Haiti disaster. With few journalists on the ground, social media forms such as Twitter, Facebook, blogs and cell phone text messages and images became crucial sources of information about, and images of, the damage and relief efforts in Haiti for the outside public and news organizations.

We begin with this description of television coverage of the aftermath of the Haiti earthquake because it illustrates many of the trends and tensions that exist in television news and current affairs today and which are addressed in this chapter. Television coverage of the Haiti disaster was not unique; other natural disasters (Typhoon Haiyan in 2013, Tropical Storm Sandy in 2012), breaking news events (the killing of Osama bin Laden by US forces in May 2011; the Sandy Hook Elementary School massacre in which 26 persons were gunned down in Newtown, CT in December 2012) and political developments (the Arab Spring that began in December 2010 and continued throughout 2013; the integration of social media such as Tumblr, Twitter and Facebook into key events of both the Obama and Romney campaigns during the 2012 US presidential election) all demonstrate similar reporting patterns. Specifically, we address how

- in a commercialized, neoliberal environment, traditional television news norms, codes and practices are exposed, undermined and used to maximize audience numbers and profit, all at once;
- 'hard' and 'soft' television news are distinguished from one another, with deference granted to the latter and its tabloidization tendencies, while decision-making is oriented to the 'bottom line' and television news and current affairs programming are commodified.
- new media technologies alter the professional news environment and make possible the participation of the 'professional amateur' known as the 'citizen journalist';
- the increase in industrial and technological convergence and the ascendancy of a new digital media environment affect approaches to news labor and the conditions and practices of newswork.

We conclude by considering how, from a scholarly perspective, studies which combine political-economic and ethnographic approaches have the advantage of providing grounded, empirical, theoretically-informed answers to questions of how television and current affairs newsworkers fare in the current environment.

TRADITIONAL JOURNALISTIC NEWS PRACTICES, IDEOLOGIES AND APPROACHES

American television journalism produces what it calls news and has defined news over the years as a particular kind of knowledge – that is, a knowledge with a singular importance. In journalism's self-description it is *the* form of knowledge necessary for the enlightenment of the public. In telling this story, journalism employs an abstract, specialized vocabulary that includes concepts like objectivity, balance, fairness, 'getting the story', sourcing, neutrality, etc. Out of this it forges its current self-understanding as 'the fourth estate' – an institution that produces 'objective' and 'balanced' opinions to assist the public in the process of decision-making. As an autonomous 'estate', journalism serves as a check on power – particularly the power of the state. Television attempted to further distinguish itself as a fifth estate as announced in *Broadcasting Magazine* (Briggs & Burke, 2005, p. 154). This was based on television's supposed unique ability to present 'live' news.

News is a set of social practices that take form in the practice of journalism – in its mode of gathering information, structure of writing and editing, approach to televisual construction, etc. These social practices serve to assert that this form of journalism is the one and only way of practicing journalism (Mattelart, 1979, p. 37). This mode of journalism makes normative claims about its practice. Supposedly, it: (1) is free of outside interference; (2) serves the public's right to know; (3) seeks to learn and present the truth; and (4) reports facts objectively and fairly (Altschull, 1995, p. 59).

The emphasis on objectivity sits uneasily with the view, common in both early and current studies of television journalism, that journalism functions in the role of 'opinion-maker'. Many early, and current, studies of TV journalism emphasize story selection and the overall effect of news on audience opinions. Gatekeeping studies have looked at how news organizations prioritize news stories, selecting the order and length of stories (White, 1950). From this foundation, gatekeeping has evolved into a way to study newsroom organizations and decision-making processes. This sort of research recognizes that news is often a collaborative process: multiple individuals with different levels of influence contribute to decisions as to 'whether a news item is permitted to pass through the gate' (Shoemaker, 1996, p. 86). This is especially true in television, where news gatekeeping is 'a group process based on discussion which began with a morning story conference and continued in the newsroom throughout the day' (Berkowitz, 1990, p. 62). In Berkowitz's model, the gate itself has several 'keys' unique to television in determining which stories will be allowed to pass through, including interest, importance and the visual quality of the story (p. 66).

A related concept is the idea of agenda-setting, first introduced by McCombs and Shaw in their 1972 study of television's influence on North Carolina voters. They found that the media do not tell viewers what to think, but rather tell them what to think about – thus setting the agenda, leading people to make decisions and prioritize events based upon how an issue or individual is framed by the news media. Viewers are more likely to see a topic as important if it has received extensive television exposure; topics are judged less important if they receive less exposure (McCombs et al., 1997). Critics note that a problem with this approach to news is that news researchers themselves pre-set the agenda by studying only topics ranked highly in Gallup polling. Agenda-setting researchers themselves engage in a type of 'issue-centered bias' which never investigates the substance of news content (Takeshita, 1997, p. 20).

What distinguishes this approach to television news is its focus on how content is prioritized, or 'stacked', within a newscast or even an individual news story. Kerbel (2000), for instance, dissects, and then mashes together, network and local news choices, demonstrating how the lurid and titillating can gain traction over more important (although perhaps duller) news items. Others turn to the individual story: specifically looking at why a story was covered or what choices led to it being covered in the manner it was (Crowley and Potter, 2005; Hess, 1996). Even the faux news program *The Daily Show* has been investigated for its function in setting a news agenda (Baym, 2005).

Since its introduction several decades ago (e.g., Goffman, 1974), framing analysis converges with the basic agenda-setting approach to describe, not only *what* we come to think about, but also *how* we come to think about various issues (McCombs, 1992). As Entman (1993, p. 52) describes, '... framing essentially involves selection and salience. To frame is to select some aspects of a perceived reality and make them more salient in a communicating text'. Several scholars consider framing to be 'second-level agenda setting' (McCombs & Ghanem, 2001, p. 69), creating layers of meaning that agenda-setting alone cannot measure and developing into an approach devoted to understanding not just what media cover but how they do so. Frames expose persistent patterns of news coverage, for example, how news topics are presented in a problem-solution format by framing devices that rule out other possible ways of thinking about the issues under discussion (Gitlin, 1980).

The problem with the majority of these studies is that they sidestep issues surrounding the status of news as a commodity, the particularity of audience readings of stories, the specificities of the televisual medium, and with the exception of some framing analyses, the ideologies of journalism. As a so-called fourth estate, journalism wields a network of norms formed in particular historical contexts, and researchers then evaluate the extent to which these norms are upheld. Among these norms is the traditional separation of entertainment from news, both of which are presented as if they require special modes of perception to properly understand them. Media has been produced as if there is such a thing as a 'current-events' sense and a 'news' sense. For the journalist these senses are intuitive yet empirical (like a sense of smell); they are unique skills that separate the journalist off as a distinct breed of professional. For the audience, these senses are constructed through the medium's stance toward the audience. Thus, there is a programming aspect to television news. Instead of the audience member relating to the news, their viewing is pre-programmed through sensing gauges including 'how much has a certain program cost ... [and] which topoi and important personalities (stars, politicians) appear today on the viewing screen ...' (Negt & Kluge, 1993, p. 153). Accordingly, 'the techniques of journalism have come to consist in the skillful filling of pre-defined genres, each of which stands for a certain definition of the audience's needs' (Smith, 1978, pp. 167–168).

Traditional approaches to journalism, at least in principle, understand the news to be composed of singular 'events'; yet the whole apparatus of news sets up schemas for constructing these singular events within very circumscribed categories. For example, early on, the news developed an entire beat structure and an organizing of time and space into specific categories of news. The sheer number of these categories from hard to soft news, from courthouse to sports to White House, signals how the sense of news is turned toward sameness.

The story of television is told fondly as one of intrepid reporters who spoke truth to power in Vietnam and brought the Watergate hearings to the public, its reporters/anchors changing history. But the other side of this has always been that television news has reacted slowly and given voice to elite sources. Almost unquestioning coverage of the Bush administration's spurious claims that Iraq had 'weapons of mass destruction', meager coverage of the civilian death toll following from the Obama administration's increase in drone airstrikes in countries such as Pakistan, and the continued tendency to take horserace approaches to elections, have made clear that these trends continue. At one time, the network news claimed autonomy – they existed in news divisions, separate from other programming divisions. Now, they are integrated into the goals of the overall network. The television journalists who attempt to work against all of this have a tougher job than ever.

COMMERCIALISM IN TELEVISION NEWS AND CURRENT AFFAIRS

Television news is almost thoroughly integrated into the overall goals of the networks, caught in a fierce battle for ratings and inundated with stories produced by public relations firms and spinmeisters. Further, the universalizing exclusion of news is shown in its construction of a punditocracy. This punditocracy is epitomized by the cable political talk shows which include various political elites, columnists, bloggers, party strategists and (usually male) self-promoting 'talking heads' who refer to themselves as 'news anchors'. Journalists working in leisure conglomerates find that news can become nothing more than promotion of other company ventures, with its value determined by virtue of its potential for cross-promotion. Campbell, Martin and Fabos note that this is precisely the problem with ABC News, which has 'increasingly been filled with promotional pieces for other Disney products, particularly films', but the problem is endemic throughout television (2005, p. 74). The problem is further complicated when these same media conglomerates produce entertainment shows that are marketed as 'news' and whose talent often appear on network or cable news programming such as *SportsCenter* (sports news, ESPN/ABC), *Access Hollywood* (entertainment news, NBC/MSNBC), *Sanjay Gupta, MD* (medical news, CNN/CBS) and *Huckabee* (political commentary, Fox).

Hallin (1996) argues that the relationship of professionalism to commercialism in journalism is a complex interaction. Professionalism is rooted in 'an ethic of "public service" ... The journalist was supposed to serve the public as a whole, and not particular interests such as the partisan causes journalists had championed in the nineteenth century, or the narrow commercial interests of advertisers and owners' (p. 220). This contrasts with commercialism, which emphasizes market position, advertising revenue and audience shares (Mosco, 1996). In television, increased competition, deregulation, the rise of local news, reality-based programming, and large media corporation mergers have pushed away from professionalism.

In respect to commercialization, anyone who believes that there was once an unquestionable 'golden age of journalism' should recall that, in the 1950s, the NBC evening newscast was called *The Camel News Caravan*, a rip and read headline service sponsored by tobacco giant R.J. Reynolds, the maker of Camel cigarettes (Bliss, 1991). The crass commercialization was a way for the network to put news on the air. For news purists, there was eventually vindication. *The Camel News Caravan*, initially popular, eventually fell in the ratings to CBS's less commercial *Douglas Edwards with the News* (Bliss, 1991, p. 239). Contemporary television news programs are rarely directly sponsored and news sets contain no blatant logos. However, commercial subject matter more and more is placed directly into the news, with stories focusing on products and industries. Commercialism is said to intrude on professional values, but professionalism too can lead to questionable practices such as accepting the government's official version of events, covering international news in terms of national security, covering news more important to elites, etc. Thus, market-oriented news and professional values interpenetrate each other, as is evident in the CBS newsmagazine *60 Minutes*. The program sets hard-hitting investigative stories side-by-side with celebrity celebrations, but always with an eye to audience entertainment values.

60 Minutes proved that news could make a profit. Further, increased competition (both cable and network) and the rise of profitable local news programming moved news from being something stations and networks aired for the good of the public to a for-profit venture (Auletta, 1992). This pattern has intensified in the 21st century due to declining audiences and increased online news sources (Pew Project for Excellence in Journalism [PEJ], 2009). Even as local newspapers are seeing circulation and revenues decline, local television news in the United States is experiencing a golden age of sorts, at least in terms of hours of news on the air. The average amount of television news produced by local stations increased by 35% from 2003 to 2009 (currently at about five hours per weekday); 'in 2009, despite the depressed economy, 28.6 percent of all local stations – and almost 40 percent in the largest markets – *added* newscasts' (Waldman et al., 2011, p. 77). Instead of replacing legacy media such as television, the internet and mobile devices tend to supplement it (at least in the US), with users 'increasingly integrating new technologies into their news consumption habits' (Pew Research Center, 2010).[1]

The norms and professional codes of journalism are formed within the logic of the market. Today, journalistic norms are wrenched out of their original contexts, employed precisely against their original intent. For instance, freedom of speech, once thought of as defending the journalistic professional against the absolutist state is now understood by media conglomerates as a mechanism that allows them to operate freely in a neoliberal environment (Negt & Kluge, 1993, p. 16). The professional codes of journalism serve as shorthand for reporters and editors in selecting and framing stories that fit within the common sense of the social formation such that these stories can cater to and reinforce mainstream audience values (McChesney, 2004, p. 68). Thus, the codes that regard official sources as legitimate, that avoid contextualization, and that canonize the practice of objectivity may result in the creation of a comfortable relationship with the audience. Cynically, it pays to practice objectivity. By avoiding taking sides the audience can be maximized. Nevertheless, and even if objectivity were possible, television news practices often are such that, while upholding the virtues of objectivity in principle, the driving motive is to make revenue by blatantly undermining the notion of objectivity. This tendency is illustrated by television news' preference for 'soft' news, the spread of 'tabloidization', and the exigency of branding. Below, we address these proclivities in turn while attempting to complicate each of these common criticisms of the news.

Hard and Soft News

News stories can be considered of a serious nature (hard news) or lighter and less important (soft news). This traditional two-part categorization dates back to 1952, when Mott defined hard news stories as those about significant events; popular interest, curiosities and other diversions were considered soft news (Mott, 1952, p. 41). Hard news stories can be seen as those which the news organization decides must be reported immediately, often about current events in the government or natural disasters (Tuchman, 1978). Soft news has less immediacy.

The challenge is to understand exactly how journalists (and the audience) define 'hard' news: as Bird and Dardenne note, the distinction includes the assumption that 'certain types of news simply "are" hard, others soft' (Bird & Dardenne, 1997, p. 335). It fails to acknowledge that the 'significance' of an event can often be determined by a variety of aspects of the life context of the citizens. As a result, definitions of 'hard' and 'soft' cannot account for complex phenomena that at a simplistic level appear 'soft', but at another level open up important philosophical and psychological questions. One example of the slippage between 'hard' and 'soft' news is the regular use, in cable network news, of video clips from *The Daily Show*, which defines itself as 'fake news' and *The Colbert Report*, which features 'fake conservative' political talk show host Stephen Colbert. While both shows are satirical, and thus 'soft', they offer information and commentary that may be more conducive to the kind of critical thinking necessary to a thriving, democratic public sphere than the 'real news' shows that use the video clips as news entertainment. As Sparks argues, one cannot support a division between serious news versus entertainment news, for 'one provides the substance for the other and the form of presentation of even that news which is not, substantially, entertainment, is that of entertainment itself' (1991, p. 70).

The Project for Excellence in Journalism categorizes stories as 'hard' or 'soft' by subject matter: international events and public policy debates are hard news, while lifestyle, entertainment and celebrity crimes are soft (Pew Project for Excellence in Journalism, 2004). Another approach divides content by the affective level of its storytelling: hard news is objective and lacking emotion, and soft news is humanized to generate emotions (Djerf-Pierre & Lofgren-Nilsson, 2003). Others argue that the 'ideal' of objectivity in hard news is a cultural value, with neutrality being esteemed in English-language media but less valued in other cultures (Thomson et al., 2008).

Modern television, embedded in entertainment conglomerates and immersed in ratings wars is oriented to soft news content because of its attraction to widely varied audience members and its economic efficiency. In US television, soft news stories made up just 60 seconds of content on the evening newscast in 1974; by 1985 that amount had doubled (Scott & Gobetz, 1992, p. 410). Fast forward to the 21st century, where US networks cover soft news across the board in news programming and cable channels fixate on scandals and celebrity news (Pew Project for Excellence in Journalism, 2008). These programming decisions are motivated in part by understanding the audience as malleable. Some academic studies mirror this view, seeing newscasts which close with an uplifting soft news story as potentially countering the impact of negative, threatening hard news stories on the audience (Zillmann & Gibson, 1994).

Tabloidization

The increase of soft and sensationalized news in television has led many critics to bemoan the 'tabloidization' of the news. Tabloidization can be defined as 'a shift away from the public toward the private in the news agenda and/or a shift toward emphasis on softer news like sports coverage or cultural discourse' (Sparks & Tulloch, 2000). Inherent in any discussion about tabloidization is the assumption that somehow news reporting was 'better' previously, and there is a current shift away from 'real' reporting toward something more salacious, sensational and prurient (Turner, 1999, p. 60). This very assumption is questionable given journalism history, which is shot through with episodes of sensationalism and human interest. Further, tabloidization is a culturally specific term and may be problematic when applied to global media; as some regions and countries may have a different standard for reporting which is only seen as tabloidization to outside eyes. Gans (2009) suggests that 'popularization' is a less pejorative term that communicates the need to present the news in such a way that it is consistent with the different aesthetic criteria and standards of those who may not be elites. However, he offers the caveat that news must be judged on the basis of empirical accuracy. Inaccurate information can hurt individuals and the larger society; a democracy needs informed citizens who, although they may prefer human interest stories, also must be familiar with 'hard' news and foreign news stories.

Nonetheless, the spread of tabloidization is a global concern. Turkish television news is critiqued for its emphasis on celebrities (specifically women) and ordinary people, who are framed as victims (Bek, 2004). In India, journalists report using tabloid-style approaches to report the news, including using hidden cameras as a tool for newsgathering and to catch public officials or celebrities in compromising situations (Rao & Johal, 2006). A study of British television comparing 1975 and 2001 found that reports on politics and foreign affairs dropped over the 25-year period, while reports on crime and human interest stories increased (Winston, 2002). The overwhelming concern by scholars is not necessarily that these stories are totally harmful, but rather that news organizations fail to place the stories within a cultural and social context. Instead, the stories are played as sensation.

But, there are complications to the simple explanation that television news is undergoing 'tabloidization'. In the UK study, for instance, stories about the economy increased over the 25-year period, which seems to run counterintuitive to a notion that news is becoming more sensationalized (of course, as the author notes, this depends upon how economic stories are being reported). Winston in fact argues that perhaps the trappings of tabloidization, such as animated graphics and contextualized discussions of a topic, can actually lead to greater understanding of a subject than a 'straight' news format (2002, p. 18). Turner warns that researchers need to consider how journalism, and specifically television journalism, functions as a form of entertainment within a specific culture (1999, p. 75). As Conboy (2004) notes, 'tabloidization may be the extended working out of the process of commercial logic with journalism as it enters its next discursive phase' (p. 181). Bird (2009, p. 44) makes the important point that the flippant use of the pejorative 'tabloidization' obscures the fact that we live in a 'digital environment in which news organizations struggle to maintain independence and profitability'; thus 'the cheap, easy, and popular story often wins out over expensive, difficult, and less popular ones'. In other words, using 'tabloidization' as a catch-all critique of media production may be too simplistic and fail to recognize other pressures which cause a news organization to report stories in the way that it does.

Branding and the Commodification of News

Situating television itself as the reference point is part and parcel of the networks' (broadcast and cable) attempts to establish brand identity. Bellamy and Traudt (2000) say this instant recognition of the 'brand' (ABC, CBS, CNN, Fox, NBC, etc.) is the network's hidden strength; a way to grow audiences and markets. This isn't just a US phenomenon. Of the five top global television networks (satellite and cable), two, CNN and BBC, focus on news (Thussu, 2009, p. 51).[2] There has also been an explosion of national and regional 24-hour news outlets, from one (US based CNN) in 1980 to well over one hundred in 2007 (Thussu, 2009, p. 51). These include English-language channels outside of the English-speaking world, such as Russia Today, France 24, Star News Asia and Central China Television, as well non-English language options, such as the pan-Latin America network Television del Sur (Spanish and Portuguese language), and the Qatar-based Al-Jazeera (p. 70), which also has an English-language channel and in 2013 debuted Al-Jazeera America on US cable and satellite systems. There are 40 24-hour news networks in India alone (p. 72).

In television, branding means carving out an identity different from other channels offering similar programming (Freeman, 1999). In the contemporary scene, where people looking for news have options of network TV, cable, newspaper, cell phone text delivery, online news, etc. establishing brand has in many cases resulted in niche marketing. So, in the US television milieu, you have conservative and liberal viewpoints occupying specific channels (Fox and MSNBC, respectively), each drawing a certain audience which trusts and believes in its reporting. Republicans see Fox as more credible than any other television news source; 34% of Republicans give Fox the highest credibility rating, compared to 19% of Democrats (Pew Research Center for the People and the Press, 2010). Democrats, meanwhile, are more likely to give MSNBC high ranks for credibility, while CNN maintains a sort of middle ground, drawing support from both parties (Pew Research Center for the People and the Press, 2010). Thus, the modern concept of branding involves creating a 'relationship', a supposed conversation with the audience – democracy through marketing.

Local television news, which is likely to cover much the same stories, is most heavily dependent on branding. In the 1970s, local stations discovered that they could make money by developing market-driven news programs that meld together news, advertising and entertainment. As Hallin points out, local TV news is influential on the entire field, because it is now the training ground for network news personnel. It is fitting into the overall context of entertainment that determines the success or failure of the journalist.

Local television news is the extreme model for the commodification of news: 'its emphasis is on technically uniform, visually sophisticated, easy-to-understand, fast-paced, people-oriented stories that are produced in a minimum amount of time' (quoted in Campbell, 2004, p. 75). What is essential is the ratings, not the journalistic content or form. As a result, local journalism strives to create a consistent product, one that induces 'comfort'. This model is extended to all forms of journalism, as is evident in the Gannett newspapers 'and their successful experiments with translating the image and reality of television news onto the printed page, recognizing that a target clientele rather than the public interest will maximize profits' (Hardt, 1998, p. 208). The extreme realization of the commodification of news is the practice of accepting stories produced by various public relations organizations as if they were news. News therefore actively participates in promotional campaigns for consumer products, political policies and cultural values.

The use of on-air talent is a crucial way that newscasts achieve this integrated approach and brand it. Anchors and reporters are framed as people viewers trust via a parasocial relationship, or pseudo-friendship (Perse, 1990). One way to develop the friendship is through the 'market-driven innovation' known as cross chat, or 'happy talk', where talent chit-chats on set in a move to make viewers feel comfortable with those on set (Hertsgaard, 1996). News consultants embraced the practice, which first evolved in local news, as a way to help promote the news team to audiences as a 'family', developing a sense of intimacy with viewers (Campbell, Martin & Fabos, 2005; Rich, 2004). At its worst, happy talk is less about promoting legitimate news, and more about increasing ratings and station profitability (Rich, 2004; Laurence, 2005).

Television news and current affairs programming are components of the overall marketization of information, with success measured in market criteria including efficiency, profit-making and customer maximization. As a result, audiences are targeted as consumers. In the following section, we address the potentialities for new media and developing journalistic forms to provide a way to question the commercialization of journalism and lead to a return to forms of active debate.

CITIZEN JOURNALISM AND NEW MEDIA

The very definition of what a journalist is has been thrown into question. Nerone's (1995, p. 157) 'journalist speaking to and for citizens' may face a divided audience – one that is more attracted to popularization and is to fuller participation. Besides professional journalism, there is an evolving labor source that has been referred to as 'citizen', 'collaborative', or 'participatory' journalism. The United States' PBS network describes citizen journalists as 'people without professional journalism training [who] can use the tools of modern technology and the global distribution of the Internet to create, augment or fact-check media on their own or in collaboration with others' (Glaser, 2006). But not everyone is satisfied with the terminology: are 'real' journalists somehow not citizens? What about people who don't live in the country that is the topic of their reporting (Glaser, 2006)? Other terms have also emerged, such as bottom-up journalism, unfiltered journalism or collective journalism. CNN, in reference to the user-generated content in its iReport platform, warns audiences, 'the stories [in this section] are not edited, fact-checked or screened before they post' (CNN, 2009). In other words, proceed into practicing or consuming citizen journalism at your own risk.

According to Schaffer (2005), writing for Harvard University's *Nieman Reports*, some mainstream journalists react to citizen journalism with skepticism and hostility, questioning whether they do 'real journalism', their accuracy, credibility and ethics, and their ability to make money from their journalistic endeavors – or, even worse, whether citizen journalists are going to siphon off money from mainstream journalistic outlets. At the same time, Schaffer (2005, p. 24) claims 'citizen journalists don't particularly aspire to be called "journalists"'. Instead, they see themselves as reacting to, or contributing to, the news.

The explosion of these citizen reporters has coincided with a development of highly portable handheld technology. Initially, it was the user- (and YouTube-) friendly Flip Mino camera; now, smart phones, which allow users to quickly use text messaging or post high-definition video content to the web over cellular phone signals, or rugged Go Pro cameras are the personal portable video systems of choice. Much of the early reporting offered a response to a tragedy or calamity, such as the personal accounts posted on the web in the wake of the September 11, 2001 attacks on the United States. The reporters initially offered 'sincere, unapologetically subjective expression to their experiences … the sending of such messages had something of a cathartic effect', a stark contrast to the reporting done by 'professional' media in similar events (Allan, 2009, p. 24). But by the time of the 2005 subway bombing in London, and the devastation wrought by Hurricane Katrina a month later, video citizen journalism came into its own. The BBC received more than 20,000 e-mails, 1,000 photos and 20 videos in just the first six hours after the subway bombing (Schaffer, 2005). Mobile telephones captured images of the events unfolding quicker and arguably more effectively than traditional broadcast journalists. 'Video clips taken with cameras were judged to be all the more compelling because they were dim, grainy and shaky, and – even more important – they were documenting an angle to an event as it was actually happening' (Allan, 2009, p. 29). The video was generated by ordinary people, covering an aspect of the story that could not be covered by professionals due to spatial constraints, delays in reaching the site of the news event or lack of access to the site, and then forwarded and replayed by professional news organizations such as the BBC or CNN.

Because television news (and often its associated web content) rely on the visual, photographs and video, often gathered through cell phone technology, have become especially important to coverage when access is delayed or denied. However, e-mail messages and text- and photo-based posts known as 'tweets' on Twitter have become increasingly important to reporting TV news by helping to provide information about an event to which the professional media do not have immediate access, as well as by helping to create a bond between reporters and anchors and their viewers.

For the most part, Western-based media use Facebook, Twitter, iReport and the like as a way to reach out to the public. News organizations such as the US-based ABC have multiple Twitter accounts where individual reporters and programs offer periodic story updates or promotions about upcoming programming. Even local stations promote their Facebook pages and ask viewers to 'like' the page; the pages offer news content but also provide a space where viewers can post video and photographs of news events. Moreover, citizen journalists are useful to news outlets because they enhance the value of the human interest angles that accompany 'hard' news.

In some countries, however, social media may offer a more resistant function. Social media can act as a watchdog, such as in China where it is used to report information which is not forthcoming from government-sponsored organizations or to criticize reporting about China by non-Chinese media (Reese & Dai, 2009). It can act as a way to transform a top-down government-driven news agenda to one that more reflects the interests and concerns of the public. An example is India's embrace of social media and citizen journalism, allowing the public to determine story content and order on nightly newscasts (Sonwalker, 2009).

In a time of such profound change it can be difficult to prognosticate about the future of news. But 2009's presidential election in Iran offers some hint of what the new news media landscape might look like. After people began voting in the June 12th election, the Iranian government began to restrict media reporting of the event, raiding news offices, confiscating footage and even attempting to jam broadcast signals and block internet access. Circumstances surrounding the 2009 Iranian election have been described as 'the perfect setup for spotlighting the strengths of Twitter, Facebook and YouTube, particularly when the professional media [were] hobbled' (Paiser, 2009, p. 42). Protestors took to the internet, using Twitter and proxy internet servers located outside of Iran to post short (140 characters or less) updates about their experiences and thoughts on the contested vote. The term '#Iran election' was the Twitter's top trending topic of 2009, with Iran and Tehran in spots four and five (Twitter Blog, 2009). Television networks worldwide reported on the election, making use of specific Twitter correspondents and reporting on the trend of Twitter as a reporting tool (Potter, 2009). But the vote was also tracked by hundreds of thousands of those who didn't work for a major television news organization via nontraditional news sources. CNN, criticized by professional journalists for an overreliance on iReporters during the election aftermath, employed Farsi speakers to help vet the veracity of citizen postings. The relationship between CNN and its (unpaid) citizen journalists 'complicates the static maps of the world that professional journalism attempts to construct' (Palmer, 2013, p. 383), specifically because of the seeming need to 'vet' amateur content while not applying those same standards to professionally-produced video. These nontraditional sources are changing the way the media reports – and the public consumes – news.

The Iran election offers a case study of a work in progress: the evolution of a movement away from the gatekeeping model of news production and to a new formula that Bruns refers to as gatewatching. In gatewatching, collaboration is the key, with the audience having access to the same sources and information as traditional news outlets and often contributing to the newsgathering, reporting and editing process (Bruns, 2008). Muddiman and Meier (2013) argue that this type of citizen involvement will help television news organizations to be more responsive to the public, increase the quality of the news product and issues covered, and contribute to the public sphere.

There are, of course, instances of irresponsible or erroneous reporting by 'average' citizens. For instance, CNN's iReport portal featured a story in late 2008 that Apple's (now late) CEO Steve Jobs had suffered a major heart attack. The story was untrue (and was never picked up by the network or listed as a network-recommended story), but nonetheless it caused significant damage, both to Apple stock prices (which plunged to a 17-month low) as well as to CNN's reputation (Allan & Thorsen, 2009, pp. 2–3). But for every negative example, there are examples of how citizen reporters helped to advance the public understanding of a news story. At CNN, iReporters contributed video about California wildfires, Midwestern floods and the Virginia Tech shooting. At Virginia Tech, a student's cell phone video footage caught the sound of gunshots from inside a nearby building (Allan & Thorsen, 2009, p. 2). At the BBC, user-generated content has become part of its newsgathering process, which the BBC actively solicits from its audience. But perhaps one of the most important stories influenced (and some would argue dramatically improved) by nonprofessional video is the ongoing Arab Spring. As Miladi observes, 'The camera-bearing citizen or the citizen journalist became the witness and sometimes a marker of the history of our time' (2011, p. 118). While in the past control of state-run media and seizing buildings were necessary for political coups, the Arab Spring largely has been an online revolution, run 'remotely via YouTube, Facebook, Twitter and satellite TV' (Miladi, 2011, p. 118). Similarly, during the same year, the gruesome assassination of Muammar Gaddafi was captured by non-professionals on the scene (Andén-Papadopoulos, 2013), and the 2012 protests in Nigeria would have gone largely uncovered were it not for citizen journalists (King, 2012). Citizen reporters offer more than just the 'accidental journalism' performed by individuals 'in the wrong place at the wrong time compelled to adopt the role of reporter' (Allan & Thorsen, 2009, p. 7). The trend is also 'a positive opportunity for journalism to improve in a way that reinforces informed citizenship' (p. 5).

Of course, the individual contribution has been a constant in the modern news gathering process, from letters to the editor to citizen 'tip lines' for story ideas. 'What perhaps is "new" about it is that it happens in the bloodletting of creativity, or, in other words, of outsourcing salaried labor in newswork to unpaid volunteers, and interpellating the aforementioned citizen-consumer as a citizen-colleague without necessarily investing in training, monitoring, moderating, or protecting arrangements' (Deuze, 2009, p. 259). This raises an important question: how does news gathering change in a society in which professional media continue to decline and reporting is left more and more to citizen journalists whose main credentials may consist of having access to new media devices for gathering and disseminating information?

THE CURRENT SITUATION OF NEWS LABOR

Political economy has spent a great deal of time describing and analyzing the concentration of ownership of the media. Bagdikian (2004) and McChesney (1999) have monitored the processes of concentration, conglomeration and integration in the media industry. For instance, the number of corporations dominating the United States media market shrank from an estimated fifty in 1983 to six in 2002, resulting in an oligopolistic market pattern where large vertically and horizontally integrated firms 'compete against each other, sell to each other and have co-ownership, revenue sharing, co-production and co-purchasing deals with each other' (Curran, 2002, p. 229). But as Mosco (1996, p. 198) puts it, 'political economy includes recognizing that, though important, corporate size and concentration are just starting points for understanding the transformation of the communications business'. The suggestion that all we need to do is decentralize the sources of the news has not been thought through clearly. Instead, we might focus on how the present structure and organization of television news affect the positions and practices of working journalists.

Television news is in the conflicted position of claiming or attempting to serve the public good at the same time that it must produce commodities for the market. A reporter is supposed to apply professional skills in producing a story that has use value for the reader and for society in general. The reporter, however, is a wage laborer and must sell their labor power on the market. 'Capital turns that labor power into a news package which, along with other stories and advertising, forms a packaged product' (Mosco, 1996, p. 146).

The intensified concentration of news media ownership is accompanied by corporate and technological convergence in which once separate news organizations increasingly are integrated with the objective of achieving harmony in both 'journalistic and business realms, in different media markets, even across distinct media (especially newspapers, TV, and internet)' (Skinner, Compton & Gasher, 2005, p. 8). Corporate and technological convergence demands that reporters engage in multitasking, repackaging and filing their stories to broadcast, print and internet news outlets.

Deuze (2008, p. 861) calls for an understanding of how, in digital culture, the roles, identities and activities of people are constantly shifting, thus calling into question what 'being a journalist' means and what this ambiguity, in turn, means in respect to the lives and labor of newsworkers. Reporters and crews have been known to use mobile phones and other portable video cameras (e.g. Flip, Go Pro) in reporting on breaking news stories in place of larger and more obtrusive traditional cameras (Palmer, 2013, p. 373). The result is that reporters may have less time to work on the newsgathering, interviewing, researching and writing that go into reporting, and this may affect the quality of the news product. A reporter who resists multitasking may be risking her or his career (Compton, 2004, p. 124).

Since the 1990, especially in Western cultures, downsizing of news labor has been the norm. The trend is particularly noticeable in coverage of foreign affairs, specifically by American television networks, which have cut the number of foreign correspondents operating from international offices. Maintaining foreign correspondents is an expense that most television networks cannot afford. Instead of maintaining correspondents in locations, the practice is now to send in reporters when an 'event occurs'. This, of course, reduces what is covered to particular types of events and it reduces the chances that the reporter will understand the deeper meaning of the region covered. This has been dubbed 'parachute journalism'.

The resulting news labor force is often made up of a decreased number of permanent staffers and an increased per diem staff (Gall, 2000; Deuze, 2007). The 'daily hire' (or per diem) worker was officially acknowledged in US news labor contracts in 1987, when television news unions agreed that networks and major market stations could hire a small number of freelancers; that number is currently about 60% of union members (Ryan, 2008). What began as a way to help cut news budgets in a time of financial stress has evolved into what the Dean of Columbia University's Graduate School of Journalism calls a 'culture of freelance' in markets of various sizes (Goldstein, 1999). This trend isn't limited to the United States. News freelancers are found in most Western countries, including the UK, Germany and Australia (Ursell, 2003; Ertel et al., 2005; Das, 2007). The model is similar to one employed by the film industry, where temporary workers are hired on a per project basis, which can last a day or several months (Ryan, 2009).

On the one hand, like most workers, freelancers are vulnerable to changing economic conditions, which can affect their pay, job security and benefits (Deuze, 2007). Freelancers and stringers are sometimes employed as a 'cheaper solution for news organizations to maintain their … coverage [generally of international stories] in the face of increasingly tight editorial budgets' (Hess, 1994). On the other hand, workers develop strategies to cope with the unpredictability of freelance life (Storey, Salaman & Platman, 2005; Ryan, 2009). A large number of freelancers acknowledge the precariousness of their work situation, but they find it 'preferable, both fiscally and psychologically, to "being paid not to leave" a miserable staff job' (Ryan, 2009, p. 658). An example of this is in the notion of being 'booked', where a freelancer reports having another assignment when called to cover an unpleasant or inconvenient story. In other words, freelancers negotiate their work, transforming instability (lack of day-to-day work) into a positive (freedom to pick and choose assignments).

A similar type of negotiation can be found in staff correspondents. For example, one study illustrates how the aforementioned Ann Curry uses her position as news anchor of NBC's *Today* as a form of leverage to encourage the network to cover socially conscious issues. She describes covering less weighty news stories during the course of her job, such as profiles of the McCaughey septuplets or the 'Octomom', as 'earning coupons' (Lake, 2008). Curry later 'redeems' these coupons to cover issues the network might otherwise ignore, such as her extensive in-depth reporting of the genocide in Darfur. This explanation potentially complicates the notion that news always is driven by the 'bottom line', only covering those stories which appeal to the largest possible audience.

CONCLUSION

The study of television news and current affairs is replete with different approaches, from mainstream research to cultural/critical studies. Gatekeeping, agenda-setting, framing and content analysis continue to be popular modes of conducting research, while critical scholars have pursued analyses based in semiotics, discourse analysis, ideological analysis and audience ethnography, and have conducted political economic studies of media corporations and labor. Research into the changing practices of newswork and the experiences of newsworkers would seem to require some combination of ethnography and political economy. Nevertheless, with a few exceptions, there are few studies that use both methodologies to illuminate newsworkers' lives and perspectives in the context of the contemporary television news landscape. Only a handful of researchers have spent extended time in newsrooms (including in Denmark, Japan and the United States); for the most part, ethnographic approaches are a rarity when it comes to studying television news. Instead, researchers rely on quantitative data, considering things like the job satisfaction of television newsworkers through surveys (Price and Wulff, 2005; Ryan, 2009).

Combining ethnography with political economy would have the advantage of providing grounded, empirical, theoretically-informed answers to questions of how newsworkers have fared given the reduction of the number of correspondents by television news organizations, the decrease in coverage of world news, the increase in spectacle and human interest in journalism, the decline of audiences for network news, and cable news' tendency to compete for audiences by creating current affairs programming based in ideological sparring. Asking such questions also would push us beyond simplistic statements about the 'dumbing down' of television news and shed some much-needed light on the experiences of the workers who create our news.

NOTES

1 Much of this expansion has come as local organizations are consolidating newsrooms, sharing stories but using different anchor talent on air (Topper and Turner, 2013).
2 The rankings are based on number of households, in millions.

REFERENCES

Allan, S. (2009). Histories of citizen journalism. In S. Allan & E. Thorsen (eds), *Citizen journalism: Global perspectives* (pp. 17–31). New York, NY: Peter Lang.

Allan, S. & Thorsen, E. (2009). Introduction. In S. Allan & E. Thorsen (eds), *Citizen journalism: Global perspectives* (pp. 2–3). New York, NY: Peter Lang.

Altschull, H. (1995). *Agents of power.* White Plains, NY: Longman.

Andén-Papadopoulos, K. (2013). Media witnesses and the 'crowd-sourced video revolution'. *Visual Communication*, 12(3): 341–357.

Auletta, K. (1992). Three blind mice: How the TV networks lost their way. New York, NY: Vintage Books.

Bagdikian, B. (2004). *The new media monopoly.* Boston, MA: Beacon Press.

Baym, G. (2005). The Daily Show: Discursive integration and the reinvention of political journalism. *Political Communication*, 22(3): 259–276.

Bek, M.G. (2004). Research note: Tabloidization of news media, an analysis of television news in Turkey. *European Journal of Communication*, 19(3): 371–386.

Bellamy, R.V., Jr. & Traudt, P.J. (2000). Television branding as promotion. In S.T. Eastman (ed.), *Research in media promotion* (pp 127–159). Mahwah, NJ: Lawrence Erlbaum.

Berkowitz, D. (1990). Refining the gatekeeping metaphor for local television news. *Journal of Broadcasting and Electronic Media*, 34(1): 55–68.

Bird, S.E. (2009). Tabloidization: What is it, and does it really matter? In B. Zelizer (ed.), *The changing faces of journalism: Tabloidization, technology, and truthiness* (pp. 40–50). London, England: Routledge.

Bird, S.E. & Dardenne, R.W. (1997). Myth, chronicle and story: Exploring the narrative qualities of news. In D. Berkowitz, *Social Meanings of News: A Text Reader* (pp. 333–350). London: Sage Publications.

Bliss, E. (1991). *Now the news*. New York, NY: Columbia University Press.

Briggs, A. & Burke, P. (2005). *A social history of the media: From Gutenberg to the internet*. Cambridge, UK: Polity Press.

Bruns, A. (2008). The active audience: Transforming journalism from gatekeeping to gatewatching. In C. Paterson & D. Domingo (eds), *Making online news: The ethnography of new media production* (pp. 171–184). New York, NY: Peter Lang.

Campbell, R., Martin, C.R. & Fabos, B. (2005). *Media and culture 5: An introduction to mass communication* (5th edn). Boston, MA: Bedford/St Martin's.

Campbell, V. (2004). *Information age journalism*. London, England: Arnold.

CNN (Producer) (2009). About CNN iReport. *CNN iReport*. Retrieved from http://www.ireport.cnn.com/about.jpsa (accessed 31 July 2014).

CNN (Producer) (January 24, 2010 Sunday). *Reliable sources: State of the union with John King* Retrieved from www.lexisnexis.com/hottopics/lnacademic (accessed 31 July 2014).

Compton, J.R. (2004). The integrated news spectacle: A political economy of cultural performance. New York, NY: Peter Lang.

Conboy, M. (2004). *Journalism: A critical history*. London, England: Sage Publications.

Crowley, C. & Potter, D. (2005). An experiment changed: TV coverage of the 2004 campaign. *Journalism Studies*, 6(2), 233–236.

Curran, J. (2002). *Media and power*. London, England: Routledge.

Das, J. (2007). Sydney freelance journalists and the notion of professionalism. *Pacific Journalism Review*, 13(1): 142–160.

Deuze, M. (2007). *Media work*. Cambridge, UK: Polity.

Deuze, M. (2008). The changing context of news work: Liquid journalism and monitorial citizenship. *International Journal of Communication*, 2, 848–865.

Deuze, M. (2009). The future of citizen journalism. Introduction. In S. Allan & E. Thorsen (eds), *Citizen journalism: Global perspectives* (pp. 253–263). New York, NY: Peter Lang.

Djerf-Pierre, M. & Lofgren-Nilsson, M. (2003). Gender-typing in the newsroom: The feminization of Swedish television news production, 1958–2000. In M. De Bruin & K. Ross (eds), *Gender and newsroom cultures: Identities at work* (pp. 81–106). New York, NY: Hampton Press.

Entman, R.M. (1993). Framing: Toward clarification of a fractured paradigm. *Journal of Communication*, 43(3): 51–58.

Ertel, M., Pech, E., Ullsperger, P., von dem Knesebeck, O. & Siegrist, J. (2005). Adverse psychosocial working conditions and subjective health in freelance media workers. *Work and Stress*, 19(3): 293–299.

Freeman, L. (1999). On TV, image is everything. *Electronic Media*, 18 (23): 18.

Gall, G. (2000). New technology, the labor process and employment relations in the provincial newspaper industry. *New Technology, Work and Employment*, 15(2): 94–107.

Gans, H. (2009). Can popularization help the news media? In B. Zelizer (ed.), *The changing faces of journalism: Tabloidization, technology, and truthiness* (pp. 17–28). London, England: Routledge.

Gitlin, T. (1980). *The whole world is watching: Mass media in the making and unmaking of the new left*. Berkeley, CA: University of California Press.

Glaser, M. (2006). Your guide to citizen journalism. *Media shift: Your guide to the digital media revolution*, 27 September 2006. Retrieved from http://www.pbs.org/mediashift/2006/09/your-guide-to-citizen-journalism270.html (accessed 22 December 2009).

Goffman, E. (1974). *Frame analysis: An essay on the organization of experience*. Cambridge, MA: Harvard University Press.

Goldstein, T. (1999). Speech to the Society of American Travel Writers. *O'Dwyer's PR Services Report*, December: 41.

Hallin, D. (1996). Commercialism and professionalism in the American newsmedia. In J. Curran & M. Gurevitch (eds), *Mass media and society* (pp. 243–264). London, England: Arnold.

Hardt, H. (1998). *Interactions: Critical studies in communication, media, and journalism*. Lanham, MD: Rowan and Littlefield.

Hertsgaard, M. (1996). Media matters. *Nation*, 263(4): 10.

Hess, S. (1994). The 'cheaper solution.' *American Journalism Review*. Retrieved from http://ajrarchive.org/article.asp?id=1377 (accessed 31 July 2014).

Hess, S. (1996). Media mavens. *Society*, 33(3): 70–78.

Kerbel, M.R. (2000). *If it bleeds, it leads: An anatomy of television news*. Boulder, CO: Westview Press.

King, L. (2012). Vetting citizen journalism. *Neiman Reports*, Summer. Retrieved from http://www.nieman.harvard.edu/reports/article/102767/Vetting-Citizen-Journalism.aspx (accessed 31 July 2014).

Lake, H.A. (2008). Gender, race, commercialism, and news values in television: An ethnographic study of NBC News anchor and correspondent Ann Curry at work. Unpublished PhD dissertation, University of Oregon.

Laurence, R.P. (2005, January 7). Which is worse, the devastation or the shameless self-promotion? *The San Diego Union-Tribune*. Retrieved from http://legacy.utsandiego.com/uniontrib/20050107/news_mz1c7remote.html (accessed 31 July 2014).

Mattelart, A. (1979). For a class analysis of communication. In A. Mattelart & S. Siegelaub (eds), *Communication and class struggle: 1. Capitalism, imperialism* (pp. 23–70). New York, NY: International General.

McChesney, R. (1999). *Rich media, poor democracy.* New York, NY: The New Press.

McChesney, R. (2004). *The problem of the media.* New York, NY: Monthly Review Press.

McCombs, M. (1992). Explorers and surveyors: Expanding strategies for agenda setting research. *Journalism and Mass Communication Quarterly* 69: 813–824.

McCombs, M. & Ghanem, S. (2001). The convergence of agenda setting and framing. In S.D. Reese, O.H. Gandy & A.E. Grant (eds), *Framing public life: Perspectives on media and our understanding of the social world* (pp. 68–81). London, England: Lawrence Erlbaum Associates.

McCombs, M.E. & Shaw, D.L. (1972). The agenda-setting function of mass media. *Public Opinion Quarterly*, 36: 176–187.

McCombs, M.E., Shaw, D.L. & Weaver, D.H. (1997). *Communication and democracy: Exploring the intellectual frontiers in agenda setting.* Mawhah, NJ: Lawrence Erlbaum Associates.

Miladi, N. (2011). New media and the Arab revolution: Citizen reporters and social activism. *Journal of Arab & Muslim Media Research*, 4(2–3): 113–119.

Mosco, V. (1996). *The political economy of communication*. Thousand Oaks, CA: Sage.

Mott, F.L. (1952). *News in America*. Cambridge, MA: Harvard University Press.

Muddiman, A. & Meier, M.R. (2013). Deliberative television: Encouraging substantive, citizen-driven news. *Journal of Public Deliberation* 9(2, Article 12). Retrieved at http://www.publicdeliberation.net/jpd/vol9/iss2/art12/.

Negt, O. & Kluge, A. (1993). *Public sphere and experience*. Minneapolis, MN: University of Minnesota Press.

Nerone, J. (ed.) (1995). *Last rights: Revisiting four theories of the press*. Urbana, IL: University of Illinois Press.

Paiser, P. (2009). Amateur content's star turn. *American Journalism Review*, 31(4): 42.

Palmer, L. (2013). 'iReporting' an uprising: CNN and citizen journalism in network culture. *Television and New Media*, 14: 367–385.

Perse, E.M. (1990). Media involvement and local television news effects. *Journal of Broadcasting and Electronic Media*, 34: 17–36.

Pew Project for Excellence in Journalism (2004). *The state of the news media: An annual report on American journalism: Network content analysis*. Available online at http://stateofthemedia.org/2004/network-tv-intro-2/content-analysis/

Pew Project for Excellence in Journalism (2008). *The state of the news media: An annual report on American journalism.* Retrieved from http://stateofthemedia.org/2008/ (accessed 15 November 2009).

Pew Project for Excellence in Journalism (2009). *The state of the news media: An annual report on American journalism.* Retrieved from http://stateofthemedia.org/2009/ (accessed 15 November 2009).

Pew Research Center for the People and the Press (2010). *Ideological news sources: Who watched and why, Americans spending more time following the news.* Retrieved from http://www.people-press.org/2010/09/12/americans-spending-more-time-following-the-news/ (accessed 21 December 2013).

Potter, N. (2009 June 19). Iran election: Online support for dissidents. *ABCnews.com.*

Retrieved from http://abcnews.go.com/Technology/story?id=7845377 (accessed 19 December 2009).

Price, C.J. & Wulff, S.S. (2005). Does sex make a difference? Job satisfaction of television network news correspondents. *Women's Studies in Communication* 28(2): 207–234.

Rao, S. & Johal, N.S. (2006). Ethics and news making in the changing Indian mediascape. *Journal of Mass Media Ethics*, 21(4): 286–383.

Reese, S.D. & Dai, J. (2009). Citizen journalism in the global news arena: China's new media critics. In S. Allan & E. Thorsen (eds), *Citizen journalism: Global perspectives* (pp. 221–231). New York, NY: Peter Lang.

Rich, F. (2004). Happy talk news covers a war. *New York Times*, 153(52914): 1–7.

Ryan, K.M. (2008). Temporary resistance: Strategies of freelancers in American news news. In R. Sickels (ed.), *The business of entertainment: Television* (pp. 125–142). Westport, CT: Praeger Press.

Ryan, K.M. (2009). The performative journalism: Job satisfaction, temporary workers and American television news. *Journalism*, 10(5): 647–664.

Schaffer, J. (2005). Citizens media: Has it reached a tipping point? In *Nieman Reports*, 59(4): 24.

Schultz, I. (2006). *Bag nyhederne. voedier, idealer og praksis [Behind the news. values, ideals, practice]*. Copenhagen, Denmark: Samfundslitteratur.

Scott, D. & Gobetz, R.H. (1992). Hard news/Soft news content of the national broadcast networks, 1972–1987. *Journalism Quarterly*, 69: 406–412.

Shoemaker, P.J. (1996). Media gatekeeping. In M.B. Salwen & D.W. Stacks (eds), *An integrated approach to communication theory and research* (pp. 79–92). Mahwah, NJ: Lawrence Erlbaum Associates.

Skinner, D., Compton, J.R. & Gasher, M. (2005). *Converging media, diverging politics: A political economy of news media in the United States and Canada*. Lanham, MD: Rowman and Littlefield.

Smith, A. (1978). The long road to objectivity and back again: The kinds of truth we get in journalism. In G. Boyce, J. Curran & P. Wingate (eds), *Newspaper history: From the 17th century to the present day* (pp. 153–171). London, England: Constable.

Sonwalker, P. (2009). Citizen journalism in India: The politics of recognition. In S. Allan & E. Thorsen (eds), *Citizen journalism: Global perspectives* (pp. 75–84). New York, NY: Peter Lang.

Sparks, C. (1991). Hildy Johnson: The vanishing 'serious press'. In P. Dahlgren & C. Sparks (eds), *Communication and citizenship: Journalism and the public sphere* (pp. 58–74). London, England: Routledge.

Sparks, S. & Tulloch, J. (2000). *Tabloid tales: Global debates over media standards*. Lanham, MD: Rowman and Littlefield Publishers.

Storey, J., Salaman, G. & Platman, K. (2005). Living with enterprise in an enterprise economy: Freelance and contract workers in the media, *Human Relations*, 58(8): 1033–1054.

Takeshita, T. (1997). Exploring the media's roles in defining reality: From issue-agenda setting to attribute-agenda setting. In M.E. McCombs, D.L. Shaw & D.H. Weaver (eds), *Communication and democracy: Exploring the intellectual frontiers in agenda setting* (pp. 15–28). Mawhah, NJ: Lawrence Erlbaum Associates.

Thomson, E.A., White, P.R. & KItley, P. (2008). 'Objectivity' and 'hard news' reporting across cultures: Comparing the news report in English, French, Japanese, and Indonesian journalism. *Journalism Studies* 9(2): 212–228.

Thussu, D.K. (2009). *News as entertainment: The rise of global infotainment*. Thousand Oaks, CA: SAGE Publications.

Topper, J. & Turner, S.D. (2013 November 6). Something's happening to local news. *The Baltimore Sun*. Retrieved from http://articles.baltimoresun.com/2013-11-06/news/bs-ed-sinclair-20131106_1_cunningham-broadcasting-fcc-rules-allbritton-communications (accessed 31 July 2014).

Tuchman, G. (1978). *Making news: A study in the construction of reality*. New York, NY: The Free Press.

Turner, G. (1999). Globalization, journalism and the possibility of critique. *International Journal of Cultural Studies*, 2(1): 59–76.

Twitter Blog (December 15, 2009). Top twitter trends of 2009. Retrieved from http://blog.twitter.com/2009/12/top-twitter-trends-of-2009.html (accessed 19 December 2009).

Ursell, G. (2003). Creating value and valuing creation in contemporary U.K. television: Or 'dumbing down' the workforce. *Journalism Studies*, 4(1): 31–46.

Waldman, S. & the Working Group on Information Needs. (2011, July). *The Information Needs of Communities*. Federal Communications Commission. Retrieved from www.fcc.gov/infoneedsreport (accessed 31 July 2014).

White, D.M. (1950). The gatekeeper: A case study in the selection of news. *Journalism Quarterly*, 27: 383–390.

Winston, B. (2002). Towards tabloidization? Glasgow revisited, 1975–2001. *Journalism Studies*, 3(1): 5–20.

Zillmann, D. & Gibson, R. (1994). Effects of upbeat stories in broadcast news. *Journal of Broadcasting and Electronic Media*, 38(1): 65–78.

20

Music on Television

Matthew Delmont

YouTube offers many examples of music on television, but two of my favorite clips are from *Seventeen*, a teen dance show broadcast by WOI-TV to central Iowa in the late 1950s. The videos feature awkward teenagers gamely dancing to 'The Stroll' by the Diamonds and to 'Betty and Dupree' by 'The King of the Stroll', Chuck Wills (The Original Stroll – February 1958, 2013; The Stroll, 2013). The Stroll, a new take on swing-era line dances where boys and girls faced each other in two parallel lines and couples took turns strutting down the aisle, was inspired by Wills's version of the popular blues song 'C.C. Rider' and popularized by the Diamonds, a white vocal group that frequently recorded cover versions of black R&B songs. In dancing The Stroll the teens on *Seventeen* were emulating their peers in Philadelphia who popularized the dance on the nationally broadcast *American Bandstand*. Less obviously, the Iowa teens were also emulating teens on *The Mitch Thomas Show*, a black teen dance show that broadcast locally to the Philadelphia area, whose version of The Stroll influenced the *American Bandstand* dancers. Unbeknownst to the teens on *Seventeen* and *The Mitch Thomas Show*, they shared a connection across regional and racial lines as television music performers and viewers.

Iowa may not be the first place that comes to mind when thinking about music television, but understanding how Iowa teenagers learned to dance The Stroll via the local *Seventeen*, the national *American Bandstand*, and the local (but far from Iowa) *Mitch Thomas Show* highlights several of the important issues related to music on television this chapter addresses, such as the relationships among local, national and transnational sites of production and reception; the shifting aesthetic and economic demands of music television production; the frequent intersections of performance and viewership in music television; and the centrality of YouTube and other online video sites to engagements with the history, present and future of music television. I also start this chapter with *Seventeen* to emphasize the challenge of surveying the vast and often precariously maintained history of music television. Iowa State University's Special Collection holds the only known kinescope of *Seventeen*, but most examples of music on television were not preserved or archived and, as a result, scholars need to explore an array of sources to study the intersections of music and television. Fortunately, music television scholars are making use of a creative and interdisciplinary array of sources. The following pages highlight several examples of this innovative scholarship, the frameworks they offer for studying music on television, and potential new areas for research.

Throughout the essay I use 'music on television' and 'music television' to describe live and recorded musical performances on television, music variety programs, dance shows, music videos, and the circulation of these forms on the Internet and mobile screens in the 'post-television' era. Referring to music 'on' rather than 'in' television signals my focus on diegetic music, rather than the program underscoring, network/station music, or advertising music examined by scholars such as Ron Rodman (2010) and James Deaville (2011). On the ordering of the terms 'music' and 'television', Norma Coates (2007b) argues persuasively for 'a conceptual shift to the study of *television music*, not music television', in order to force 'scholars to think about television music as television, to foreground the corresponding industrial, economic, programming, production, and business aspects of the medium as informed by the engagement with popular music' (p. 23). Like Coates, this chapter attends to the history and present of music on television without reading this complex range of examples through a single network (MTV) or a single form (music video). Since television scholars are this collection's primary audience, however, I have elected to use 'music television' to give the terms equal analytic standing, which I take to be central to Coates's argument.

While Elvis Presley's performances on *The Ed Sullivan Show* in 1956 and 1957 are the best-known examples of popular music on early television, music has figured prominently on television since the medium's earliest days. Murray Forman (2012) explores how, as television infrastructure took shape in the 1930s and 1940s, producers experimented with a range of musical programming. Analyzing corporate correspondence, memoranda, program scripts and photographs, Forman shows how television executives, programmers and engineers struggled to develop visually and commercially appealing musical performances that could sell the idea of television to the public. In addition to emphasizing artists' visual appearance, early television producers also paid close attention to the use of camera placement, shot composition, and lighting to frame these musical performances. 'Although television had yet to fully mature', Forman argues, 'by 1935 the conceptual framework and representational principles that eventually came to dominate were already discernible and many of the general qualities inscribed in this early "TV laboratory" trial remained more or less intact when television was finally introduced on a massive commercial scale at the end of the 1940s' (2012, p. 26). Like Forman, Keith Negus (2006) demonstrates that musical performances were also central to early programming in the UK, as jazz pianists Fats Waller and Art Tatum, vaudevillian Sophie Tucker, and over a dozen operas appeared on BBC Television in the late 1930s (pp. 311–312).

As television sales took off in the late 1940s, musicians participated in television's commercial ascent with a mix of optimism and anxiety. The opportunity to reach vastly larger audiences appealed to musicians like jazz trumpeter and trombonist Kirby Stone, who told *Down Beat* magazine: 'We could have knocked around in clubs for 10 years and never have been seen by the number of people who have seen us on television. One night on TV is worth weeks at the Paramount' (as cited in Forman, 2012, p. 57). At the same time, 'doubling' on television and nightclub stages proved taxing for musicians, and television's promotional power, its ability to elevate careers and break new artists, was never accessible to the majority of musicians. Still, early television viewers saw and heard a wide variety of musical performances that encompassed and gave shape to several genre categories, such as 'amateur talent shows, big band "cavalcades" or "revues", DJ and teen dance shows, "hillbilly"/western variety, hit parade shows, musical quiz shows, and musical variety' (Forman, 2012, p. 115). Broadway, as Kelly Kessler (2013) argues, was among the major sources for early television, with over 100 original musicals and Broadway adaptations appearing on US television by the end of the 1950s. Like the musical medleys in many of these Broadway adaptations, Gary Burns (1998) discusses how *Your Hit Parade* (1950–1959) featured a cast of vocalists who performed the most popular songs of the day, and Heather McIntosh (2004) offers examples of similar programs, in which actors pantomimed skits to contemporary hits, that originated on Cincinnati's WCPO-TV and were picked up for national broadcast by ABC and DuMont.

These histories of early television show how conventions for showing popular music on television, and myths surrounding these presentations, took shape. Singers and musicians had to be taught how to project their performances, often without a live audience, to television viewers hundreds or thousands of miles away. Television producers, for their part, had to figure out how to use lighting, staging, camera framing and visual effects to maximize these performances. Despite numerous errors and miscues in these early, and usually live, productions, this early history of television demonstrates that music television was well established before Elvis Presley broke into television with several appearances on the Dorsey Brothers' *Stage Show* in 1956. The most popular story about early music television production practices is that Ed Sullivan only allowed Presley to perform on *The Ed Sullivan Show* if the cameras stayed above waist level and did not show Presley's gyrating hips. Norma Coates (2007a) demonstrates that this story, which ignores Presley's numerous television engagements in 1956 and the fact that his full body was on show in two previous appearances on *The Ed Sullivan Show*, developed as a foundational music television myth to explain the mainstreaming of rock and roll in the late 1950s. Coates show how this myth supports a separation of 'authentic' rock and roll (favored by male rock critics) from 'commercial' rock and roll on television (favored by teenage girls) that, despite historical evidence to the contrary, continues to frame many popular histories of rock and roll.

Among the limitations of this masculinist rock and roll discourse is that it obscures the more nuanced ways television made music meaningful for audiences. *American Bandstand* (1952–1957 local; 1957–1989 national), one of the television shows most often blamed for sanitizing rock and roll, used extended close-ups on dancers' feet so that viewers at home could follow along and learn the dances steps (Delmont, 2012). *American Bandstand* depended on the creative energies of black teenagers, such as the aforementioned *The Mitch Thomas Show* dancers who popularized The Stroll, but *American Bandstand* discriminated against black teenagers during its years in Philadelphia, counter to host Dick Clark's claim to the contrary. Far from trivial, scholars like Tim Wall (2009) and Julie Malnig (2006) have argued that these televised teen dances were important sites of cultural sharing and appropriation and were crucial to making rock and roll and rhythm and blues part of the daily lives of teenagers. Locally broadcast black teen dance shows, like Bob King's *Teenarama Dance Party* (1963–1970) in Washington, DC, and J.D. Lewis's *Teenage Frolics* (1958–1983) in Raleigh, NC, offered black teens a rare opportunity to appear on television and blurred the boundaries between music television performers and viewers. Similarly, Christine Acham (2005) and Christopher Lehman (2008) have shown how Don Cornelius's *Soul Train* (1970–2006) was an important commercial and political step toward securing more black-oriented, black-produced and black-owned media. Among other locally broadcast black music television programs, Carrie Allen Tipton's (2011) work on *Parade of Quartets* (1954–present), a gospel music program in Augusta, Georgia, highlights a musical genre that is not frequently discussed in music television scholarship.

The false dichotomy between authentic musical artists and commercial music television also conceals the experiments in televisual style in 1960s music television. Aniko Bodrogkozy (2001) shows how *The Monkees* (1966–1968) used unorthodox production techniques to playfully subvert the group's bubblegum image. 'Scenes were underlit or overlit, and the cameras kept running after a scene was supposedly over', Bodrogkozy notes (2001, p. 70). The producers hired young, inexperienced directors who would embrace the show improvisatory and untraditional style. Coates's research on Jack Good, who produced *Six-Five Special* (1957–1958) and *Oh Boy!* (1958–1959) in the UK and *Shindig!* (1964–1966) in the US, shows how Good brought new levels of movement, energy and theatricality to music television. 'For Good', Coates argues, 'musicians were little more than actors, characters in the scenes that he envisioned in his mind, to be brought to life with costumes, lighting, acting, sets, and blocking' (Coates, 2013, p. 303; see also Hill, 1991). As Gayle Wald (in press) shows, Ellis Hazlitt's PBS variety program *Soul!* (1968–1973) filmed artists like Labelle, Curtis Mayfield, Stevie Wonder, and Ashford and Simpson in a naturalistic style, without lip synching and usually with all-live musicians to convey an aura of 'realness' and emphasize the show's theme of black pride.

The introduction of lower cost and more portable video equipment in the late 1960s led to new engagements among musicians, experimental filmmakers and visual artists. Barbara London (2010) locates several pre-MTV music video experiments, such as Captain Beefheart airing an album promo on public-access cable television in Los Angeles in 1970 and the Residents shooting film shorts that screened in art house theaters and later played on MTV and USA Network's *Night Flight* (1981–1986). The British *So It Goes* (1976–1977) on Granada Television broadcast several emerging punk bands, despite the fact that both the bands and the audience shared, as producer Steve Hawes notes, a 'healthy disregard for television' (Hawes, 1999, p. 60). In New York, intersections between music and visual art took place on *Glenn O'Brien's TV Dance Party* (1978–1982), a live, weekly, public access, avant-garde variety show, that George Clinton of Parliament Funkadelic described as 'anarchy *Howdy Doody*' (Galvin, 2013). Kristen Galvin notes that the show's regular guests included Debbie Harry and Chris Stein of Blondie, the painter and graffiti artist Jean-Michel Basquiat, and hip-hop artist and later host of *Yo! MTV Raps* (1988–2004) Fab 5 Freddy (Fred Braithwaite). The rest of the country got to see and hear New York's Downtown scene when all of these artists appeared in Blondie's video for 'Rapture' (1981), the first rap video aired on MTV. Music videos, as Maureen Turim (2007) and Antje Krause-Wahl (2010) show, continued to influence and be influenced by avant-garde high art well after 1980.

Taking a long view of the developments in music television from the 1930s to the 1970s makes it possible to see the development of MTV and other music video channels as part of the longer history of music on television. Through an analysis of the video for Busta Rhymes's 'Gimme Some Mo' (directed by Hype Williams), Kay Dickinson (2007) explores the limitations of picking sides between two of the canonical texts in early music video scholarship, E. Ann Kaplan's *Rocking Around the Clock* (1987) and Andrew Goodwin's *Dancing in the Distraction Factory* (1992). Rather than focusing on video or music, Dickinson argues that scholars must attend to 'image and music synchronously' in order to explore 'how the body and its citizenship is outlined, colored in, and erased in media mergings' (2007, pp. 19, 27). Lori Burns and Jada Watson's (2010) essay on the Dixie Chicks' video 'Top of the World' (directed by Sophie Miller) demonstrates an example of this methodology as they use several diagrams to illustrate how the musical, lyrical, and visual aspects of the video foreground different subject positions. Along similar lines, Carol Vernallis (2004) offers a detailed analysis of the music video conventions and the various modes of continuity and discontinuity that developed in MTV's first twenty years. 'It might be helpful', Vernallis suggests, 'to imagine the various elements of music video's *mise-en-scène* as separate tracks on a recording engineer's mixing board: any element or combination of elements can be brought forward or become submerged in the mix' (2004, p. x). Mark Fenster (1993) explores how the country music industry adapted the rock and pop music video conventions Vernallis describes to fit the established conventions of country music. While these and other studies offer detailed and nuanced analysis of music videos, more work needs to be done to connect music videos to concurrent televisual representations in other genres, such as news, sports, comedy and drama. For example, how did images of blackness in hip hop videos aired on Black Entertainment Television's (BET) *Rap City* (1989–2009) and *Yo! MTV Raps* intermingle with 'urban crisis' news reports or the increasing television visibility of black athletes? In taking up questions like these, scholars can emphasize the breadth of music television programming and the ways in which music television is regularly enlisted in debates around race, class, gender, sexuality, youth and citizenship in other television formats.

In 1987, MTV International began operating distinct from the US MTV channel, and by 1997 MTV affiliates outside of the US (MTV Europe, MTV Mandarin, MTV India, MTV Latin America, MTV Australia, MTV Japan, MTV New Zealand, MTV Brazil and MTV Asia) reached 300 million households (Kun, 2002, p. 102; see also Klanten, Meyer & Jofré, 2005). Since its inception MTV International executives have answered accusations of cultural imperialism by focusing on how the different stations celebrate local music cultures, and the best work on the global reach of MTV probes the tensions among the local, national and transnational aspects of music video production, circulation and reception. Kip Pegley (2008) offers a comparative analysis of how MTV's US station and the Toronto-based MuchMusic construct racial, gender and national identities. Both Pegley and Antti-Ville Kärjä (2007) look at the Finnish music video program *Jyrki*, which operated under the MuchMusic brand and format, and MTV Nordic, a regional subdivision of MTV Europe, finding hybrid cultural formations that are shaping new definitions of Finnishness. Josh Kun (2002) examines Tijuana NO!, a pro-anarchy punk group signed to the German-owned multinational record label BMG International, who have had several videos appear on MTV Latin America (see also Hanke, 1998). The group hails from the border city of Tijuana, Mexico and many of their songs and videos focus on the tensions, paradoxes and energies of the politics and history of the US–Mexico border. Their video for 'Transgresores de la Ley' is a tribute to the Ejército Zapatista de Liberación Nacional and its leader Subcomandante Marcos who issued a Declaration of War against the Mexican government in 1994 (p. 112). Featuring footage of Marcos, the video offers up sounds, lyrics and images that depict 'Mexican nationalism as a corrupt, weak face hiding behind a mask of U.S.-financed lies' (p. 114). This is not the local music culture MTV International expected to promote.

Music videos in China offer a different example of localization, as Anthony Fung (2008) shows in his analysis of how state broadcasting interests, China Central TV and Shanghai Media Group, have worked closely with Channel V (owned by Rupert Murdoch's Star TV) and MTV Mandarin to produce local popular culture. 'The political and economic forces of the state have combined with global capital in a symbiotic relationship', Fung argues. 'The state works to accelerate the localization of transnational culture while the latter promotes it' (2008, p. 85). In practice, this often leads to musical and visual styles from abroad being combined with local performers and styles, such as integrating 'hip-hop music and dance with Chinese icons, including local composers, local bands, and Chinese folksingers' (2008, p. 100). While privately owned broadcasting plays a larger role in Indonesia than in China, R. Anderson Sutton (2003) identifies a similar dialogic relationship between local and global styles in looking at Indonesian pop music videos broadcast on MTV Southeast Asia, *dangdut* videos (Indonesia's de facto national music) and *Dua Warna* (Two Colors), a music performance program that combined Indonesian pop music with more traditional ethnic and regional musical styles. Sutton finds that in its effort to produce 'Indonesian' national music television, *Dua Warna* 'is much more in the spirit of Western taste for an exotic tinge', but that this has as much to do with the elite status of the show's producers and the dynamics of class, ethnicity and regional identities within Indonesia as it does with the influence of MTV (Sutton, p. 333). Like Kun, Sutton highlights how global cultural flows lead to unexpected connections in music videos, such as *dangdut* programs which, despite their identification as 'very Indonesian', portray 'dancers with futuristic spacesuit-inspired costumes, planets, asteroids, galaxies, and other space references', as well as a Mexican themed show that interspersed *dangdut* songs with performers dressed in sombreros and Mexican peasant outfits with an 'educational segment on varieties of Mexican cacti and cooking tips on chicken fajitas by the owner of a Mexican restaurant in Jakarta' (p. 329; see also Hayward, 1998).

Tony Langlois's (2009) ethnographic research on music video consumption in the Moroccan cities of Fez and Oujada is a particularly good example of local music video culture that is not framed by MTV. Video Compact Discs (VCDs), which are more affordable than DVDs but have lower image and sound quality, proliferate in markets in Morocco and other developing countries. These discs have facilitated the development of a robust music video culture in Morocco with low-cost videos of local musicians and illegally ripped musical performances from Arabic satellite TV programs among the most popular offerings. Western products play a limited role in this market, with the exception of mash-up videos in which 'Excerpts from Michael Jackson or Shakira videos, dance scenes from Jim Carrey's *The Mask* (1994), or from Disney's *The Jungle Book* (1967), are manipulated to fit the song structures of popular cha'abi songs' (Langlois, 2009, p. 78). Since VCDs are most commonly watched on home TVs and social spaces are commonly segregated along gender lines, Morroco's VCD music video culture is geared toward female viewers. More recently, Moroccan music videos appear on YouTube and other websites, uploaded primarily by Moroccans living outside of the country (Langlois, 2009, p. 79).

Questions surrounding ethnic, regional and national identities are central to much of the work surrounding music videos and MTV International, and similar questions are applicable to singing competitions like the *Eurovision Song Contest* (1956–present), as well as the numerous versions of *Idol* (2001–present), *Got Talent* (2006–present), X *Factor* (2004–present), and similar programs that developed in the 2000s and 2010s. According to Freemantle Media, the international production company that distributes the show, there have been 199 *Idol* series in 46 territories (Idols, 2013). The scope of the *Idol* franchise poses obvious research challenges, but Katherine Meizel's (2010) work on *American Idol* offers one productive methodological approach. '*American Idol* and its identity politics,' Meizel argues, 'play out across multiple places and in multiple spaces – onstage, onscreen, online, and in line waiting for *Idol* events' (2010, p. 10). Meizel's research draws not only on analysis of *American Idol* broadcast footage, but also ethnographic interviews and participant observation at the show's local cattle-call auditions, and analysis of online commentary related to the program in official, critical and fan venues. Other scholars could take up these diverse research methods to analyze multiple national versions of *Idol* in comparative perspective, how the shows' appeals to voting and national identity function in different countries, or how the shows' relatively low-cost 'reality' format fits into different national media environments.

Since the mid-2000s, YouTube, Dailymotion, Vimeo, Vevo, Metacafe, MyVideo, and similar video websites have made music related videos available instantaneously on a proliferation of screens, such as computers, cell phones, and tablets. Roger Beebe and Jason Middleton identified these new means for disseminating and viewing music videos as one of the inspirations for their 2007 edited collection, *Medium Cool: Music Videos from Soundies to Cellphones*, and this new media landscape is even more important today. Web video platforms have fueled a renewed market for professionally made music videos, as well as billions of amateur and fan videos that take advantage of digital video cameras and webcams. In her analysis of music videos in the YouTube-era, Carol Vernallis (2010, 2013) traces several aesthetic differences from earlier genres in the broadcast or cable television eras, such as repetition, digital weightlessness, visual images scaled to smaller screens, layering of intertextual references, and soliciting viewer participation. Despite the lack of historical context, the presence of advertisements, and the routine removal of copyrighted material, YouTube and other online video sites have become important repositories for music television history. Among the most well organized examples are the 120 Minutes Archive (http://120minutes.tylerc.com/), an online archive of playlists and links to YouTube videos for nearly 893 episodes of MTV's college/alternative rock program *120 Minutes* (1986–2003), and Kumalo (www.kumalo.com), which features contemporary music videos from 33 African countries. The dubious copyright status and subsequent instability of many historical recordings on YouTube is a new iteration of a problem television scholars have always faced, locating primary sources. This challenge is particularly acute for scholars of music television because many of the programs were regarded as ephemeral or may be subject to copyright claims by both the television broadcaster and the record company. In light of this, music television scholars should be proactive in maintaining the sources about which they write through sites such as Critical Commons (www.criticalcommons.org), a public media archive and fair use advocacy network.

While scholars are offering new ways to understand how music and images are intersecting in the contemporary media environment, the YouTube era should also encourage more research on the commonalities and discontinuities among web video, MTV-era music videos, and earlier eras of recorded music performances. Amy Herzog's (2010) work on visual music jukeboxes (Soundies in the 1940s and Scopitone films in the 1960s), for example, offers television studies scholars a model for thinking about how people engage with recorded musical performances outside of the home, as does Anna McCarthy's (2001) analysis of 'ambient television' monitors in bars, airports and stores. The present convergence of television and the internet should also prompt renewed attention to the technological aspects of producing, watching, and hearing the sights and sounds of music television in different eras. Screen size and portability, image resolution, sound quality and compression, and systems of distribution have changed dramatically since the 1930s and scholars could build on work by Tim Anderson (2006), André Millard (2012), and Jonathan Sterne (2012) to examine what these technological changes have meant for music television.

Like many of the works cited in this chapter, scholars can also profitably make use of archival documents, newspapers, trade journals, magazines, advertisements, memorabilia, oral histories and ethnographies to fill-in, complement and complicate the available video footage of music television. By foregrounding the evidentiary challenges and archival precarity of music television research, scholars can make these gaps and silences part of the larger history of music television. Such an approach destabilizes a neat progression across iconic performances and eras, so that rather than moving from Elvis Presley on the *Ed Sullivan Show* in 1956, to the debut of Michael Jackson's 'Thriller' video on MTV in 1983, to PSY's 'Gangnam Style' or whatever replaces it as the most viewed video on YouTube, we might instead focus on the vast, messy, and in many cases lost, array of shows and clips that have been crucial to every era of television.

The clips of Iowa teens dancing The Stroll on WOI-TV's *Seventeen* are instructive in this regard. WOI-TV was the first commercial broadcast station in the US owned by a university, and Iowa State University started the station with money left over from federal grant given to the school for research on the Manhattan Project, the World War II initiative that led to the first atomic bombs ('Vote to Sell TV Station Splits Iowans', 1992). At the time of writing, these two YouTube videos have been viewed nearly two million times, which, accounting for repeat viewers, is still many more people than would have seen the original video when it aired in the Ames-Des Moines region in 1958. At least one of these YouTube viewers was inspired to create a video that intercuts the footage of Iowa teens dancing The Stroll with footage from the well-known *Soul Train* dance line (Soul Train vs. The Stroll, 2013). The juxtaposition of the clips highlights the most obvious differences between the shows – the conservatively attired and physically restrained white teens on the black and white 1950s *Seventeen*, dance in sharp contrast to the flamboyantly dressed and dynamic black performers on the vibrantly colorful 1970s *Soul Train*. Less obvious are the points of connection between the clips. Both programs, for example, use the lines of dancers to frame full shots of the couples dancing up the aisle, a single camera technique that *American Bandstand* popularized. While *Soul Train* came to national attention after it moved to Los Angeles and became syndicated in 1971, the show started a year earlier as local dance program like *Seventeen* on WCIU-TV, a UHF station in Chicago (a local Chicago version of *Soul Train* continued until 1976, with reruns airing until 1979). By 1970, Chicago already had a history of music television shows with black hosts, the *Al Benson Show* and Richard Stamz's *Open the Door Richard* both had brief periods of success in the 1950s, as well as teen dance programs, *Kiddie-a-Go-Go* and *Red Hot and Blues*, the latter of which was hosted by black DJ Big Bill Hill and featured black pre-teen dancers (Lehman, 2008, p. 28; Stamz, 2010). As the first nationally broadcast black dance show, *Soul Train* built on these Chicago programs, while also responding to the dance program genre that had long welcomed black music more than black dancers. One does not need to know this history to appreciate the mash-up of *Seventeen* and *Soul Train*, but the video, uploaded by a YouTube user in the Ukraine, hints at the frequent intersections among the history, present and future of music television.

REFERENCES

Acham, C. (2005). *Revolution Televised: Prime Time and the Struggle for Black Power*. Minneapolis, MN: University of Minnesota Press.

Anderson, T. (2006). *Making Easy Listening: Material Culture and Postwar American Recording*. Minneapolis, MN: University of Minnesota Press.

Beebe, R. & Middleton, J. (2007). *Medium Cool: Music Videos from Soundies to Cellphones*. Durham, NC: Duke University Press.

Bodrogkozy, A. (2001). *Groove Tube: Sixties Television and the Youth Rebellion*. Durham, NC: Duke University Press.

Burns, G. (1998). Visualising 1950s Hits on Your Hit Parade. *Popular Music*, 17(2), 139–152.

Burns, L. & Watson, J. (2010). Subjective Perspectives through Word, Image and

Sound: Temporality, Narrative Agency and Embodiment in the Dixie Chicks' Video 'Top of the World'. *Music, Sound, and Moving Image*, 4(1), 3–37.

Coates, N. (2007a). Elvis from the Waist Up and Other Myths: 1950s Music Television and the Gendering of Rock Discourse. In R. Beebe & J. Middleton (Eds) *Medium Cool: Music Videos from Soundies to Cellphones* (pp. 226–251). Durham, NC: Duke University Press.

Coates, N. (2007b). Filing in Holes: Television as a Recuperation of Popular Music on Television. *Music, Sound, and Moving Image*, 1(1), 21–25.

Coates, N. (2013). Excitement is Made, Not Born: Jack Good, Television, and Rock and Roll. *Journal of Popular Music Studies*, 25(3), 301–325.

Deaville, J. (2011). A Discipline Emerges: Reading Writing about Listening to Television. In J. Deaville (Ed.) *Music in Television: Channels of Listening* (pp. 7–33). New York, NY: Routledge.

Delmont, M. (2012). *The Nicest Kids in Town: American Bandstand, Rock 'n' Roll, and the Struggle for Civil Rights in 1950s Philadelphia*. Berkeley, CA: University of California Press.

Dickinson, K. (2007). Music Video and Synaesthetic Possibility. In R. Beebe & J. Middleton (Eds) *Medium Cool: Music Videos from Soundies to Cellphones* (pp. 13–29). Durham, NC: Duke University Press.

Fenster, M. (1993). Genre and Form: The Development of the Country Music Video. In S. Frith, A. Goodwin, & L. Grossberg (Eds.) *Sound and Vision: The Music Video Reader* (pp. 109–128). New York, NY: Routledge.

Forman, M. (2012). *One Night on TV Is Worth Weeks at the Paramount: Popular Music on Early Television*. Durham, NC: Duke University Press.

Fung, A. (2008). *Global Capital, Local Culture: Transnational Media Corporations in China*. New York, NY: Peter Lang.

Galvin, K. (2013). TV Party: Downtown Music Scenes Live on Your TV Screen. *Journal of Popular Music Studies*, 25(3), 326–348.

Goodwin, A. (1992). *Dancing in the Distraction Factory: Music Television and Popular Culture*. Minneapolis, MN: University of Minnesota Press.

Hanke, R. (1998). 'You Quiero Mi MTV!': Making Music Television for Latin America. In T. Swiss, J. Sloop & A. Herman (Eds) *Mapping the Beat: Popular Music and Contemporary Theory* (pp. 219–245). Malden, MA: Blackwell.

Hawes, S. (1999). I Was There: Putting Punk on Television. In A. Blake (Ed.) *Living Through Pop* (pp. 60–62). London, England: Routledge.

Hayward, P. (1998). The New Tradition: Titus Tilly and the Development of Music Video in Papua New Guinea. In P. Hayward (Ed.) *Sound Alliances: Indigenous Peoples, Cultural Politics and Popular Music in the Pacific* (pp. 142–157). New York, NY: Cassell.

Herzog, A. (2010). *Dreams of Difference, Songs of the Same: The Musical Moment in Film*. Minneapolis, MN: University of Minnesota Press.

Hill, J. (1991). Television and Pop: The Case of the 1950s. In J. Corner (Ed.) *Popular Television in Britain: Studies in Cultural History* (pp. 90–107). London, England: British Film Institute.

Idols (2013). Retreived May 10, 2013, from www.fremantlemedia.com/Production/Our_Brands/Idols.aspx

Kaplan, E.A. (1987). *Rocking Around the Clock: Music Television, Postmodernism, and Consumer Culture*. New York, NY: Methuen.

Kärjä, A-V. (2007). Visions of a Sound Nation: Finnish Music Videos and Secured Otherness. In R. Beebe & J. Middleton (Eds) *Medium Cool: Music Videos from Soundies to Cellphones* (pp. 174–199). Durham, NC: Duke University Press.

Kessler, K. (2013). Broadway in the Box: Television's Infancy and the Cultural Cachet of the Great White Way. *Journal of Popular Music Studies*, 25(3), 349–370.

Klanten, R., Meyer, B. & Jofré, C. (2005). *On Air: The Visual Messages and Global Language of MTV*. Berlin, Germany: Die Gestalten Verlag.

Krause-Wahl, A. (2010). 'Why Artists Make Clips': Contemporary Connections between Art and Pop. In H. Keazor & T. Wübbena (Eds) *Rewind, Play, Fast Forward: The Past, Present, and Future of the Music Video* (pp. 207–224). Bielefeld, Germany: transcript Verlag.

Kun, J. (2002). 'The Sun Never Sets on MTV': Tijuana NO! and the Border of Music Video. In M. Habell-Pallán & M. Romero (Eds)

Latino/a Popular Culture (pp. 102–116). New York, NY: New York University Press.

Langlois, T. (2009). Pirates of the Mediterranean: Moroccan Music Video and Technology. *Music, Sound, and Moving Image*, 3(1), 71–85.

Lehman, C. (2008). *A Critical History of Soul Train on Television*. Jefferson, NC: McFarland.

London, B. (2010). Looking at Music. In H. Keazor & T. Wübbena (Eds) *Rewind, Play, Fast Forward: The Past, Present, and Future of the Music Video* (pp. 59–65). Bielefeld, Germany: transcript Verlag.

Malnig, J. (2006). Let's Go to the Hop: Community Values in Televised Teen Dance Programs of the 1950s. *Dance & Community: Proceedings of the Congress on Research in Dance*, August, 171–75.

McCarthy, A. (2001). *Ambient Television: Visual Culture and Public Space*. Durham, NC: Duke University Press.

McIntosh, H. (2004). Music Video Forerunners in Early Television Programming: A Look at WCPO-TV's Innovations and Contributions in the 1950s. *Popular Music and Society*, 27(3), 259–272.

Meizel, K. (2010). *Idolized: Music, Media, and Identity in American Idol*. Bloomington, IN: Indiana University Press.

Millard, A. (2012). *Beatlemania: Technology, Business, and Teen Culture in Cold War America*. Baltimore: The Johns Hopkins University Press.

Negus, K. (2006). Musicians on Television: Visible, Audible and Ignored. *Journal of the Royal Musical Association*, 131(2), 310–330.

The Original Stroll – February 1958 (2013). Retrieved May 10, 2013 from www.youtube.com/watch?v=UrGLNtZ0rEg

Pegley, K. (2008). *Coming to You Wherever You Are: MuchMusic, MTV, and Youth Identities*. Middletown, CT: Wesleyan University Press.

Rodman, R. (2010). *Tuning In: American Narrative Television Music*. New York, NY: Oxford University Press.

Soul Train vs. The Stroll (2013). Retrieved May 10, 2013 from www.youtube.com/watch?v=4Wwlz1FGNu8

Stamz, R. with Roberts, P. (2010). *Give 'Em Soul, Richard!: Race, Radio, & Rhythm & Blues in Chicago*. Urbana, IL: University of Illinois Press.

Sterne, J. (2012). *MP3: The Meaning of a Format*. Durham: Duke University Press.

The Stroll w/ The King of The Stroll: Chuck Willis - Betty and Dupree ('Seventeen') (2013). Retrieved May 10, 2013 from www.youtube.com/watch?v=PajBJOnRYBI

Sutton, R.A. (2003). Local, Global, or National? Popular Music on Indonesian Television. In L. Parks & S. Kumar (Eds) *Planet TV: A Global Television Reader*. New York: New York University Press.

Tipton, C.A. (2011). 'I Got That Something That Makes Me Want to Shout': James Brown, Religion, and Gospel Music in Augusta, Georgia. *Journal of the Society for American Music*, 5(4), 535–555.

Turim, M. (2007). Art/Music/Video.com. In R. Beebe & J. Middleton (Eds) *Medium Cool: Music Videos from Soundies to Cellphones* (pp. 83–110). Durham, NC: Duke University Press.

Vernallis, C. (2004). *Experiencing Music Video: Aesthetics and Cultural Context*. New York, NY: Columbia University Press.

Vernallis, C. (2010). Music Video and YouTube: New Aesthetics and Generic Transformations: Case Study – Beyoncé's and Lady Gaga's *Video Phone*. In H. Keazor & T. Wübbena (Eds) *Rewind, Play, Fast Forward: The Past, Present, and Future of the Music Video* (pp. 233–259). Bielefeld, Germany: transcript Verlag.

Vernallis, C. (2013). *Unruly Media: YouTube, Music Video, and the New Digital Cinema*. New York: Oxford University Press.

Vote to Sell TV Station Splits Iowans: Does a Commercial Venture Fit into an Educational Mission? (1992). *New York Times*, July 22, pp. A17.

Wald, G. (in press). *'It's Been Beautiful': Soul! and Black Power Television*. Durham, NC: Duke University Press.

Wall, T. (2009). Rocking Around the Clock: Teenage Dance Fads from 1955 to 1965. In J. Malnig (Ed.) *Ballrooms, Boogie, Shimmy, Sham, Shake: A Social and Popular Dance Reader* (pp. 182–198). Urbana, IL: University of Illinois Press.

21

Reality Television

Mark Andrejevic

Reality TV is a television genre that, broadly construed, is as old as TV itself, but which developed into a significant global programming trend in the 1990s and 2000s. Reality TV continues to develop new formats and to explore permutations of existing ones while borrowing elements from other programming formats including game shows and documentaries. Even incomplete lists of reality formats since the 1990s number in the hundreds; it is difficult to pinpoint a number because new variations, spinoffs, and imitations are generated on an ongoing basis. In the United States, the Emmy awards added three new categories to address the growth of reality programming in the early 2000s and global reality TV franchises like *Big Brother* and the various *Idol* (singing contest) formats have proliferated alongside original and derivative local formats.

Because the genre continues to re-invent itself, and because it comprises a wide range of formats, including dating shows, talent contests and makeover shows, there is no clear-cut thematic definition that does justice to it while excluding non-reality shows. In general, however, reality TV refers to unscripted entertainment programming that features events in the lives of its cast members, who, depending on the format, include both celebrities and 'real' people – those who have no prior claim to star status. Any attempt to define a form of programming that ranges from found-video shows like *World's Dumbest Criminals* to competitions like *The Amazing Race*, and makeover and documentary style programming like *Celebrity Rehab* in terms of formal similarities or generic attributes is doomed to failure. The reality TV boom of the turn of the millennium is perhaps best characterized by its promise to reconfigure the boundary between producer and consumer, celebrity and audience in the name of a mediated version of authenticity. One of the defining characteristics of reality programming in the digital era is access to the once restricted realm of celebrity: the promise that TV will feature not just professional actors and celebrities following plotlines crafted by Hollywood insiders, but that the doors have been thrown open to the audience to participate in and become a part of the programming they consume (and, relatedly, that celebrities have daily lives, traumas, and joys that make them just as 'real' as non-celebrities).

A characteristic recurring element of this incarnation of reality TV is its promise of access to behind-the-scenes reality in the form of situations or aspects of the production usually limited to actors and producers. This promise of backstage access, and in some cases even audience participation, is part of the overarching claim of reality TV to erode the traditional boundary between media insiders and media outsiders, celebrities and producers on the one hand and audiences and 'real' people on the other. The flip side of the promise of audience access to the once restricted realm of celebrity might be described as the 'humanization' or 'realization' of celebrities whose lives have provided the content for a celebrity themed shows including *The Osbournes* and *I'm a Celebrity ... Get Me Out of Here!* Reality shows that provide behind-the-scenes access to the backstage lives of rock stars, actors and other celebrities complement the formats that transform audience members into celebrities: both provide access to once rarefied realms, and both provide opposite halves of the equation in which celebrities are revealed to be, as the entertainment tabloid *US Weekly* puts it in one of its regular photo features: 'just like us'.

In economic terms, reality TV is a popular format with producers because it is relatively inexpensive and flexible. That is to say it aligns itself with production imperatives in an era of burgeoning bandwidth, increasingly comprehensive audience measurement, and multiplying channels. As Bondebjerg (2002) puts it, 'reality TV can be analyzed not just as a typical time-bound hybrid form of factual entertainment but also as a metaphor or symptom of the development of a commercialized and globalized media-culture' (p. 159). Thus, the development of reality TV can be situated within a broader set of cultural and economic logics as a genre that, in keeping with the rise of interactive media, reconfigures the boundary between producer and consumer, 'talent' and audience, while addressing production needs in a multi-channel, multi-platform, globalized TV market, and simultaneously catering to media-saturated audiences savvy about the contrived character of the media products they consume.

On the one hand, reality TV uses the promise of participation to allow members of the audience into the rarefied celebrity realm. On the other it offers backstage access as a means of transforming celebrities into 'real people'. In both cases, there is an underlying promise to level the distinction between celebrity and everyday life: to demonstrate that celebrities are in some respects, just like us, and on the other that we, the members of the audience, have the potential to become celebrities. Turner (2010) has described this leveling promise as a manifestation of the 'demotic' turn: not necessarily a form of political or social empowerment (that is not 'democratizing'), but the mobilization of popular appeal and the promise of democratization as a marketing strategy.

Successful reality formats have been built around celebrities like Ozzie Osbourne (*The Osbournes*) and Joseph Simmons (*Run's House*; Simmons was a founding member of the seminal hip-hop group Run-D.M.C.) as well as people drawn from the ranks of the audience, as in the case of formats like *The Real World*, *Survivor* and a range of dating and makeover shows. There is some debate as to whether quiz show formats like *Who Wants to be a Millionaire* or talent shows like the *Idol/Pop Star* franchise fit into the category of reality TV proper. On the one hand, such shows hearken back to well-established formats that include scripted elements and are built around celebrity talent including hosts and judges. On the other hand, they feature central elements of the promise of the 'demotic turn'. Whereas quiz shows have historically featured the wide-ranging knowledge of their contestants, thereby distinguished them from the typical audience member, *Who Wants to Be a Millionaire* relies on the promise that any viewer is a potential contestant, and opens the phone lines after the show for applicants. With a little bit of luck and the help of some 'lifelines', the message is that anyone might win.

Although talent shows do rely on a hierarchy of abilities, their reality show versions emphasize the discovery of undiscovered talent in the populace, the paradigmatic case perhaps being that of Susan Boyle, an unprepossessing church choir member from Scotland who became internationally famous overnight when she surprised skeptical judges and thrilled the audience on the talent show *Britain's Got Talent*. Boyle neatly captured the promise of reality TV and highlighted its synergistic relationship with the internet, where clips of her performance went viral. The result was a multi-platform spiral of publicity: the more the media covered her performance, the more they promoted the online clips, and the more these were viewed, the more the media reported on the phenomenon of her online popularity.

Reality talent formats tend to emphasize the discovery of talent in undiscovered places (relative unknowns, or, in the case of celebrity talent shows, the aptitudes – or lack thereof – of people who are famous for other things, as in the case, for example, of *Dancing with the Stars*). At the same time such shows cater to the promise of behind-the-scenes access by focusing on details of the cast members' personal lives and providing glimpses of production details, including, for example, audition tapes or practice sessions and backstage interviews.

More than a programming fad or trend, reality TV represents a mode of production (McCarthy, 2004a) that replaces writers with story editors and, professional actors with often unpaid or lower paid 'real' people or celebrities past their prime. Even in the case of reality shows focused on the lives of celebrities, who are paid in accordance with their fame, the costs of soundstages, sets and production studios are kept in check by on-location filming. Because many reality formats are less expensive than other programming genres, such as scripted dramas and comedies, they allow for programming flexibility: they can be accelerated (aired several days a week) or discontinued (if ratings are low) and replaced with another rapidly produced and relatively inexpensive program. Moreover, the localizable character of reality shows has led to the multiplication of successful international franchises like *Big Brother* and the *Idol* shows. Such formats allow for the marketing of an internationally successful format that can be populated with local cast members: the success of a format builds its name, whereas localization allows for the customization of the format to local audiences. The logic of mass customization, like that of participation, aligns the genre with more general trends in the information economy, which relies on data processing to create more flexible supply chains and customized marketing and sales campaigns. Although reality TV is not unique to the multi-channel, internet era, it is uniquely suited to it, which may explain why it became a programming phenomenon at the turn of the millennium.

HISTORY

Entertainment programming based on unscripted interactions of non-celebrities long predates television. For example, *Candid Camera*, often cited as a reality TV precursor and influence, was a carry-over from the radio, based on the success of *Candid Microphone*. Similarly, the US radio show *Nightwatch*, which followed police on patrol in California might be considered a radio-era precursor to police reality shows like *Cops* in the US, *Emergency 999* in the UK, *Highway Patrol* in Australia, and other similar shows.

Given the wide range of formats encompassed by the genre it should not be surprising that the genealogy of reality TV is a complex and hybrid one that overflows the disciplinary field of television studies thanks to its roots in the history of cinema (and, even further back in the novel), its connection to the online world, and its reliance on the influence of the social psychology experiments of the 1960s (for more on the connection between reality TV and social psychology, see McCarthy (2004b)). Considered within the context of the mass media, reality-based films that record the rhythms of daily life were, as documentary filmmakers Ilisa Barbash and Lucien Taylor (1997) argue, one of the first manifestations of the moving picture medium. The inventors of film-based motion pictures, Louis and Auguste Lumière 'churned out film after film about apparently inconsequential moments of daily life: workers leaving their factories; a train arriving at the station … and lumberjacks going about their work' (Barbash and Taylor, 1997, p. 15).

This early slice-of-life cinema anticipated the mundane character of shows like *Big Brother* and *The Real World*, which often focus on the details of daily life: cast members sweeping the yard, washing the dishes, and sitting around the table with nothing much to say. With the development of documentary film, recording the rhythm of daily life became not just a historical genre, but, as Barbash and Taylor (1997) note, an anthropological one – a way of presenting the lives of other people, often in far off lands. In this sense, the medium served as a means of overcoming spatial and temporal boundaries, recreating either historically or geographically distant lives.

The advent of the reality genre takes place when documentary techniques are used not to document the daily life of geographically and culturally remote peoples but to document (for entertainment purposes) the lives of proximal, contemporary figures as representatives of typical – hence real – people. It is this representativeness – this 'reality' – that works to level the distinction between celebrity and audience. The fictional sitcom and family drama anticipated this reality TV by relying upon scenarios based on the daily, domestic routines of fictional families like the Cleavers and the Ricardos. It provided a glimpse behind the façade of suburbia into the domestic lives of American families. Insofar as such shows focused on private rather than public life, they helped draw back the curtains on the internal dynamics of family life and provided a 'behind-the-scenes' look at gender and family relationships, albeit fictional and often idealized. As Meyrowitz (1986) argues, the content of TV programming, in its reliance on intimacy, often took viewers behind the scenes, serving as a demystifying machine that, for example, portrayed to children what the world of adults looked like when they were not around. Rather than reinforcing distanced images of authority, it got up close to reveal moments of intimacy, cracks in the façade, and the details of daily life. In this regard, the logic of reality TV represents the culmination of the promise of TV as what Kavka (2008) describes as a 'technology of intimacy, a machine that functions by drawing viewers *close*' (p. 5).

In this regard, the question becomes not so much which forms of programming anticipated and contributed to the development of reality TV, but which *didn't*. Scholars have, for example, located genetic influences of 21st-century reality TV in the British documentary television tradition, the rise of sensationalist, human-interest-oriented tabloid television, the creation of new hybrid forms of documentary entertainment, game shows, talk shows, DIY instructional shows and soap operas (see, for example, Corner, 2002; Hill, 2005; Kilborn, 1994; Murray, 2004). The fact that the list might well go on is suggested by Hill's thumbnail summary of influences: 'popular factual television [the term she uses to describe reality TV] has developed during a period of cross-fertilization with tabloid journalism, documentary television and popular entertainment' (Hill, 2005, p. 23). The breadth of 'popular culture' and the range of the formats it encompasses suggest the promiscuity of influences, which continues to grow as the genre develops and experiments with new formats. We might, for example, include in the list of influences situation comedies such as *The Brady Bunch* and *The Beverly Hillbillies* that helped influence subsequent reality formats (*My Fair Brady* and *Amish in the City*), police procedurals with their various attempts at portraying gritty veracity that anticipate shows like *Cops* and various border security and policing formats, and the range of shows that have generated 'whatever-happened-to' celebrities whose fame can be recycled for reality TV.

On the one hand, then, we can trace the reality lineage back to shows like *Candid Camera* and *An American Family* that might be considered early versions of reality TV – although the term does not come into general usage until the 1980s. On the other hand, precursors can be located in a broad range of formats that represent elements and tendencies later picked up on and adapted to reality TV proper. It is not hard, in this light, to consider the whole range of TV formats prior to the reality TV boom of the early 2000s as raw material for reality TV producers to pick over, reference, and otherwise mine for ideas.

The history of the genre cannot be separated from the economic and technological logics that facilitate its development. The manufacture of lightweight portable video equipment that made it possible to leave the soundstages and shoot relatively inexpensively on the fly contributed to the development of both documentary filmmaking and its forays into television production, such as the PBS series *An American Family* and its UK successor *The Family*. *An American Family*, which documented the daily life of the Loud family and famously chronicled the parents' divorce and the eldest son's decision to come out of the closet anticipated subsequent shows devoted to the daily lives of cast members over a period of time, including such formats as *The Real World*, *Road Rules* and *The Osbournes*. Promotions for *An American Family*, which compressed seven months worth of filming (three hundred hours of raw footage) into 12 weekly one-hour installments, presciently billed the Louds as 'TV's First Real Family' (Alexander, 1973, p. 28). The show triggered many of the responses that have characterized the media coverage of later reality formats: laments over the decline of the private sphere and the role played by television in promoting an inordinate fascination with celebrity, as well as musings on the educational – indeed anthropological – character of the show (with frequent quotes from Margaret Mead, who described the advent of this form of reality TV as very possibly 'as important for our time as were the invention of drama and the novel for earlier generations') (McCarthy, 1973, p. 72). The show also triggered a debate that has long characterized the practice and reception of documentary film: how does the artifice of filmmaking impact the 'reality' of what it seeks to portray. The response of *An American Family*'s producer Craig Gilbert anticipated those of later reality producers: 'If reactions were modified because of the camera, those reactions were still valid' (McCarthy, 1973, p. 75). The conditions might have been artificial, according to such an account, but the responses were real. This notion of validity lays claim to what might be described as the Petri-dish character of reality TV and its description, harking back to the social psychology experiments of the 1960s, as a form of social experimentation. The logic is a familiar one: sometimes creating contrived circumstances is necessary for isolating and studying interactions, whether they are chemical or social. Picking up on the social psychology lineage, both psychological and sociological approaches have drawn upon reality TV as a means of studying interpersonal interactions (see, for example, Beattie, 2003; Godard, 2003). Geoffrey Beattie, who served as an on-air psychologist and commentator for the UK version of *Big Brother*, published a book on non-verbal communication based on theories derived from his observation of the show, which, he argued, showed, 'behaviour in sufficient detail in a long enough context so that we can begin to understand the individuals and to get some hint as to why they are doing what they are doing' (2003, p. 3). The show allowed for observation and interpretation, according to Beattie, 'in a way that no psychology experiment that I know of has ever allowed before' (2003, p. 3).

It is perhaps not insignificant that *An American Family*, as the first manifestation of a 'reality' TV format that provided comprehensive documentation of the rhythm of proximal daily life, occurred during the period characterized by both theorists and business consultants as the transition to post-Fordist, or flexible capitalism (Harvey, 1990; Pine, 1993). Reality TV emerged during a period in which the destabilization of mass society was accompanied by a reformulation of the boundaries that helped maintain its social and cultural hierarchy. Media theorist Josh Meyrowitz (1995), for example, interprets the 'question-authority' mentality of the 1960s generation as a result, at least in part, of the role that television played in eroding the boundaries that enclosed the private lives of the powerful and famous – bringing them to the level of 'real' people: 'Most people who step forward into the television limelight and attempt to gain national visibility become too visible, too exposed, and are thereby demystified. The more we see them, the more ordinary they appear' (Meyrowitz, 1995, p. 48). The hit reality show *The Osbournes*, based on the daily life of heavy metal rocker Ozzy Osbourne and his endearingly eccentric family, provides one example of the demystification facilitated by the intimacy of the medium. Once feared and reviled as another 'satanic' element in popular music, Ozzy Osbourne had become the bumbling, harmless hero of a family show, the addled Ward Cleaver of the dawn of the millennium.

THE PROMISE OF THE REAL

If it is possible to point to a tangled lineage for different categories of reality TV – the game show, which features contrived if unscripted antics of audience members as well as celebrities (in their 'just-like-us' mode), the quiz show, the hidden camera show, the TV documentary – it is also possible to locate a shared logic that organizes these disparate influences into the reality TV boom at the turn of the millennium. It is tempting to interpret the promise of reality TV as one that juxtaposes reality to 'fakeness', contrivance, or artifice, as if what reality TV provides is an antidote to mediation: direct access to the real. This temptation has resulted in ongoing discussions in the popular press and some academic circles as to just how real reality TV is. If the situations are contrived, does that undermine the promise of access to reality? What if scenes are re-shot or cast members are told to behave in a particular way or are coached to say or repeat certain lines? As Kavka among others points out:

> The criticism that reality TV is not in fact 'real' because the shows are heavily manipulated ... is also dependent on the construct of a dumbed-down viewership that conflates what plays out on one side of the screen – framed as spectacle – with what happens on the other – grounded in the experiential world. (Kavka, 2008, pp. 22–23)

Typically it is a criticism directed by pundits or academics toward an imagined audience who, in reality, are well aware that the genre is described not simply as 'reality' but as 'reality *TV*' with full emphasis on the fact of its mediation. As Hill notes, 'Most viewers expect reality TV to be artificial' (2007, p. 140). However, authenticity and artifice are not necessarily opposed; in the experimental model of the sciences and social sciences, artifice can serve as a technique for extracting a moment of insight or knowledge. In this respect, it is not necessarily contradictory to accuse reality TV of being the product of artifice and at the same time to scan it for traces of authenticity.

In the case of the reality TV formats associated with the genre's rapid growth in the 1990s and 2000s it may be more accurate to think of the promise of reality as construed not in opposition to artifice, but rather to *abstraction* – that is, the forms of rationalization and differentiation that come to characterize industrial society and the mass media products it produces. The 'reality' of reality TV in this context is reframed as a challenge to the top-down forms of centralized media production reliant upon remote decision-makers serving up pre-scripted commodities to a mass audience. In this regard, the reality TV promise aligns itself with broader transformations in the media environment that characterize the turn of the millennium. Just as the internet mobilizes the promise of interactivity as an antidote to the depredations of top-down, one-way forms of mass produced culture, reality TV offers the promise of participation either directly or by proxy as a means of countering the non-participatory character of industrially produced culture. Reality TV, in this sense, is presented as *real* not because of an absence of mediation, but because of the ways in which this mediation promises to overcome the forms of differentiation associated with mass societies and their media. With the advent of industrialization, the critique of the forms of de-skilling associated with the abstraction of the mental work of planning from the mechanical routine of labor on the factory floor carries over into the critique of mass media for a kind of de-skilling of the audience. The notion of a true popular culture as that which is created by and for the people as a participatory endeavor is, according to this critique, replaced by mass-produced commodity culture in which a clear distinction develops between performer, creator and audience.

The promise of reality TV then is to overcome the distinctions of mass media from within. Reality in this context refers not to the lack of artifice, but to the participation of the public in the artifice they consume – and help to create. In this regard, the appeal to reality and the related notion of authenticity promises to overcome the estrangement of mass culture, to allow viewers to recognize their own contributions in the culture they consume, whether directly or by proxy. Direct forms of participation are characteristic of shows that allow viewers to shape the outcomes, as in the case of the *Idol* and *Big Brother* formats. Participation by proxy refers to the use of cast members who are selected from the audience, which is then invited to identify with them as representatives of the 'real' people who comprise the viewership. Research has indicated that viewers tend to assess the authenticity of cast members and to critique those who are considered to be phony or playing for the camera (see, for example, Andrejevic, 2004; Hill, 2005). Perhaps this behavior is critiqued because the intrusion of performance reinstates the boundary between audience and talent, and undoes the promise to counter the abstraction of mass culture. Indeed, the genre puts a premium on cast members who are willing to be themselves, warts and all, for the camera.

THE POLITICAL ECONOMY OF REALITY TV

Reality TV is more than a genre; it is, as McCarthy (2004a) has noted, a mode of production, and this is why its lifespan is not to be measured solely in terms of trend cycles or programming fads. In many countries reality TV relies on a different set of labor relations than scripted programming insofar as it replaces unionized actors and writers with (in many cases) lower paid or unpaid flexible labor. In keeping with the broader promise of interactivity in the digital era, reality TV offers up as its own reward – the promise of participation in a realm from which the audience has long been excluded. In so doing it offloads many of the costs of TV production onto selected members of the audience. Consider, for example, the case of the Montour Spartans, a high school football team who served as the subject of a reality show called *Bound for Glory* that documented their rather dismal 2005 season. The show's producers obtained a full cast, complete with football players, coaches, parents, friends, teachers, and more, plus a complete 'stage' set including the high school, the football field, and so on, in exchange for locker room renovations, weight room equipment and a few other amenities.

Because of such savings, reality TV shows have been estimated to cost, on average half as much or less than scripted formats, but like many averages, this figure can be misleading (Hirschorn, 2007). Formats like *Survivor* which film in remote locations with a full production crew and sets can be comparable in cost to scripted shows, whereas found video shows like the funniest home video formats cost only a tiny fraction of a comparable length fictional format. In general, however, reality formats are cheaper, more flexible formats that can be used by broadcasters to counter the threat of actors' and writers' strikes. With respect to key members of the production crew, producers in the US have been able to sidestep collective bargaining agreements:

> From the outset, owners have insisted that producers and editors are not 'writers' who pen scripts and dialogue, and so the WGA [the writers' union] was shut out of reality programming. As a result, the sector teems with substandard conditions – eighteen-hour work days, chronic job instability, no meal breaks, no health benefits, and employer coercion to turn in time cards early ... When employees of *America's Top Model* voted to join the WGA in summer of 2006, they were summarily fired. (Ross, 2009, pp. 136–137)

As for cast members, reality TV's ability to use unpaid or low-paid labor has hinged on its claim that the cast is not working, but rather participating in a voluntary game show or documentary. Compensation is much higher for some celebrity participants in formats like *The Osbournes* or *The Girls Next Door* (which documented the life of Playboy mogul Hugh Hefner and his three girlfriends).

The need for low-cost programming was exacerbated by the advent of satellite, cable and, somewhat later, digital TV in many markets as well as the rise of commercial networks in regions once reliant on public or state broadcasting. In the United States, for example, Raphael (2004) argues that reality TV served as a response to the growing number of video channels, the high levels of corporate debt taken on by networks in the 1980s, and changes in ratings measurement techniques that, in some cases, lowered ratings and hence ad revenues for the networks. Cost-cutting aspects of the genre included its reliance on non-union labor and its cultivation of relatively low production value aesthetics (especially in the found video and *Cops* formats), and its ability to be sold both in its original form and as a franchise format in international markets. Internationally, reality TV catered to the need of public broadcasters in other parts of the world to compete with growing competition from commercial broadcasters as the public service model in many nations gave way to a more commercially driven system.

The fact that reality TV can be a relatively low-cost genre helps explains its international character. As Hollywood producers, long reliant on the creation of costly, high-production-value formats as a means of staving off international competition sought cheaper fare, they turned to formats pioneered abroad, often in smaller, less well-financed markets. Thus, for example, the funniest home video format, which became a staple of low-budget reality TV in the US and elsewhere, traces its roots back to a segment in a Japanese variety show that capitalized on the growing popularity of portable home video recorders in the late 1980s. Formats that swept the US market in the early 2000s, including *Survivor* and *American Idol* were first developed in Sweden and the UK respectively.

Catering to local markets can mean making significant format changes, as in the case of the US version of *Big Brother*, which switched from an audience elimination vote to a cast elimination vote after it became apparent that US audiences were evicting the cast members who created the most controversy and thus the most drama and ratings appeal. The Chinese adaptation of a *Survivor*-style format famously focused not on individual competition but on collective team building (Turner, 2010), and in the Middle East, the *Star Academy* talent show format has been re-invented as a competition in reciting the Koran (Kraidy, 2009). The synergies of such global-local formats help build an international brand reputation for localizable formats.

Despite the fact that reality TV cast members are, in many cases, contributing to the production of successful commercial formats, they have not, for the most part, been accorded the legal status of workers. The question has, however, been raised, especially with respect to children's participation, whether cast members should be subject to the protections of labor laws and collective bargaining agreements. In the US, the popular reality show *Jon & Kate Plus Eight*, triggered an investigation into whether producers were complying with state labor laws, and another US reality format with a child-dominated cast, *Kid Nation*, raised similar questions. In the United Kingdom, the government has expressed concern about the treatment of children on shows like *Britain's Got Talent* after a 10-year-old broke down in tears during the show's semi-final competition. In response to concerns over the treatment of reality TV cast members, the French courts have ruled that some reality TV participants are governed by French labor laws and are thus entitled to receive minimum wage, vacation time and overtime. Given the genre's reliance on cheap labor, it is not surprising that the French court ruling led to predictions of the demise of French reality TV.

For the most part, however, reality shows seem to operate in a loophole opened up by game shows and documentaries. Even though cast members spend long hours in front of the cameras to generate revenues for commercial organizations and often sign draconian contract agreements ceding control over the rights to their life stories, their portrayal, and even their ability to speak to media outlets or to profit from their appearance on reality TV, they are nonetheless treated not as workers but as volunteers. The producers reap the benefits of a vast surplus casting pool: they can get participants to sign such one-sided contracts because there are plenty of others in line who will sign if they do not. As Turner has pointed out, reality TV celebrities, unlike other performers, negotiate their contracts before they have received any of the publicity that might give them bargaining power (2010, p. 36). This is the advantage of the model of dispensable celebrity (Collins, 2008) exploited by reality TV producers: they do not need to pay a premium for cast members who are, for the most part, unknown (or, in the case of celebrity reality TV, well past the peak of their fame). Moreover, reality TV fame tends to have a short shelf-life. A tiny fraction of cast members go on to more lucrative forms of fame, whereas the vast majority discover that their celebrity was a result of the show's success, rather than the other way around. In the case of reality TV, the formats and not the fungible cast members are the real stars.

The reality TV mode of production lends itself to the sensibility of a media savvy era, one in which reflexivity about the media publicity machine is taken for granted. The construction of the spectacle is pushed to the fore as an integral part of the spectacle itself. If the content of reality TV is, at one level, widely variable, ranging from pranks to competitions, to behind-the-scenes views of domestic and professional lives, it is in another respect quite uniform: it is fascinated with the productive power of the media as a celebrity-generating apparatus. This fascination peaks in a self-stimulating cycle when the media marvels at its own power, which it describes as the audience's own in a hyper-interactive era.

The marriage of reality TV and the internet is built in part upon their simultaneous embrace of the promise of interactivity. Reality TV exemplified the promise of the internet (or perhaps, vice versa) – that everyone could become a star. The popularity of webcam sites, which 'netcast' images of the daily lives of their subjects to anyone who might be interested, anticipated the reality TV boom of the turn of the millennium. Like reality TV, the internet promised to reconfigure the boundary between audience and performer, consumer and producer. Given this affinity, it is not surprising that the reality TV shows of the early 2000s found a variety of ways to leverage their interactive appeal across multi-media platforms including the internet and mobile phones. Versions of the *Big Brother* format, for example, provided online access to live video feeds of the *Big Brother* house, allowing viewers to follow a live version of the show online.

Many other reality formats created their own online presence, complete with video clips, bulletin boards, chat rooms, and other supplementary online content. The web presence was not limited to official sites, however. Several formats spawned viewer created sites that fostered everything from ardent fandom to mockery. Early on in the run of the shows in the US, for example, SurvivorSucks.com and BigBrotherBlows.com were popular sites that, even when critical about the shows, helped to generate both audience and media attention. Viewers went to these websites not just to share their reactions to the show, but also to share their research about the show and to speculate about its outcome. Long discussion threads were devoted to unearthing inadvertent clues dropped by producers and editors about the finalists for *Survivor*. Producers played along by providing false leads or indeterminate information: leaving some cast members, for example, out of preview footage to suggest that they may have been eliminated in subsequent episodes. Some websites like *Television Without Pity* devoted forums to particular shows that featured episode recaps and discussion boards. Other sites sprung up to categorize reality TV shows and provide news about forthcoming formats.

In an era in which TV producers started to worry about competition from the internet, reality TV helped demonstrate some of the ways in which the old medium might adapt to and embrace the new. Producers discovered that the internet was a way to allow viewers to stay in touch with the show between episodes – it served both a promotional and a feedback function, providing information about which characters received the most online attention and how viewers reacted to developments in the show. For some shows, the forums came to serve as an online focus group and even a site for cast members to interact with viewers. With the advent of YouTube and other online video sites, the internet became a place where viewers could catch up on missed episodes and view and share highlights. In an era in which TV producers emphasize the crucial importance of a multi-platform presence for their programming, reality TV provided ample opportunities for tapping into the interactive capability of the internet and for benefiting from the activities of online fans.

The rapid spread of mobile telephony during the same period also fit well with the multi-platform strategy of voting-based reality shows. People's choice shows like the *Idol* format and some versions of *Big Brother* could generate revenues by allowing viewers to vote by phone. With the development of mobile phones with Internet capability, these interactive platforms converged, and it became possible to call in a vote to a reality show with the same device that provided access to online episodes and to the show's web presence.

As TV undergoes transformations associated with the changing media environment, reality TV has served as a suitably flexible and adaptable format. It has, for example, provided ample opportunities for addressing the threat to interstitial advertising posed by video recorders and time shifting, proving itself amenable to new forms of sponsorship and product placement. In the US, the *Idol* format has signed lucrative sponsorship agreements with companies including Ford, Subway and Coke, regularly incorporating the products and brands into the show's content. The makeover formats have proven particularly adept at product placement, not least because they tend to rely heavily on consumer goods and services. Even the successful *Survivor* series, which takes place in locales notable for their lack of consumer goods, relies heavily on product placement. As *Survivor* producer Mark Burnett put it, the show 'is as much a marketing vehicle as it is a television show. My shows create an interest, and people will look at them, but the endgame here is selling products in stores – a car, deodorant, running shoes. It's the future of television' (as quoted in Sager, 2001).

A DEMOCRATIC GENRE?

Because reality TV has styled itself as the people's genre – one which allows audiences access to realms previously limited to a small group of professionals – it has frequently been referred to as a democratic genre. As a researcher for *American Demographics* put it, summarizing the results of an audience survey, 'The popularity of this format with youth also has a lot to do with their growing up in a democratized society, where the Internet, Web cams and other technologies give the average Joe the ability to personalize his entertainment' (Gardyn, 2001, p. 37). The ready assimilation of personalized entertainment to political democratization via new media is a recurring theme in the celebratory musings of the promoters of mass-customization in the interactive era. Folding reality TV into the mix has led some critics and pundits to discern in the genre's persistent and global popularity evidence of a universal yearning for participatory democracy. In its account of *Big Brother: Africa*, for example, *The New York Times* implied that part of the format's popularity derived from the fact that it modeled democracy for a 'strife-torn' continent:

> *Big Brother* defenders are many. Though it may be subtle, one theme they point to is democracy. The contestants are nominated for eviction by their housemates and then voted off by viewers on the Internet or by cellphone text messaging. The will of the people decides how the show unfolds. (Lacey, 2003, p. 8)

The media reception of reality TV as emblematic of democratic desire is not limited to the US. A Malaysian media commentator explored at length the 'political undertones' of successful reality TV, understood as 'an exercise in participatory democracy, of sorts' that stimulated (or simulated) a vibrant public sphere:

> Viewers get to vote to help determine the winners. People debate the merits of their candidates in office corridors, chat rooms, restaurants. They rejoice when their favourites win, and grouse about 'rigged' voting when they get booted out ... You cannot get any more democratic: the people have the power here and they are itching to use it. When you have emotions as the driving force, and mobile phones at the ready, you can bet people will continue to participate. (Samat, 2004, p. 1)

Writing about the Australian reception of the genre, one journalist observed that, 'Reality TV represents the democratisation of television ... Today anyone with a phone and 55 cents can take charge' (Ferguson, 2004, p. 11). In the UK, the head of the company that brought the popular reality *Big Brother* show to the UK, Peter Bazalgette, received widespread news coverage for his claim that reality TV might serve as model for boosting public interest in the political process – suggesting that the House of Commons might benefit from secret voting and regular online votes 'to make Parliament more interactive' (Hinsliff, 2003, p. 15).

Also drawing on the UK success of *Big Brother*, Coleman's (2006) study of the show's viewers suggests that reality TV might serve as the model for a form of vernacular politics that overcomes the boundary between high politics and low culture. Shows like *Big Brother*, Coleman (2006) suggests, provide a kind of compensatory participation directed not toward distant political representatives and the issues they debate, but, rather, toward more accessible and popular aspects of social life with which viewers can directly identify. As he puts it:

> *Big Brother*, although substantially incomparable to a general election, is characterized by the very qualities that *political* participation lacks. *Big Brother*, in this sense, represents another side of democracy; a counterfactual democratic process in which conspicuous absences in contemporary political culture are played out. (Coleman, 2006, p. 458)

In a similar vein, Hartley has described reality TV as a paradigmatic 'plebiscitary format' – one in which the core democratic practice of voting has been revitalized, 'as if someone had pressed the refresh button on one of the oldest technologies of democracy' (2008, p. 126). While acknowledging that the shows are not necessarily political in the narrow sense, he insists that, 'the plebiscitary format *is* an experience of democracy' (2008, p. 126). This is a generously expansive, not to say slightly vague definition of democracy as, 'doing something together, not just being told what to do' (2008, p. 126). Coleman has argued that politicians have much to learn from the plebiscitary appeal of *Big Brother*:

> If politicians really want to reconnect (or, most likely, connect for the first time) with broad sections of the public that have come to regard them as irrelevant, malevolent or worse, they may need to come to terms with approaches to representation that capture the symbolic, dramatic and banal aspects of human experience. (Coleman, 2006, p. 458)

Perhaps the advent of political blogging and 'tweeting' has already started to realize the reality-TV version of politics anticipated by Coleman. Whether or not the result ends up being as politically 'invigorating' as he suggests remains subject to debate, but the reality TV industry has already capitalized on the political comparison with a sub-genre devoted to politics. If reality TV is a democratic genre – a model for the democratization of entertainment, according to such accounts – why not take it as a model for making democracy itself more entertaining? In 2004 and 2005, three political reality shows were introduced – one each in Australia, Great Britain and the US – that featured a political competition, voting, and the selection of a winning candidate by viewers. In each case, the premise of the show was that the winner could draw on the prize money and publicity to mount a political campaign, though the US version eventually backed off and switched the timing of the show to make it impossible for the winner to meet the deadline for securing a place on the ballot.

In each case, moreover, the stated objective of the producers was to jumpstart political engagement by drawing on the popularity of reality TV. As the head of current affairs for England's version of *Vote for Me* put it, 'We know about the power of television ... and here we're trying to use it for something good and worthwhile' (as quoted in Showtime Public Relations Release, 2004). In this regard, it might be possible to argue that reality TV was hearkening back to (one of) its roots in the documentary tradition devoted to the promotion of civic participation and democratic citizenship.

The end result was indecisive at best: *American Candidate* failed to generate more than a half-hearted buzz and fared poorly in the ratings. The candidates fielded by both versions of *Vote for Me* did run for office, but made hardly a dent in the election returns, the British candidate finishing not only far behind the winner, but also behind a perennial candidate who called himself Lord Toby Jugg of the Monster Raving Looneys (Freeman, 2004, p. 1). It would be hard to make the case for the significance, in its own right, of a show like *American Candidate*, despite the fact that it was, for the most part, an intelligent, well-constructed and even engaging series – certainly more so than many of its ratings-grabbing cousins in the reality TV world. After 2005 the political sub-genre largely faded from the scene, along with the dream of R. J. Cutler (producer of the show) that reality TV could serve as an alternative route to public office: one that bypassed both the machine of the political party and the increasingly costly campaign process that, in the United States, has made it difficult for all but the very wealthy and well-connected to mount a nationwide campaign. The political promise of reality TV was that, as in the case of other talent-search shows, the media could level the playing field by making it possible for a talented unknown to gain the recognition that might serve as the basis for a successful campaign. As in the case, of the entertainment industry, the genre was offered as a way of bypassing existing pathways to success and creating a new one that might give anyone a chance.

As for other forms of reality TV, these can only stand as examples of a 'democratic' turn in entertainment programming if one subtracts at least one key element from any recognizable definition of what the term might mean: the element of self-rule. Democracy that isn't about deliberating over and shaping the policies that govern our lives is hardly deserving of the name. This is not to denigrate the cultural and personal significance for viewers, but to suggest that some difference, at least conceptually, remains between the realms of commercial entertainment and politics. As Turner (2010) points out, 'there is no necessary connection between on the one hand, a broadening demographic in the pattern of access to media representation, and on the other hand, a democratic politics' (p. 134).

In the end, the connection between reality TV and democracy can be read through the assimilation of the marketplace to democracy: the notion that consumer signalling is tantamount to – or perhaps preparation for – direct participation. Much of what passes for participation on interactive reality TV formats – voting, posting on message boards, e-mailing producers or cast members – might be better described as focus group activity: the generation of consumer preferences and reactions for producers. The *Pop Idol* format, for example, doubles as nation-wide market research and promotion for products that will be sold at season's end: albums and concerts by the show's top performers. This is participation, and it is a way of aggregating and taking into account viewer preferences, but only to sell them a product that they have helped to create – that is, to enlist them in the process of marketing to themselves.

A PARTICIPATORY SPECTACLE: REALITY TV AND GOVERNANCE

The 'demotic' (Turner, 2010) promise of reality TV implies a critique of mass media culture, whether produced by the state, public service entities or commercial institutions, for being engineered from above and thus non-participatory. In this regard it embraces the more general promise of interactivity. When *The New York Times* can refer to the broadcasting era, viewed against the background of new interactive technology, as 'a Stalinist plot erected, as it has been, on force from above rather than choice from below', it sounds like it is channelling the arguments of critical theory (Lewis, 2000, p. 36). Debord's (1995) famous description of the society of the spectacle, for example, describes a world in which the mass media have taken on an autonomous and oppressive logic of their own: 'The spectacle is the nightmare of imprisoned modern society which ultimately expresses nothing more than its desire to sleep. The spectacle is the guardian of sleep' (p. 22). Within such a system, 'The concentration of "communication" is thus an accumulation, in the hands of the existing system's administration, of the means which allow it to carry on this particular administration' (Debord, 1995, p. 27). The notion of the spectacle is associated with the passivity of the populace.

The political promise of reality TV is predicated on its challenge to placation and passivity. If lack of participation is authoritarian, oppressive and demobilizing, the story goes, then participation may be empowering, democratic, and invigorating. Critical media studies, however, has raised the possibility of a 'third way' – one which does not reproduce the binary logic of passivity as authoritarian and participation as empowering. Drawing on the Foucault-inspired governmentality literature, scholars have considered what it might mean to think of a participatory spectacle – one in which the conduct of a populace is managed in large part by their own increasingly active participation (see, for example, McCarthy, 2007; Ouellette, 2004; Ouellette and Hay, 2008; Palmer, 2003).

This scholarship has focused largely on the 'makeover' sub-genre of reality TV programming: shows that instruct participants and, by extension, viewers, in technologies and practices for managing their lives, and taking care of their careers, homes, children, bodies and selves. Like many reality formats, makeover shows mix entertainment with pedagogy. Even as they are presented with egregious examples of failures of self-management or personal tragedy, such shows instruct viewers in techniques and technologies for taking responsibility for an increasing range of career-, relationship-, or health- and happiness-threatening liabilities, ranging from therapeutic intervention to plastic surgery. The message of personal responsibility in such shows takes shape against the background of the dismantling of the welfare state associated with the rise of neo-liberalism and the forms of competition that characterize the global triumph of capitalism. Against the background of privatization and market processes, the wager of neo-liberal forms of government is that public conduct can be regulated through the identification and internalization of economic and social imperatives that maximize chances and manage the risks of the marketplace. The concept of governmentality, in this context, serves as, 'a powerful model for understanding the cultural and political manifestations of neoliberalism, in which state policies synchronize with cultural practices to apply market-based individualism as a governmental rationale across the institutions and practices of everyday life' (McCarthy, 2007, p. 20). What such an approach brings to makeover shows is an understanding of the way in which they model the regulation of conduct in the name of freedom, participation and self-realization. As Ouellette puts it in her analysis of the US reality show *Judge Judy*, in which a real judge adjudicates disputes in a mock TV courtroom, such shows, 'construct templates for citizenship that complement the privatization of public life, the collapse of the welfare state, and most important, the discourse of individual choice and personal responsibility' (Ouellette, 2004, p. 232). Makeover shows, like *Judge Judy*, then, can be described as modeling forms of '"cultural training" that govern indirectly in the name of "lifestyle maximization," "free choice," and "personal responsibility" in the context of the mobilization of a widening array of risks' (Ouellette, 2004, p. 232, internal quotes are from Rose, 1996, pp. 55, 58–9). This is not to say that such shows, in themselves, serve as behavior templates for viewers, but rather to highlight the way in which they portray and embody emerging regimes of participatory governance – control in the name of the maximization of freedom – and fit within a widening array of self-help media and the galloping privatization of social services and social life.

REALITY TV AND THE SURVEILLANCE SOCIETY

The element of voyeurism on one hand and exhibitionism on the other has been a recurring theme in popular criticism of reality TV for helping to erode the private sphere and promoting a culture of self-display and self-disclosure. Again, reality TV needs to be situated in the broader context of new forms of self-display and 'ego-casting' (Rosen, 2005, p. 51) enabled by the internet. The progression from personal web pages to webcams, blogs, MySpace, YouTube and Facebook fits neatly alongside the varied forms of self-disclosure that take place on reality TV. If participants have been accused of an inordinate tendency to self-disclosure, the charge of voyeurism has typically been used to denigrate both the genre and its audiences: 'Voyeurism or peeping tom syndrome, links with the other two topics of harm and "dumbing down", as voyeurism implies watching reality TV is a form of socially deviant behaviour' (Hill, 2005, p. 84). By contrast, Hill argues, based on her interviews with audience members, that reality TV is better understood as a form of 'people watching' that may serve an educational function despite the fact that, 'audiences consider much reality TV to be entertaining rather than informative' (Hill, 2005, p. 71).

Hearkening back to the social psychology experiments of the 1960s, many reality TV shows invoke an implied informational or educational function, describing themselves directly as social experiments. Some formats enlist the contributions of psychologists or other experts to interpret the show for audiences, providing lessons in people watching and social dynamics. Makeover shows and intervention shows (in which friends and family members confront someone about their addiction or their wardrobe) in particular incorporate an instructional component, providing 'expert' commentary and analysis to guide cast members and explain the interventions to audiences. Viewers are invited to learn not just from the experience of others but also from the guidance of the expert consultants.

If such programming highlights the educational and experimental value of reality TV, it also often emphasizes a related range of benefits for cast members. The cast members on shows like *Big Brother* and *The Real World* (a format that places several young people in a house to live and work together) repeatedly invoke the therapeutic benefit of being watched as an opportunity to learn about themselves and to work through various personal and professional issues (Andrejevic, 2004). The opportunity of participating as a cast member on a reality show is portrayed as a way of learning how to maximize one's personal assets and thereby to get the most out of life. In this regard, such shows frame a willing submission to monitoring not as a form of oppressive surveillance, but as a liberating opportunity for self-expression via self-disclosure.

This framing of the role of monitoring and surveillance should be viewed within the larger context of the proliferation of state and commercial forms of monitoring ushered in by digital media at the turn of the millennium. More than ever before in human history, the wired populace finds itself subject to comprehensive forms of monitoring, to the collection and storage of information about the details of their daily lives. Much of this monitoring is done by commercial organizations for marketing purposes, and the information collected serves as a resource or asset that can be bought and sold.

The offloading of some aspects of production onto consumers and viewers reflects a more general trend in the information economy. In both the case of reality TV and that of the emerging interactive economy, the contribution of the audiences comes in the form of willing or unknowing submission to increasingly comprehensive forms of monitoring. In both cases this activity is portrayed as a form of participation, an opportunity to contribute by expressing oneself in the form of self-disclosure.

Reality TV may not *create* a productive surveillance society, but it stages and partakes of its logic. It reframes willing submission to monitoring in the post-Cold-War era as a form of democratic empowerment, participation, self-expression – in short, the exact opposite of what surveillance stood for in the era of Orwell's Cold War Big Brother. At the same time, from the perspective of the viewer, monitoring is framed as a strategy for gauging authenticity as well as a form of experimental and instructional observation. Hill has argued that reality TV, insofar as it prompts questions and debates over the authenticity of cast members' performances, 'invites a critical viewing mode': 'The genre invites the viewer to engage in debate, to question what is authentic and what is staged, to judge the actions of non-professional actors faced with challenging situations' (Hill, 2007, p. 140). In so doing, it invites viewers to adopt the monitorial mode: to embrace surveillance as a means for getting at the truth of the behavior of others. In keeping with an attitude of reflexive savviness, it fosters a fascination with behind-the-scenes access as an avenue for debunking misleading appearances and arriving at underlying truths.

Transposing this logic into the realm of power and politics, Coleman (2003) has claimed that this critical mode is a potentially empowering one, claiming that a show like *Big Brother* is political in the sense that, 'it shifts control toward the receiver of messages and makes all representations of reality vulnerable to public challenge and disbelief' (p. 35). By contrast, Palmer (2003) has argued that surveillance themed shows – and he has in mind CCTV shows that allow viewers access to the images videotaped by authorities – foster a form of identification with authorities. Rather than questioning the monitoring power of those who operate the surveillance cameras, he claims that such shows, 'invite us to share the perspective of other watchers, to be in the viewing booth with the other instruments of authority. The very idea of taking a critical distance on the institution doing the surveying is becoming untenable' (Palmer, 2003, pp. 35–6). The paradox of the critical attitude modeled by reality TV, then, emerges in the way the promise of access to authenticity via monitoring invites both a savvy skepticism toward contrivance but also an uncritical identification with surveillance strategies themselves: a paradoxically uncritical skepticism, or naïve savviness.

THE FUTURE OF REALITY TV

The popular press has tended to treat reality TV as simply one genre among many, and thus its popularity as a fad or passing trend whose life cycle will mimic those of other popular genes like medical dramas or police procedurals. Understood as something more than a trend – as a mode of production suited to an era of flexible, customized, globalized, multi-channel and multi-platform media, reality TV is not likely to fall out of favor with programmers any time soon. Viewers may well tire of particular formats over time, but reality TV is not a genre in the standard sense of being circumscribed by formal conventions: it has the ability to continue to develop new sub-genres that have very little in common content-wise with preceding formats. Rather, the trajectory of reality TV has been one of multiplication and permutation triggered by occasional innovation. Reality TV is a parasitic and imitative format, one that builds on developments in daily life and past formats – both fictional and reality – to create new shows.

The economic viability of reality formats will depend upon the ready availability of inexpensive labor, whether in the form of unpaid cast members, non-union story editors, and faded celebrities willing to work cheaply for another shot at fame. This means much depends on a regulatory structure that continues to allow producers to claim that cast members are not employees and therefore not subject to workplace regulations or collective bargaining agreements.

Reality TV, like other forms of 'interactive' culture that cater to mass audiences and welcome their participation, will continue to serve as a target for cultural critics, who have accused it of debasing the culture, eroding privacy and fostering an unhealthy obsession with celebrity. There is a moralizing tone to such critiques, which treat the desire for reality-TV fame as one more get-rich-quick scheme: an attempt to claim the desserts of celebrity without having done the work or truly deserving it.

From a media studies perspective, reality TV will continue to be not just one more genre, but rather a form of media production that stages the rise of digital interactivity more broadly, along with the economic imperatives of a changing political and technological environment. As contrived as it may be, reality TV reflects back to us the promise of participation refracted through the lens of contemporary media culture in the era of global capitalism.

REFERENCES

Alexander, S. (1973, January 19). The silence of the Louds. *Newsweek*, pp. 23–24.

Andrejevic, M. (2004). *Reality TV: The Work of Being Watched*. Boulder, CO: Rowman and Littlefield.

Barbash, I. and L. Taylor (1997). *Cross-Cultural Filmmaking*. Berkeley: University of California Press.

Beattie, G. (2003). *Visible Thought*. London: Routledge.

Bondebjerg, I. (2002). The mediation of everyday life: Genre, discourse, and spectacle in reality TV. In Anne Jerslev (Ed.), *Realism and 'Reality' in Film and Media*. Copenhagen: Museum Tusculanum Press, pp. 159–192.

Coleman, S. (2003). A tale of two houses: The House of Commons, the *Big Brother* house. *The Hansard Society*. http://www.hansardsociety.org.uk/wp-content/uploads/2012/10/A-Tale-of-Two-Houses-2003.pdf.

Coleman, S. (2006). How the other half votes. *International Journal of Cultural Studies*, 9, pp. 457–479.

Collins, S. (2008). Making the most of 15 minutes: Reality TV's dispensable celebrity. *Television & New Media*, 9(2), pp. 87–110.

Corner, J. (2002). Performing the real: Documentary diversions. *Television & New Media*, 3(3), pp. 255–269.

Debord, G. (1995). *The Society of the Spectacle*. New York: Zone Books.

Ferguson, T. (2004, July 12). Democracy finally gets chance of stardom. *Courier Mail* (Queensland), p. 11.

Freeman, S. (2004, May 6). Election night: How it unfolded. *The Times* (London), p. 1.

Gardyn, R. (2001, September). The tribe has spoken. *American Demographics*, pp. 34–40.

Godard, E. (2003). Reel life: The social geometry of reality shows. In Matthew J. Smith and Andrew F. Wood (Eds), *Survivor Lessons: Essays on Communication and Reality Television*. Jefferson, NC: McFarland & Co., pp. 73–96.

Hartley, J. (2008). *Television Truths*. Malden, MA: Blackwell.

Harvey, D. (1990). *The Condition of Postmodernity*. Cambridge: Blackwell.

Hill, A. (2005). Reality TV: *Audiences and Popular Factual Television*. London: Routledge.

Hill, A. (2007). *Restyling Factual TV: Audiences and News, Documentary and Reality Genres*. London: Routledge.

Hinsliff, G. (2003, June 1). Is *Big Brother* key to winning next election? *The Observer*, p. 15.

Hirschorn, M. (2007, May). The case for reality TV. *The Atlantic*, pp. 32–33.

Kavka, M. (2008). *Reality TV, Affect and Intimacy: Reality Matters*. London: Palgrave Macmillan.

Kilborn, R. (1994). How real can you get? Recent developments in 'reality' television. *European Journal of Communication*, 9(4), pp. 421–439.

Kraidy, M. (2009). *Reality Television and Arab Politics*. Cambridge: Cambridge University Press.

Lacey, M. (2003, September 4). Reality TV rivets Africa, to the churches' dismay. *The New York Times*, p. 8.

Lewis, M. (2000, August 13). Boombox. *The New York Times Magazine*, pp. 36–52.

McCarthy, A. (2004a). Reality TV and its implications for television studies. Presentation to the annual convention of the Society for Cinema and Media Studies in Atlanta, GA, March 5.

McCarthy, A. (2004b). 'Stanley Milgram, Allen Funt, and Me': Postwar Social Science and the 'First Wave' of Reality TV, in Susan Murray and Laurie Ouellette (Eds), *Reality TV: Remaking Television Culture*. New York: NYU Press, pp. 19–39.

McCarthy, A. (2007). Reality television: A neoliberal theater of suffering. *Social Text*, 25(4) (Winter), pp. 17–41.

McCarthy, B. (1973, July). An American Family & The Family of Man. *Atlantic Monthly*.

Meyrowitz, J. (1986). *No Sense of Place: The Impact of the Electronic Media on Social Behavior*. Oxford: Oxford University Press.

Meyrowitz, J. (1995). Mediating communication: What happens? In J. Downing et al. (Eds), *Questioning the Media: A Critical Introduction*. London: Sage, pp. 47–56.

Murray, S. (2004). I think we need a new name for it: The meeting of documentary and reality television. In S. Murray and L. Ouellette (Eds), *Reality TV: Remaking Television Culture*. New York: NYU Press.

Ouellette, L. (2004). Take responsibility for yourself. In S. Murray and L. Ouellette, *Reality TV: Remaking Television Culture*. New York: NYU Press, pp. 231–250.

Ouellette L. and J. Hay (2008). *Better Living Through Reality TV*. London: Blackwell.

Palmer, G. (2003). *Discipline and Liberty: Television and Governance*, Manchester: Manchester University Press.

Pine, J. (1993). *Mass Customization: The New Frontier in Business Competition*. Cambridge: Harvard University Press.

Raphael, C. (2004). The political economic origins of Reali-TV'. In S. Murray and L. Ouellette, *Reality TV: Remaking Television Culture*. New York: NYU Press, pp. 119–136.

Rose, N. (1996). Governing 'advanced' liberal democracies. In A. Barry, T. Osborne and N. Rose (Eds), *Foucault and Political Reason*. University of Chicago Press, pp. 37–64.

Rosen, C. (2005). The Age of Egocasting. *The New Atlantis*, fall/winter, pp. 51–71.
Ross, A. (2009). The political economy of amateurism. *Television & New Media*, 10, pp. 136–7.
Sager, M. (2001, July). What I've learned: Mark Burnett. *Esquire*, 136: 94.
Samat, H. (2004, July 18). Reality takes a deep bite. *New Straits Times* (Malaysia), p. 1.
Showtime Public Relations Release (2004, January 8). Showtime to introduce *American Candidate*.
Turner, G. (2010). *Ordinary People and the Media: The Demotic Turn*. London: Sage.

22

Television Drama

Jason Jacobs

From the beginning of the medium, television drama has been held up – rightly or wrongly – as an emblem of its artistic and aesthetic possibilities, although it has always suffered at the same time by perceived 'contamination' because of its proximity to other narrative forms such as film, radio and theatre. Hence in the earliest days, TV drama practitioners attempted to articulate (in keeping, implicitly, with the predominant modernist discourses of the time) television's distinction as a medium, its separateness from other creative presentations of dramatic material. This insistence on the purity of a medium has continued to maintain a grip on producers and critics alike, producing a continually renewable series of assertions about the emergence and loss of various 'Golden Ages' of drama, such as the early to mid-1950s US network anthologies, the innovations of the British single television play in the 1960s, up to the approbation heaped on the current HBO/cable signature dramas of the past fifteen years. Given the emblematic status of drama as part of the definition of what television is capable of, and the striving for various kinds of purity in the estimation of drama's contribution to the distinction of the medium, it is illuminating to track the various ways in which the evaluation of the significance of drama as a form of television output has been employed as an object in ongoing debates about the medium's value.

Of course television drama meant, and means, something different from television fiction in general. In the 1950s and 1960s, certainly in the UK, drama referred to either single 'plays' (the word reflecting the sedimented inheritance of theatrical tradition) – usually one hour or 90-minute one-off narratives – or episodic serials, but did not normally include other clearly dramatic forms such as soap opera. That is, television drama was a kind of elite genre separate from, say, the cop show or medical drama. By the end of the 1950s in the US 'drama' was much more quickly subsumed within ongoing generic series and it was not until recently that signature dramas, although still episodic, emerged as prominent artistic contributions in their own right, resembling the strongly authored television plays of the UK, and indeed there has been some resentment from British cultural nationalists at the supposed failure to acknowledge this (see, for example, Brunsdon, 2008, 133–4). But there is a deeper aspect to this than just industrial differentiation. For, as I've suggested above, the ability of television to produce *elite* dramatic instances rather than 'merely' innovating within existing generic staples, was and is seen as an indicator of the aesthetic health of the medium as a whole, not just its nationally relevant storytelling abilities. Maintaining and policing this boundary between the merely popular and the authentically artistic (authentic because it both exemplified the possibilities of the medium and could participate in the wider history of valorized aesthetic form) became a central critical activity and one that was put under considerable scrutiny as television studies developed its critical momentum.[1] As one-off anthology dramas gave way to the hybrid serialized-serial in the 1980s and 90s, that desire to keep a separation between elite drama and the rest continued. What I want to do here is (briefly) track what I see as the *consequences* of this development of critical and academic responses to elite television drama of this kind. But rather than pointing to an ambient modernist critical stance as a justification for identifying the possibilities of the medium, I will argue that the suppression of a nurturing a critical ambition that can articulate the achievement of fine television drama has become extremely problematic, despite the fact that it is and should continue to be essential for strengthening and calibrating our judgement of television fiction as a whole.

To make this case clear I will reframe current and past events through the lens of a (schematic, but helpful) binary that seeks to capture two intersecting and dominant trends in the scholarly and critical accounts of television drama. Since the early days of staging excerpts from current West End plays for television (as the BBC did in the late 1930s) there has been a debate about the function of television drama, one that draws thinking toward two paradoxically overlapping poles.

First, there is the approach to drama as a means to represent, promulgate and distribute content of national or cultural importance to the wider mass audience, and which maintains that study or analysis of such content can tell us a lot about the way an author, an institution and a culture thinks and feels about itself. To adapt Charles Barr's account of Ealing Studios we might call it a 'reflection/projection' dialectic, where the mediation of cultural values and content is understood as a mechanism whereby the lived experience of a time and place is recruited for both ideological and aesthetic purposes by highly organized media institutions and their creative employees (Barr, 1993). Or, to put it even more crudely, and using one of the earliest technologically inflected senses of live television, television drama is a *relay* of cultural matter, say of national and local significance, transmitting 'the DNA of its time'.[2] The problem with this, as V.F. Perkins has argued, is that tempts the critic or producer to 'redeem the work by giving it undue credit for being what it must always be, a product of its time' (1992: 203); equally we should acknowledge that undue *criticism* is typically a product of television studies scrutiny of works that *fail* to relay, in adequate ways, the socio-historical content of the time. This critical approach, exemplified by the a recent collection of scholarly work on *Mad Men* (Goodlad et al., 2013) seeks to subordinate the drama to its processes of mediating and articulating (often with illusionist omissions, suppressions, masking and a general failure to be congruent with whatever the current liberal-left orthodoxy is at the time of writing) current social and cultural attitudes and structures of feeling/anxiety/etc. The problem here is that the scholar/critic tends to assume a position of authority over the object of their analysis: the drama is something to be assayed, sifted, 'worked through' until its core socio-historical content is laid bare.

There is another way to take drama, which has historically played a minor role in academic scholarship (see, for example, Brandt, 1981) and is far more prevalent in popular and journalistic television criticism. This is an approach to television drama as an expression or articulation that has aesthetic value in itself, quite apart from its putative function as a socio-historical weathervane. Taking it this way allows those critics to access and evaluate its currency as part of a dramatic tradition, in dialogue not only with theatrical and cinematic and radio drama forms, but also with its competitors. In this case television drama is seen primarily as consisting of *expressive* artefacts, allowing us to calibrate the value of the artwork, and their creators' contribution to an ongoing tradition. The dominant comportment here is one of *submission* to the power and achievement of the work; critical activity is directed to illuminating both.

Of course the two 'poles' of this binary are in fact intimately interrelated, and this is not a strict binary division at all but instead a distribution of emphasis with gradations and shadings along a critical continuum. For those engaged at the relay end the content is clearly for them also expressive – of a national or cultural kind – while for those who wish to drill into the form and style of expression itself it cannot be insignificant that such content is relayed, transmitted and distributed in certain ways, or that socio-historical content is more often than not a prominent feature. Indeed, decades of critical theory have taught us that the aesthetic can be taken as an epistemology, a way of knowing the world beyond its appearances, in those appearances. Nonetheless, for the expressivists, the possibility of deep artistic achievement is often in view; for the relayers, not so much. And it is in the tension between these different evaluative stances that the most important debates take place.

The orientating canopy stretching over both attitudes is a concern with the medium that I have previously characterized as drawing on some traditions of modernist criticism where form and content were, at the deepest core of the artwork, in self-reflexive engagement with the medium in which it inhered. It is no surprise that television drama has been the one of the key foci for the calibration of the aesthetic health of television as a medium, for the simple reason that dramatic forms have a long historical tradition of artistic achievement against which it can be assessed. As Sergio Dias Branco notes:

> Ted Cohen suggests that if philosophy of art devoted attention to television it would concentrate first on dramatic and comedy series. They seem likely to him to be of more interest than other programmes and other aspects of television. That is probably because within the televisual context they are more easily identified as artworks. Fiction television series resemble fiction films in the sense that they are both produced with expressive and narrative ends in sight. This leads us to put them in the same category. This similarity does not exist in regards to quiz shows, for example, which are programmes that present a rewarded competition for entertainment purposes. (2010: 5)

Sometimes however this has been taken as an illegitimate distortion of the medium's folk and demotic character, a striving for legitimacy through the import of 'literary' values and other impurities which undermine the assessment of the real connective opportunities television affords for articulating and reflecting the popular; hence the medium operates by these lights as a bulwark against bourgeois canons (see Lury, 2007; Newman and Levine, 2012). There is perhaps some truth in this, but the question has a remainder, which is why such striving for artistic status, and its critical assessment should be problematic at all? A lot of art may reflect or reject bourgeois or elite ideology, and interact fluently with several artistic forms and traditions, but that, in itself, does not abrogate the universal reach of such art, the ways in which its peculiar ability to transcend the conditions (and ideology) of its making dislocates it from a necessarily strong association – or imprisonment – in its time.[3] To be sure, not all art achieves such exalted status, and the particular embeddedness of television output within national and popular systems of broadcasting and distribution might be taken to further limit its escape velocity. But as the work of Lukacs on the historical novel and his later writing on aesthetics teaches us, the work of art is a special kind of entity, 'characterized by the organic unity of its form and content allowing a truth of universal significance to appear in a directly apprehensible, sensuous-concrete embodiment' (Markus, 1998: 170). The fact is that the vast majority of critical and scholarly attention to television drama has grappled with the immediacy of the concrete contemporary embodiment – its significance for its time and place – rather than matters of 'organic unity' or universality. This is puzzling at first sight because television drama of all objects seems to offer a unique path beyond the contemporaneity of most television output, which is to say, however engaged its narratives are with the here and now its necessary location within a tradition of dramatic output ineluctably situates it as a candidate, however unsuccessful in particular cases, that can speak beyond its time of making. But to grant that as a possibility and a criterion of greatness equally pits such drama and such evaluations against the entire momentum of television studies as it has so far developed, where the emphasis tends to degrade and instrumentalize aesthetic issues into questions of whether or not it is a 'progressive text' or whether it exploits the 'possibilities of the medium'. Hence medium specificity is recruited to police the potential of television drama to participate alongside the assessment of other drama, confined by the insistence on the limits of television itself as *ipso facto* illegitimate as a contender for greatness alongside other art traditions.

There is a historical background to this odd refusal. Before the academic field of television studies rose to prominence in the 1970s there were many attempts to theorize the medium and its works by practitioners, critics and academics. William Boddy (1986), in his deconstruction of the assertion of the 'Golden Age' (his scare quotes) of US television drama in the 1950s, is highly critical of what he deems as the 'frankly amateur' (p. 125) theorizing of early television criticism that accompanied the rise of live anthology series such as *Studio One* and *Kraft Television Theatre*. It is important to see this 'amateur' response as part of that larger trend, exemplified in the art criticism of Clement Greenberg, to stress test the medium against essentially modernist criteria of its fitness to express the 'possibilities of the medium'.[4] Indeed this was a central trope in the UK investigation into the future of broadcasting during the late 1950s and early 60s, which was published as the *Report of the Committee on Broadcasting, 1960* (more commonly referred to as the Pilkington Report). This was peppered with references to the 'triviality' of commercial television which was deemed to fail in its responsibility to 'realize its potentialities' (1962: 34). Hence, as scholars of UK early drama history have shown, there was an engagement in carving out a distinctive identity for television drama because of its very similarity to narrative forms of theatre, film and radio – especially important in institutions like the BBC where radio was privileged and the organizational thirst for identity was strong among its fledgling staff (see Jacobs, 2000).[5]

The first significant aesthetic crisis at depth in television drama occurred in the US during the move of production from the East to the West Coast during the 1950s. The output of Rod Serling exemplifies the difficulties this shift represented: his work in Emmy-winning single dramas like *Patterns* (1955) and *Requiem for a Heavyweight* (1956) was typical of the concentrated character study that had come to be seen to epitomize the essence of television drama's attention, which had its coronation with Paddy Chayefsky's celebrated *Marty* (1953). But Serling's later work on *The Twilight Zone* used the generic lens of the telefantasy anthology within which political and social themes could be explored. By contrast, in the UK a different industrial dynamic was at work; the advent of a commercial television network competitor in 1955 into what had been until then a BBC monopoly on all broadcasting including television, inaugurated a new trend in dramas that engaged with the local and the particular (making them hard to export). By the 1960s US drama was almost exclusively shot on film and was dominated by the episodic series form, supplemented by imports from the UK as showcased by, among others NET and later *Masterpiece Theatre* (see Hilmes, 2012). In the meantime the UK continued to valorize, and experiment within, the single play format with anthologies such as *Play for Today* establishing individual auteur-writers such as Dennis Potter, Jack Rosenthal and Trevor Griffiths. Such 'plays' (as they were still quaintly referred to) were either studio bound and shot on videotape or, increasingly in the 1970s, shot on film or a blend of both to achieve the finished product (this created some critical debate about the appropriateness of using one medium in another, see Taylor (1990) and Sutton (1982)). Already the idea that live television drama was a pure form of the medium had been shown to be an extrapolation of technological and institutional necessity into an aesthetic virtue (and there were indeed some of these); now the ingress of film usage seemed to threaten the erasure of the distinction between film and television. With the beginning of Channel 4 in 1982 and its commitment to produce films that would receive their premiere on television, this synthesis of media seemed to be complete.

As these institutional, industrial and technological shifts unfolded, one of the unfortunate critical tendencies that began its domination of British debate about television drama was conducted fully on the relay side. That is, although many of the debates were engaged with textual and narrative questions, they tended to subordinate issues of form and style to matters of address and mass reception. The handling of content in discussions about naturalism and realism was used instrumentally to assess the extent to which television drama as a relay of content had the potential to 'agitate' the viewer out of a complacent absorption in the flow of television content, establishing 'distance' which in the anti-illusionist tradition of *Screen* studies passed for a radical position (see, for example, Williams, 1977; Potter, 1984; Caughie, 2000). What mattered here was not so much the content *per se* but its ability to unmask ideologically hidden truths about the worlds depicted in their fiction, and that content was assessed in regard to this criterion more or less exclusively. This avoidance of the artistic force and achievement of such dramas meant that the assessment of their value turned on a judgement about their efficacy as political and ideological stimulants. While professional television critics were keenly aware of the necessity of making judgments about artistic value at however a low critical altitude of analysis, scholarly responses spent decades avoiding this fundamental issue. Instead the emphasis has been on the effectiveness of television drama as an instrument of social and cultural engineering which, over the decades, has been denuded of any radical content and is now simply congruent with ambitions of the State. When Lez Cooke concludes his book *British Television Drama: A History* with the assertion that the 'strength of British television is its diversity and its pluralism', we can see these instrumental consequences in action; innovation and experimentation are not offered as aesthetic values but as part of a 'radical and progressive tradition' which constitutes television drama's role in the construction of 'cultural diversity' (2003: 196). Such a degraded, programmatic notion of artistic output which sees programme controllers and directors as agents of social and cultural change or custodians of putatively progressive traditions inserts a stencil of contemporary prejudices where a critical engagement with, and evaluation of, television drama's artistic achievements should be.

This is one of the consequences of relay-criticism, its subordination of television drama to categories and criteria (often derived from the social sciences or cultural studies) already established outside of the shows themselves, but which they supposedly transmit, however partially. The Goodlad collection about the show *Mad Men* (2013) exemplifies the current health of relay criticism: many of the essays therein are predicated on the assumption that the show tells us things about the world and about the falseness or partialness of its rendering that only critical study can reveal to us.

Expressive criticism, by contrast, waits patiently to be told and taught by the object with which it is engaged. It grants the possibility – nearly unfeasible for relay critics, for whom the aesthetic is merely an epistemology for grasping the social and historical – that the dramas engaged with are not subsumable beneath a critical canopy of explanation. To be sure, part of their achievement might well be an articulation or partial articulation of a view or perspective on society, history or people. But there is a significant remainder, not least the embeddedness and responsiveness of such shows to their own artistic intentions and aspirations, not to mention the wider, longer and deeper dramatic traditions of which they are a part.

The reasons for the scholarly friendship with the relay side in television studies are complex and largely connected to wider and deeper shifts in the humanities as a whole (such as the influence of structuralism, the absorption of the methods and perspectives deriving from the social sciences), and television studies, an amalgam of various disciplinary modes, happened to emerge and mature largely after these battles seemed over.[6] Since many of the new theories of society and culture that television studies drew upon were derived from a denigration of human subjectivity and agency, it is little surprise that television's elite dramatic output, which necessarily foregrounds creative intentionality, has been neglected in its own terms.[7] This has resulted in a relatively poor account to date of the richness of the output of television drama over the past three decades, the content of which has barely been addressed head-on. In particular, the critical allegiance to the putative 'death of the author' (which is to say the denial of what any sane person would recognize as the artistic intentionality underpinning any great art) has meant some desperate wriggling in the face of the success of overtly signature forms of television drama from the late 1990s onwards. The necessarily collaborative nature of television production and writing means that decanting the results to a single person, *à la* the auteur theory, is hardly tenable. But what is crucial is the significance of artistic *authority* as an element in such output, whether it be the 'show runner' or another formation of creative personnel. Equally, the artistic contribution of performers themselves, in the unique position of playing characters who do not know their final destinations in the narrative architecture, has been little remarked upon. Amidst the contemporary debates and subsequent revisionist histories of television fiction in the 1970s, few have considered in depth the sheer range of dramatic achievement by performers on television. John Caldwell (1995) characterized the 35mm filmed output of US genres in the 1970s (such as *CHiPs*) as a 'zero-degree style', therefore utterly eliding the extraordinary contributions of those show's casts – think of the light-hearted wit of the heavy-footed and strangely asexual John Bosley in *Charlie's Angels* as played by David Doyle, or the scruffy-dog Satanic tenacity of Peter Falk's Columbo. Caldwell's account of the what he calls the stylistic exhibitionism of television in the 1980s is snootily dismissive (in a Hoggartian sense) of its shiny industrialized packaging, and takes expression (such as the gorgeous evolution in the colour palettes of Michael Mann's *Miami Vice* – the stage on which Crockett and Tubbs dance their camp version of police detection) as a mere relay of highly industrialized semiotic content.

This is not to say that character and performance were totally occluded in television studies' engagement with drama, but that they tended to be subordinate to what were considered more important matters: fiction as a transmitter of cultural content – say identity, or postmodernism, or assertions of 'progressive' or 'non-progressive' content.[8] Intersecting with these but also at some theoretical distance are the many dedicated edited collections that emerged during the early to mid-2000s around critically fashionable television drama, such as the I.B. Taurus series *Reading Contemporary Television Series*. Here one can see the orthodoxy of relay criticism beginning to loosen, although there remain significant elisions and absences. Steven Peacock and my criticisms of the approaches in many of the scholarship to be found here is apposite:

> … there has been a shift from considerations of television in general (as a medium, in terms of 'flow' for instance) to particular programmes … Yet such studies remain, for the most part, informed by approaches through which theory is mapped onto the television 'text' to decipher its so-called coded meanings. Equally, despite many writers' assertions that, within these readings, close textual analysis will be employed, there is a key conflation of terms. Too often, such analysis becomes systematic, determined to 'solve' the text's engagement with a specific subject, rather than employing critical principles to feel through its tensions and complexities, keeping them in play. Further, such work resists a dedicated and sustained scrutiny of television style, attempting to undertake 'close textual analysis' without getting close to the text's integral compositional elements. (Jacobs and Peacock, 2013: 2)

In the face of the undeniable amplification of the creative and artistic achievement of US signature drama, British television scholarship has been thoroughly engaged with a revisionist revisiting of British television drama, cultivating deep analysis of past achievement. Since the early 1990s there has been a significant augmentation of our knowledge of the institutional, technological and cultural history of British television drama (see, for example, Bignell et al., 2000; Corner, 1991; Jacobs, 2000; Wheatley, 2006), as well as specific work on significant creative individuals (see, for example, Rolinson, 2005; Vice, 2009). As Donald Sassoon (2006) has demonstrated, such cultural nationalism tends to emerge and coalesce in view of putative threats from other cultures' successful purchase on the global imagination. Perhaps it is no coincidence that as US drama emerged as a distinctive and diverse artistic force in the late 1990s, there were several attempts to discipline this output: either as 'quality' (effectively this is a generic and industrial category not an evaluative term), or as fodder for a legitimizing aesthetic manoeuvre, vitiated further by its inheritance of other artistic traditions and tropes. Sudeep Dasgupta (2012) has convincingly argued that attempts by scholars to denote such output as merely an elite artefact in fact reproduces at a theoretical level the corporate pitch of those that fund and market the product ('This is not TV…') while hegemonically owning and wielding the folk category of the popular. A similar move to negate the force of US drama can be seen in scholarship that insists on seeing television drama's power through the lens of a nationally inflected folk sense of the medium (Buonanno, 2012). Beyond Europe and the US, the role of drama in reflecting and projecting local and national stories, especially in those nations undergoing internal cultural and economic change, has lent further energy to those on the relay side, since the function of narrative in making such changes legible to the wider population has immediate sociological relevance. This was clear during the 1980s in Australia when various television mini-series (such as those made by the Kennedy-Miller production company) revisited emblematic events in that nation's history, but with an eye on shaping and re-shaping the ambient cultural understanding of that history; in a similar way emergent drama in Singapore, Indonesia, China and India has captured the attention of cultural studies scholars in that region and beyond (see, for example, Huat, 2008). The dominance of cultural-studies-inflected accounts of television in the Asian region almost inevitably means the emphasis is on the relay side, with matters of narrative and style subordinate to questions of mediation and distortion and control of national stories by large entertainment conglomerates. In view of the prominence of artistic achievement in contemporary television dramas, we now witness such academic traditions – once new and 'rebellious' – struggling to incorporate the 'aesthetic' into their ossified theoretical frameworks; but so deep was the excision of traditional humanities models of artistic intention, of harmony, unity and coherence – in favour of minimizing the role of human agency and in praise of fragmented subjectivity, the discursive structuring of response, and the withdrawal of engagement with evaluation – that it seems unlikely such approaches will be illuminating or productive.

Although I have emphasized the significant mismatch between the astonishing achievements of television drama across the world, particularly in the US during the past three decades, and the paucity of expressive writing on it by the academy, there are traces and signs of productive engagement with it, both within and outside the academy. Popular journalism in print and online forms are lifting some of the evaluative load, with critics such as Matt Zoller Seitz and Clive James (a veteran of television criticism whose witty style has matured in line with the content he now engages with), who have been particularly eloquent in their strong engagement with innovation and style in contemporary television drama. In the academy, Sarah Cardwell has pioneered what we might call the new expressionist criticism of television drama (even where the objects of her attention are not quite worthy of it (Cardwell, 2005)); and Sean O'Sullivan's work on television narrative has dared to draw on literary antecedents with novel and engaging results (O'Sullivan, 2010, 2011; see also Branco, 2010, on serial narrative). Even veteran film style analysts famed for their eloquence when handling a Hitchcock or a Capra have been drawn by the remarkable presence of recent drama (Rothman, 2013; Toles, 2013) and they have been joined by new voices, also expressivists interested in mining and illuminating the profound moral, ethical and aesthetic matters that such art brings to our attention (see Clayton, 2013; Logan, 2013).

Finally, given the importance of intentionality to the understanding of such dramas, as a next step I would insist that ideas of the death of the author be rejected for what it always was – an attempt to demote human artistic agency. In terms of television drama we are clearly not often dealing with single authors (although that may happen, and certainly, given the industrial investment in signature and auteur television, individuals do have a lot of say), but collaborative intentionality. One suggestion by a prominent television screenwriter, Ted Mann (*Deadwood*, *Hatfields & McCoys*) is that we see the influence of authorship on others as akin to the development of 'schools' or movements in art history (such as the Hudson River School); that is, groups of artists working on a show, each lending their skills and distinct manner of creative shaping to a larger, agreed, sensibility. Certainly this means, as Matt Hills (2011) correctly puts it, a 'pre-structuralist' response, but this is not the danger, or the recidivism he characterizes it as. Instead we need new ways of reconciling relay and expression, of connecting the social and historical to the granular, metabolic stylistic and aesthetic details that capture in gesture, mood and extended narrative forms, the ongoing mediation of art and society in television drama.

NOTES

1 For examples of critical manoeuvring that seeks to maintain and police the authority and distinction of television as a popular/folk medium, see Newcomb (1974) and Lury (2007).
2 This notion is taken from an essay by Martin Scorsese (2013) on the language of cinema, where he notes, '[t]he American movie critic Manny Farber said that every movie transmits the DNA of its time'.
3 Rosen (1998) and Cavell (1976), in very different ways, explore this unique aspect of the artwork.
4 The classic and influential articulation of this tendency is of course Greenberg (1978).
5 This frantic struggling to carve out an institutional identity as well as a medium-specific one amongst activity that is not, on the face of it, dissimilar, has its analogue in the world of human culture as popularized by Freud's term the 'narcissism of small differences' (Freud, 2007: 268).
6 See Brunsdon (2004) for a history of the formation of television studies.
7 For a theoretical account and history of the denigration of subjectivity in the 20th century see Heartfield (2006).
8 See Harris (2006) for an exemplar of this tendency.

REFERENCES

Barr, Charles [1977] (1993). *Ealing Studios*. London: Studio Vista.

Bignell, Jonathan, Stephen Lacey & Madeleine Macmurraugh-Kavanagh (Eds) (2000). *British Television Drama: Past, Present and Future*. Houndmills: Palgrave.

Boddy, William (1986). The Shining Centre of the Home: Ontologies of Television in the 'Golden Age'. In Phillip Drummond and Richard Paterson (Eds), *Television in Transition*. London: British Film Institute.

Branco, Sergio Dias (2010). *Strung Pieces: On the Aesthetics of Television Fiction Series* (Doctoral dissertation, University of Kent).

Brandt, George (Ed.) (1981). *British Television Drama*. Cambridge: Cambridge University Press.

Brunsdon, Charlotte (2004). Television Studies. In Horace Newcomb (Ed.), *The Encyclopedia of Television* [second edition]. Chicago: Fitzroy Dearborn, 2299–304.

Brunsdon, Charlotte (2008). Is Television Studies History? *Cinema Journal*, 47(3), 127–37.

Buonanno, Milly (2012). *Italian TV Drama and Beyond*. Bristol: Intellect.

Caldwell, John Thornton (1995). *Televisuality: Style, Crisis, and Authority in American Television*. New Brunswick, NJ: Rutgers University Press.

Cardwell, Sarah (2005). Television Aesthetics and Close Analysis: Style, Mood and Engagement in *Perfect Strangers* (Stephen Poliakoff, 2001). In John Gibbs and Douglas Pye (Eds), *Style and Meaning: Studies in the Detailed Analysis of Film*. Manchester: Manchester University Press.

Cavell, Stanley (1976). *Aesthetic Problems of Modern Philosophy: Must We Mean What We Say?* Cambridge: Cambridge University Press.

Caughie, John (2000). *Television Drama: Realism, Modernism, and British Culture*. Oxford: Oxford University Press.

Clayton, Alex (2013). Why Comedy is at Home on Television. In Jason Jacobs & Steven Peacock (Eds), *Television Aesthetics and Style*. New York: Bloomsbury.

Cooke, Lez (2003). *British Television Drama: A History*. London: British Film Institute.

Corner, John (Ed.) (1991). *Popular Television in Britain: Studies in Cultural History*. London: British Film Institute.

Dasgupta, Sudeep (2012). Policing the People: Television Studies and the Problem of 'Quality'. *NECSUS European Journal of Media Studies*. Retrieved from www.necsus-ejms.org/policing-the-people-television-studies-and-the-problem-of-quality-by-sudeep-dasgupta/

Freud, Sigmund (2007). The Virginity Taboo. In *The Psychology of Love*. London: Penguin Books, 262–278.

Goodlad, Lauren M.E., Lilya Kaganovsky & Robert A. Rushing (Eds) (2013). *Mad Men, Mad World: Sex, Politics, Style and the 1960s*. Durham: Duke University Press.

Greenberg, Clement (1978). Modernist Painting. In Richard Kostelanetz (Ed.), *Esthetics Contemporary*. Buffalo: Prometheus Books, 198–206.

Harris, Geraldine (2006). *Beyond Representation: Television Drama and the Politics and Aesthetics of Identity*. Manchester: Manchester University Press.

Heartfield, James (2006). *The 'Death of the Subject' Explained*. London: BookSurge Publishing.

Hills, Matt (2011). Television Aesthetics: A Pre-Structuralist Danger. *Journal of British Cinema and Television*, 8(1), 99–117.

Hilmes, Michele (2012). *Network Nations: A Transnational History of British and American Broadcasting*. New York: Routledge.

Huat, Chua Bong (2008). Structure of Identification and Distancing in Watching East Asian Television Drama. In Chua Bong Huat & Koichi Iwabuchi (Eds), *East Asian Pop Culture: Analyzing the Korean Wave*. Hong Kong: Hong Kong University Press, 73–90.

Jacobs, Jason (2000). *The Intimate Screen: Early British Television Drama*. Oxford: Oxford University Press.

Jacobs, Jason & Steven Peacock (Eds) (2013). *Television Aesthetics and Style*. New York: Bloomsbury.

Logan, Elliott (2013). Flashforwards in *Breaking Bad*: Openness, Closure and Possibility. In Jason Jacobs & Steven Peacock (Eds), *Television Aesthetics and Style*. New York: Bloomsbury.

Lury, Karen (2007). A Response to John Corner. *Screen*, 48(3), 371–6.

Markus, György (1998). Lukács, György. In Michael Kelly (Ed.), *Encyclopedia of Aesthetics*. New York: Oxford University Press.

Newcomb, Horace (1974). Toward a Television Aesthetic. In *TV the Most Popular Art*. New York: Anchor Books, 243–264.

Newman, Michael Z. & Elana Levine (2012). *Legitimating Television*. New York: Routledge.

O'Sullivan, Sean (2010). Broken on Purpose: Poetry, Serial Television, and the Season. *Storyworlds: A Journal of Narrative Studies*, 2, 59–77.

O'Sullivan, Sean (2011). Space Ships and Time Machines: *Mad Men* and the serial condition. In Gary R. Edgerton (Ed.), *Mad Men: Dream Come True TV*. London: Tauris Press, 115–30.

Perkins, V.F. (1992). The Atlantic Divide. In Richard Dyer & Ginette Vincendeau (Eds), *Popular European Cinema*. London: Routledge, 194–205.

Potter, Dennis (1984). Some Sort of Preface ... In *Waiting for the Boat: On Television*. London: Faber and Faber, 11–35.

Report of the Committee on Broadcasting, 1960 (1962). London: HMSO.

Rolinson, Dave (2005). *Alan Clarke*. Manchester: Manchester University Press.

Rosen, Charles (1998). The Ruins of Walter Benjamin. *Romantic Poets, Critics, and Other Madmen*. Cambridge: Harvard University Press, 131–81.

Rothman, William (2013). Justifying *Justified*. In Jason Jacobs & Steven Peacock (Eds), *Television Aesthetics and Style*. New York: Bloomsbury.

Sassoon, Donald (2006). *The Culture of the Europeans: From 1800 to the Present*. London: HarperCollins.

Scorsese, Martin (2013). The Persisting Vision: Reading the Language of Cinema. *The New York Review of Books*, 15 August. Retrieved from www.nybooks.com/articles/archives/2013/aug/15/persisting-vision-reading-language-cinema/?pagination=false DOA 20

Sutton, Shaun (1982). *The Largest Theatre in the World: Thirty Years of Television Drama*. London: BBC.

Taylor, Don (1990). *Days of Vision*. London: Methuen.

Toles, George (2013). Don Draper and the Promises of Life. In Jason Jacobs & Steven Peacock (Eds), *Television Aesthetics and Style*. New York: Bloomsbury.

Vice, Sue (2009). *Jack Rosenthal*. Manchester: Manchester University Press.

Wheatley, Helen (2006). *Gothic Television*. Manchester: Manchester University Press.

Williams, Raymond (1977). A Lecture on Realism. *Screen*, 18(1), 61–74.

23

Sperm Receptacles, Money-Hungry Monsters and Fame Whores: Reality Celebrity Motherhood and the Transmediated Grotesque

Brenda R. Weber and Jennifer Lynn Jones

In current academic and industrial usage, transmediation is the persistence of a story world across platforms, with each iteration a unique narrative extension. Although based in intertextuality and emerging from contemporary convergence culture, transmediation differs from other common forms of convergence, like sequels and franchises, in that it's not a continuation or adaptation, but an expansion of the narrative. Transmediation is thus less diachronic than synchronic in form, and the transmediated narrative has the potential to emerge simultaneously across the mediascape.

Marsha Kinder initially defined what she called 'transmedia intertextuality' as 'an expanding supersystem of entertainment' organized through 'relations across different narrative media' (1991: 1, 2). In Henry Jenkins' description, 'A transmedia story unfolds across multiple media platforms, with each new text making a distinctive and valuable contribution to the whole' (2008: 95–6). Ideally, writes Jenkins, 'each medium does what it does best – so that a story might be introduced in a film, expanded through television, novels, and comics; its world might be explored through game play or experienced as an amusement park attraction' (p. 96). Jenkins further argues that each level of transmedia storytelling is ideally self-contained, so that a viewer need not play the game to enjoy the film, or vice versa. The key to transmedia storytelling, explains Jenkins, is perpetual newness, since 'redundancy burns up fan interest and causes franchises to fail. Offering new levels of insight and experience refreshes the franchise and sustains consumer loyalty'. Jenkins further contends that transmedia storytelling is dictated by the horizontal integration of the entertainment industry, in which 'a single company may have roots across all of the different media sectors' and thus, this company is able to control and determine 'the flow of content across media' (p. 96).

In both 'transmediation' (which in the above definition is audience-centered) and 'transmedia storytelling' (which in the above account is industry-centered) there is a coherence and integrity to each respective art form, so that, for example, the TV story is different from the video game story, which is, in turn, different from the film story. And each of these forms exists independently from the other. As so defined, there is also a causal chronological linearity between mediated formats, so that even while the amalgamated telling comes to represent the full story and consequent text, there is a clear first telling of the story that leads to continuing extensions of the narrative. The metamorphic aspects of transmediation may thus help to disguise how the process still remains structured in dominance. In describing the work of the supersystem, a corollary concept to transmediation, for instance, Marsha Kinder notes that it may enable a 'protean malleability' but still functions hegemonically, 'as a social apparatus that transmits and reproduces the dominant ideology ... through widely accepted cultural practices' (1991: 37). As such, despite its transgressive elements, transmediation still has the potential to sustain hierarchies of both media and identities throughout its various proliferations.

In this chapter we use a new term, the transmediated grotesque, to stand for a transaction that messily confuses its genre, agent/s of creation, and moment of birth, signifying the grotesque in both its fluidity and polyvalence. In the case study we discuss here, the transmediated grotesque occurs simultaneously, created by both cultural producers and audiences, distributed through multiple levels of the media industry, including union shops, major networks, obscure independents, and individual bloggers and computer users. The liquidity of the transmediated grotesque muddies linear provenance in terms of causality. Instead of creating new-ness, transmediated continuity allows for a perpetual 'now-ness' in which ideas circulated at one moment continue to have relevance, blending with new forms and new times to structure an eternal present. In so doing, the transmediated grotesque recrafts notions of provenance and coherence. In many ways, this idea very much describes the postmodern palimpsest theorized by such scholars as Jean Baudrillard or Frederick Jameson, in which history flattens into a perpetual field. But where palimpsest theory suggests that all elements are still present in their basic form but hidden under layers, like an elaborate archaeological puzzle, the transmediated grotesque, as we are using the term, denotes a transformation at the atomic level. What existed before mutates through the transmediated exchange.

Gender is clearly critical in this analysis, since within the complex fusion of transmediation, gender modalities are apparent (if not always stable) reference points for producers, consumers, and reality television participants. But gender is also critical to transmediated continuity for its ties to the classed and raced body. Indeed, we consider the form and function of the body and its position in a field of signs as another element of media that plays within the complex folds of the transmediated grotesque. In this respect, the body is not only located within the flesh and blood of a real person but also represents ideas bigger than and uncontained by 'realness'.

Notions of the female grotesque help us understand why. As theorized by literary scholar Mikhail Bakhtin, the grotesque body is that which threatens to explode, perforating the abstract, spiritual, noble and ideal within its 'unruly biological and social exchange'.[1] For Bakhtin, the grotesque body was about the sheer materiality of embodiment – the shit, urine, blood and sweat of the 'bodily lower stratum' (1984: 20). And while Bakhtin allowed for a 'comic' and carnivalesque communal function of the grotesque in the premodern world, its meanings have become more defused within modernity. Mary Russo extends Bakhtin's reading, for instance, offering a version of the female grotesque that mindfully pushes a feminist agenda able to allow for the 'heterogeneous, strange, polychromatic, ragged, conflictual, incomplete, in motion, and at risk' (1995: vii).[2] The female grotesque, writes Russo, is a bodily category in the 'anxious and hopeful point in between' (p. viii). It can thus be located not in its association with the 'low' or as distant from the high but in the insistent interstitiality where surfaces and interiors are both refuted and fused. Thus, even from its position in the 'in between', the grotesque threatens rupture and destabilization.[3]

Much-despised reality star Kate Gosselin offers a helpful example of the transmediated grotesque. Or rather, the *idea of Kate* is a good case in point. Since 2006, when she, her husband Jon, and their family of eight children, emerged into celebrity, the idea of Kate has gone from being a working-class mother whose pregnant fantastic belly full of six children expanded the capacities of what the female body might do, to a shrewish 'bitchy' wife who emasculated her 'Asian looking'[4] husband, to a wronged and sympathetic divorcee (when that hen-pecked husband cheated on her), to a glamorous role model as a celebrity single mother, to a public failure, 'destroyed by greed' for fame and plastic surgery.[5] This is the chronological version of her star text, rich with sexist assumptions about women's worth and behavior. But Kate is also the uneven sum of these parts, her 'Before' body never fully leaving, her 'After' body always tethered to these different iterations of her. Kate's reality shows were cancelled in 2011, but in late 2013 and early 2014, the Gosselins drew new attention and new controversy through a series of public battles with one another, played out in court, Twitter, tabloids and various television shows, including Oprah's *Where Are They Now?* and the *Katie Couric Show*.[6] For Jon Gosselin, the resurgence of his private trials in a public arena reinforced a sense that he is a burnt-out loser who now waits tables for a living, reinforcing the 'racial castration' that often adheres to Asian men.[7] When he appeared on VH1's *Couple's Therapy* in 2014 with a girlfriend whose nastiness far surpassed Kate's, the 'idea of Jon' easily reinforced this emblem of failed masculinity. Partly because she was always more famous and more controversial than Jon, but also because her public image conformed to a notion of the female grotesque, the 'idea of Kate' took on more of the atemporal palimpsest marked by transmediation. She was 'the good reality mom' (as reinforced by a *People Magazine* feature in January 2014) and the 'fame whore' (as insinuated by an *In Touch Weekly* cover from November 2013), a real mother worried about how cold it was at the bus stop that morning or where to find the best price on back-to-school clothes (as her January 2014 Twitter feeds indicate) and a heartless monster who mistreats her children (as reported in an unauthorized tell-all book from 2013, *Kate Gosselin: How She Fooled the World*).

As activated through the accelerant of 21st-century transmediation, there is perhaps no better Petri dish for the female grotesque than the reality celebrity mother, who keens for recognition and reward. As a sign, she fuses real and reality and brings together a potent combination of narcissism, sexualization, pathology and demonization, all fueled by low-level (shame-saturated) reality celebrity and playing out in the context of a multi-platform mediascape that invites (and requires) individual anonymous reactions as agents of disciplinary sanction. In this chapter, we consider Farrah Abraham, as a way of best demonstrating the meanings of the transmediated grotesque, since her reality story illustrates the degree to which misogyny is used to police the 'gender crimes' committed under the sign of celebrity motherhood.

FARRAH: 'I MADE A PORNO BUT I'M NOT A PORN STAR'

Las Vegas, ever the fabled home of scandal and sin, has a new feather in its cap: as the putative place where female reality celebrity goes to renew itself (or die trying). Head toward the airport on I-115 south of the strip and you will be arrested by a gigantic billboard emblazoned with the image of Farrah

Figure 23.1 Farrah Abraham, reality star, as featured attraction at Crazy Horse III in Las Vegas

Abraham (à la *Teen Mom*) in a pink bikini. She began stripping at a Vegas nightclub, Crazy Horse III, in the summer of 2013 (Figure 23.1).

Farrah's fellow 'reality star' strippers at the Crazy Horse include Heidi Montag (*The Hills*), Josie Stevens (*Married to Rock*) and Heather Chadwell (*Rock of Love*), each having gained extra-diegetic notoriety for using copious amounts of plastic surgery to write a version of big-breasted hyperfemininity on her body. It isn't the fact of Abraham smiling down from the billboard, her dark hair blowing back over one shoulder, or even her own conspicuous aesthetic augmentations that startle as much as it is the fact of the billboard itself in its intermedial fusion of reality celebrity, the sex industry, and mainstream entertainment (the CBS logo perched conspicuously under the ad). Indeed, the composite image strikes us as a cultural meme rich in information about gender, celebrity and reality television, and the perfect example of the transmediated grotesque announced through the reality celebrity mom.

Farrah Abraham first found reality celebrity on the MTV anthology program *16 and Pregnant* (2009), moving on to be one of four featured mothers in the first iteration of MTV's weekly series *Teen Mom* (2009–2012). As reality texts, both *16 and Pregnant* and *Teen Mom* work according to the logic of the cautionary tale, where the demands of diaper changing, reduced finances, emotional tension and a child's endless needs work in tension with more systemic issues, such as familial dysfunction, social-class stresses and substance abuse. As Laurie Ouellette (2014) puts it, the shows function more as birth control than entertainment, offering, in turn, a finely tuned pedagogy in governmentality that works to condition viewers to make putatively better choices than those depicted on the screen. These choices are fully in line with state ideologies that seek to create citizens who accept, and even embrace, an economy of individualism, can-do optimism and strategic self-management.

Farrah became pregnant in 2008 by her then-boyfriend Derek, who died in a car crash two months before their daughter Sophia was born in early 2009. If MTV reader boards and our own responses can be trusted, her segments on *Teen Mom* often elicited extreme discomfort in viewers, both due to the rage that percolates just under the surface with her parents and because of the saccharine-like façade that passes for intimacy within the family. Both Farrah and her mother talk in a nasally whine that is particularly annoying when they switch to baby talk with dogs and toddlers. In contrast to the other women on *Teen Mom* who visibly had to cope with financial difficulty, Farrah and her family seemed to be more economically prosperous. Farrah took pictures of Sophia with her iPad and iPhone, she owned her own car, she lived in her own charming Victorian home (provided by her parents), and she expressed aspirations to attend college and to be both a chef and a model. While fan blogs seem puzzled by Farrah's ethnicity (her dark hair and perpetual tan have many respondents convinced that she is Middle Eastern or Latina), no posts that we have read question her middle-class residency, even while many deride her for being class-less. Yet, these middle-class aspirations, themselves markers of socioeconomic stability, seemed to fall by the wayside in favor of a career as a famous person, or as she calls it, a television personality and entertainment entrepreneur. With cancellation of *Teen Mom* eminent, Farrah released several music singles and an autobiography, *My Teenage Dream Ended*, neither of which did well in sustained sales or publicity.

Since the show's end in 2012, Farrah's 'where is she now' check-ins have involved the requisite elements of the starlet in distress: in March 2013, she was arrested for driving under the influence; in May 2013, she was 'forced' to release a sex tape, using Vivid Entertainment, a mainstream porn company that also released Kim Kardashian's 'private' sex tape; in July 2013 she entered rehab for alcohol abuse, only to be 'kicked out' early;[8] in December 2013 she posed for the cover of *Girls and Corpses*, a fetish magazine in which 'hot girls cuddle cold corpses', and she started 'causing trouble' on a new reality show, VH1's *Couple's Therapy*, on which she appears as a single person trying to understand why she has such difficulty in relationships.[9] Throughout her rise to ignominy, she has engaged in feuds with other faltering celebrities, including a beleaguered Charlie Sheen, who called her 'a desperate guzzler of stagnant douche agua' when she leaked their flirtatious texts to TMZ, a celebrity gossip website.[10] Radar online announced a 'celebrity smackdown' between Abraham and former *Real Housewives of New York City* star, Bethenny Frankel, whom Farrah termed 'rude' and 'against women', for inviting her to appear on *The Bethenny Frankel Show* and then exposing her to audience critique for waxing her daughter's unibrow. Farrah gained further controversy in the waning months of 2013 by posting and then tweeting her fans an Amazon Christmas wish list that included such items as a $1,000 pair of glitter pumps and a $1,500 bookcase. In all, she asked for 30 items with a combined price tag of $13,000. When realitytea.com posted her wish list, the response was more than vicious. 'mamazog' offered her 'a supply of Valtrex [an antiviral drug to treat herpes] and a burner phone for Sophia with CPS on speed dial' while 'Norma Jeane' preferred a more violent expedient, 'I would buy her a gun so she can all do us the pleasure of blowing her brains out and putting us out of this misery'.[11]

For our purposes Farrah Abraham's star-text demonstrates a number of perplexing elements about transmediation, gender and celebrity, primarily the tacit and overt authorization of a spectacularized sexualization that gains currency in the cross-platform misogynist policing of female behaviors. Much recent scholarship has elucidated the way that postfeminist television culture often involves – and indeed, calls for – 'the pathologization of women – single women especially – through a disciplinary gaze that revels in their moral failings, vulnerabilities, and excesses' (Heller, 2014: 133; see also Holmes and Negra, 2011: 11). Postfeminism, notes Diane Negra (2008), is also marked by several telling factors, including female empowerment through sexualized visuality, hyperdomesticity and the idealization of motherhood, and new norms through which work might be understood, all of which plays into a neoliberal logic in which 'good choices' bring about 'success' in a supposedly free market, where larger impediments of racism, sexism and classism are no longer considered to be relevant. The 'Farrah Abraham Brand' is now recognizable across a broad range of mostly low-status intermedial platforms: reality television, tabloid journalism, talk shows, Facebook, Twitter, celebrity-related blogs, Keek and pornography, both print and video.

By this logic, Farrah Abraham has been playing by the rules of transmediation, postfeminism, and neoliberalism beautifully, using medical technology to alter her body in ways that reinforce normative, even excessive, codes of white female beauty, and then parlaying this body-as-commodity into a free market of celebrity that requires one earn attention through any means possible. In their study of the Kardashian family brand, Maria Pramaggiore and Diane Negra describe the postfeminist cultural turn as combining 'commercial enterprise and frank sex talk', both of which often provide the very terms for which female reality celebrities are censured (2014: 88). True to this script, Farrah's 'free to be me' behaviors have also generated enormous popular invective, unleashed in social media like blogs, Facebook and Twitter. In so doing, Farrah neatly fits within two reality celebrity and postfeminist typologies: the ambitious mother, such as Kate Gosselin or Kris Kardashian Jenner, using children to gain a toehold in a competitive media market, and the celebutante, such as Kim Kardashian and Paris Hilton, using sex and scandal to perpetuate popular interest. In all cases, these celebrities are considered to be poseurs – in that they can make no 'legitimate' claim to fame through talent or achievement but instead offer up their private selves for public consumption. As Erin Meyers (2013) phrases it, particularly with the advent of reality television, 'Talent is no longer a pre-requisite for entrance into the public eye (if it ever really was), but rather [fame now requires] the willingness to "be yourself" at all points within the circuit of celebrity production'. Conventionally, talent as used in this context refers to long-standing masculinist notions of value stemming from achievement in the public sphere. What's striking about the transmediated grotesque in the reality television celebrities we consider is precisely the talent they exhibit for spinning their celebrity across a variety of mediated platforms. Keeping themselves relevant, even if reviled, constitutes not only a form of talent but of labor in this current neoliberal, postfeminist economy.

Importantly, such talent is also a part of transmediation, as it is not coherent unto itself but bleeds across platforms, uniting diverse media under a banner of affective outrage and audience disgust, an element that ties reality celebrity to the leaky body of the female grotesque. Indeed, much as in the case of Gosselin, respondents to websites, user boards, YouTube videos, and Facebook pages use misogynist and trans-phobic punishing language to censure Farrah with impunity, not only demonstrating a salacious pleasure in calling her stupid or ugly because 'her lips look like worms' and she has 'horse teeth'[12] and 'tranny boobs',[13] but also furthering the reality star's celebrity brand through their strong negative affect. 'The unmitigated gall of this sperm receptacle' writes one poster to the *US Weekly* website.[14] When cultural commentators make comparisons across forms – saying that Abraham is simply mimicking the Kardashian model of lowbrow salaciousness – fan boards are doubly invective. Writes '411adrianne' in response to a Hollywood Scoop YouTube video in which Farrah boasts that she outshines Kim Kardashian in every way, 'I'm not a fan of Kim by any means, but … at least in Kim's life the men WANT to stay, this fugly monster can't get any guy to stay around her longer than they get in her nasty pants, what a dumb ugly ho, I can't figure out where the arrogance comes from, other than the fact she was spoiled rotten, she's so nasty looking, and always thinks she's better than EVERYONE, LOL!!!!!'[15] Interestingly, public rebuke rarely uses homophobic language in relation to Abraham, calling her same sex erotics 'hot' girl-on-girl or even 'lesbian' behavior that clearly codes as exciting for an interpellated heterosexual male viewer.[16]

In her simultaneous repudiation and fusion of these mediated typologies, Farrah furthers a concept of the transmediated grotesque, a point that becomes all the more evident when the matter of her sex tape/porn video gets a bit more scrutiny. The video, named *Farrah Superstar: Backdoor Teen Mom*, features Abraham having oral, vaginal and anal sex with a seemingly anonymous man, later revealed to be porn star James Deen. The polymorphous erotics of the tape themselves speak to an eruptive sexuality that queers the body and its pleasures, but it is the pointed lack of eroticism within the *Backdoor Mom* video that strikes us as the richest element for study within the context of the transmediated grotesque. The tape, later released by porn distributor Vivid, reportedly earned Abraham $1.5 million (and has been so popular that viewer demand crashed Vivid's website). Claiming that she wanted a 'personal video' that would offer her visual proof of her 'banging body'[17] someday in the future and also 'show her feminine side',[18] Farrah also argued, in vacillating and bafflingly contradictory fashion, that she and Deen were dating, that she was single and wanted to memorialize her body and so hired a professional company who, in turn, hired Deen, and that she and Deen made the video on their own. In an interview with *Entertainment Tonight*, she played an interesting version of the victim card, arguing that reality celebrity had robbed her of a personal life and any chance of having a boyfriend, and so hiring an actor was her only option for sex or intimacy.[19] Her move to go public and sell the video, she argues, was preemptive, to keep remuneration where it was due. As she announced to Kirk Fox on *The Test*, a talk show where participants engage in a technologized or biomedical test to settle a dispute, 'I just see people trying to make money off of me. I need to grab the bull by the horns. I don't care if everyone sees my *** [bleeped out]. I don't care if anyone sees my boobs. I'm proud of how I look and how I feel about myself, and I want to remember that moment'.[20] These statements clearly echo Negra's contention that pornography has become a mainstream element of postfeminist culture that must negotiate a complicated divide between healthy erotic expressions and sexual exploitation. Within this, Negra (2008: 100) notes, it is the sex worker who emerges as 'one of popular culture's most regular archetypes of paid labor ... particularly in those genres and forms that play to a younger demographic', and so it is perhaps no surprise that Farrah's attempts at self-expression and economic reward have moved her into the adult entertainment world.

Yet, Farrah's new career as a sex worker is unsettling, even for pro-sex feminists like us, largely because heretofore Farrah's celebrity brand has been predicated on offering access to a personal, even if not a particularly private, selfhood. Those who have been following her since 2009 know what her bedroom looks like, each phase of her daughter's development, how her parents address her, the kind of car she drives, where her puppy sleeps. The intermedial residue that clings to her star text thus makes it impossible to depersonalize her in the context of pornography, to make her only a vagina or anus receiving a cock. Access to her private life has also provided ample coverage of her private parts, the 'lower bodily stratum' that to Bakhtin epitomizes the grotesque. In researching this article, for instance, we have seen Farrah give birth, we have seen James Deen put his fist in her anus, we have seen her drunk and crying and vomiting (but we have not seen pubic hair or cellulite, taboos of greater resonance than her non-private parts). Combined with this sense of access to her widely mediated body is an affective discomfort due to the fact that the sex video is itself incredibly unsexy, for many of the same reasons that her woodenness and abrasiveness on the MTV shows made her annoying. She is a bad actor, and even for the hyperbolic theater of porn, her expressions seem hackneyed and overdone (Figure 23.2). In pointed contrast to both Kim Kardashian and Paris Hilton, both of whom released 'private' sex tapes that are blurred, darkened and fuzzy, Farrah's hyperbolic affect and the crystalline appearance of the footage itself code this 'backdoor mom' as fully on stage.

CONCLUSION

Our point in laying out this controversy is not to flay Farrah on some sort of charge of disingenuous behavior, suggesting, as did both

Figure 23.2 Farrah's face of faux pleasure

Dr Phil and Kirk Fox on their respective talk shows, that if she would just 'come clean' then America wouldn't find fault with her. Nor is it exclusively to point out how perfectly Farrah's declarations against body shame and for self-love find expression in the neoliberal, postfeminist métier that demands sexualized visibility in exchange for improving one's market appeal. Instead, we have tried to work through the complicated dynamics at play in the transmediated grotesque that moves in mercurial fashion in ways both rhyzomatic and teleological. Because both the body and scandal help constitute the intermedial domain of continuity that we address here, Farrah's reality celebrity – her currency as a 'reality star' – secures the attention capital that makes her, quite literally, a billboard for postfeminist sexual politics. But it is her disbursement across these other forms of media that continues to offer her brand its purchasing power. Her image advertises a hybridized self-as-product (and body-as-self) dependent on sexualized display, lowbrow scandal and mediated saturation, in this case served up through a thriving career in the sex industry. Never has the mediascape been more literal about the relation between neoliberalism and postfeminism, which tasks good citizens with self-management strategies that make a viable (read sellable) commodity of the self, and which reinforces a norm of hypersexualization and display as measures of personal value for women. If the grotesque, as Mary Russo tells us, stands in pointed display to conventional conceptions of gender, race, class and sexuality, then Farrah's behavior and body explode these categories of value in their simultaneous instantiation of ideal and material, inside and outside, real and reality, normative and transgressive.

In Farrah's case her self-promotion brands her as a bad self-manager, and her poor choices have made her callous to herself, her daughter and her family. 'Farrah Abraham has made a number of questionable choices in her short time on earth', writes *The Huffington Post*, 'but it was the porno she agreed to do with adult film star James Deen that she says "ruined" her life.' Within hours of that article's posting, respondents rained critique: "No, porn didn't ruin her life ... her never-ending bad choices have ruined her life. Leave porn's good name out of this!" (James H.); "That's her life-ruining mistake? THAT? Not the poor treatment of her mother/father, child, and self on national television? Everyone despised her before the video, but that's the one she's going with?" (Mary Jump)'.[21]

Misogyny in the public forums of new media has become so common that it is almost unremarkable. Writes online journalist Amanda Hess, who herself has been threatened with rape and decapitation, 'Making quick and sick threats has become so easy that many say the abuse has proliferated to the point of meaninglessness, and that expressing alarm is foolish' (2014). But Hess rightly notes that 'casual' hatred, violence, and intimidation exact a painful emotional toll on women, both individually and collectively. While the term 'trolling' has been used to typify a broad range of anti-social internet behavior, increasingly trolling has taken a turn to the misogynist, used against those who identity as female or those who speak from feminist perspectives. Comments and threats, generated by both male-identifying and female-identifying social media users often sexualize and silence women and girls through condemnation, critique or intimidation.

When the troll meets the succubus in the brave new world of transmediated continuity, all manner of the grotesque is let loose, yielding what Patricia Yaeger (2000) describes as a body where the 'volume is turned up' (p. 10), thus obscuring the field around it. 'Monster tropes', says Yaeger, 'become switching points in a stratified culture overloaded with change: a culture whose categories have short-circuited precisely because they carry too much information' (p. 27). In this, contends Yaeger, tropes of the monstrous and the grotesque, the trolling fan/viewer/respondent and the reality momster, allow for the kind of 'social explosions' that enable 'entire populations to resist social change' (p. 27). And while it may never be possible for the reality celebrity mom/monster to turn down the volume and carry less information, it is possible for feminist media theory to increase its bandwidth by engaging in the transmediated sites that allow misogyny to flow rapidly through the twenty-first century mediascape.

NOTES

1 Bakhtin, Mikhail, 'Perforations: Grotesque Corpus', www.pd.org/topos/perforations/perf3/grotesque_corpus.html (accessed December 22, 2013).
2 For a substantive reading of theories of the grotesque from ancient Rome to 20th-century performance art, see the introduction of Russo's *The Female Grotesque* (1995).
3 Kristen Warner (2013) nicely ties this idea of excess to the black woman's body on reality TV.
4 Jon Gosselin is an American of Korean and European decent. EthniCelebs reports that the 'Korean gene is very strong', and 'Kate said that she wanted her kids to look like Jon and the she always wanted little "China dolls"', www.ethnicelebs.com/jon-gosselin (accessed January 5, 2014).
5 *In-Touch Weekly*. November 2013.
6 In a nutshell, Kate accused Jon of stealing her diary from the hard drive of her computer and later giving that information to Rob Shutter, who authored a book called *Kate Gosselin: How She Fooled the World* (2013). Initially available on Amazon, the book was pulled from circulation in the wake of the lawsuit.
7 For a particularly snarky profile of Jon, see Van Zuylen-Wood (2013). See also Eng (2001).
8 'Farrah Abraham "forced to leave rehab for disruptive behaviour": Report', 9 July 2013, http://pagesix.com/2013/07/09/farrah-abraham-forced-to-leave-rehab-for-disruptive-behavior-report/ (accessed December 1, 2013).
9 Farrah's singleness on the show was apparently another of her faux scenarios. She claimed to producers, participants, and audiences that she was dating Brian Dawe, but

after Dawe failed to arrive on the show, celebrity blogs leaked that he and Farrah had lied about their relationship from the beginning as a 'business deal' to get them on the reality show. 'Farrah Abraham's Couples Therapy Lies Exposed!', Radar Online, January 6, 2014. http://radaronline.com/exclusives/2014/01/farrah-abraham-fake-couples-therapy-boyfriend-brian-dawe-real-girlfriend-victoria-stokes/ (accessed January 8, 2014).

10 'Charlie Sheen SLAMS Farrah Abraham', Gossip Cop, June 14, 2013, http://www.gossipcop.com/charlie-sheen-farrah-abraham-fight-feud-letter-texts-lobotomy-porn-douche-tranny-boobs/ (accessed November 14, 2013).

11 'Farrah Abraham Tweets Amazon Wish List, Asks Fans to Buy Her Gifts', Reality Tea, August 21, 2013, http://www.realitytea.com/2013/08/21/farrah-abraham-tweets-amazon-wish-list-asks-fans-to-buy-her-gifts/#/slide/1 (accessed December 6, 2013).

12 'Farrah Abraham Causes Trouble on "Couples Therapy" Proving Everyone Right', The Stir, December 1, 2013, http://thestir.cafemom.com/entertainment/164834/farrah_abraham_causes_trouble_on. (accessed December 2, 2013).

13 Megan Morris, 'Farrah Abraham Takes her Tranny Boobs to Rehab', July 1, 2013, http://www.sheknows.com/entertainment/articles/1004853/farrah-abraham-goes-into-rehab (accessed December 7, 2013).

14 'Farrah Abraham Amazon Wish List: Teen Mom Star Asks Fans for Expensive Gifts', *US Weekly*, August 28, 2013, http://www.usmagazine.com/celebrity-news/news/farrah-abraham-amazon-wish-list-teen-mom-star-asks-fans-for-expensive-gifts-2013288

15 'Farrah Abraham Says She Outshines Kim Kardashian in Every Way', Hollywoodscoop TV (no date), http://www.youtube.com/watch?v=vlwy9q8LUBY (accessed December 8, 2013).

16 Abraham Tanksley (nd), 'Farrah Abraham is a Lesbian', http://www.youtube.com/watch?v=nhdKfMSJj_Y (accessed December 13, 2013. Interestingly, one of the stronger lines of punishing critique came when Farrah, being interviewed by Allie Conti of the *Miami New Times*, answered a question about whether or not she was a feminist by saying, 'I'm pretty feminine'. When the reporter corrected her understanding, Farrah responded: 'What does that mean, you're a lesbian or something?' (Conti, 2013).

17 *The Test*.

18 'Teen Mom Farrah Exposed'. *Dr Phil*, April 19, 2013.

19 Her words: 'I have no relationships and I'm, like, sad sometimes. So, taking all this into consideration, which some find it hard to, that's what brought me here today. I felt this was my way of embracing my sexuality and being happy for me', Farrah Abraham on *Entertainment Tonight*, 8 May 2013, http://www.bing.com/videos/search?q=entertainment+tonight+farrah+abraham&FORM=VIRE1#view=detail&mid=054BD84F1A9CE6C48EC8054BD84F1A9CE6C48EC8 (accessed December 1, 2013).

20 'Teen Mom, Sex Tape Mishap', *The Test*, CBS TV Distribution, Stage 29 Productions, Jay McGraw, Executive Producer, September 23, 2013.

21 'Farrah Abraham Admits Doing Porn "Ruined" Her Life', *The Huffington Post*, January 8, 2014, http://www.huffingtonpost.com/2014/01/08/farrah-abraham-porn-ruined-life_n_4563292.html (accessed January 9, 2014).

REFERENCES

Bakhtin, M. (1984). *Rabelais and His World*. Bloomington, IN: Indiana University Press.

Conti, Allie (2013). 'Farrah Abraham on Feminism: "I'm Pretty Feminine"'. *Miami New Times*, September 3. http://blogs.miaminewtimes.com/cultist/2013/09/farrah_abraham.php. (accessed November 2, 2013).

Eng, David L. (2001). *Racial Castration: Managing Masculinity in Asian America*. Durham and London: Duke University Press.

Heller, Dana (2014). 'Wrecked: Programming Celesbian Reality'. In Brenda R. Weber, ed., *Reality Gendervision: Sexuality and Gender on Transatlantic Reality TV*. Durham: Duke University Press.

Hess, Amanda (2014). 'Why Women Aren't Welcome on the Internet'. *Pacific Standard Magazine*, January 6. http://www.psmag.com/navigation/health-and-behavior/women-arent-welcome-internet-72170/ (accessed January 8, 2014).

Holmes, Su and Diane Negra (2011). 'Introduction: In the Limelight and Under the Miscroscope – The Forms and Functions of Female Celebrity'. In Su Holmes and Diane Negra, eds, *In the Limelight and Under the Microscope: Forms and Functions of Female Celebrity*. New York: Continuum.

Jenkins, H. (2008). *Convergence Culture: Where Old and New Media Collide*. New York, NY and London: New York University Press.

Kinder, M. (1991). *Playing with Power in Movies, Television, and Video Games: From Muppet Babies to Teenage Mutant Ninja Turtles*. Berkeley, CA: University of California Press.

Meyers, Erin (2013). 'The Reality of Contemporary Stardom'. FLOWTV, November 18. http://flowtv.org/2013/11/the-reality-of-contemporary-stardom/ (accessed December 1, 2013).

Negra, Diane (2008). *What a Girl Wants: Fantasizing the Reclamation of Self in Postfeminism*. London: Routledge.

Ouellette, Laurie (2014). '"It's Not TV, It's Birth Control": Reality TV and the "Problem" of Teenage Pregnancy'. In Brenda R. Weber, ed., *Reality Gendervision: Sexuality and Gender on Transatlantic Reality TV*. Durham: Duke University Press.

Pramaggiore, M. and Negra, D. (2014). 'Keeping Up with the Aspirations: Commercial Family Values and the Kardashian Brand'. In Brenda R. Weber, ed., *Reality Gendervision: Sexuality and Gender on Transatlantic Reality TV* (pp. 76–96). Chapel Hill, NC: Duke University Press.

Russo, M. (1995). *The Female Grotesque: Risk, Excess and Modernity*. New York, NY and London: Routledge.

Van Zuylen-Wood, Simon (2013). 'Jon Gosselin in the Wilderness'. *Philadelphia*, December 27. http://www.phillymag.com/articles/jon-gosselin-wilderness/?all=1 (accessed January 1, 2014).

Warner, Kristen (2013). '"They Gon' Think You Loud Regardless": Ratchetness, Reality Television and Black Womanhood'. Presentation at Reality Gendervision, Bloomington, Indiana, April 25–27.

Yaeger, Patricia (2000). *Dirt and Desire: Reconstructing Southern Women's Writing, 1930–1990*. Chicago: University of Chicago Press.

PART IV

Audiences, Reception, Consumption

24

From the Networks to New Media: Making Sense of Television Audiences

Laura Grindstaff

INTRODUCTION

For the past half-century and more, television has occupied a central place in American domestic and national culture. Unsurprisingly, much time and energy has been devoted to studying television audiences. And yet, paradoxically, the more we know, the less coherent the concept of the audience becomes. Particularly in the multimodal, multiplatform, convergent, digital-interactive, 'new media' era, what constitutes 'television' let alone the 'television audience' is by no means self-evident. What we now call television is an inseparable part of media streams that people encounter everyday across a wide variety of contexts. In her 2004 article 'The challenge of changing audiences: Or, what is the audience researcher to do in the age of the internet?' Sonia Livingstone characterizes the television audience as a 'moving target' for scholars. Surely it is that. This movement hasn't so much deterred research as posed new questions and challenges, especially in terms of methodology.

This chapter explores some key movements and moments in the study of television audiences, which necessarily means exploring concepts such as commodification, reception, consumption, and participation. This exploration is far from exhaustive, of course. The bodies of work focused on television audiences, broadly conceived, stretch wide and deep. They encompass different theoretical traditions, countless methodological choices and competencies, varied national and international industrial systems, local and global interdependencies, and different cultural/intellectual priorities. Mine is one perspective, shaped by the specificities and idiosyncrasies of my own training and social location, including the American context.

In what follows, I first provide a brief summary and overview of key theoretical traditions that grapple with broad questions regarding the role and place of television in American society. Conceptualizations of the television audience are implicit rather than explicit in much of this work, often couched in a language of influences or effects. I then examine three different empirical approaches to researching audiences – three different interventions – that represent distinct ways of thinking about audiences. The first is industrial, in which the audience is a commodity measured and sold to advertisers by media companies. The main focus here is on measurement and ratings. The second intervention reflects a body of academic work that typically goes by the unsexy label 'reception studies', in which viewers/readers actively interpret television texts under specific socio-cultural conditions. Included here are ethnographic approaches to television audiences, by which I mean approaches that demonstrate various degrees of interest in and attention to material context (under what circumstances do people watch/use television?) instead of or in addition to symbolic content (what meanings do viewers make of what they watch?). The third is the scholarly study of fans and fandoms, which could be subsumed under reception and/or ethnographic approaches but which I treat separately because fandoms hold open the promise of moving us away from notions of 'audiences' toward notions of 'publics', an important distinction I borrow from Daniel Dayan (2001). These three approaches do not so much reflect chronological developments in the study of television audiences (note, for instance, that industrial ratings are more sophisticated and influential than ever before) as different methodological choices stemming from different assumptions about presumed passivity versus activity on the part of viewers/users. Finally, I conclude with a brief discussion of some key challenges associated with studying television and television audiences in the new media era. In their recent book *Spreadable Media*, Henry Jenkins and his colleagues ask 'what constitutes meaningful participation' in our contemporary media environment? (Jenkins et al., 2013, pp. 153–194). Their insights are particularly relevant for the study of television, and for understanding persistent inequalities surrounding media access and use.

SOME THEORIES OF TELEVISION AND SOCIETY

Historically, scholarly interest in the social role and impact of television has been driven by the medium's pervasiveness, as well as its ability to bring the outside world into the home and thereby connect disparate individuals to one another in virtual space via notions of 'the audience'. The fact that millions of people watched the same program simultaneously formed the basis for caring and theorizing about television. As Leo Bogart wrote in 1956, 'with no other form of impersonal communication has the sharing of experience been possible on so universal a scale and to so intense a degree as with television' (p. 2). In the early years as today, television – and mass media more generally – prompted considerable debate about the changing nature of society, the public sphere and the public good.

In a positive vein, social scientists such as John Dewey, William James and Robert Park believed that mass media, if managed well, could strengthen democracy by socializing people into a common set of norms and values (Grindstaff and Turow, 2006). Marshall McLuhan (1964) posited an even more optimistic (some say celebratory) view of electronic media as enabling a global village transcending time and place, a thesis later explored in a more detailed and historicized fashion by Joshua Meyrowitz (1985). Less optimistically, the rise of mass communications, in concert with industrialization and technological change, was said to breed cultural mediocrity (according to mass society critiques) and/or inhibit revolutionary class consciousness (according to Marxist critiques). Adorno (1957), for example, in concert with other scholars of the Frankfurt School, denounced television for insinuating the capitalist mode of production into everyday leisure, including into people's psychic lives. Baudrillard (1983) also took a pessimistic/deterministic stance, suggesting that the primary effect of television was to substitute a representation of reality (simulacrum) for reality itself.

In her now-classic essay 'Audience Control' Muriel Cantor (1980) notes how both the mass society and Marxist critiques of television, which mirrored concerns about forms of mass media preceding television, contained implicit assumptions of audiences as powerless and manipulable – either by technology or capitalist ideology or both. Television had negative 'effects' on society because audience response was said to be determined in large measure by the industrial nature of the medium. Unsurprisingly, this stood in stark contrast to early industry discourse about the television audience, which positioned viewers as 'in control' of television content in the form of ratings. As Cantor points out, the industry perspective didn't necessarily position the audience as active, but nor did it position the audience as a passive, undifferentiated mass; rather, the audience was understood to be a market of specific demographic characteristics, some subset of which was said to shape programming through ratings (more on this topic shortly).

Antonio Gramsci's concept of hegemony complicated Marxist critiques of the media and paved the way for another set of influential theories of television, beginning in the 1970s. In arguing that particular relations of ruling prevail not because they are imposed on people against their will but because they are accepted as common sense by the rulers and the ruled alike, Gramsci offered a more nuanced theory of power that posited ideology as collectively held and needing to be continually re-secured. These ideas influenced Todd Gitlin (1979), for example, who was interested in how prime-time television could be simultaneously appealing to audiences and sustain class hegemony. Although not passive, the audience in Gitlin's view had limited power to influence content because the commercial system is able to absorb and harmonize conflicting demands and definitions of reality in ways that ultimately reaffirm the status quo. The hegemony concept also found expression in the work of Raymond Williams (1974), Horace Newcomb (1974), Gaye Tuchman (1974), Stuart Hall (1980) and Douglas Kellner (1981), among others, all of whom emphasized television as a site of contradiction where meaning is struggled over and not simply given or assumed. Newcomb in particular helped shift the discussion of television from a discourse of 'mass communication' to a discourse of 'popular culture', with an attendant shift in the degree of agency accorded audiences. In *TV: The Most Popular Art*, Newcomb emphasized the complexity of television entertainment with regard to plot, character and genre, and the multiple levels of meaning available to viewers in making sense of television narratives.

It was Raymond Williams (1974) and Stuart Hall (1980), however, who were most influential in shaping the study of television within the context of the emerging field of cultural studies in the UK and abroad. I will discuss Hall in a later section, for it was his encoding-decoding model that inspired much of what we now call reception studies, including the empirical study of television audiences. Williams' influence was in some ways more mobile and wide-ranging, inspiring a new generation of scholars in the humanities (particularly those trained in film analysis) who welcomed ways of thinking about television as something other than discrete programs to be analyzed or a capitalist institution to be condemned (see Spigel, 1992). His 1974 book, *Television: Technology and Cultural Form*, elucidated the concept of 'flow' – the movement of fragmented text across time and space – and forced a consideration of television as a mode of address that structured experience apart from specific questions of content or message. Williams argued that television is both an intention and an effect of the social order, including relations of power and inequality, and as such it offers people a kind of language or grammar for understanding and negotiating those relations. The idea that television has an industrial mode of address – an experiential aesthetic – was fruitfully employed and debated by other scholars (see Ellis, 1982; Kaplan, 1983; Newcombe and Hirsch, 1983; Browne, 1984; Fiske, 1987; Caldwell, 1995; Lembo, 2000) and remains an important touchstone for rethinking televisual aesthetics in light of recent industrial and technological change (see Boddy, 2004; Caldwell, 2004; Parks, 2004; Uricchio, 2004; Wood, 2007).

PAINT-BY-NUMBERS TELEVISION: THE AUDIENCE AS COMMODITY

Far removed from discussions of hegemony and flow, the television industry from the very beginning has had its own preferred way of thinking about audiences: in terms of ratings, as commodities to be measured and sold to advertisers. Indeed, the vast majority of time and money devoted to researching audiences occurs outside of academia. Ratings research is important to discuss here, not because it accurately assesses what television audiences are up to, but because it animates a critical discourse among scholars with an imperative for operationalizing audiences differently. Currently, Nielsen Media Research retains its monopoly over television ratings production, even as the Nielsen company itself has changed hands (it is now owned by the Dutch media conglomerate VNU). The function of the company is to put a value on advertising time, as determined by the size of a program's audience and other audience demographics such as age, sex and income. According to Nielsen statistics, Americans spend more than 34 hours per week watching TV, plus another 3–6 hours per week watching recorded programs. The average household has access to more than 100 channels and several different television sets (Hinckley, 2012).

Dallas Smythe (1977) is typically credited with formulating a theory of the audience as commodity. Although not uncontested (see Caraway, 2011), this theory suggests that the activity of watching television represents a form of wageless labor that audiences engage in on behalf of advertisers. Audiences get rewarded with programming – what Smythe calls a 'free lunch' – in exchange for doing the work of constituting themselves as a potential market for advertised goods. For Smythe, audiences are simultaneously doing productive work for the capitalist (the advertiser) and reproducing their own labor power as viewers of programming. Jhally and Livant (1986) argue something similar, substituting 'programmer' for 'advertiser' in the formulation. For them, the viewing audience, having already received its 'wage' in the form of programming, is working on behalf of the television programmer rather than the advertiser; the programmer then converts surplus watching time into additional advertising revenue.

As Brett Caraway (2011) notes, the industry construction of the audience as commodity is, in a very real sense, fictitious, because no one knows whether viewers exposed to specific advertising messages actually purchase the products advertised. Advertisers are thus not buying audience power but the ratings companies' promises about viewers' future purchasing behavior. Networks and cable companies pay ratings firms – predominantly, Nielson – to help them reliably predict the realization of surplus value in the form of the consumption of goods, 'but the whole system of commodity exchange is speculative – the networks are acquiring credit based on surplus value which has yet to be realized' (Caraway, 2011, p. 701). For this reason, Caraway (2011) believes the economic transaction described by Smythe is better characterized as rent: the media owner rents the use of the medium to the advertiser who is interested in gaining access to an audience, and speculation on the size and quality of the audience determines the rent charged.

In her trenchant critique of the commodity audience, Eileen Meehan (1990) makes a related but different point when she notes that the measurement techniques used in ratings research construct the very thing being measured. Her careful historical account of the development of ratings systems in the US, beginning first with radio and extending to television, demonstrates that different methods produce different ratings for the same program, partly as an artifact of the methods themselves. Because the specter of different ratings for the same programs threatens to disrupt the established business of buying and selling the commodity audience, networks and advertisers agree to accept a monopoly in ratings production if this monopoly can balance out discontinuities in demand (networks want to charge advertisers as much as possible for delivering audiences, advertisers want to pay as little as possible for those audiences), while satisfying the need for a single, agreed-upon measure of viewing in the form of the commodity audience.

The commodity audience is not viewers writ large, of course, but the subset of viewers who are sampled – until recently, mostly by paper diaries and electronic people meters installed in selected homes. The meters record what is being watched and who is watching, provided viewers remember to push log-in buttons (each member of the household has a button associated with her demographic information). Consequently, ratings do not represent the wishes of the television audience *qua* audience because most members of the viewing public are not measured and therefore literally don't count; rather, ratings reflect 'the forced choice behavior of the *commodity audience* within limitations set by continuities in demand, market conditions, production costs, and changing conditions in the general economy' (Meehan, 1990, pp. 126–127). Ratings are forms of measurement selected on the basis of economic goals, Meehan reminds us, not according to the rules of social science. 'The difference between the commodity audience and the public viewership, between manufacturing the commodity audience through ratings and measuring the public taste through social research cannot be overemphasized' (Meehan, 1990, p. 127).

In the new media environment with the rise of digital television and the dispersal of television programming across multiple interfaces and delivery systems, the search for 'reliable' audience measures has taken some interesting turns. Nielsen has begun to track time-shifting on DVRs (where viewers can shift when they watch their chosen programs) and is considering the use of cable set-box data (STB data), a transmission from the cable signal back to the cable operator that gives a complete picture – not just a sample – of what viewers in a particular place are tuned to at any given time (www.nielsen.com/us). According to the *Hollywood Reporter*, Nielsen has partnered with Twitter to measure TV-related tweets, and is poised to install new hardware and software in its 23,000 sample homes to capture viewership not only on cable, satellite and over-the-air broadcasts but also devices that deliver streaming video services provided by companies like Netflix and Amazon (Block, 2013). The portable people meter (PPM), initially developed by Arbitron (a Nielsen rival-turned-acquisition), promises to extend the boundaries of media consumption to outside the home. Although not yet widely adopted, the PPM is a pager-sized device that monitors the individual viewer rather than the television set by picking up a unique digital code embedded in the audio tracks of all the radio and television channels that a PPM-wearer is exposed to throughout the day. Theoretically, with the cooperation of entertainment companies, it could detect everything from DVDs to video games to MP3 music files and even whether a person drives by a particular billboard or electronics store (Gertner, 2005). Of course, whether or not people are actually paying attention to the channels and signals registered by their PPM devices – or any of the in-home measurement tools, for that matter – is an open question. But the question may not matter much in the long run. The Nielsen-Arbitron experiment in PPMs has a twist: 70,000 PPM-wearers are being tracked, not for the sake of ratings, but to match all the advertisements and messages they hear to the actual purchases they make using bar-code technology (Gertner, 2005).

If the new media era represents challenges for audience measurement, it also changes the nature of the 'labor' performed by audiences for media owners and advertisers. Indeed, as television itself proliferates across the digital landscape, so does the potential 'work' of television audiences. Philip Napoli (2010) notes that the notion of the audience-as-worker, which may not have been entirely persuasive when what was being monetized was the act of watching television programs, becomes decidedly more concrete in the new media environment where audiences not only watch/receive but create/use content. Today, 'the creative work of the audience is an increasingly important source of economic value for media organizations' (Napoli, 2010, p. 511). The industry itself clearly recognizes this, even if individual users – and some of the academics writing about them – do not. The wealth of scholarly work on fandom notwithstanding, scholars lag behind industry stakeholders in thinking about audiences as producers as well as consumers of content (see also Turow, 2005).

What is the nature of audience productivity, from an industry perspective? The way Napoli describes it, Web 2.0 applications such as Facebook and YouTube enable people to communicate in a community of sorts, with the advertising revenues they generate being derived from audience attention captured with content produced by members of that user/audience community. In other words, 'aggregating or providing a common platform for user-generated content, and then selling advertising on these platforms, represents the core business model of most Web 2.0 applications' (Napoli, 2010, p. 512). User 'content' extends beyond selfies and homemade videos, of course. User-generated content comes in the form of comments, ratings and reviews for products and services, which represent an important source of monetized value for organizations involved in the production and distribution of media. Audiences further create value for advertisers when they assist with the actual marketing of products – producing their own commercials, engaging in word-of-mouth endorsements online (sharing, liking, recommending), and/or integrating brand messages into their own Facebook or MySpace pages (Napoli, 2010, p. 512). And then there is the work of audiences in helping to generate popularity and buzz for specific programs on fan pages, chat rooms and message boards, not to mention the 'work' of voting people off an island or fashion runway in the latest reality program.

Consequently, as Napoli observes, the old distinction between scholars who claimed audiences are working for advertisers (Smythe, 1977) and those who claimed audiences are working for programmers (Jhally and Livant, 1986) has collapsed, because, in the new media era, audiences are clearly working for both. For Napoli, what is so remarkable about this development is the extent to which people (1) engage in the production of media content absent any expectation of financial compensation, and (2) appear willing to allow others – notably media organizations – to capture the revenue generated by their aggregated effort. Mark Andrejevic (2004) and others have extended this argument in important ways beyond audiences to the on-camera participants of reality-based programming, whose flexible, insecure, non-union and largely uncompensated labor generates enormous profits for the television industry. The 'work of being watched' and the willingness of people to engage in this work – including but not limited to reality-TV participants and those seeking visibility on YouTube, Facebook, Instagram, etc. – is consistent with the push toward new forms of celebrity and a new culture of surveillance in which watching and being watched is increasingly normalized and monetized (Andrejevic, 2004; Ouellete and Hay, 2008).

TELEVISION RECEPTION: BEYOND PEOPLE METERS

The 'television audience' as measured by industry stakeholders operates within a closed feedback system informed by industrial logic. 'Institutional knowledge is not interested in the social world of actual audiences ... [but] in an objectified category of users to be controlled', writes Ien Ang. '[This construction] enables television institutions to develop strategies to conquer the audience so as to reproduce their own mechanisms of survival' (1991, p. 154).

Reception studies emerged partly in response to the industrial logic as defined above, and partly in response to traditions of social science research focused on media 'effects'. The results of laboratory experiments, content analyses and large-scale attitudinal surveys published in mainstream academic journals from the 1960s onward (typically by psychologists, social psychologists, or mass communication scholars trained in statistical methods) formed a significant core of research on the topic of television audiences, although practitioners generally did not claim membership in something called television studies or audience studies. In contrast to effects researchers who saw themselves as scientists testing hypotheses, reception studies scholars saw themselves as analysts exploring/theorizing an interactive process. To paraphrase James Halloran (1970), the question in reception studies is not what the media does to people, but what people do to the media.

When examining qualitative traditions of television audience research, it is difficult to disentangle media studies from cultural studies, particularly in the UK where a focus on media developed in tandem with cultural studies. Both Stuart Hall's (1980) encoding-decoding model of media reception and the interview-based studies of television audiences by Morley (1980) and Hobson (1982) were foundational in shaping growing scholarly interest in the qualitative, experiential dimensions of television reception and use (as well as engagement with other forms of popular culture). Reception theory, reader-response theory, the text-reader model, ethnographic studies of audiences, and even 'uses and gratifications' research – all are interventions in the reception studies tradition that, in different ways and to varying degrees, move us away from an understanding of 'the audience' as an effect of the text or production process toward the notion of audiences as active makers of meaning. For this reason, Fiske (1987, p. 16) prefers the term 'reader' over 'audience' in referring to television viewers, as the latter implies 'a homogeneous mass of people who are all essentially identical, who receive the same messages, meanings and ideologies from the same programs, and who are essentially passive'. He later coined the term 'audiencing' to make much the same point, believing the verb form of the noun better captures the active, participatory quality of television consumption (Fiske, 1992). John Hartley (1999), too, has been a strong proponent of the 'active audience' paradigm. This is not a more objective conceptualization of the audience, only a different one. As Fiske (1989) insists, there is no such thing as 'the television audience' apart from the methods used to study it (see also Allor, 1988; Dayan, 2001). The main contribution of reception studies is to demonstrate the meaning-making capacity of audiences within particular cultural and historical contexts, underscoring the diversity of meanings, the diversity of interpretive practices and the diversity of audiences, while still retaining notions of textual structure, industrial practice and social location. Although not focused on television, Janice Radway's (1984) important study of romance readers is clearly an early intervention along these lines.

In his essay 'Encoding/decoding', Hall (1980) theorized the media-audience circuit as reciprocal but not equal: the ideology of the culture industries may be hegemonic and work to secure social and political consensus, but people may respond to and interpret media texts in a variety of ways. For Hall, there is a necessary correlation between people's social positioning and the meanings they generate. This introduces a potential tension into the circuit, between the meaning encoded at the point of production (which necessarily bears the imprint of dominant ideology) and the meanings decoded at the point of reception by viewers whose social location may position them against that ideology. Viewing television thus involves negotiation between reader and text, with some readings being preferred but no reading being imposed. Hall offered three generalized reading strategies for characterizing viewers: dominant (the reader agrees with and accepts the dominant ideology); negotiated (the reader accepts the dominant ideology for the most part but has to customize it to fit her local circumstance); and oppositional (the reader opposes the dominant ideology).

Obvious problems exist with the model (why only three reading strategies? How do we know which readings are preferred? Do oppositional readings matter in the real world or only in readers' heads?) and, naturally, the notion of the active, resisting audience can be carried too far, especially if presumptions of semiotic resistance are accorded great social or political significance. The main contribution of the encoding/decoding approach was to provide theoretical justification for conceptualizing television audiences differently: not as an irrational mass manipulated by ideology on the one hand, and not as an assemblage of rational individuals strategically consuming media for identifiable and measurable reasons on the other, but rather as complex, messy subjects embedded in cultures and communities. Dayan (2001, p. 748) aptly describes it as 'a framework that abandons individual psychology and the study of the structural coherence of a text to concentrate on the nature of the relationship between text and reader'. He outlines four main assumptions of the framework : (1) the meaning of a text is not pre-given but is produced in the context of reception; (2) the analyst does not have privileged knowledge of the text; (3) readers/viewers are varied, as are contexts of reception; and (4) meanings, rather than the text itself or the industrial system that produces it, are the starting point for the study of 'effects' (p. 749). Texts and anthologies devoted to the study of television and its audiences testify to the centrality of this perspective (see Allen, 1987; Fiske, 1987; Seiter, 1990; Morley, 1992; Hay et al., 1996; Abercrombie and Longhurst, 1998; Newcomb, 2000; Tulloch, 2000; Gorton, 2009; Briggs, 2010; Seiter et al., 2013).

In television audience scholarship, the contours of reception continue to shift as more studies are carried out and more types and levels of context are considered. One strand of reception study has focused on viewer interpretations of specific programs, genres or sets of programs (Morley, 1980; Hobson, 1982; Ang, 1985; Liebes and Katz, 1990; Livingstone, 1990; Jhally and Lewis, 1992; D'Acci, 1994; Livingstone and Lunt, 1994; Gripsrud, 1995; Manga, 2003; Hill, 2005; Skeggs and Wood, 2012; Sender 2012). Liebes and Katz (1990), for example, asked groups of people from five different cultures to watch and discuss the prime-time soap opera *Dallas*, revealing the importance of distinct national/cultural repertoires to interpretations of the show. Jhally and Lewis (1992) interviewed viewers of *The Cosby Show* and concluded, among other things, that the program encouraged 'enlightened racism'. Press (1991) and Manga (2003) explored class differences among women viewers of prime-time programming and daytime talk shows respectively, while Hill (2005), Skeggs and Wood (2012) and Sender (2012) all focus their attention on viewers of reality television.

A second strand of reception research examines the broader domestic (and sometimes public) contexts of television use/consumption in everyday life (Hobson, 1982; Morley, 1986; Palmer, 1986; Lull, 1990; Gray, 1992; Buckingham, 1993; Brown, 1994; Gillespie, 1995; Gauntlett and Hill, 1999; Lembo, 2000; McCarthy, 2001; Fisherkeller, 2002; Bird, 2003; Mayer, 2003). This work examines who watches television, the various conditions under which watching occurs (when, where, why, how), and how television use intersects and overlaps with other aspects of daily life. Topics include the gendered use of technology within the family (Morley, 1986; Gray, 1992), the unique ways that children relate to television (Palmer, 1986; Buckingham, 1993), the sociality of television use among people of different occupational backgrounds (Lembo, 2000), the deployment/reception of television in public settings such as waiting rooms, airports, bars and retail spaces (McCarthy, 2001), the use of television and video in building community and recreating cultural traditions across ethnic diasporas (Gillespie, 1995; Mayer, 2003), and the meanings and uses of television culture in the lives of American adolescents as they play out in the varied contexts of family, school and peer group (Fisherkeller, 2002).

Together with a limited subset of fan studies, this second strand is most often identified as ethnographic, despite the fact that extended interviews and short-term encounters with specific groups or individuals are more common than is sustained fieldwork within a culture or community (Fisherkeller [2002] is a notable exception). As Lotz (2000) reminds us, classifying one's object of study as an 'audience' versus a 'culture' remains a key difference between media studies and anthropology when investigating media consumption. Classic ethnographic immersion is more easily accomplished in the relatively bounded spaces of television production (e.g. Grindstaff, 2002) than reception, reception being a more fluid, geographically dispersed and privatized phenomenon (see Radway, 1988; Morley and Silverstone, 1990; Moores, 1993; Ang, 1996; Seiter, 1999). Not only is it is difficult to hang out in people's homes (or cars or offices or dorm rooms) and watch them watch/use television, but television use cannot easily be separated from the rest of everyday life, as it unfolds either on- or off-line. In the words of Ang (1996, p. 68) '"watching TV" is no more than a short-hand label for a wide variety of multi-dimensional behaviors and experiences implicated in the practices of television consumption ... [consequently] it becomes difficult to demarcate when we are not part of the television audience'. This is dilemma is only magnified in the contemporary media environment by the dispersion of television texts across multiple mediums and platforms.

FANS AND FANDOMS: THE PARTICIPATORY AUDIENCE

Studies of fans and fandoms partially sidestep the problem of how to locate the when and where of television consumption because fans often create and sustain self-consciously identified communities and subcultures. The study of fans has been one of the signature contributions of media studies generally and reception studies specifically. This is because, in part, fans crystallize both what is concerning and what is promising about television consumption in the modern era. According to Jensen (1992), early critics saw fans as lonely, isolated individuals whose affinity to a media figure or text is either pathetic (the fan as nerd or geek) or dangerous (the fan as psychopath), or a member of a hysterical crowd (the screaming/fainting Beatles fan) or uncontrolled mob (drunken, destructive soccer hooligans). All four tropes, Jensen argues, reflect anxieties about the decline of local familial and community-based ties and their substitution by impersonal, mediated forms of sociability. On the more optimistic side, scholars recognized the promise of active, creative, 'producerly' engagement with media texts for the purposes of building new forms of community – the dominant characterization of fans that held sway in what Gray et al. (2007) call the 'fandom is beautiful' phase of fan studies.

The analytic framework for this initial phase of fan studies came from French anthropologist Michele de Certeau via Henry Jenkins. De Certeau's theory of 'poaching' offered media scholars a way of understanding fan activity as productive and participatory within an overall context of inequality and institutional marginalization (de Certeau, 1984). As peasants and not proprietors in the media landscape, the power of fans is the power of appropriation and consumption rather than production, even as consumption is understood to have a productive dimension. According to de Certeau, there corresponds to the rationalized, spectacular production of the culture industries another type of production, called 'consumption'. Consumption is 'devious' and 'disperse', he says, '... it insinuates itself everywhere, silently and almost invisibly ... it does not manifest itself through its own products, but rather through its ways of using the products imposed by a dominant economic order (1984, p. xii). The power of appropriation doesn't level the playing field; there is no equivalence in the production-consumption relation. Rather, poaching articulates a struggle over meaning that both reflects and constitutes unequal power relations in late modernity.

De Certeau usefully distinguishes between strategies and tactics to reinforce this point. Strategists are the people who get to make and enforce the rules; they have institutional power. They are the politicians, the lawmakers, the policymakers, the CEOs, the educators, the movie moguls and television producers – they are the cultural capitalists, or what Fiske (1989) calls 'the power bloc'. Tacticians, on the other hand, are producers with a small 'p'. Lacking an institutional power base, they are the ones for whom the rules are made. Their power is the power of appropriation, of making do with what they have. Tactics are thus more ephemeral and fleeting; as de Certeau would say, they are opportunities 'seized on the wing' (1984, p. xix) by those 'already caught in the nets of "discipline"' (p. xv). To the extent that the average person's relationship to the culture industries is on the consumption rather than the production side of the equation, we are all tacticians rather than strategists – we don't own the land, but we can poach on it and potentially recraft it to better suit our interests and desires.

Henry Jenkins (1992) famously applied these ideas in his study of *Star Trek* (and other media) fans in his now-classic book *Textual Poachers*. Jenkins saw fandom as a particularly good example of poaching because fans were both persistent and inventive in their efforts to reclaim media imagery for themselves. Fans refuse the high-culture mode of reception in which audiences are expected to be passive and worshipful, maintaining a distance between artist and audience. For Jenkins, this refusal to pay homage to authorial control is important because it challenges the ability of media producers to determine the creation and circulation of meanings: once characters become part of popular discourse they become the property of the fans who fantasize about them, not of the industry executives who produce and merchandise them. In this formulation, fandom goes beyond being a regular viewer of a favorite program because it translates viewing into some kind of cultural activity: sharing thoughts and opinions with others, joining a community of fans with common interests, even generating original art work, poems, novels, screenplays, zines and videos. Indeed, some fan activities go beyond poaching in that people not only poach on the property of others, they make their own property, their own productions. 'Fans possess not simply borrowed remnants snatched from mass culture, but their own culture built from the semiotic raw materials the media provides' (Jenkins, 1992, p. 49).

Television fans thus differ from other categories of viewers because they approximate what Dayan (2001) calls a 'public' rather than merely an 'audience'. In Dayan's view, reception studies in the reader-response or text-reader tradition, although an improvement over earlier effects models of research, nevertheless create the audience as an artifact of the method; in eliciting statements that viewers would never make if not for the provocation of the researcher, and in analyzing reactions whose nature is typically private and non-discursive, scholars incorporate viewers into an invented discourse that would not otherwise exist. Ethnographic studies of viewers as 'interpretive communities' only partially resolve this problem. Unlike an audience, a public, according to Dayan (2001), has a milieu that sustains sociability, a self-reflexive sense of itself as a public, and the capacity for self-representation. For Dayan (2001), 'true' publics do not form around a medium (television or any other), but in relation to a social problem and with respect to other publics. That being said, he sees fandoms as approximating publics. He considers them to be 'quasi-publics', excluding them from full-fledged membership because they are, in his words, 'ephemeral' and 'non- serious', focused on mimicry and play rather than real socio-political issues (Dayan, 2001, p. 752).

Setting aside this problematic re-inscription of the very cultural hierarchies that fandom works to challenge (high/low, serious/trivial, information/entertainment, etc.), we can see that the first wave of fan studies, of which *Textual Poachers* was a part, varied in topic and focus but generally confirmed the image of fans as active consumers who worked within and against commercial culture to create media publics (see Bacon-Smith, 1992; Lewis, 1992; Harrington and Bielby, 1995; Tulloch and Jenkins, 1995; Penley, 1997). Gray et al. (2007) note that these studies do not so much deconstruct the binary in which fans are positioned as 'other' to the 'normal' (detached) media consumer, as attempt to differently value the fan's place in the binary. Valuing fans differently does not mean projecting onto them oppositional tendencies, of course, and for the most part scholars have avoided this. As Jenkins (1992) reminds us, not all readings are oppositional, not all readers are resistant, and not all resistance is progressive; for the most part, fans gravitate toward particular media texts (presumably inflected with dominant ideology) because of some compatibility or affinity between the text and fans' pre-existing cultural beliefs and commitments. Some studies in fact, revealed how fan activity works to maintain rather than challenge existing systems of classification and thus existing cultural and social hierarchies (see Thornton, 1995; Harris and Alexander, 1998; Jancovich, 2002; Jancovich et al., 2003).

For Gray et al. (2007), the chief shortcoming of early fan studies is not its celebratory tone, although there is some of that, but its tendency to exclude from systematic study the most common or typical exemplar of fandom – the person who loves a show, watches it religiously and talks about it enthusiastically, but does not otherwise engage in fan activities. In other words, there is a bias toward organized, active, highly-visible groups or subcultures (for an important exception, see Harrington and Bielby, 1995). Subsequent studies began to right this imbalance, situating organized fandom on a continuum from regular viewing to amateur content-production and widening the field of fan studies to encompass greater conceptual, theoretical and methodological diversity (see Barker and Brooks, 1998; Adden, 1999; Brooker, 2002; Hills, 2002; Thomas, 2002; Juluri 2003; Sandvoss, 2003, 2005).

Arguably, the new media environment has made fandom more relevant than ever before. Far from existing on the fringe of media consumption, the DIY practices associated with fandom have emerged as central features of television and media consumption in the digital age. In *Convergence Culture*, Jenkins (2006) describes a moment when 'fans are central to how culture operates … the concept of the active audience, so controversial two decades ago, is now taken for granted by everyone involved in and around the media industry' (2006, p. 1). Media companies act differently today – generating new kinds of content and forming new relationships with consumers – because they have been shaped by the increasing visibility of participatory culture, once associated primarily with fandom (Jenkins, 2006, 2007; Jenkins et al., 2013). For Jenkins, the interactive audience of participatory media culture is more than a marketing concept and less than a democracy: media industries have to accommodate the interests of consumers even as they seek to bend consumers to their interests (Jenkins, 2006). In his afterward to the anthology *Fandom* edited by Gray et al. (2007), Jenkins argues we should avoid celebrating a process that commodifies fan cultural production and sells it back to us, but we must also acknowledge new trends that make companies more responsive to committed consumers and that extend the influence fans exert over the media to wider publics (2007, p. 362). We are witnessing a new kind of cultural power, he says, 'as fans bond together within larger knowledge communities, pool their information, shape each other's opinions, and develop a greater self-consciousness about their shared agendas and common interests' (p. 363). Likening these new knowledge communities to 'collective bargaining units for consumers', he speculates that 'as fandom becomes part of the normal way the creative industries operate, then fandom may cease to function as a meaningful category of analysis' (p. 364).

CONCLUSION

The new media environment, characterized by digitalization, convergence, choice, interactivity, intertextuality and extraterritoriality, presents both opportunities and challenges for the study of television and television audiences. As Livingstone (2004) aptly notes, it turns out that the 'television' of media theory was a temporary, particularistic phenomenon and not a timeless, universal one; scholars have mostly attended to mass-broadcast, non-interactive television along with the sit-on-the-couch domestic audience. Today, in the post-network era, television is present in multiple locations and on multiple platforms not only in the home but in all manner of public and private spaces; it is used not only for entertainment/leisure but for surveillance and social control; it allows people to watch their favorite shows but also shop, bank, vote, and shift programming to the internet; people not only receive television via cable, satellite and the internet, they carry it around on cell phones, tablets and personal video recorders (PDVs) – breathing new life into Raymond Williams' characterization of viewing as a form of 'mobile privatization' (Williams, 1974) and prompting Grindstaff and Turow (2006) to prefer the term 'video cultures' to 'television'.

The economic, industrial and technological changes in the production and distribution of television are more easily documented and better understood than are commensurate changes in reception and use. To quote Spigel (2004: 6), 'as images multiply on a variety of delivery systems and platforms, who knows what audiences are seeing – much less thinking – any more'. The challenge for audience studies of television is understanding *how* people are engaging with video cultures (contexts, patterns and practices of reception), *why* people watch/use/interact with these cultures (to what purposes), and *what* people are watching, in terms of the meanings television/video texts convey, given the varied and multiple modes of engagement. For exploring the *how* question, Lotz (2009) sees three developments as key: (1) the emergence of 'on demand' technologies, which represent a fundamental break from the programming schedules of the network era; (2) the existence of extradomestic viewing contexts, which free programming from the TV set in the living room; and (3) the increasingly individualized organization of the medium's use – made possible largely by digitization and cross-platform delivery. The *why* and *what* questions are proving more difficult to study in a qualitative manner. Simply tracking the programs or genres people watch, or monitoring discrete, user actions such as clicking, linking, liking or favoriting do not tell us much about interpretive processes at work or how the very meaning of television/video texts might hinge on their context of use. More promising are online spaces that encourage viewer commentary, response and discussion because such spaces potentially tell us something about what some viewers think, even as the form of communication shapes its expression. User actions from 'liking' to 'commenting' do indicate activity – and interactivity, of a sort – but within frameworks established by the classificatory systems being deployed. As Jean Burgess and Joshua Green (2009) note in their analysis of YouTube (without doubt one of the key new-media sites for watching TV/video), the different ways of measuring the popularity of videos posted on YouTube – 'most viewed', 'most responded', 'most discussed', 'most favorited' – constitute different versions of what YouTube is, and what it is for.

Lotz (2009) reminds us that the extradomestic and individualized use of TV has not entirely replaced older modes of viewing, rather old and new coexist. Moreover, even within the new-media environment, there is a blend of old and new content. Burgess and Green (2009) call attention to the existence of 'two YouTubes' – the YouTube consisting of user-generated content (garage-band music videos, home movies, fan-generated mash-ups of favorite programs and characters, vlogs and user-generated news and information) and the YouTube consisting of traditional media content (clips and occasionally whole episodes from news and entertainment programming, trailers for television shows and Hollywood films, advertisements, sporting events, etc.). In reality, the two versions co-exist and collide (not always harmoniously, as lawsuits over copyright indicate), but the larger point is that Web 2.0 applications and digital platforms for 'television' are not free from commercial pressures and industrial participation and indeed represent new opportunities for industrial colonization. Although user-generated content exceeded traditional-media generated content on YouTube by a slight margin at the time when Burgess and Green conducted their research in 2006–2007, the overall trend since then, predictably, is toward commercial use (see Kim, 2012). At the same time, the patterns of use revealed by the popularity measures on YouTube suggest important differences in viewer/user engagement. Whereas the 'most viewed' category was dominated by traditional-media content, the 'most responded' and 'most-discussed' categories were dominated by user-generated content, indicating that although viewers are certainly watching 'television' on YouTube, they are also using the site to view, respond to and discuss other content that is not commercially – or industrially – generated (Burgess and Green, 2009). And since responding and commenting are themselves forms of content production, it's fair to say user-generated content begets more user-generated content at higher rates than does traditional-media content. In other words, users appear more interested in and willing to engage in a participatory and producerly way with other users, as would be predicted by fan studies.

This blurring of production and use/consumption, what Axel Bruns (2008) calls 'produsage', is characteristic of the new media era, although its prevalence can be over-stated. Acknowledging the importance of the blurring does, however, push back against the tendency to valorize new media only for its productive capacity. Burgess and Green (2011) as well as Jenkins et al. (2013) caution against recreating a hierarchy in which production is the ultimate goal and consumption its poor relation. In the words of Burgess and Green (2011, p. 82), 'continuing to value only those who produce replicates the politics of the previous system. It's important to consider the possibility that forms of participation requiring original content creation are potentially less inclusive than forms of participation that combine a range of modes of engagement'. That being said, however one counts participation, there is still the problem of what Jenkins et al. (2013) call 'the participation gap'. In describing our culture as becoming more participatory over time, we're speaking in relative and not absolute terms, they remind us. Even if we value, in the spirit of de Certeau, consumption as part of and not separate from production, we do not live in a society where communicative capacity is equally distributed. 'Insofar as the [capacity] to meaningfully participate … [is] linked to educational and economic opportunities, then the struggle over the right to participation is linked to core issues of social justice and equality' (Jenkins et al., 2013, p. 194). This seems to me the underlying issue at stake in any consideration of how and why people consume television, whether the 'television' under consideration is old or new, broadcast or narrowcast, fragmented or unified, celebrated or condemned. Media consumption differs from the consumption of other goods and services precisely because media texts are symbol systems that connect interior and exterior worlds, and as such they enable and constrain the production and circulation of meaning, and even our very imaginations.

REFERENCES

Abercrombie, N. and Longhurst, B. (1998). *Audiences: A sociological theory of performance and imagination*. Thousand Oaks, CA: Sage.

Adden, R.C. (1999). *Popular stories and promised lands: Fan cultures and symbolic pilgrimages*. Tuscaloosa AL: University of Alabama Press.
Adorno, T. (1957). Television and the patterns of mass culture. In B. Rosenberg and D.M. White (Eds) *Mass culture: The popular arts in America* (pp. 474–88). New York: Free Press.
Allen, R. (1987). *Channels of discourse: Television and contemporary criticism*. Chapel Hill: University of North Carolina Press.
Allor, M. (1988). Relocating the site of the audience. *Critical Studies in Mass Communication* 5: 217–233.
Andrejevic, M. (2004). *Reality TV: The work of being watched*. New York/Toronto/Oxford: Rowman & Littlefield Publishers, Inc.
Ang, I. (1985). *Watching Dallas: Soap opera and the melodramatic imagination*. London: Methuen.
Ang, I. (1991). *Desperately Seeking the Audience*. London: Routledge.
Ang, I. (1996). *Living room wars: Rethinking media audiences for a postmodern world*. London/New York: Routledge.
Bacon-Smith, C. (1992). *Enterprising women: Television fandom and the creation of popular myth*. Philadelphia: University of Pennsylvania Press.
Barker, M. and Brooks, K. (1998). *Knowing audiences: Judge Dredd, its friends, fans, and foes*. Luton, UK: University of Luton Press.
Baudrillard, J. (1983). *In the shadow of the silent majorities*. New York: Semiotext(e).
Bird, E. (2003). *The audience in everyday life: Living in a media world*. New York: Routledge.
Block, A.B. (2013). Nielsen agrees to expand definition of TV viewing. *The Hollywood Reporter*. www.hollywoodreporter.com/news/nielsen-agrees-expand-definition-tv-422795 (posted 2/20/2013).
Boddy, W. (2004). Interactive television and advertising form in contemporary US television. In L. Spigel and J. Olsson (Eds) *Television after TV: Essays on a medium in transition* (pp. 113–132). Durham, NC: Duke University Press.
Bogart, Leo (1956). *The age of television*. New York: Ungar.
Briggs, M. (2010). *Television, audiences, and everyday life*. Berkshire, UK: Open University Press.
Brooker, W. (2002). *Using the force: Creativity, community, and Star Wars fans*. New York: Continuum.
Brown, M.E. (1994). *Soap opera and women's talk: The pleasure of resistance*. Newbury Park, CA: Sage.
Browne, N. (1984). The political economy of the television (super) text. *Quarterly Review of Film Studies* 9(3): 174–182.
Bruns, A. (2008). *Blogs, Wikipedia, Second Life and beyond: From production to produsage*. New York: Peter Lang.
Buckingham, D. (1993). *Children talking television: The making of television literacy*. London: Falmer.
Burgess, J. and Green, J. (2009). *YouTube: Online video and participatory culture*. Cambridge: Polity Press.
Caldwell, J. (1995). *Televisuality: Style, crisis, and authority in American television*. New Brunswick, NJ: Rutgers University Press.
Caldwell, J. (2004). Convergence television: Aggregating form and repurposing content in the culture of conglomeration. In L. Spigel and J. Olsson (Eds) *Television after TV: Essays on a medium in transition* (pp. 41–74). Durham, NC: Duke University Press.
Cantor, M. (1980). Audience control. In H. Newcomb (Ed.) *Television: The critical view*, 4th edition (pp. 361–381). New York: Oxford University Press.
Caraway, B. (2011). Audience labor in the new media environment: A Marxian revisiting of the audience commodity. *Media, Culture & Society* 33(5): 693–708.
D'Acci, J. (1994). *Defining women: Television and the case of Cagney and Lacey*. Chapel Hill, NC: University of North Carolina Press.
Dayan, D. (2001). The peculiar public of television. *Media, Culture & Society* 23: 743–765.
De Certeau, M. (1984). *The practice of everyday life*. Berkeley: University of California Press.
Ellis, J. (1982). *Visible fictions: Cinema, television, video*. London: Routledge.
Evans, E. (2011). *Transmedia television: Audiences, new media, and daily life*. New York/Abingdon: Routledge.
Fisherkeller, J. (2002). *Growing up with television: Everyday learning among young adolescents*. Philadelphia: Temple University Press.
Fiske, J. (1987). *Television culture*. London/New York: Routledge.
Fiske, J. (1989). *Reading the popular*. London: Unwin Hyman.

Fiske, J. (1992). Audiencing: A cultural studies approach to watching television. *Poetics* 21: 345–359.

Gauntlett, D. and Hill, A. (1999). *TV living: Television, culture, and everyday life*. London/New York: Routledge.

Gertner, J. (2005). Our ratings, ourselves. *New York Times Magazine* (April 10): 34–41.

Gillespie, M. (1995). *Television, ethnicity, and cultural change*. London: Routledge.

Gitlin, T. (1979). Prime-time ideology: The hegemonic process in television entertainment. *Social Problems* 26 (February): 251–266.

Gorton, K. (2009). *Media audiences: Television, meaning, and emotion*. Edinburgh: Edinburgh University Press.

Gray, A. (1992). *Video playtime: The gendering of a leisure technology*. London/New York: Routledge.

Gray, J., Sandvoss, C. and Harrington, C.L. (Eds) (2007). *Fandom: Identities and communities in a mediated world*. New York/London: New York University Press.

Grindstaff L. (2002). *The money shot: Trash, class, and the making of TV talk shows*. Chicago: University of Chicago Press.

Grindstaff, L. and Turow, J. (2006). Video cultures: Television sociology in the 'new TV' age. *Annual Review of Sociology* 32: 103–125.

Gripsrud, J. (1995). *The Dynasty years: Hollywood television and critical media studies*. London: Comedia.

Hall, S. (1980). Encoding/decoding. In S. Hall, D. Hobson, A. Lowe and P. Willis (Eds) *Culture, Media, Language* (pp. 128–139). London: Hutchinson.

Halloran, J.D. (1970). *The effects of television*. New York: HarperCollins.

Harrington, L. and Bielby, D. (1995). *Soap fans: Pursuing pleasures and making meaning in everyday life*. Philadelphia: Temple University Press.

Harris C. and Alexander, A. (Eds.) (1998). *Theorizing fandom: Fans, subculture, and identity*. Cresskill, NJ: Hampton.

Hartley J. (1999). *The uses of television*. London/New York: Routledge.

Hay, J., Grossberg, L. and Wartella, E. (Eds.) (1996). *The audience and its landscape*. Boulder, CO: Westview Press.

Hill, A. (2005). *Reality TV: Audiences and popular factual television*. London/New York: Routledge.

Hills, M. (2002). *Fan cultures*. London/New York: Routledge.

Hinckley, D. (2012). Americans spend 34 hours a week watching TV, according to Nielsen numbers. *New York Daily News* (Sept. 19). www.nydailynews.com/entertainment/tv-movies/americans-spend-34-hours-week-watching-tv-nielsen-numbers-article-1.1162285

Hobson, D. (1982). *Crossroads: The drama of a soap opera*. London: Methuen.

Jancovich, M. (2002). Cult fictions: Cult movies, subcultural capital, and the production of cultural distinction. *Cultural Studies* 16(2): 306–322.

Jancovich M., Reboll A.L., Stringer, J. and Willis, A. (Eds.) (2003). *Defining cult movies: The cultural politics of oppositional taste*. Manchester, UK: Manchester University Press.

Jenkins, H. (1992). *Textual poachers: Television fans and participatory culture*. London/New York: Routledge.

Jenkins, H. (2006). *Convergence culture: Where old and new media collide*. New York: New York University Press.

Jenkins, H. (2007). Afterword: The future of fandom. In J. Gray, C. Sandvoss, and C.L. Harrington (Eds.) Fandom: Identities and communities in a mediated world (pp. 357–364). New York/London: New York University Press.

Jenkins, H., Ford, S. and Green, J. (2013). *Spreadable media: Creating value and meaning in a networked culture*. New York/London: New York University Press.

Jensen, J. (1992). Fandom as pathology: The consequences of characterization. In L. Lewis (Ed.) *The adoring audience: Fan culture and popular media*. New York: Routledge.

Jhally, S. and Lewis, J. (1992). *Enlightened racism: The Cosby Show, audiences, and the myth of the American dream*. Boulder, CO: Westview.

Jhally, S. and Livant, B. (1986). Watching as working: The valorization of audience consciousness. *Journal of Communication* 36: 124–143.

Juluri, V. (2003). *Becoming a global audience: Longing and belonging in Indian music television*. New York: Peter Lang.

Kaplan, E.A. (Ed.) (1983). *Regarding television*. Los Angeles: American Film Institute.

Kellner, D. (1981). Network television and American society: Introduction to a critical theory of television. *Theory and Society* 10(1): 31–62.

Kim, J. (2012). The institutionalization of YouTube: From user-generated content to professionally generated content. *Media Culture & Society* 34(1): 53–67.

Lembo, R. (2000). *Thinking through television*. Cambridge, MA: Cambridge University Press.

Lewis, L. (Ed.) (1992). *The adoring audience: Fan culture and popular media*. New York: Routledge.

Liebes, T. and Katz, E. (1990). *The export of meaning: Cross-cultural readings of 'Dallas'*. New York: Oxford University Press.

Livingstone, S. (1990). *Making sense of television*. London: Pergamon.

Livingstone, S. (2004). The challenge of changing audiences: Or, what is the audience researcher to do in the age of the internet? *European Journal of Communication* 19(1): 75–86.

Livingstone, S. and Lunt, P. (1994). *Talk on television: Audience participation and public debate*. London/New York: Routledge.

Lotz, A. (2000). Assessing qualitative audience research: Incorporating feminist and anthropological theoretical innovation. *Communication Theory* 10(4): 447–467.

Lotz, A. (2009). What is US television now? *Annals of the American Academy of Political and Social Science* 625 (September): 49–59.

Lull, J. (1990). *Inside family viewing: Ethnographic research on television audiences*. London: Routledge.

Manga, J. (2003). *Talking trash: The cultural politics of daytime TV talk shows*. New York: New York University Press.

Mayer, V. (2003). *Producing dreams, consuming youth: Mexican Americans and mass media*. New Brunswick, NJ: Rutgers University Press.

McCarthy, A. (2001). *Ambient television*. Durham, NC: Duke University Press.

McLuhan, M. (1964). *Understanding media: The extensions of man*. New York: McGraw-Hill.

Meehan, E. (1990). Why we don't count: The commodity audience. In P. Mellencamp (Ed.) *Logics of television* (pp. 117–137). Indiana and Bloomington: Indiana University Press.

Meyrowitz, J. (1985). *No sense of place: The impact of electronic media on social behavior*. New York: Oxford University Press.

Moores, S. (1993). *Interpreting audiences: The ethnography of media consumption*. London: Sage.

Morley, D. (1980). *The 'Nationwide' audience: Structure and decoding*. London: British Film Institute.

Morley, D. (1986). *Family television*. London: Comedia.

Morley, D. (1992). *Television, audiences, and cultural studies*. New York: Routledge.

Morley, D. and Silverstone, R. (1990). Domestic communications: Technologies and meanings. *Media Culture, & Society* 12(1): 31–55.

Napoli, P. (2010). Revisiting 'mass communication' and the 'work' of the audience in the new media environment. *Media, Culture & Society* 32(3): 505–516.

Newcomb, H. (1974). *TV: The most popular art*. New York: Anchor Press.

Newcomb, H. (Ed.) (2000). *Television: The critical view*. New York: Oxford University Press (6th edition).

Newcomb, H. and Hirsch, P. (1983). Television as cultural forum: Implications for research. *Quarterly Review of Film Studies* 8(2): 45–55.

Ouellette, L. and Hay, J. (2008). *Better living through reality TV: Television and post-welfare citizenship*. Oxford: Blackwell Publishing.

Palmer, G. (1986). *The lively audience: A study of children around the TV set*. Sydney: Allen & Unwin.

Parks, L. (2004). Flexible microcasting: Gender, generation and television-internet convergence. In L. Spigel and J. Olsson (Eds.) *Television after TV: Essays on a medium in transition* (pp. 133–156). Durham, NC: Duke University Press.

Penley, C. (1997). *NASA/TREK: Popular science and sex in America*. London: Verso.

Press, A. (1991). *Women watching television: Gender, class, and generation in the American television experience*. Philadelphia: University of Pennsylvania Press.

Radway, J. (1984). *Reading the romance: Women, patriarchy, and popular literature*. Chapel Hill, NC: University of North Carolina Press.

Radway, J. (1988). Reception study: Ethnography and the problems of dispersed audiences and nomadic subjects. *Cultural Studies* 2(3): 359–376.

Sandvoss, C. (2003). *A game of two halves: Television, football, and globalization*. New York: Routledge.

Sandvoss, C. (2005). *Fans: The mirror of consumption*. Malden, MA: Polity.

Seiter, E. (1990). Making distinctions in TV audience research: the case of troubling interview. *Cultural Studies* 4 (1): 61–84.

Seiter, E. (1999). *Television and new media audiences*. New York: Oxford University Press.

Seiter, E., Borchers, H., Kreutzner, G. and Warth, E.M. (Eds.) (2013). *Remote control: Television, audiences, and cultural power* (3rd edition). London/New York: Routledge.

Sender, K. (2012). *The makeover: Reality television and reflexive audiences*. New York/London: New York University Press.

Skeggs, B. and Wood, H. (2012). *Reacting to reality television: Performance, audience, and value*. Abingdon/New York: Routledge.

Smythe, D. (1977). Communications: Blindspot of Western Marxism. *Canadian Journal of Political and Social Theory* 2(2): 120–129.

Spigel, L. (1992). Introduction. In R. Williams, *Television: Technology and cultural form*. Hanover and London: Wesleyan University Press [first published 1974].

Spigel, L. and Olsson, J. (Eds.) (2004). Television After TV: Essays on a Medium in Transition. Durham, NC: Duke University Press.

Thomas, L. (2002). *Fans, feminism, and 'quality' media*. New York: Routledge.

Thornton, S. (1995). *Club cultures*. Cambridge UK: Polity.

Tuchman, G. (Ed.) (1974). *The TV establishment*. Englewood Cliffs, NJ: Prentice-Hall.

Tulloch, J. (2000). *Watching television audiences: Cultural theories and methods*. London: Arnold.

Tulloch, J. and Jenkins, H. (1995). *Science fiction audiences: Watching 'Dr. Who' and 'Star Trek'*. London/New York: Routledge.

Turow J. (2005). Audience construction and culture production: marketing surveillance in the digital age. *The Annals of the American Academy of Political and Social Science* 597:103–21.

Uricchio, W. (2004). Television's next generation. In L. Spigel and J. Olsson (Eds) *Television after TV: Essays on a medium in transition* (pp. 163–182). Durham, NC: Duke University Press.

Williams, R. (1974). *Television: Technology and cultural form*. London: Wesleyan University Press.

Wood, H. (2007). Television is happening: Methodological considerations for capturing digital television reception. *European Journal of Cultural Studies* 10(4): 485–506.

25

Effects and Cultivation

Michael Morgan, James Shanahan and Nancy Signorielli

In 1952, the Federal Communications Commission lifted the four-year 'freeze' it had imposed on the allocation of new licenses for broadcast television stations; the moratorium had been instituted to allow the Commission to work out how best to allocate frequencies and increase channel assignments for the new medium that was about to burst onto the scene. In the sixty-plus years since, television has become not only the most prominent of all mass media, but also one of the defining elements of modern life. It has also been the focus of a vast amount of scholarly research and public debate, neither of which shows any sign of subsiding, even as various newer technologies (the internet, social media, videogames, cell phones, etc.) enter the fray.

Over the decades, the institutional structure and the technology of television have seen immense changes, as has its programming. Yet, television still garners more of the public's time, and more of the advertisers' dollars, than any other medium. We are a long way from 1952, but television still remains the source of the most widely-shared cultural stories, images and lessons. New technologies (on-line streaming, on-demand, DVRs, smartphones and tablets) that make it easier to watch whenever, and wherever, we want, only augment its centrality in our lives. Concerns about the 'effects' of its portrayals of violence, sex, gender and racial/ethnic stereotypes, the family, medicine, politics, health, nutrition and weight, drugs and alcohol, mental illness, and much more, continue to motivate and frustrate researchers, parents, teachers, health professionals, and others.

This chapter explores one particular way those effects may be explored, through an approach known as *cultivation analysis*, developed by George Gerbner (1919–2005). Growing up in pre-war Hungary, Gerbner spent his summers in the countryside, and was struck by the way the folk stories, songs and dances of peasants in the rural Hungarian villages seemed to both reflect and recreate their particular ways of seeing the world. Later, as a scholar starting out in mid-20th-century

America – amidst the post-war explosion of mass media – his attention turned to the fact that the cultural process of story-telling had undergone a profound transformation. No longer hand-crafted and passed on face-to-face over the generations, Gerbner pointed out that *our* folktales were now mass-produced by commercial conglomerates and mass-consumed to a historically unprecedented degree. He devoted his career to investigating how those manufactured stories both reflect and shape the culture that produced them.

Gerbner's goal was to understand the consequences of growing up in a cultural environment dominated by mediated, profit-driven mass communication (Gerbner, 1973). He devised a three-pronged approach, called Cultural Indicators, to investigate: (1) the institutional processes underlying the selection and production of media content ('institutional process analysis'); (2) the most prevalent, aggregate images in that content ('message system analysis'); and (3) the contribution of that content to audience beliefs and behaviors ('cultivation analysis'). While this model, its conceptual assumptions, and its methodological procedures can be applied to any dominant medium of communication (or any system of story-telling), cultivation analysis has been largely focused on television, since no other medium exceeds its hold on our time and attention. In its simplest form, cultivation analysis asks if those who watch more television ('heavy viewers') have conceptions of social reality that are more reflective of what they see on television compared to people who have similar demographic characteristics but who watch less television. Cultivation analysis, then, examines how the stories we – as a culture – watch on television contribute to our beliefs and attitudes about the 'real' world.

Cultivation was conceived as a critical alternative to traditional approaches to the study of media 'effects'. While the vast bulk of media effects research had been predicated on finding some sort of attitude or behavior change resulting from exposure to some type of media message, Gerbner was not concerned with 'changes' in outlook that were due to viewing in the short-term. Those were questions for persuasion and marketing research. Rather, he saw shared cultural outlooks and assumptions as being nurtured, maintained and reinforced by television's messages in large communities over long periods of time. All communication, he argued, cultivates the assumptions, points of view and relationships on which it is premised. With mass communication, what is cultivated is standardized and more widely shared than was ever before possible. The common symbolic environment in which we grow up and live gives shape and meaning to all that we do; the more we 'live' in that synthetic but coherent world, the more it cultivates our conceptions of social reality.

To some, this was a common-sense idea. However, it raised the hackles of many social scientists who saw a too-simplistic criticism of television or even a misuse of social science research methods; as a result, it provided an opportunity for innumerable colloquies, critiques, revisions, and reappraisals (see Shanahan and Morgan, 1999, for a review). But despite, or perhaps because of, the enormous amount of criticism that has been leveled at cultivation, it has been one of the most frequently cited theories of mass communication or media effects since the 1970s. Various studies of the literature regularly place cultivation, along with agenda setting and uses and gratifications, as one of the three most cited theories of media effects (Bryant and Miron, 2004). One analysis of media effects research articles published between 1993 and 2005 found that cultivation was the *most* cited theory (Potter and Riddle, 2007). By 2014 more than 600 relevant studies had been published (two-thirds of which are extensions, replications, reviews and critiques conducted by independent researchers not associated with Gerbner and the original research team).[1] Cultivation studies have been carried out in Argentina, Australia, Belgium, Brazil, China, England, Germany, Hungary, Israel, Japan, Mexico, Russia, South Korea, Sweden, Thailand and elsewhere. Clearly, such a large body of work – and all the complex issues and implications it raises – cannot be exhaustively treated here; this chapter attempts to provide only a general introduction to this area of research. (For a more extensive examination of the cultivation literature, see Morgan, Shanahan and Signorielli (2009, 2012) and Shanahan and Morgan (1999).)

CULTIVATION ANALYSIS METHODS

Cultivation analysis typically begins by identifying the most common and stable patterns in television content, emphasizing the consistent images, portrayals and values that cut across program genres. This is accomplished either by conducting a message system analysis (a content analysis) or by examining existing content studies. Typical examples of content patterns that have been identified include the high levels of violence in prime-time TV programs (e.g., Gerbner, 1970) or under-representations of specific social groups (e.g., Signorielli, 1993). Once those patterns are identified, the goal is to ascertain if those who spend more time watching television are more likely to perceive the real world in ways that reflect those particular messages and lessons. For example, do heavy viewers (having been immersed in the violent world of television) also see the world as a more violent place?

Cultivation hypotheses are most often tested using survey procedures to examine relationships between amount of television viewing and conceptions of social reality. Several different types of questions are asked. Some juxtapose answers reflecting the statistical 'facts' of the television world with those more in line with reality (often referred to as 'first-order' cultivation measures). As an example, compared to light viewers, heavy viewers would be more likely to think that there are more law enforcement professionals in the workforce (Gerbner and Gross, 1976). Other questions examine symbolic transformations and more general implications of the message system data (often referred to as 'second-order' cultivation measures). For example, questions about peoples' fear of victimization or the extent to which they think others can be trusted fall into this category. Finally, some questions simply ask about beliefs, opinions, attitudes or behaviors. For instance, in many studies, heavy viewers have been shown to harbor more traditional beliefs about gender roles (Morgan, 1982; Signorielli, 1993).

Studies use different types of samples (national probability, regional, convenience) of children, adolescents or adults and assess the amount of viewing by asking how much time the respondent spends watching on an 'average day' (although many investigators have operationalized amount of viewing in different ways). These data may be used in their original form (a ratio scale) or may be grouped by level of exposure ('light', 'medium' and 'heavy' viewing), on a sample-by-sample basis. People who regularly consume a great deal of television differ from light viewers in many ways besides simply how much time they spend watching. To deal with this, differences between the responses of light, medium and heavy viewers are examined within specific demographic subgroups, and the effects of other variables are statistically controlled.

As with many media effects studies, cultivation analyses typically generate small effect sizes. Even those who watch relatively little television may watch 7–10 hours a week – and these viewers certainly interact with those who watch more. Thus, the cards can be statistically stacked against finding evidence of cultivation. Nevertheless, even small differences between light and heavy viewers may indicate far-reaching consequences. With very large audiences, a difference of a fraction of a percentage point in ratings can indicate the success or failure of a program (and be worth millions of advertising dollars), and a difference of a few percentage points in an election will determine who wins or who loses. Similarly, small relationships across large populations of heavy and light viewers may indicate profound social effects, especially as they cumulate over time.

CRITIQUES AND REFINEMENTS

Few theories of media effects (and perhaps few areas of social research in general) have been critiqued as heavily and fiercely as cultivation. (For some examples, see Doob and McDonald, 1979; Hirsch, 1980; Hughes, 1980; Potter, 1993, 1994; Wober, 1990; see also Shanahan and Morgan, 1999.) These critiques have focused on many issues, including cultivation's emphasis on overall amount of television exposure (as opposed to specific types of programs), the way television viewing is measured and divided into relative levels of exposure, justifications for interpreting television-world answers, the linearity of cultivation associations, and much more. Some of the most heated issues of contention have revolved around questions of spuriousness and the proper use of statistical controls.

Early analyses typically applied a single control variable at a time. But various critics – reanalyzing the same data – argued that when multiple controls were applied at the same time, the relationships between television viewing and attitudes mostly disappeared. This led them to conclude that there was essentially no evidence to support the cultivation hypothesis.

Gerbner and his colleagues countered that even if a relationship disappears under multiple controls, significant patterns may still exist within specific subgroups, often reflecting a pattern they called 'mainstreaming' (Gerbner, Gross, Morgan and Signorielli, 1980). Among light viewers, people who differ in terms of background factors such as age, education, social class, political orientations and region of residence tend to have sharply different conceptions of social reality regarding violence, interpersonal mistrust, gender-role stereotypes, and a broad range of political and social outlooks. Yet, among heavy viewers across those same groups, those differences tend to be much smaller or even to disappear entirely.

For example, Gerbner et al. (1980) found that low-income respondents were more likely than those with higher incomes to say that 'fear of crime is a very serious personal problem'. Among low-income respondents, amount of television viewing was not related at all to perceptions of crime. In contrast, although higher-income respondents as a group were less likely to think of crime as a serious personal problem, the heavy viewers with higher incomes were much more likely than the light viewers to be especially worried about crime. Heavy viewers with higher incomes had the same perception as those with lower incomes; among heavy viewers, the difference stemming from income was sharply diminished. For another example, among lighter viewers, those with more education have been found to express more 'progressive' attitudes about gender roles, but this difference disappears among heavy viewers; more educated heavy viewers express the same 'traditional' beliefs as do those with less education.

A second process also emerged to explain and predict how viewing may impact different viewers. Specifically, direct experience may be important for some viewers; the phenomenon called *resonance* was proposed to illustrate how a person's everyday reality and patterns of television viewing may provide a double dose of messages that 'resonate' and amplify cultivation. For example, those who live in high crime urban areas often show stronger relationships between amount of viewing and self-reported fear of crime. The idea here is that both the 'TV world' and the real world reinforce the impression that violence is widespread.

Much of the back-and-forth dialogue took place in the early to middle 1980s and from it emerged a few important theoretical outcomes. Overall, cultivation theory and its basic research approaches flourished, even as the critiques also gained some level of acceptance among researchers. From these early debates, cultivation research grew and expanded. The overall focus of cultivation was broadened from its beginnings in violence to issues related to sex roles, aging, politics, health, science, religion, the family, the environment, and many other topics.

Also, the study of cultivation was taken up by many individual researchers who had not worked or trained with the original Cultural Indicators research team; it became an international research phenomenon. This also led to important extensions and new strands of research, such as those focusing on the cognitive processes underlying cultivation.

At its outset, cultivation may have been an easy target because it blended critical theory about mass media with empirical research techniques, cutting across the typical divisions between critical and empirical research that were so common at that time. Nevertheless, the amount of research continued to grow. Shanahan and Morgan (1999) reviewed over 20 years' worth of studies in a meta-analysis. This analysis found that the body of cultivation research as a whole exhibited a small but persistent relationship between television exposure and beliefs about the world.

In short, hundreds of studies conducted over the past four decades have (mostly) found that there are relationships between media exposure and people's worldviews. But important questions remain: how large are these relationships, are they real, are some people more vulnerable to them than others, do they vary across different topics, and will we continue to find them as our media environments keep evolving?

RECENT EXTENSIONS OF CULTIVATION

Today, in addition to the questions noted above, there are many and various views about how cultivation should be studied, and much ongoing exploration. This section briefly discusses some of these new developments and perspectives (for an extended treatment, see Morgan et al. (2012)). Most of the studies and reviews reported here are supportive of the basic idea of cultivation, though many propose extensions and modifications to the methods originally suggested by Gerbner, and there has been continued growth in the ways researchers conceptualize the activity of watching television and its consequences.

Cultivation studies have continued in the more traditional vein by updating and extending some of the classic dependent variables including images of crime (Custers and Van den Bulck, 2011) and sex roles (Harrison, 2003; Scharrer, 2012). Some of these studies have looked at the impacts of specific programs and genres, or the fear of events such as the 9/11 attacks. In addition, as perceptions and views about minorities have changed greatly during the past forty plus years, the number of cultivation studies in this area has also grown (e.g., Mastro and Tukachinsky, 2012). For instance, Nisbet and Myers (2012), looking at the relationship between television viewing and views on homosexuality, utilized an innovative multi-level methodology that illustrates how social change, television depictions and opinions can be analyzed over time. Along the same lines, Hardy (2012), bringing a cultivation perspective to political communication, found that television viewing still tends to cultivate middle-of-the-road ideological self-perceptions.

One of the most fruitful and active areas of development in recent cultivation research has illuminated the cognitive mechanisms that 'explain' the cultivation process. In an extensive series of studies, Shrum (see Shrum and Lee (2012) for a comprehensive summary) has examined television's role in how people make first-order judgments (estimates of frequency and set size in the real world) and second-order judgments (attitudes, opinions and feelings). Shrum and Lee (2012) posit that very different cognitive processes are at work when television viewing influences first- and second-order judgments. First-order cultivation judgments are memory-based, and formed through heuristic processes. Television images influence these judgments when someone recalls relevant information often seen in television programs. This occurs when people take cognitive short-cuts when forming a judgment; for heavy viewers, television-based information is more accessible, available, recent and vivid. On the other hand, second-order cultivation judgments are formed in an 'online' process, in which 'the influence of television on judgment occurs during viewing, as information is processed' (Shrum and Lee, 2012, p. 162). Although 'active' processing can counteract first-order cultivation (by making people aware of television as the source of their images), deep involvement or 'transportation' in a story can enhance second-order cultivation.

Another issue in more recent cultivation research is whether the 'realism' of TV content matters for cultivation. Many scholars have examined the notion of 'realism', particularly 'perceived realism', as a critical component in understanding cultivation. While many researchers have assumed that only programs perceived as 'realistic' would influence beliefs about the world, recent research shows that the situation is not that simple. For example, the research of Busselle and Bilandzic (2012) suggests that perceived realism is a default condition and that most content, even that recognized as fantasy, has some element of perceived realism and thus that viewers carry impressions from television to their understanding of the real world.

'Classic' cultivation has always been concerned with the long-term and long-range effects of living with television in general. As Shanahan and Morgan (1999) note, it 'is the pattern of settings, casting, social typing, action and related outcomes that cuts across program types and viewing modes and defines the world of television' (p. 30). Consequently, 'orthodox' cultivation studies predict that those who watch more in general, television's heavy viewers, see more of these images and messages on a regular basis and that their views of social reality are influenced by the messages and images they see day in and day out, regardless of their individual selections or favorite programs.

Nevertheless, many researchers see genre as an important element in cultivation research. Some scholars, for example, believe that cultivation research should focus on different genres of programs because they assume that viewers watch more selectively than ritualistically (especially now, with many more channels and many more options), and that different types of programs present diverse and distinct views of the world and social reality. Cohen and Weimann (2000) note that different genres, although typically driven by their own specific formulas, present viewers with diverse views about the world; crime dramas or action adventure programs, for example, focus on public life and the social order, while situation comedies, family dramas and soap operas focus on domestic issues, family relationships and friendships. Interestingly, Grabe and Drew (2007) suggest that one reason scholars examine possible genre-related differences in cultivation is that content analyses typically show variation in portrayals among genres. Situation and romantic comedies present different types of stories than crime dramas, with the former genres having considerably less violence than the latter. In addition, Bilandzic and Busselle (2012) argue that genre's status as an important component of narrative implies that it is useful to look at cultivation within genres (such as crime programs). Among the many genres recently investigated are 'sexually objectifying media' (Aubrey, 2006), music videos (Buellens, Roe and Van den Bulck, 2012) and hospital dramas (Hetsroni, 2009).

Some researchers postulate that viewers' conceptions about fear and violence are specifically the result of viewing crime-related programming. Despite the knowledge that television's crime programs typically distort statistics about crime in society, several studies show relationships between conceptions about crime and the viewing of crime programs. For example, Holbert, Shah and Kwak (2004), using a national probability sample, found that viewing television news and police reality programs was related to measures of fear of crime but that viewing crime dramas was not. This analysis also found that nonfiction viewing (such as police reality programs) tended to be more strongly related to conceptions of social reality than the viewing of fictional crime dramas. They interpreted this difference as due to the increased perceived realism of the police reality programs. (Many other genre-based studies have produced a variety of genre-related findings.)

It may not be surprising that there are many instances in which exposure to very specific media, particularly specific genres of television programs, is related to having very specific views about the world and social reality. The cultivation perspective, however, focuses more on people's general viewing habits. Even in today's fragmented viewing world, viewing more television tends to mean greater exposure to what Gerbner conceived as the message 'system', not just a collection of disparate and idiosyncratic viewing experiences. Patterns and similarities across all viewing genres are what matter most for the idea of cultivation. Even if people state that they have a preference for watching a specific genre (as we all do), overall media use normally encompasses viewing many different types of programs (as well as using other media such as magazines and newspapers). Although someone may profess to have a television diet consisting primarily of viewing crime programs, such as the *Law and Order* or *CSI* series, the images to which they are exposed may go far beyond just crime and violence. For example, while *Law and Order* and *CSI* are based primarily on the investigation of crime, these programs also offer images of interpersonal and family relationships as well as other thematic elements showing how the world works. Consequently, heavy viewers of these programs will not only learn about crime and violence, they will also learn important life lessons about how people interact with each other.

An interesting innovation in cultivation analysis comes from Riddle (2012), who has developed a measure to assess viewing across a person's lifetime. Her research posits that because cultivation is concerned with systematic, *long-term* exposure to television, a measure is needed to assess how people have watched television in the past, regardless of their current viewing habits. Indeed, Riddle has found that measures of cumulative, lifetime exposure to television are more strongly related to conceptions and beliefs than are measures of current exposure levels. Consequently, the historical over-time dimension has emerged as a very important component of cultivation research.

Among the more recent developments in cultivation research are interesting linkages of cultivation with other theories of mass communication. For example, Hetsroni and Lowenstein (2012) suggest a model to link cultivation with agenda-setting research. They look at both theories in terms of first- and second-order media effects, hypothesizing a single process that might unite the two. In addition, Diefenbach and West (2012) look at the intersection of cultivation with the Third Person Effect. They posit that because cultivation is an example of a media effect that people can think about (thus generating possible expectations about how others might be affected), looking at these two approaches in tandem makes considerable sense, and they present data to support this view. Shanahan and Scheufele (2012) examine theoretical and empirical intersections between cultivation and the spiral of silence. Cultivation is also being examined through the lens of the critical cultural theorist. Ruddock (2012) links cultivation with ritual theory within the focus of digital sports media and fandom. Interestingly, Ruddock's work suggests that the idea of cultivation may become even more relevant as new media emerge.

Ruddock's argument notwithstanding, change in the media environment since cultivation was developed has been dramatic, and may seem to challenge or invalidate many fundamental assumptions of cultivation analysis. While the way we now receive our 'stories' has changed, many aspects of their content has not and, in some ways, they have become even more homogeneous. Today's media are dominated by transnational, global conglomerates whose goal is to maintain a large share of the audience and increase profits. Indeed, as there are more and more channels and more and more outlets, the number of transnational companies continues to become smaller and smaller. Consequently, the smaller number of production venues may result in a more homogeneous than heterogeneous media environment. Cultivation has always been concerned with the broad underlying elements of content and how audiences interact with these messages. Indeed, as Shanahan and Morgan (1999) note, 'the content of messages is more germane than the technology with which they are delivered' (p. 201). Watching a television program on a laptop, tablet, smartphone or any other technology is still 'watching television'.

Cultivation continues to be a theory that is frequently tested, cited and scrutinized, as it continues to evolve. Morgan et al. (2012) summarize the evolution of cultivation during the past twenty years. Many of the ideas for the future of cultivation suggested by Signorielli and Morgan (1990) have come to fruition (and many issues and problems they noted seem to be not so important anymore). Yet, as with any workable and viable theory, there are still important insights to be gained as to where cultivation research can go in the future. We believe that cultivation's stature as one of the most well-known and well-used theoretical approaches in communication is well justified, and we hope that this chapter provides a useful view of where the research stands at this juncture. We look forward to seeing where research on cultivation will lead us in the next twenty years.

NOTE

1 A complete bibliography of Cultural Indicators work is located at: http://people.umass.edu/mmorgan/CulturalIndicatorsBibliography.pdf

REFERENCES

Aubrey, J.S. (2006). Effects of sexually objectifying media on self-objectification and body surveillance in undergraduates: Results of a 2-year panel study. *Journal of Communication*, 56(2), 366–386.

Beullens, K., Roe, K. and Van den Bulck, J. (2012). Music video viewing as a marker of driving after the consumption of alcohol. *Substance Use & Misuse*, 47(2), 155–165.

Bilandzik, H. and Busselle, R. (2012). A narrative perspective on genre-specific cultivation. In M. Morgan, J. Shanahan and N. Signorielli (Eds), *Living with television now: Advances in cultivation theory & research* (pp. 261–285). New York, NY: Peter Lang.

Bryant, J. and Miron, D. (2004). Theory and research in mass communication. *Journal of Communication*, 54, 662–704.

Busselle, R. and Bilandzic, H. (2012). Cultivation and the perceived realism of stories. In M. Morgan, J. Shanahan and N. Signorielli (Eds), *Living with television now: Advances in cultivation theory & research* (pp. 168–186). New York, NY: Peter Lang.

Cohen, J. and Weimann, G. (2000). Cultivation revisited: Some genres have some effects on some viewers. *Communication Reports*, 13(2), 99–114.

Custers, K. and Van den Bulck, J. (2011). Mediators of the association between television viewing and fear of crime: Perceived personal risk and perceived ability to cope. *Poetics*, 39, 107–124.

Diefenbach, D. L., and West, M. D. (2012). Cultivation and the third-person effect. In M. Morgan, J. Shanahan and N. Signorielli (Eds), *Living with television now: Advances in cultivation theory & research* (pp. 329–346). New York, NY: Peter Lang.

Doob, A. and Macdonald, G. (1979). Television viewing and fear of victimization: Is the relationship causal? *Journal of Personality and Social Psychology*, 37(2), 170–179.

Gerbner, G. (1970, March). Cultural Indicators: The case of violence in television drama. *The Annals of the American Academy of Political and Social Science*, 388, 69–81.

Gerbner, G. (1973). Cultural Indicators: The third voice. In G. Gerbner, L. Gross and W. Melody (Eds.), *Communications technology and social policy* (pp. 555–573). New York: John Wiley & Sons.

Gerbner, G. and Gross, L. (1976). Living with television: The violence profile. *Journal of Communication*, 26(2), 173–199.

Gerbner, G., Gross, L. Morgan, M. and Signorielli, N. (1980). The 'mainstreaming' of America: Violence profile No. 11. *Journal of Communication*, 30(3), 10–29.

Grabe, M.E. and Drew, D. (2007). Crime cultivation: Comparisons across media genres and channels. *Journal of Broadcasting & Electronic Media*, 51(1), 147–171.

Hardy, B. W. (2012). Cultivation of political attitudes in the new media environment. In M. Morgan, J. Shanahan and N. Signorielli (Eds), *Living with television now: Advances in cultivation theory & research* (pp. 101–119). New York, NY: Peter Lang.

Harrison, K. (2003). Television viewers' ideal body proportions: The case of the curvaceously thin woman. *Sex Roles: A Journal of Research*, 48, 255–264.

Hetsroni, A. (2009). If you must be hospitalized, television is not the place: Diagnoses, survival rates and demographic characteristics of patients in TV hospital dramas. *Communication Research Reports*, 26(4), 311–322.

Hetsroni, A. and Lowenstein, H. (2012). Cultivation and agenda-setting: Conceptual and empirical intersections. In M. Morgan, J. Shanahan and N. Signorielli (Eds), *Living with television now: Advances in cultivation theory & research* (pp. 307–328). New York, NY: Peter Lang.

Hirsch, P. (1980). The 'scary world' of the non-viewer and other anomalies: A reanalysis of Gerbner et al.'s findings of cultivation analysis, Part I. *Communication Research*, 7(4), 403–456.

Holbert, R.L., Shah, D.V. and Kwak, N. (2004). Fear, authority, and justice: Crime-related TV viewing and endorsements of capital punishment and gun ownership. *Journalism & Mass Communication Quarterly*, 81(2), 343–363.

Hughes, M. (1980). The fruits of cultivation analysis: A re-examination of the effects of television watching on fear of victimization, alienation, and the approval of violence. *Public Opinion Quarterly*, 44(3), 287–302.

Mastro, D. and Tukachinsky, R. (2012). Cultivation of perceptions of marginalized groups. In M. Morgan, J. Shanahan and N. Signorielli (Eds), *Living with television now: Advances in cultivation theory & research* (pp. 38–60). New York: Peter Lang.

Morgan, M. (1982). Television and adolescents' sex-role stereotypes: A longitudinal study. *Journal of Personality and Social Psychology*, 43(5), 947–955.

Morgan, M., Shanahan, J. and Signorielli, N. (2009). Growing up with television: Cultivation processes. In J. Bryant and M. Oliver (Eds), *Media effects: Advances in theory and research* (3rd edn, pp. 34–49). Hillsdale, NJ: Erlbaum.

Morgan, M., Shanahan, J. and Signorielli, N. (2012). *Living with television now: Advances in cultivation theory and research*. New York: Peter Lang.

Nisbet, E. C. and Myers, T. E. (2012). Cultivating tolerance of homosexuals. In M. Morgan, J. Shanahan and N. Signorielli (Eds), *Living with television now: Advances in cultivation theory & research* (pp. 61–80). New York, NY: Peter Lang.

Potter, W.J. (1993). Cultivation theory and research: A conceptual critique. *Human Communication Research*, 19, 564–601.

Potter, W.J. (1994). Cultivation theory and research: A methodological critique. *Journalism Monographs*, 147, 1–35.

Potter, W.J. and Riddle, K. (2007). A content analysis of the media effects literature. *Journalism & Mass Communication Quarterly*, 84(1), 90–104.

Riddle, K. (2012). Developing a lifetime television exposure scale: The importance of television viewing habits during childhood. In M. Morgan, J. Shanahan and N. Signorielli (Eds), *Living with television now: Advances in cultivation theory & research* (pp. 286–306). New York, NY: Peter Lang.

Ruddock, A. (2012). Cultivation analysis and cultural studies: Ritual, performance, and media influence. In M. Morgan, J. Shanahan and N. Signorielli (Eds), *Living with television now: Advances in cultivation theory & research* (pp. 366–385). New York, NY: Peter Lang.

Scharrer, E. (2012). Television and gender roles: Cultivating conceptions of self and others. In M. Morgan, J. Shanahan and N. Signorielli (Eds), *Living with television now: Advances in cultivation theory and research* (pp. 81–100). New York, NY: Peter Lang.

Shanahan, J. and Morgan, M. (1999). *Television and its viewers: Cultivation theory and research*. Cambridge: Cambridge University Press.

Shanahan, J. and Scheufele, D. (2012). Cultivation and the spiral of silence: Theoretical and empirical intersections. In M. Morgan, J. Shanahan and N. Signorielli (Eds), *Living with television now: Advances in cultivation theory and research* (pp. 347–365). New York, NY: Peter Lang.

Shrum, L. J. and Lee, J. (2012). Multiple processes underlying cultivation effects: How cultivation works depends on the types of beliefs being cultivated. In M. Morgan, J. Shanahan and N. Signorielli (Eds), *Living with television now: Advances in cultivation theory & research* (pp. 147–167). New York, NY: Peter Lang.

Signorielli, N. (1993). Sex roles and stereotyping on television. *Adolescent Medicine: State of the Art Reviews*, 4(3), 551–561.

Signorielli, N. and Morgan, M. (Eds) (1990). *Cultivation analysis: New directions in media effects research*. Thousand Oaks, CA: SAGE.

Wober, J. (1990). Does television cultivate the British? Late 80s evidence. In N. Signorielli and M. Morgan (Eds), *Cultivation analysis: New directions in media effects research* (pp. 207–224). Newbury Park, CA: SAGE.

26

Active Audience and Uses and Gratifications

Helen Wood

It is not often that these two approaches to audiences are written about together in television studies books, and yet they share the basic premise that audiences are not the passive dupes characterized by some accounts from popular imagination and reinforced by effects research. The dualism 'active/passive' was framed in public debate and in communication studies around the spectre of television as the mass medium par excellence. Early critics and commentators of television, mostly elites, were worried about its role in generating an atomized mass of individuals whose passive 'heavy viewing' represented a pathological failure of character associated with women, children and the lower classes. One influential text, *Living With Television* by Ira Glick and Sidney Levy from 1962, locates the image of the couch potato with the working classes who are 'less motivated to act in an energetic, censoring or selective fashion' (cited in Butsch, 2008, p. 131). Against this grain, any conversation about audiences actively using television in their own interests seems an entirely sensible proposition with which both traditions would concur.

Nevertheless, the two approaches to understanding the 'active' role that audiences take in television consumption have differed significantly and they have emerged as separate fields of inquiry: the first, 'active audience', is most usually associated with a qualitative cultural studies approach to audience, and the second, 'uses and gratifications', is applied more often by quantitative psychological approaches. Any good audience studies undergraduate course should, and usually does, include mention of both, even if the temptation is to give one or the other short shrift. The account that follows is an attempt to discuss their various merits and limitations in tandem, with the intention that holding them in tension might prove ultimately productive. For reasons of a linear narrative it begins with an outline of uses and gratifications research followed by a summary of the active audience tradition. This chapter then takes up three areas of discussion around questions of power, the ethnographic tradition, and the dynamic between method and meaning-making.

USES AND GRATIFICATIONS TRADITION

We begin with uses and gratifications by means of a historical narrative which describes how this field of inquiry emerged from early communication studies and concerns about the mass media in general. Despite years of concentration on the negative effects of the mass media, characterized by the post-war interest in propaganda and the influence of media violence on children, it was the uses and gratifications tradition that first pushed the idea of audiences as active. It is largely credited with shifting the focus from 'what does the media do to people?' to 'what do people do with the media?' The earliest approach to analysing gratifications, in terms of the media's ability to satisfy certain social and psychological needs, came from Harvey Cantril in 1942 on the reception of radio quiz shows. It set in motion a functional approach to the media and television which ultimately argued that, 'people bend the media to their needs more readily than the media overpower them; that the media are at least as much agents of diversion and entertainment as of information and influence [and] that the selection of media and content, and the uses to which they are put, are considerably influenced by social role and psychological predisposition' (Katz, Gurevitch and Haas, 1973, p. 166).

The earliest uses and gratifications research therefore considered the *motives* behind media consumption, in order to understand the various functions of particular media. So for instance, studies suggested that soap operas help to reassure oneself of one's own role in life (Herzog, 1941), or that newspapers satisfy a need for information which assists with the trials of daily life (Berelson, 1949), or that classical music on the radio helps people's endeavours in upwards mobility (Suchman, 1942). Rather than starting with media content (as was the preoccupation with research on media propaganda in effects research) a focus on the personal functional uses of mass media explains *why* it is taken up and begins with the audience itself. Uses and gratifications therefore presumed that the mass media does not have a role if the audience does not have a use for it and thus the effects and the uses and gratifications traditions evolved in tandem. They were both well funded (particularly during the growth of US Communication Studies) and were deemed to explain both sides of the mass communication phenomenon.

Since the early brokering of uses and gratification research in conversation with the 'effects' tradition, research ebbed away but went through something of a revival in the 1970s and 1980s. During that period television studies as a field was also establishing itself, although the uses and gratifications tradition of reception research does not often figure as central to its formative narratives. In this later period the account of needs and uses that has become most prevalent to media studies in general comes from Denis McQuail in 1987.[1] Uses are complexly differentiated and cannot be folded into a homogeneous concept of media effect. Rather they span four realms of individual personal need that are fulfilled by the mass media:

Information

- Finding out about relevant events and conditions in immediate surroundings, society and the world
- Seeking advice on practical matters or opinion and decision choices
- Satisfying curiosity and general interest
- Learning; self-education
- Gaining a sense of security through knowledge.

Personal identity

- Finding reinforcement for personal values
- Finding models of behaviour
- Identifying with valued others (in the media)
- Gaining insight into one's self.

Integration and social interaction

- Gaining insight into the circumstances of others; social empathy
- Identifying with others and gaining a sense of belonging
- Finding a basis for conversation and social interaction
- Having a substitute for real-life companionship
- Helping to carry out social roles
- Enabling one to connect with family, friends and society.

Entertainment

- Escaping, or being diverted, from problems
- Relaxing
- Getting intrinsic cultural or aesthetic enjoyment
- Filling time
- Emotional release
- Sexual arousal. (McQuail, 1987, p. 73)

The primary concern of uses and gratifications was with the various uses of particular media, rather than with the *content* or aesthetics of particular media texts, which is why the approach has not always been central to television studies curricula but has rather thrived in the Mass Communications administrative tradition. However, it is easy to see how much of the premises or motives for media use outlined above easily apply to the quotidian experiences and pleasures of television viewing: for example as a 'friend' in the home, or as a way of reassuring ourselves about the world at large.

THE ACTIVE AUDIENCE TRADITION

In contrast to the well-funded uses and gratifications audience paradigm, what has been labelled the 'active audience' tradition in Cultural Studies is usually credited as stemming from within a critical position from the more modestly funded Centre for Contemporary Cultural Studies (CCCS) in Birmingham, UK, in the 1980s. That tradition was in conversation with the Mass Communications tradition, but it was more concerned with issues related to the ideological and social power of the media, rather than with interests derived from understanding personal and psychological need. This tradition took its starting point from concerns over the ideologies of the mass media taken from a Marxist perspective which tries to understand the extent to which the ideas of the ruling classes become the dominant and taken-for-granted ideas of any political epoch. In that sense this tradition was precisely concerned with struggles over the *content* of television. The active audience tradition held meaning-making at the centre of its attempt to understand contemporary socio-cultural formations. This meant that, at least at the outset, television programmes were kept in view, whilst the uses and gratifications tradition, in taking the audience as its starting point did not seek to compare analyses of content with analyses of audience reception.

The cultural studies tradition began with Stuart Hall's (1980) 'encoding and decoding' model which owed a great deal to the legacy of European work on semiotics in the cultural analysis of meaning. In that essay Hall was concerned with the idea that messages are coded at one end in the production of media content, and ultimately decoded in the very 'social' contexts of reception at the other end. This complicated the older communication models of the effects tradition which privilege a linear model of 'sender-message-receiver', and created a space for the acknowledgement that media messages are not necessarily consumed as they are intended, since the audience must deconstruct those messages from within their very specific social locations. In trying to understand the media's role in constructing cultural meaning both processes were central and not necessarily symmetrical.

Hall's original thesis was entirely theoretical, but it is worth pointing out that like the uses and gratifications approach it saw the audience as differentiated, not as a 'mass'. It follows Umberto Eco's work that discussed the television viewer's understanding of programmes as being dependent upon his/her 'general framework of cultural reference … his ideological, ethical religious standpoints [his] tastes, his value systems' (1972, p. 115). Notice that what is different here from the uses and gratifications model is the language of differentiation: taste, value systems and ideology are inherently *social* phenomenon that belong to an understanding of social structure, not individual in the sense that they are prioritized by understanding personal psychological needs. The sense that personal needs are often also structured by social formation is a point we return to later.

David Morley (1980) first tested Hall's theoretical approach on audiences in attempting to assess the ways in which audience members may or may not be co-opted into the dominant ideological messages of media according to their social position:

> What is needed is an approach which links differential interpretations back to the socio-economic structure of society – showing how members of different groups and classes, sharing different cultural codes, will interpret a given message differently, not just at the personal/idiosyncratic level, but in a way systematically related to their socio-economic position. (1980, p. 88)

Morley's investigation of different socially oriented responses to *Nationwide* (a British early evening magazine current affairs television programme) demonstrated how audience members incorporated, negotiated or resisted, the preferred reading of the text according to their socio-economic position thereby supporting Hall's thesis. Many other studies have followed this tradition, utilizing the encoding-decoding paradigm: Corner et al.'s (1990) *Nuclear Reactions* which focuses on audience readings of programmes about nuclear power; Jhally and Lewis's (1992) audience research on the reception of *The Cosby Show* and Schlesinger et al.'s (1992) work in *Women Viewing Violence.* But the focus on establishing a 'preferred reading' in the interests of understanding the ideological message of the television text has meant that it has often been easier to apply the model to audience research on news and non-fiction programming than to entertainment programming, 'where it is much more difficult to identify a single message, or even set of propositions with which audience members could agree or disagree' (Seiter, 1999, pp. 20–21). Indeed, it still remains that the question of the relative power of 'messages' over audiences is one of the most contested terrains in audience research and one which recurs in discussions of new media and participatory culture.

QUESTIONS OF POWER

The uses and gratifications tradition was criticized mostly by the cultural studies tradition for its lack of interest in questions of power, for the 'tendency on the part of gratificationists to overplay "audience freedom" and ignore issues of ideology completely' (Moores, 1993, p. 7). This is primarily because the theoretical underpinning of uses and gratifications owed a great deal to the prominent perspective of functionalism from social science, which was popular in the 1940 and 1950s, particularly in the US. Functionalist sociology has its roots in the work of Emile Durkheim who described the way in which the elements of a society worked together to maintain stability. In this case this philosophy worked to hold an understanding of the media in balance, to *explain* rather than to *challenge* its workings. Uses and gratifications approaches sometimes problematically assume that people are ultimately free to choose the media that they want as individuals (White, 1994). This means that researchers come to the research with a set of preformed assumptions which are themselves problematic and not tested: 'a) media selection is initiated by the individual, b) expectations for media use are produced from individual predispositions, social interaction and environmental factors and c) active audiences [have] goal oriented media behaviour' (Wimmer & Dominick 1994, cited in Ruggiero, 2000, p. 11).

Let us take an example related to a presumed need for interaction that media consumption can fulfil. Rubin, Perse and Powell argue that, 'a person's reliance upon a mass medium for need fulfilment and the availability of functional alternatives are related negatively. As the perceived number of functional alternatives decreases, the more reliant a person is upon a mass medium for the fulfilment of basic needs' (1985, p. 159). The methodology applied to this type of hypothesis was to try and correlate individuals' use of local news in relation to a psychological 'loneliness scale' via 329 questionnaires. Viewers rated statements like, 'I feel in tune with the people around me', or 'I lack companionship' alongside statements about the news like, 'I look forward to hearing my favourite newscaster' and 'I sometimes make remarks to my favourite newscaster during the newscast'. The actual findings were rather inconclusive, suggesting that there is no relationship between loneliness and television interaction per se, but they suggest that there is a correlation between those who rely on TV when lonely and television interaction. This is then further qualified by a rendering of earlier research which suggests that the correlation may be stronger for 'game show and daytime serial viewing than [for] news viewing' (Rubin, Perse & Powell, 1985, p. 173). Uses and gratifications research therefore largely undertakes large-scale surveys in searching for various correlations between need and media use. The emphasis is upon individual human behaviour which brackets off both the social context of the viewer and the content of television.

Philip Elliott's (1974) well-known critique of the tradition suggests that uses and gratifications is overly empiricist and lacks a prior social theory. For example, the participants in the para-social interaction study outlined above are 'viewers', but there is no record of their social or cultural background which would certainly have a direct influence on the features of loneliness. Similarly, the assumption that there is a weak correlation between loneliness and news tells us nothing about the genre or textual features of local news itself – despite the ultimate comparison with other genres. Therefore the relationship between television forms themselves and the social fabric of the viewership are left as untouched territory in the methodological emphasis upon measurement and scale. In summary, uses and gratifications' focus upon individual behaviour circumvents the very *social* processes that are central to the sensibilities of the cultural studies television audience researcher.

One problem with any functionalist perspective of social phenomena is that its findings might serve to ultimately justify the status quo. If individuals can use the media to meet certain ends in any number of ways, and their needs are being met, what impetus is there for social change? For cultural studies the point of research was always about challenging socio-political situations, especially for particular marginalized groups in society. However, it is this imperative that has set in motion a particular turn to audience activity that has also come under fire. The growth of what has often been termed the active audience paradigm was sustained by researchers who also wanted to defend the audience groups which they studied. Television studies itself owes a great deal to feminist academics working with an interest in a popular medium and popular texts like soap opera that many women favoured. They wanted to rescue the medium, programmes and, ultimately, women viewers from the kind of over-easy derision that has persisted since the early years of television.

Ellen Seiter et al.'s (1989) classic essay '"Don't treat us like we're so stupid and naïve": Towards an ethnography of soap opera viewers' encapsulates a research tradition which gave credence to the active viewing strategies of women soap opera viewers as critical and engaged, rather than accepting the picture of passive dopes that had held sway in popular criticism. A body of work in feminist television studies defended the female viewer as adopting a resistive position in relation to the text which offers pleasures that may indeed afford certain kinds of freedoms in otherwise restrictive lives. Much of this research took its lead from Janice Radway's (1984) research on women readers of romance fiction, who found value and nurture from popular texts in an environment where they were themselves mostly the nurturers. For instance, Dorothy Hobson's (1982) work demonstrated the critical faculties of women viewers who were fans of the much-maligned low-budget soap in the UK, *Crossroads*. Ien Ang's (1984) research on Dutch television viewers of the American soap opera *Dallas*, argued that the women found value in the series for its 'emotional realism' as a particular structure of feeling rather than through interrogating narratives for empirical truth. Andrea Press (1991) discussed the generational and class differences of women viewers and found that middle-class women were more likely to adopt a critically distanced viewing strategy, whilst working-class women quite often applied a 'relational' viewing style appropriating texts to their own lives.

In a similar vein, research into concerns over the dominance of the Western media market over the rest of the world and thus the threat of cultural imperialism, also actually found 'resistance' from audiences. Studies on the global exports of soap opera found that viewers incorporated the dominant meanings of soap opera in ways dependent upon their own localized meaning systems, rather than adopting the text as 'straight'. Middle Eastern viewers of the television programme *Dallas*,[2] for example, saw the show as evidence that capitalist ideology is morally corrupt, rather than as a celebration of American culture (Liebes and Katz, 1990). Similarly, Daniel Miller, in his work on the soap opera *The Young and the Restless* in Trinidad, found that despite fears over growing Americanization on the island, large numbers of viewers used the soap in order to think through local issues (Miller, 1993). Thus, rather than confirming the media imperialism thesis, empirical work on the ground tended to find that even imported media forms were used to confirm localized senses of meaning.

According to many television and media studies textbooks these traditions were characterized as looking for the value of audience pleasures as resistance. It is for this particular combination of 'pleasure and resistance' that the active audience paradigm has gained its name, features which, according to Gray (1999), have been seriously over-played in crude summaries of the field. Here 'action' refers to a reactionary position taken against television texts but also against the restrictive conditions of existence. For instance Mary Ellen Brown in her work on soap opera audiences suggests that, 'the pleasure that women experience when talking about soap operas and constructing their own spoken texts is often a resistive pleasure' (Brown, 1994, p.11). What made this position so potent was the idea that television texts (particularly fictional ones) often do not contain a very determined 'preferred reading' with which to understand any dominant ideological framework of meaning.

The major theorist credited with securing the active audience theoretical paradigm is John Fiske (1987) who seized on the idea that texts are often polysemic in nature; that is that they are open to multiple readings depending upon one's social location. This position then was furthered by subsequent studies of television fandom, where fans openly critique and re-construct texts for themselves, also embracing the possibilities opened up by participatory culture (Bacon-Smith, 1992; Jenkins, 1992; Baym, 2000; Hills, 2002). Whilst it is clear that the arguments made in most active audience studies are much more complex and detailed, characterizing them as about pleasure as resistance means that they have moved somewhat away from the original interests of the critical tradition into understanding the political workings of ideology. Instead, some of these studies were deemed to over-celebrate audience agency and freedom and came under serious attack for 'pointless populism' (Seaman, 1992). Where meaning is indeterminate, audiences are free to shape it to suit their own ends, which is a move that rather ironically echoes the cultural studies critique of the analysis of need in the uses and gratifications paradigm.

ETHNOGRAPHIC TRADITIONS AND QUESTIONS OF USE

Where there is a less clear distinction between the two sets of interests is possibly in the ethnographic tradition of audience research. The very turn to the audience at all in television studies was sometimes characterized as the 'ethnographic turn' – an interest in people. In *Family Television* David Morley (1986) proposed an increased need to understand the social contexts through which people received the media, arguing that the focus group (and by inference perhaps still less the questionnaire) could not take into account the way television is stitched into the fabric of the everyday and the domestic. Echoing a broader interest in ethnographic traditions from anthropology into the study of 'alien' cultures, sociologists and media researchers took to more embedded methods for understanding the appropriation of television. This involves spending time in households and observing use and interaction around the television set. Morley's study describes the way in which family power dynamics impact upon the *use* of the television set and choices of programming. Similarly, James Lull's (1990) research on the *social* uses of television, based on ethnographic research, produced a typology of both the structural and relational uses of television which is not entirely dissimilar to some of the premises of audience motivations that we saw earlier in the uses and gratifications approach from McQuail.

Social Uses of Television

Structural

- **Environmental**: Background noise; companionship; entertainment
- **Regulative**: Punctuation of time and activity; talk patterns

Relational

- **Communication Facilitation**: Experience illustration; common ground; conversational entrance; anxiety reduction; agenda for talk; value clarification
- **Affiliation/Avoidance**: Physical, verbal contact/neglect; family solidarity; family relaxant; conflict reduction; relationship maintenance
- **Social Learning**: Decision-making; behaviour modelling; problem-solving; value transmission; legitimization; information dissemination; substitute schooling
- **Competence/Dominance**: Role enactment; role reinforcement; substitute role portrayal; intellectual validation; authority exercise; gatekeeping; argument facilitation. (Lull, 1990, p. 36)

The difference between studies like these and those of the uses and gratifications tradition is that social context plays as great a role as individual need in determining use, since individual need is derived from social positioning. Studies like Dorothy Hobson's (1980) work on young mothers' use of the media in the home, address the gender politics of mothering and domestic responsibility within which media combats loneliness and structures the repetitive tasks of housework. Ann Gray's (1992) study on the use of the video cassette recorder demonstrated the way in which the technology became part of the gendered power relations of labour in the home. These studies certainly had questions of use at their heart, but they were firmly structured around the dynamic between use and power relations through small-scale ethnographic studies.

More recent studies from an ethnographic tradition have sought to make insights into television as both text and technology. These studies rely heavily on collecting intricate detail and on the rich description and ultimate interpretation of findings by the researcher. Marie Gillespie's (1995) study of schoolchildren living in Southall in London, discusses the various ways in which television as text and technology is used in the negotiation of diasporic experience and in the power struggles between generations. Thomas Tufte's (2000) work on women living in the favelas of Brazil shows how the technology itself marked the significatory and permeable boundary between home and street, whilst the narratives of the popular telenovelas were appropriated and related to personal struggles and hopes for change. Purnima Mankekar's (1999) work on the role of television in Indian women's lives demonstrates the ways in which text and socio-political context are intricately interwoven. She describes how images from traditional series like *Mhabharat* resonate so emotionally with women viewers that they were able to 'confront and theorize their own vulnerabilities as women' (1999, p. 256) from within the increasing violence of the Indian state. The richness of these findings, where structural formations, social contexts and the workings of the text come to bear upon individual lives can only really be gathered by a method which relies on intensive and localized periods of ethnographic data collection.

METHOD AND MEANING-MAKING

Despite their origins in very different traditions there are, therefore, moments when the two audience paradigms share research interests. So, for instance, some uses and gratifications researchers have pointed to the importance of the social origins of audience needs and their gratifications. Katz, Blumler and Gurevitch (1973) asked, for example, 'what needs if any are created by routine work on an assembly line, and which forms of media exposure will satisfy them?' (1973, p. 516). McQuail, Blumler and Brown (1972) suggest that those who reported 'educational appeal' as a major gratification of media had left school at the minimum age. In these ways, much like some of the arguments made by feminists working on soap opera, the media offers compensatory practices, gratifying needs that are frustrated by other structures of social life (Fiske, 1987).

But what significantly differs for the two research paradigms are the ways in which which research questions are addressed – questions of method. In posing the social question about the structure of labour cited above, Katz, Blumler and Gurevitch (1973) concede that it is harder to outline the type of hypotheses necessary for the purpose of large-scale survey type research: 'it is more difficult to conceive of a general theory that might clarify the various processes that underlie any such specific relationships' (1973, p. 517). Audience researchers from within a more cultural-studies-focused tradition do not necessarily begin with set of tightly ordered hypotheses which they aim to 'test' in order to produce a general theory. Rather information is gathered 'on the ground' and interpreted in relation to specificity rather than generalizability. This sets up the greatest divide between the two camps: the active audience tradition might be said to offer more explanatory power in terms of local detail, which is strong on validity but weak in terms of representativeness, whilst the uses and gratifications tradition offers the precise reverse, enabling broader generalizable abstractions.

Kim Christian Schroder (1999) offers a very useful account of the methodological differences between the uses and gratifications and what he calls the 'interpretative' tradition and the various arguments that have been levelled at each side in terms of their value. He goes on to outline an approach which attempts to get the best of both worlds: to figure in reliability, validity *and* representativeness to the research design. He outlines a process which attempts to make qualitative data generalizable through practices such as the peer-auditing of qualitative research in order to ensure that interpretation is accounted for more scientifically. However, there have been audience studies which have worked which large amounts of both qualitative and quantitative data which might also be said to partly bridge the divide (Jhally and Lewis, 1992; Livingstone and Lunt, 1994; Livingstone, 1998).

Questions of method and the reliability of data are of course central to any research paradigm. Even Rosengren, one of the biggest critics of the cultural studies tradition, suggests that it has produced 'an impressive number of ingenious and insightful case studies'

(1993, p. 13); although 'case studies', or what Morley (1992) might call a 'map of interpretative communities', are not enough for those searching for some over-arching explanation of more general behaviour. Television Studies is concerned with a medium that primarily captures and generates forms of immediacy that is closely tuned to moments in history, and thus we might assess the explanatory power of its research in terms of what it can tell us about any particular socio-political conjuncture. But that is not to say that personal gratifications are not, of course, intricately bound up in our immediate experiences of social formations. Television studies researchers might want to hold on to an understanding of gratifications to reveal how social and cultural structures and *meanings* help create and manage needs in the first place. Questions of use and need are also germane to elaborations of the workings of power.

In a recent research project, *Reacting to Reality Television* (2013), Bev Skeggs and I have thought about what audiences *do* with reality television in order to understand what reality television *means* in the current environment. Admittedly, this was not done with any particular investment in bridging audience research divides. Our study involves forty women from four different socially located groups across London, who took part in three different stages of audience research. By uses and gratifications standards of scale it is a case study. During the research we found practices like judging reality television participants, taking the moral high ground, using *Schadenfreude* to feel relieved at personal circumstances, and seeking out empathy and common ground with those on television. These are all, on one level, instances which meet personal and individual needs, but they are also stitched to the very fabric of contemporary social life. The ways in which our research participants felt the intensities of certain feelings was dependent upon their social location, and the types of high moral ground that they could take were entirely dependent upon the social, material and cultural resources that they had to hand. For example, in the absence of cultural capital our working-class women used motherhood as their main source of cultural value, whilst other groups could draw on taste, education and cultural difference. Having spent a good deal of time also thinking about meaning in reality television texts, we came to see this audience engagement as a process in which economic forces and cultural life have become increasingly alert to self-scrutiny and evaluation in the interests of late capitalism. For example audiences use *Schadenfreude* – 'thank goodness my life isn't like that' – to assuage anxieties about their own failings or injustices in a system which requires constant evaluation. It is this process which secures in our audiences hope for the future as a 'cruel optimism' (Berlant, 2011), which we suggest is in itself an ideological process: 'By metonymically mapping failure and lack of value reality television incited our audiences to defend and prove their *own*, and in this reaction their investment in the ideological mechanism of value struggle itself was secured' (Skeggs and Wood 2013, p. 233).

Questions of use and need are inescapably part of the way in which power is experienced. What audiences *do* with television is vital to the explanation of what television *means*. This is potentially exacerbated by television's migration onto alternative platforms and the invitation to take part in television culture – vote, phone-in, download, navigate online platforms, generate content – which provide different challenges for the audience researcher (Jermyn and Holmes, 2006; Bolin, 2010; Evans, 2011). Given all of this, it is possibly time to overcome distinctions such as those between uses and gratifications and the active audience traditions and their related legacies as we embrace a new 'participation paradigm' (Livingstone, 2013) where 'activity' will be even more central to the figuration of the television audience of the future.

NOTES

1 See Denis McQuail on the contemporary relevance of the uses and gratifications approach: http://podcasts.ox.ac.uk/people/denis-mcquail

2 *Dallas* is an American prime-time television soap opera that originally ran from 1978 to 1991. It revolved around the Ewings, a wealthy Texan family in the oil and cattle-ranching industries and was a global export.

REFERENCES

Ang, I. (1984). *Watching Dallas: Soap Opera and the Melodramatic Imagination*. London: Methuen.

Bacon-Smith, C. (1992). *Enterprising Women: Television Fandom and the Creation of Popular Myth*. Pennsylvania: University of Pennsylvania Press.

Baym, N.K. (2000). *Tune In, Log On: Soap, Fandom, and Online Community.* Thousand Oaks, CA: Sage.

Berelson, B. (1949). What 'missing the newspaper' means. In P.F. Lazarsfeld & F.N. Stanton (Eds), *Communications Research 1948–9*. New York: Duell Sloan and Pearce.

Berlant, L. (2011). *Cruel Optimism*. North Carolina: Duke University Press.

Bolin, D. (2010). Digitization, multiplatform texts and audience reception. *Popular Communication* 8(1), 72–83.

Brown, M.E. (1994). *Soap Opera and Women's Talk*. Thousand Oaks, CA: Sage.

Butsch, R. (2008). *The Citizen Audience: Crowds, Publics and Individuals*. London and New York: Routledge.

Cantril, H. (1942). Professor Quiz: A gratifications study. In P.F. Lazarsfeld & F. Stanton (Eds), *Radio Research 1941* (pp. 34–45). New York: Duell, Sloan and Pearce.

Corner, J., Richardson, K. & Fenton, N. (1990). *Nuclear Reactions: Form and Response in 'Public Issue' Television*. London: John Libbey.

Eco, U. (1972). Towards a semiotic enquiry into the television message. *Working Papers in Cultural Studies CCCS 3*.

Elliott, P. (1974). Uses and gratifications research: A critique and a social alternative. In J.G. Blumler & E. Katz (Eds), *The Uses of Mass Communications*. London: Sage.

Evans, E. (2011). *Transmedia Television: Audiences, New Media and Daily Life*. London: Routledge.

Fiske, J. (1987). *Television Culture*. London: Routledge.

Gillespie, M. (1995). *Television, Ethnicity and Cultural Change*. London: Routledge.

Glick, I. & Levy, S. (1962). *Living With Television*. Chicago: Aldine.

Gray, A. (1992). *Video Playtime: The Gendering of a Leisure Technology*. London: Routledge.

Gray, A. (1999). Audience and reception research in retrospect: The trouble with audiences. In P. Alasuutari (Ed.), *Rethinking the Media Audience*. London: Sage.

Hall, S. (1980). Encoding/Decoding. In S. Hall et al. (Eds), *Culture, Media, Language: Working Papers in Cultural Studies*. London: Hutchinson.

Herzog, H. (1941). Motivations and gratifications of daily serial listeners. In W. Schramm & D. Roberts (Eds), *The Process and Effects of Mass Communication* (Revised 1971). Champaign: University of Illinois Press.

Hills, M. (2002). *Fan Cultures*. Routledge: New York.

Hobson, D. (1980). Housewives and the mass media. In S. Hall et al. (Eds), *Culture, Media, Language* (pp.104–114). London: Hutchinson.

Hobson, D. (1982). *Crossroads: The drama of a Soap Opera*. London: Methuen.

Jenkins, H. (1992). *Textual Poachers: Television Fans and Participatory Culture*. Routledge: New York.

Jermyn, D. and Holmes, S. (2006). The audience is dead: Long live the audience! Interactivity, 'telephilia' and the contemporary television audience. *Critical Studies in Television* 1, 49–57.

Jhally, S. and Lewis, J. (1992). *Enlightened Racism: The Cosby Show, Audiences and The Myth of the American Dream*. Boulder CO: Westview Press.

Katz, E., Blumler, J. and Gurevitch, M. (1973). 'Uses and Gratifications Research', *The Public Opinion Quarterly* 37(4): 509–523.

Katz, E., Gurevitch, M. and Haas, H. (1973). 'On the Uses of Mass Media for Important Things', *American Sociological Review* 38 (April):164–181.

Liebes, T. and Katz, E. (1990). *The Export of Meaning*. Oxford: Oxford University Press.

Livingstone, S (1998). *Making Sense of Television: The Psychology of Audience Interpretation*. London: Routledge.

Livingstone, S. (2013). The participation paradigm in audience research. *The Communication Review* 16, 21–30.

Livingstone, S. and Lunt, P. (1994). *Talk on Television*. London: Routledge.
Lull, J. (1990). *Inside Family Viewing*. London: Routledge.
Mankekar, P. (1999). *Screening Culture: Viewing Politics: An Ethnography of Television, Womanhood and Nation in Postcolonial India*. Durham: Duke University Press.
McQuail, D. (1987). *Mass Communications Theory: An Introduction*. London, Thousand Oaks, New Delhi: Sage.
McQuail, D., Blumler, J.G. & Brown, J.R. (1972). The Television Audience: A Revised Perspective. In D. McQuail, (Ed.), *Sociology of Mass Communications*. Harmondsworth: Penguin.
Miller, D. (1993). The young and the restless in Trinidad: A case study of the local and the global in media consumption (pp. 163–182). In R. Silverstone & E. Hirsch (Eds), *Consuming Technologies*. London: Routledge.
Moores, S. (1993). *Interpreting Audiences: The Ethnography of Media Consumption*. London: Sage.
Morley, D. (1980). *The Nationwide Audience*. London: BFI.
Morley, D. (1986). *Family Television: Cultural Power and Domestic Consumption*. London: Comedia.
Morley, D. (1992). *Television, Audiences and Cultural Studies*. London: Routledge.
Press, A.L. (1991). *Women Watching Television: Gender, Class, and Generation in the American Television Experience*. Pennsylvania: University of Pennsylvania Press.
Radway, J. (1984). *Reading the Romance: Women, Patriarchy and Popular Literature*. Chapel Hill: North Carolina University Press.
Rosengren, K.E. (1993). From field to frog ponds. *Journal of Communication* 43(3): 6–17.
Rubin, A.M., Perse, E.M. & Powell R.A. (1985). Loneliness, parasocial interaction, and local television news viewing. *Human Communication Research* 12(2), 155–180.
Ruggiero, T.E. (2000). Uses and gratifications theory in the 21st century. *Mass Communication and Society* 3(1), 3–37.
Schlesinger, P. et al. (1992). *Women Viewing Violence*. London: BFI.
Schroder, K. (1999). 'Best of both worlds? Media audience research between two paradigms. In P. Alauutari (Ed.), *Rethinking the Media Audience*. London: Sage.
Seaman, W.R. (1992). Active audience theory: Pointless populism. *Media, Culture and Society* 14, 301–311.
Seiter, E. (1999). *Television and New Media Audiences*. Oxford: Clarendon Press.
Seiter, E., Borchers, H. & Warth, E. (1989). 'Don't treat us like we're so stupid and naïve': Towards an ethnography of soap opera viewers. In E. Seiter et al. (Eds), *Remote Control: Television, Audiences and Cultural Power*. London: Routledge.
Skeggs, B and Wood, H (2013). *Reacting to Reality Television: Performance, Audiences and Value*. London: Routledge.
Suchman, E. (1942). An invitation to music. In P.F. Lazarsfeld & F.N. Stanton (Eds), *Radio Research, 1941*. New York: Deuell, Sloan and Pearce.
Tufte, T. (2000). *Living With the Rubbish Queen: Telenovelas and Modernity in Brazil*. Luton: University of Luton Press.
White, R.A. (1994). Audience interpretation of media: Emerging perspectives. *Communication Research Trends* 14(3), 3–36.
Wimmer, R.D. & Dominick J.R. (1994). *Mass Media Research: An Introduction*. Belmont, CA: Wadsworth.

27

Raced Audiences and the Logic of Representation

L.S. Kim

The relationship between media and audiences has been at the core of Television Studies, helping to map historical shifts in the intellectual imperatives, social contexts, and cultural concerns expressed in work in the field. There are several theoretical and methodological approaches to studying television and audience, some follow one another, some are in opposition, some converge; the fact of the plethora of theories and strands marks a consistent characteristic of Television Studies: it is ever dynamic and always curious about the object-subject dialectic of television.

Anchored at one end by television's historic sibling rivalry with cinema, and buoyed at the other by defeating the perennial threat of its death at successive technological thresholds, television mediates society. The persistence of television and the regulation of its presence and broadcasting indicate a political importance. At once 'just entertainment' as well as ideologically powerful, television's relationship to its audiences is both direct and indirect in terms of reception (sometimes people acknowledge how televisual images shape their perspectives), and both mythic and material when it comes to market operations (the belief that 'green is the only color that matters' and the chase after consumers aged 18–34 remain presuppositions on the production side). Understanding this relationship more fully is key to critiquing and potentially affecting televisual cultural production.

In terms of raced audiences, a key question is: what defines raced audiences, that is, how are audiences raced? The role of television in the production of raced audiences is performed through a number of composite forces, including: the state, public policy, advertising and marketing practices, program makers, networks and pressure groups. In addition to the semiotic negotiation of racial meaning/s, the process of racialization both on television and through television as a site, involves such extra-textual dimensions.

INTRODUCTION: RACE AND AUDIENCE

The topic of raced audiences, and race and television more broadly, can be seen as part of a disciplinary trajectory, and yet racial discourse is a way to inform and innovate audience studies and research. In other words, race has and hasn't been part of the historiographic work of Television Studies. There have been specific case studies focusing on programs or genres, and studies about the representation of racial or ethnic groups; these are sprinkled into the larger repertoire. Perhaps this can be attributed to a kind of racial bypassing on the larger front of media and culture, as part of the 'race matters' debate that exists in more colloquial conversations about watching television. For some, race personally and politically does not matter in this 'post-race' period; industrially, there is a tension between the old network system[1] and the post-network system in which forms of marketization, branding and production of raced audiences engage with both racialization and disavowal.

In considering raced audiences, there are attendant compound concepts related to audience – viewing practices, ethnography – and to race – as signifier, as controlling image, as polysemic. Television Studies has rejected the theoretical framework of the passive viewer (hypodermic needle model), yet as an academic discipline it hasn't simply abdicated television's multiple meanings to postmodernist pastiche; there are spaces and phenomena in between these two poles that studies of television's racial meanings explore and uncover. While this semiotic function is crucial, so too is the production of a kind subjectivity that speaks to the project of self-making. Pleasure, as consumed and as cultivated, is generated from the subject-viewer in conjunction with the anticipatory offerings by television writers, producers, makers and advertisers.

Television Studies began with essays about the apparatus, about flow, about advertising, and about fandom before the term was fully formalized. A predominant genre that was analyzed was the daytime (and nighttime) soap opera. Female spectatorship and pleasure were key components. Female consumers were acknowledged beginning in the 1950s as programming developed to center around the themes of domesticity and the family. The television industry has always valued female viewers, and television texts have long provided points of view of, and points of identification for, women (even if ideologically managed). This focus on who in the cinematic experience and according to film theory was excluded from subjectivity (both on-screen as well as in the audience), set a path for Television Studies – and for television – to be open to possibilities for difference.

If gendered viewing is accepted as a concept, so raced viewing ought to be. But acknowledging these axes of difference has not historically been an explicit phenomenon; much of raced viewing has been made invisible, stitched back into the production of whiteness as the unmarked viewing position. While the idea that there are audiences of varying races watching television in national and international contexts is a basic one, the experience of raced viewing is less concrete. The history of this viewing project features the production of subjectivity through the figure and position of whiteness as the unmarked signifier. And it constitutes a story about television studies, television, and television's role in the project of racialization.

I see 'identification' at the crux and wish to point out a premise that has functioned as a fairly unchallenged rationale in both industry practice and academic theory: that one identifies with a character that is (racially) like-oneself. Accompanying this false premise is an underestimation about what viewers can experience and come to accept. At the same time, Television Studies has long acknowledged the uniqueness of television viewing (apart from film viewing), in that pleasure is dispersed, rather than intensely located in a psychic drive.

While psychoanalytic theory is deeply embedded in the vocabulary about the film experience, such theory is for the most part inapplicable to the televisual experience – it is not about a homogenous gaze, a 'sonorous envelope', and regression into pre-verbal childhood, but rather, television is more about a multiplicitous glance within communal spaces, and liveness within a social milieu. Nevertheless, a singular approach to identification remains in much of the way television is discussed.[2] It is not hard to realize that viewers identify with more than one character, sometimes simultaneously; furthermore, viewers are capable of identifying – aligning and allying – with a character that is not of the same gender or race,[3] or disidentifying with the 'intended' character alignment. Racial discourse via television involves raced viewing as well as racialized representations; identification is a connecting mechanism, but its operation is much more social than psychological. Race on television is less about projecting an ego ideal on screen, than about mapping out social relations in the real world.

That is, part of the logic of representation is its serving as a site for the production of identification. Especially given the alignment of media/television with political claims on social justice and equality emanating from government reports such as that of the Kerner Commission in the late 1960s[4] or from scholarly debates and analyses about television and democracy since the birth of the medium, the history of the project of television follows a trajectory of racing audiences. The Kerner Commission report in 1968 berated federal and state governments for failed housing, education and social-service policies; the report aimed some of its sharpest criticism at the mainstream media, stating, 'The press has too long basked in a white world looking out of it, if at all, with white men's eyes and white perspective'. Its most famous passage warned, 'Our nation is moving toward two societies, one black, one white—separate and unequal'. As I have written elsewhere, the representation of race is about the representation of race relations. The making of race and racing audiences is a key project of television.

A matrix of elements comprises raced viewing:

- The audience member as an individual (with her/his own psyche)
- The audience member as part of a community (self-identified or as-perceived by those not part of the group)
- The image-text which is racialized (but which may not be believed to be by all)
- The construction/production of an image-text, specifically, a character (both discrete and reoccurring)
- The expectation of 'tastes' or at least consumer practices for different racial groups (which usually boils down to marketing language)
- A basic belief about identificatory processes (that delimits how one identifies with a character or figure)

PRODUCING RACED AUDIENCES

Using the US as a model, particularly with the campaign, election and re-election of President Barack Obama, the media culture in the United States is one comprised of racial discourse, whether in debating the relevance of race and even when denying it. It is a preoccupation either way. Racing audiences is part of the practice in a nation that is exceptionally diverse, extols its diversity, and yet sustains wide swaths of segregated social life. In political speech 'Americans' are at once addressed as a uniform (if not unified) national body and also as particular pockets of the electorate (separated out by race and gender).[5] Raced audiences might be taken as an expression – or effect – of the dual practice by television (by virtue of its own conventions and logic) of address and separation. During the two most recent US presidential elections, coverage and discussions gauged election outcomes based on polls of 'Blacks' and 'Whites' and invoked the importance of 'the Latino vote', thus, designating racial positions to members of an audience.[6] Similarly, public policy on a topic such as immigration is transmitted through televisual discourses in a way that fragments audiences politically. Public policies on television, itself – for example, related to standards and practices or violence – also speak to conservative audiences generating fear and reasserting a need for control. These policy debates become moral debates, and terms such as illegal alien or violent criminal are often code for raced subjects. The state participates in racing audiences by provoking alignments, which are politicized-racialized-classed.

On the seeming opposite end of the spectrum, various media watch groups that pressure the industry and/or the government to change or halt a representational practice, can also essentialize by arguing that something is injurious to 'the' Black or Asian American or American Indian or Latino audience. Indeed, these are racial projects in the sense that they use the logic of race to construct points of identification and belonging that seek redress for racial exaggeration and exclusion. And, somewhere in-between, advertising executives target 'ethnic markets' and conjure a sense of who their audience is.[7]

The production of raced audiences comes through address: political entities, advertisers, programmers 'produce' their anticipated audiences according to categories of age, gender, region, race. At the same time, as with the concept of hegemonic hailing, there is ability on the part of the viewer to correct or defy how we are being addressed. Uses and gratification theory helps explain part of the story, in that this approach resuscitates the passive viewer as one who is not a victim but rather as one who figures out how to use media for her/his own gratification. The emphasis is on a goal-oriented choice-making; however, defining the raced audience member is less about use and more about process. A cultural studies approach is more effective in illuminating the process; indeed, Stuart Hall's analysis interpreting meanings as dominant, negotiated or oppositional echoes how audience members can accept, engage or reject the raced position they are being offered. In representational culture, there is generally an iconic shorthand that prompts people to abstract the easier racial meaning, and which naturalizes racial (and class and gender) hierarchy. Becoming aware that media producing audiences is contingent to audiences consuming media, asserts and empowers us as consumer-viewers. (That is, we can resist or contest, as a start.) And yet, what is implicit here is that the projection and contestation are different sides of the same process – racialization is expressed in the organization of markets and advertising address at the same time as they are organized as part of the national body, political interests and cultural identities. Television is a mechanism for their coproduction, and identification is part of the organizing logic of race and television.

The role of television in claims about race among various institutions, movement organizations, marketers and creative staff has been to reconcile race alternatively as a 'problem' to be resolved or as not-an-issue. For example, on the one hand, specific programs have taken criticism for a caricatured depiction of a person of color and networks have felt obliged to color-up their line-up due to activist pressuring; on the other hand, the sighting of a few faces of color broadcast on the small screen, or even balkanized channel offerings are seen as proof for the argument that there is a representational democracy.

HISTORICAL FRAMEWORK: DEBATES ABOUT REPRESENTATION

Approaches to studying race and television have focused on accounting, emphasizing the need for screen time. To be represented means having a presence in popular culture and in political society; so a core part of the campaign to address racial representations has rested on the belief that to be represented is not to be invisible.

From a historical US business standpoint, a guiding principle began that since most of the viewing audience was white, programming should not 'offend' or discomfort, white audiences; the general presumption followed that almost all characters and storylines were to be white and that interracial interactions would be as minimal and benign as possible. At the same time, television came on to the scene in the post-war, pre-civil rights 1950s accompanied by idealism about the new democratic art. Never denying its commercialism, television nevertheless had the potential to produce an image of a more equal America. The Kerner Commission report in 1968 emerged as a direct response to the civil unrest and protests against the lack of economic opportunities; it criticized the media for its failure to help reduce segregation and its privileging of white perspective. But how could life on television approximate a racial equality that was not yet achieved in life off-screen?

Very early US programming indicated ethnicity – Italian American, Irish American, Norwegian American, Jewish American and Black American, for example. From the late 1940s through the mid-1950s there was a morphing from working-class immigrant past to 'American' middle class – a transformation to a mythic middle-class whiteness. Ethnicity mostly disappeared, and while race was a conspicuous difference, it was rarely seen. As scholars have written about, television in this era helped to teach viewers how to become consumers, and it projected images of the ideal American family. Such families were portrayed in a way that contributed to the creation of a cultural whiteness.

Also early on came criticism (by some) about problematic representations of race. Derogatory or egregiously misrepresented characterizations of people of color, particularly African Americans, were examples of negotiations amongst writers/producers, sponsors/advertisers, actors/performers, viewers/consumers, media watch/activist groups. The beginning conceptual framework in media criticism and ideological struggle was – and to an extent remains – one which gauges the quantity or quality of images and the extent to which they are positive or negative. But, as media criticism developed, a question raised by Marlon Riggs in his 1991 film, *Color Adjustment* about the history of representation of African Americans on television became increasingly complex: What is the parameter used to determine: 'Is this a positive image?'

Is positive and negative always relative and shifting? Is the positive–negative dyad a limited strategy? (Yes and yes.) The conversation about race and representation must continue beyond (rather than end at) 'example x is racist'. For audiences to participate in change, it is vital to consciously reject stereotyped images and their attendant stories; thinking of ways to then demonstrate consumer demand for more creative representations of race is the next step. As television producer Loni Ding advocated for, tropes must not merely be displaced, but fully replaced (cited in Leong, 1992).

While an increased number of faces of color on screen is important and has slowly come about, the quality, or qualities, of such representations can be marked by criteria such as:

- Authorship – who is writing the role; who has the power to create new characters and the opportunity to cast actors?
- Self-representation – to what degree does a minority community influence how characters of color are produced, projected, circulated?
- Mode of production – is a mainstream commercial medium able to innovate racial representation?
- Subjectivity – it requires a conscientious skill to write and create characters of color in ways that ensure their perspective is shared and understood. Garnering a viewer's empathy would be the goal.

The dynamic between the production of images, of characters, of stories and the ways in which such programming is watched risks getting caught in a tautology. It is the 'Does television mirror society, or does society mirror television?' question. Imagining that television progresses gradually is simplistic and teleological. Furthermore, being content with 'change-will-come-with-time' is a passive non-argument.

Historically, the representation of race relations has taken two dichotomous forms: as harmonious (whether one racial group was enslaved/employed as servants to another, or whether portrayed as neighbors-friends for whom race was no matter), or as rancorous (racially divided viewers and citizens in the midst of so-called race riots, or during high-profile, highly-charged cases such as those involving O.J. Simpson or Trayvon Martin/George Zimmerman). The shifting nature of television's account of race relations, including its own stakes and investments in different moments as culturally legible and economically profitable, belies the logic of television representation as a site for the production of identification.

KEY EXAMPLES: VIEWING RACE, SIGNAL MOMENTS

In the US market context there is a resistance to change in media practices that have already proven profitable. But the system is not closed. There are examples that demonstrate that television in the abstract and discrete televisual discourses occupies a space for negotiation and can participate in a hegemonic struggle against racial hierarchy. The following are part of a constellation of programs that illustrate ways that US television history comprises a history of the representation of race relations (both on-screen and off). The emphasis on televisual race relations is a practice of racing audiences. Changes in production circumstances also explain why there have been moments when the racing of audiences was foregrounded (rather than oblique); the demise of the networks, the rise of cable, and, particularly, the emergence (however temporary) of 'Black block' programming on Fox, UPN and the WB, continued a pattern of projecting audience tastes according to race.

The review in this chapter features a majority of texts representing African Americans for several basic reasons. First, in the instances of programming that have included people of color ranging from fiction to sports (and in cinema, for that matter), African Americans, have had a more noticeable presence in comparison to other racial groups. Second, African Americans have mobilized successfully as activists and as industry professionals, and have become significant players as actors, producers, writers and directors. And, third, 'race' in US culture has long been understood as Black and as a dynamic between Black-and-white; the post-Civil Rights discourse (that racial equality has been achieved) is often blanketed over numerous groups, with achievements by African Americans serving as representative.

Contested, Contentious Texts

Perhaps the first public wrangling over the representation of race on television involved *The Amos 'n Andy Show* (1951–1953). As with the landmark protest against *The Birth of a Nation* (1915) lodged by the NAACP, the organization asserted its presence and its concern about the depiction of African Americans in one of the very first American television series. The promise of equal or better treatment after the Second World War in light of the service that African Americans, Japanese Americans, Native Americans and other minorities gave was not realized. Criticism of *The Amos 'n Andy Show* was two-fold – that it depicted a segregated world (it had an all-Black cast), and that many of the characters were shown to be incompetent, particularly in white-collar professions. The NAACP was successful in seeing *Amos 'n Andy* cancelled; along with that came an end to the employment for the program's actors and a loss of African American faces on the screens of the new medium.

Famed Hollywood film actresses faced career questions in relation to the kind of roles they would take, and could accept responsibility for. A series about a domestic, the eponymous *Beulah* (1950–1953), ran for three years and each year a different actress played Beulah. Academy-Award winning, Hattie McDaniel (well known for playing Mammy in *Gone With the Wind* (1934)) transferred the role of a slave in the antebellum South to that of a servant in the suburbs. She quit after one year, though she is known to have famously said: 'I'd rather make $700 a week playing a maid than earn $7 a day being one'[8]. Like McDaniel, Louise Beavers has a prolific filmography (often in the role of a maid or mammy archetype). She is perhaps most known for starring alongside Claudette Colbert in *Imitation of Life* (1934), a rather ground-breaking melodrama about race, class, interracial friendship, single motherhood, racial melancholy and racial pride. Beavers fell ill during the shooting of *Beulah*, and acted in just one season. The series persisted in its popularity, with seeming interchangeability among the lead actresses. Esteemed stage actress, Ethel Waters, was the third African American woman to occupy the role of Beulah. Beulah was Mammy, inasmuch as she 'served' to make her white employers and the white audience comfortable, with her subordinate position and the outwardly harmonious yet unequal race-class relations she symbolized.

Both *Beulah* and *Amos 'n Andy* made spectacle of African Americans. At the same time, the performances of the actors provided a kind of affirmation for African American viewers. Actress Diahann Carroll mentions in *Color Adjustment* that her parents did not like her to watch *Amos 'n Andy*, nevertheless, she admired the artists she saw acting on television. The concepts of 'reading against the grain' or winning performatively even if losing narratively provide ways to negotiate contested images.[9] These theories meant to wrest power away from 'the text' have moved close analysis of television forward. The three-choice frame used in understanding the media – as 'media-powerful' (focused on manufactured culture), or 'audience-powerful' (focused on consumer society) or 'in-between' – elects the obvious, third choice.[10]

The 1940s and 1950s marked a shift in media theory, embodied in the Frankfurt School's emigration from fascist Europe to capitalist North America. Their pessimism about mass culture injecting an oppressive ideology into the public lessened. It was the American researchers, however, who by the 1960s, had forged an empirically grounded 'Sociology of Mass Persuasion', which produced a more qualified notion of 'media power', in which media consumers were increasingly recognized not to be totally passive 'victims' of the Culture Industry. Such elaborations in the nuances of media culture advanced Television Studies, though they did not specifically address minority members of society. What about the fact of unequal consumer access in advocating for an audience-powerful perspective? Is the power of the audience parceled according to racial and economic status? Did audience desires on the part of African Americans count in the 1950s and 1960s?

Nat King Cole is a figure that inspired cultural pride. His talent, elegance and confidence, though, were offensive to many white viewers when he was on stage with white female singers. He was not allowed to be on stage with a white woman alone. His disembodied recording voice, on the other hand, was more widely accepted. *The Nat King Cole Show* (1956–1957) could not find regional sponsors (who were worried about their white viewership), and it was canceled.

Julia was a prime-time fiction program with an African American lead character, though without an African American man/husband/father, that gained enough viewership to last three seasons (1968–1971). Starring lighter-complexioned Diahann Carroll, the program presents a perfect case study of raced audiences in that there were basically four response positions: African Americans who liked the program; African Americans who didn't; whites who liked the program; and whites who didn't. Julia was dignified (positive) or 'an oreo'[11] (negative) according to some African American viewers; she was 'too beautiful' (a negative critique by some white viewers) or harmoniously part of a white, pink-collar world (a positive critique by whites). Letters poured in, and executive producer and writer, Hal Kanter, made a conscious effort to create a character with depth ('as an apology for *Beulah*', he states in *Color Adjustment*). The lesson is three-fold: that the same image/text elicits variable interpretations; that conscious or conscientious work can be done in the production/creative process; and that it is probably not possible to satisfy everyone and it could be a distraction to try.

Race and Class

Moving from a bourgeois aesthetic to portraying a working-class reality, US programs emerged in the early 1970s such as *All In the Family* (1971–1981), *Good Times* (1974–1979), *Sanford & Son* (1972–1977), *Chico and the Man* (1974–1978), *What's Happening?* (1976–1979) and other spin-offs or derivative series. Race, or racial difference, came to represent a less desirable class position. Viewing alliances and alignments were designed according to race, class and geography. Counter or counter-programmed to the sub-genre of 'the ghetto sitcom' were notably nostalgic series such as: *The Waltons* and *Little House on the Prairie*, and, in the second half of the decade, *Happy Days* and *Laverne & Shirley* in which whiteness was preserved and idealized. Furthermore, rather than the acceptance of one's lot in life demonstrated in the former list of texts, conformity to the Horatio Alger myth of pulling oneself up by their bootstraps in the latter grouping functioned to (re)establish hard-working, aspirational white families in these televisual tales about what it meant to be American. This was a form of racing the audience in which whiteness served as the unmarked site of mobility aspirations.

The growing awareness of the fact of plural audiences was an added layer to the continuing contentiousness about the use of stereotypes. J.J.'s oft-occurring smile and the awaited delivery of his signature line, 'dy-no-mite!' in *Good Times* overtook the memory of its beginnings as a program that revealed some of the hardships experienced by people who live in housing projects and suffer from unemployment. Getting up the rent, and getting by, were part of the characters' existence and the show's storylines and humor. *The Jeffersons* (1975–1985), one of many very popular Norman Lear productions and one of several spin-offs, featured a story of upward mobility. Wise-cracking, sometime jive-talking Mr George Jefferson had 'moved on up, to the East Side' through the success of his dry-cleaning business. His wife, Louise, no longer had to work as a maid, and, in fact, she hired one of her own, Florence. George Jefferson was a kind of counterpart to Archie Bunker, whose working-class neighborhood in Queens he moved into dramatically being the first to integrate it (an intermediary step before crossing the Hudson River to Manhattan); both were vocally opinionated about race. *The Jeffersons* enabled his perspective, of a once-poor-now-wealthy Black man, to be shared. He expressed fears about losing it all and Louise expressed anxieties about having money; they dwelled among whites, though they did not much alter their personalities. George was not shy about expressing his disapproval of his neighbors, an interracial couple (a significant 'first' on American television). The fact that *The Jeffersons* and other programs with racial characters were comedies helped to deliver the messages about class, race relations, cultural difference and integration.

The Fresh Prince of Bel Air (1990–1996) depicts a similar story of class movement through a character who declares to hold on to his (class-race) roots, at least stylistically; the class tension, however, is expressed almost exclusively within the insular terrain of Will's African American family (their surname is Banks) whose easy integration into a presumed white and elite L.A. lifestyle is not iterated on screen.[12] In some ways, more than the text, characters or *mise-en-scène*, it was the post-L.A. Uprising context of *Fresh Prince* that enabled multiple and exchangeable points of identification; the ethos of the program basked in the hopefulness and youthfulness of the Clinton Administration, gesturing toward both the discourse of color-blindness of Bush's politics and the arguments about a post-race society engendered by Obama's presidency. While outwardly more liberal and less conservatively-mobilized than *The Cosby Show* of the 1980s, *Fresh Prince* continued television's project of racing audiences through a fish-out-of-water plot device in which viewers and sponsors could choose to see or not to see race as a factor, and by proffering a 'good Black' – a younger, hipper Bill Cosby – who was both 'street' and approachable. Actually, both are from Philadelphia. Comic Bill Cosby became 'tame'[13] by donning sweaters and penning a best-seller, *Fatherhood* (Richard Pryor took a different path[14]); Will Smith was a rapper turned blockbuster action star, spending time on prime-time gaining a cross-cultural fan base who could simultaneously note and efface his and their own race-class difference.

Mediated Events

Non-fiction, real-world events as packaged into television programming are also ways that raced viewing develops. Analyses have been written about the Civil Rights Movement, the Vietnam War, the Rodney King Beating, along with the not-guilty verdict against the police officers which prompted the L.A. Uprising, and the O.J. Simpson Case, and they all illustrate my thesis that the representation of race is ultimately about the representation of race relations. Individually, these are fascinating as key events in social history, and they are rich examples of racialized television texts broadcast to a viewership that is folded into an expression and practice of racing audiences.

The news coverage of the Civil Rights Movement starting in the late 1950s acted as a generic form for the representation of African Americans. At this time, most viewers (in and beyond the US) saw African Americans as noble, peaceful activists, as well as victims of brutal discrimination. African Americans saw themselves, and saw images of themselves circulated, as important and meaningful in a society on the verge of immense Contrary to the news were prime-time comedies in which there was no acknowledgement of a racial struggle. Furthermore, the structural and moral work that news organizations and networks contributed in order to broadcast the Civil Rights Movement, took place during a formative time for the television industry and thus married television and race in a certain way. That is, we can understand television as a specific vehicle for racial meaning to enter public discourses, and racial discourse as a way that television enunciates itself.

The era of the Vietnam conflict was the first time that images of a US war were seen regularly, nightly, on televisions. In the 1960s, news film was shot, processed, shipped and edited (a thoughtful process that took time, quite the reverse of today's propensity for an endless, live stream of visual 'information'). The conflict in Vietnam was contested, messy, and not as legible as past wars in the international arena; it was the first time that the United States did not achieve a definitive victory. It was also the first time that there was a lack of support for the war effort and its troops by Americans, garnering a strong anti-war sentiment. At the same time that an anti-establishment disposition arose in response to and in protest of the violent images of this military engagement, a more deeply entrenched visual language was employed to depict 'the enemy'. Racialized images of Vietnamese peasants/Vietcong soldiers were rooted in a Yellow Peril ideology born during Western Expansion in relation to Chinese immigrant railroad workers and which grew during World War II in relation to Japan, rendering Japanese Americans suspect. Televisual discourses engaged with racial and national identity formation in the US as well as abroad through a mechanism of alterity (i.e., marked as American or non/un-American). Domestically, they functioned to inspire social change and yet also to dehumanize Asian bodies. While there was outrage against the gruesome war, the televisuality of the Vietnam War did not encourage perspectives other than the thrust of a mainstream, white middle-class American experience. Despite the ways that mainstream television news instigated resistance to the war in Vietnam, it did not account for nor allow, for example, the perspectives of Asian American soldiers who were simultaneously American and 'like' the enemy. Or of returning veterans, disillusioned and traumatized. (Who did they identify with, when news images flashed on the screens before them?)

The Rodney King beating caught on tape became a video played and re-played in news broadcasts around the world, enabling an unlikely figure for viewers of widely variant backgrounds to identify with (an intoxicated, working-class African American man with an arrest record). A great number of viewers aligned themselves with the beating victim, thus castigating the excessive force used by the L.A.P.D. officers. Furthermore, the amateur video was a vital piece of evidence that produced a paradigm shift for many: the concept that law enforcement may act with excessive force or excessive scrutiny when dealing with minority men was exposed to a mass public and accepted by a mass audience (it is something men of color have known and rightfully feared for a very long time). The 'video-low' quality of the recording made by a white man from the balcony of his hotel during a visit to L.A. was an aesthetic that lent itself to television of the early 1990s.[15] As a kind of flip-side to one of the longest-running programs in US history, the conservative reality series, *Cops*, the text of the Rodney King beating was evidentiary material for audiences; the impromptu, amateur and unofficial visual text begged a questioning of the nearly impenetrable characterization of police enforcement as heroic and trustworthy, and was effective in presenting the perspective of the disenfranchised.

While King was not completely innocent (he did lead police on a high-speed chase), the case set a different story about race and race relations (i.e., about a powerful white authority, and defenseless Black citizenry). Not since the news broadcasts of the Civil Rights Movement was there a near consensus about the reading of racial meaning. However, this shared audience position across race, class and gender was quickly lost by the time the 'not guilty' verdicts for four L.A.P.D. officers came down in a conservative white suburb of L.A. The images and stories of the civil unrest in another part of the city, Korea town, which began to be broadcast a few hours after the verdicts were announced proved to be a complex, polysemous text. In Los Angeles the members of different communities (for instance, Simi Valley, 'the three Bs': Beverly Hills, Brentwood and Bel Air, and South Central) and the viewers – local, national, and international – were divided along racial and class lines. There was a shift from the fairly unified reading of the video-taped beating to how different audiences read (alternatively): the L.A. Uprising, or the L.A. Rebellion, or the L.A. Riots. Accompanied by Rodney King's press conference plea for peace, the televisuality of the few days of civil unrest performed a kind of 'perverse ventriloquism' swapping one racialized story (of victimization) for another (of victimizing).[16] Audience realignments settled.

Coming two years later, audience alignment in the O.J. Simpson trial, preceded by the live-on-TV slow-speed car chase, was more blatantly polarized. The vast majority of white Americans thought that 'he did it', that he committed the double-murder of his ex-wife and her friend, while the vast majority of African Americans hoped for a 'not guilty' verdict. The courtroom drama was a television program of its own, ripe with colorful characters, a murder mystery plot, a romance gone awry, and the question about whether 'race mattered'. Mr Simpson was on trial, but so was the justice system. The case, and the mediated event, cannot be seen apart from the case of the Rodney King beating and the 'travesty of justice' in the (first) case against the Los Angeles Police Department. (Like the officers, O.J. Simpson was also tried a second time, and found culpable.) Simpson was a formidable character, as a Black man ('who became a whole lot darker' argues Todd Boyd)[17], a beloved sports icon and a wealthy individual who had 'crossed over' into white society, divorcing his African American first wife from Potrero Hill and taking a second wife, a European American younger, blonde beauty whom he lived with in Brentwood. The story of interracial marriage was bothersome to both whites and African Americans, and the fear of miscegenation was not too far under the surface of the trial.

Trial coverage ultimately revealed the fervor with which many whites wanted O.J. Simpson to be found guilty and punished, as well as the impassioned hope many non-whites had that this one time a Black man would be spared in the *criminal* justice system, one which had discriminated against men of color since its founding. Identification, then, was not with a particular individual (Simpson for instance), but rather, with a political position (conscious or unconscious) about race, class and justice. While there had been high-profile legal cases within popular culture before, never before the O.J. Simpson trial had there been such a mediated, public engagement with tough questions about race. Televisual discourses merged with racial discourses, and racial(ized) representations (of Black men, white women, wealth) were demonstrative of race relations in the social world adjacent to, even intermingled with, the world of television.

Bill Cosby; Oprah Winfrey

There are two important figures in the history of television, whose popularity demonstrates or mythifies race relations in a way that reconfigures race (blackness and otherness) as not-a problem and as not 'mattering'. They both are recognized as people of color and are held up by African American communities with pride. Their cases involve a question of how they amassed such unparalleled status and global success as minority stars. How have both Bill Cosby and Oprah Winfrey negotiated and enabled pleasure and identification by multiple, pan-ethnic, international audiences? How do they signify and de-signify Blackness as the same time?[18]

A near opposite of O.J. Simpson in style and temperament (though both amassed wealth in the 1980s) is actor and comedian, Bill Cosby. Though he appeared steadily in a number of popular television series beginning in the 1960s and was a Grammy-award-winning stand-up comic during the 1970s, Cosby's pinnacle of fame came from his carefully crafted performance as Dr Cliff Huxtable in the world-wide hit, *The Cosby Show* (1984–1992).[19] His role functioned inter- and extra-textually, endearing Huxtable/Cosby as everyone's ideal father; fatherhood was elevated to race-less status.[20] And within the discourse of Reaganism, blackness as class could be shed through hard work and blackness as ethnicity could be worn fashionably and acceptably in the nice-looking, family values-driven Brooklyn brownstone where the Huxtables dwelled. Not unlike the raced viewing of *Julia* 30 years earlier, *The Cosby Show* was loved by a matrix of viewers, for example, by black South Africans who were proud to see themselves reflected in characters who achieved high professional status and were treated with regard, and by white American conservatives who liked to see this 'middle-class' Black family as proof of a meritocracy (and as a case against affirmative action policies). The dialectic of sameness and difference compelled viewership, as did the strategic writing and skillful performances.[21]

Ms Winfrey likewise has navigated her career as a journalist strategically and skillfully. Finding and growing her niche in the talk show format with a compassionate feminism, her audience of Black and white women proved to be loyal, strong and expanding. While early in her position on *The Oprah Winfrey Show* (1986–2011) there was a way in which she figured as a comforter (some called Mammy-like) and less as a provocateur, and although her universe was made up of many suburban moms, Oprah Winfrey nevertheless exhibited power in her inimitable 25-year-run. So Amazonian is she, that through her production company, Harpo Productions Inc., Winfrey is involved in film production, television production, radio and print, and in 2011 she launched her own network, OWN.[22] She has been credited with championing a then lesser-known senator from the state of Illinois, Barak Obama; it has also been suggested that Winfrey run for elected office, her media identity is so deep and wide reaching.

On the one hand, Winfrey embodies the 'despite-all-odds' grit of a poor Black girl originally from the American South, and on the other, her professional success has reached the stratosphere, beyond even the accomplishments of most privileged, white males. Unlike someone like Rupert Murdoch and his dynastic empire, however, Oprah Winfrey operates as a different kind of symbol of success, one in which her gender and race are pronounced – projected upon and through her body. It is a body not of a mother or wife, but of a singular, exceptional individual who feels accessible and familiar.[23] Unlike Cosby, though, whose presence was simultaneously a signifier of race and a disavowal of racial difference, Winfrey maintains race as a marker of difference – whether as topics of discussion, subjects in her telefilms, images in her magazine, or organizing frameworks for the programming on OWN.

Moving into the next era of media production-consumption, Winfrey invites viewers to come as they are –diverse, self-appointed and self-defined – though they are identified as part of a community, a club. As a women's interest, minority-run, cable network (co-owned by Discovery), OWN is an experiment in programming, marketing, hailing; it stumbled in the beginning, with a significant financial loss in its first year. In 2013 Winfrey partnered with independent producer-writer-director-performer and personal friend, Tyler Perry, and his original series brought OWN high ratings.[24] Pooling their talents, ambition and followers, it is a potent partnership and intriguing new business model. On the face of it, programming on OWN appears more Black – with African American interview subjects and original new series featuring African American lead characters and predominantly Black casts. As of August 2013, The Oprah Winfrey Network was in 73% of US homes with televisions. In August 2014, OWN launched a digital series with YouTube Space, called *Who Am I*. Women of color in the media scheduled to appear include: celebrity personality, Nicole Ritchie, make-up artist and international cosmetics entrepreneur, Michelle Phan, and actress/singer, Brandy. From the syndicated broadcast of her talk show spanning rural-urban-suburban settings, to the narrow(er)-casting of her network, Winfrey signals a new kind of racing of audiences. She is testing her 'universal' appeal on new grounds, and with more demonstrably raced material. I think her expectation is that audiences will race themselves – identify – differently than alignment according to one's (same) race. Perhaps she, like I do, hopes that viewers can break from traditional patterns of racial identification.

CONCLUSION: A DIFFERENCE IN VISIBILITY?

There are other (digital, on-line) platforms and circuits of US television portrayals of race relations, and there are also alternative models that come back to interrupt or destabilize a US view. Raced audiences have been an effect and expression of US television's racialization process. This project is reliant on a logic of representation that produces identification along racial lines. But there is a limit to the US version and view of race relations.

Beyond the US broadcast borders, the new wave and way of watching television, on-demand or on-line, translated, adapted or dubbed, can deter and obscure the original racialized orientation of a text. The phenomenon of remakes of reality television formats (for example, _____ *Idol* or _____*'s Got Talent*) in settings where the entire cast of contestants, judges and studio audience is monoracial defies the often, racialized subtext in the US versions. International viewers do not necessarily subscribe to the politics of representing race relations as harmonious or dichotomous; in their own cultural production, ethnic, racial or national difference may be more frankly stated in class terms.

Domestically, production norms have shifted along with the structure of the network system, the regulatory environment, and sponsorship now aiming at reaching different market segments. Quoting scholar Herman Gray:

> In the initial phases of the network system, questions of racial representation were addressed in terms of assembling and integrating audiences around the logic of sameness and the lore of least objectionable programming. Hence, racing and raced audiences were structured around the normative ideal of whiteness as the universal subject of address; with the shift to narrowcasting, streaming, and other audience centered pull technologies, the issue of racing and raced audiences appears to cater to and cultivate differences, except in those cases of national address, crisis, and event based television.

In other words, visibility of racial difference is being produced in the space of the niche. Raced audiences are finding 'homes' in disparate channels, shows, characters, contexts and fan cultures. Whether pleasure comes from disidentifcation or alignment, the project of racing audiences remains but the placement of viewers has become more self-directed.

So what do the disparate offerings and choices in the televisual landscape say about the direction of race and representation? One approach is Oprah Winfrey's: acknowledging raced audiences but mixing it up in terms of *not* presuming raced tastes. The question is: what kinds of representation of race relations are there in niche-programming? And further, can audiences identify differently in the space of the niche? Are interracial relationships and multi-racial families, for example, more likely to be visible in narrowcast or broadcast (or elsewhere), and would such discourses of race make a difference? In other words, is it possible to birth a new logic of representation?

ACKNOWLEDGEMENT

My thanks to Dr. Herman Gray for his guiding comments in the preparation of this chapter.

NOTES

1 The network era can be understood as the late 1950s through the early 1990s in which 'adjustments' were made, as Marlon Riggs describes them, in what I call, a racial economy.
2 For example, feminist critics asking, 'which of the characters in *Sex and the City* do you identify with?' Or television executives arguing the difficulty of featuring an Asian American male lead character because 'audiences won't be able to identify with him'.
3 After all, what have I, or any person of color been doing most of our lives?
4 The National Advisory Commission on Civil Disorders, known as the Kerner Commission after its chair, Governor Otto Kerner, Jr. of Illinois, was an 11-member commission tasked by President Lyndon B. Johnson to investigate the causes of the 1967 riots in cities across the US including Detroit, Los Angeles, Chicago and Newark. A report was released seven months later although the recommendations were rejected by the president (who helped pass the Civil Rights Act in 1964 and the Voting Rights Act in 1965); the 400-page document became a best-seller among 2 million Americans.
5 For an interpretation of Barack Obama's mobilization of race as a tool, see Beth Coleman's analysis of his campaign speech about the 'perfectability of race' (2009).
6 As for Asian Americans, director Justin Lin (of the landmark independent film, *Better Luck Tomorrow* (2003) and of the highly popular and financially profitable *The Fast and the Furious* franchise) shared an anecdote: at a pitch meeting, the question of raced viewers came up and Lin was told that Asian Americans are not included on the pie chart because they 'count as white'.
7 The documentary, *Brown is the New Green: George Lopez and the American Dream* (2011), provides a useful study on race and marketing and peek at George Lopez's personal strategies related to story development in his prime-time series (which reached the coveted 100th episode). Lopez's performative strategies take a turn after *The George Lopez Show*, in the form of his late-night talk show on which he opened his stage to numerous minority artists, as well as his stand-up performances in which the address of race (his own identity and the identification of those in the audience) is more direct and far less tame than on the ABC-into-syndication sitcom. *The George Lopez Show* on ABC led the way for *Ugly Betty*, adapted from a Latin American telenovela, premiered on ABC in 2006 and released in about a dozen other countries in Europe, plus Japan.
8 Whether accurate or not, this famous sentiment has been attributed to Ms. McDaniel, and carries

into myth – rationalizing or justifying when actors of color take on unflattering or problematic roles.

9 I adapt Patricia Mellencamp's concept of winning performatively whilst losing narratively, which she uses to describe Lucille Ball in her essay, 'Situation Comedy, Feminism, and Freud: Discourses of Gracie and Lucy' (1996).

10 Darnell Hunt (2004), after Stuart Hall.

11 The colloquial term is a description or criticism of an individual as Black (skinned) on the outside, but (acting/feeling) white on the inside.

12 DJ Jazzy Jeff, a real-life collaborator and friend of Will Smith's from Philadelphia, appeared in the first four seasons of *The Fresh Prince of Bel Air*. His performance was a foil for Will's acclimatization to life in Bel Air; Jazzy Jeff's character lives in South Central, maintaining Will's ties/sense of authenticity to a certain kind of Blackness, with the convenience of having a room in a mansion staffed with a butler.

13 See John Fiske's chapter, 'Hearing Anita Hill and Seeing Bill Cosby', in which he discusses the notion of tame(d) Blacks versus untamed Blacks, within the context of what Herman Gray would call the 'Sign of Blackness'.

14 See Christine Acham (2005) chapter on *The Richard Pryor Show*, a case study of a stand-up comedian who refused to capitulate to network standards, literal and philosophical.

15 See John Caldwell (1995), especially Chapter, 11, 'Televisual Politics: Negotiating Race in the L.A. Rebellion'.

16 This is a term used by David Palumbo-Liu (1994) in his analysis of the use of the press image of the vigilante Korean store owner holding a gun to protect his property against 'looters' in 'Los Angeles, Asians, and Perverse Ventriloquisms: On the Functions of Asian America in the Recent American Imaginary'.

17 Commentary from *O.J.: A Study in Black and White* (1992), an HBO Sports documentary.

18 See works by Sut Jhally and Justin Lewis, and Timothy Havens (2004).

19 The series' ending came on the 2nd night of the L.A. Uprising in April 1992. This television event was invoked by another television event – live news reports of the civil unrest, accompanied by a plea from Cosby, himself, to stay home and watch the show (rather than go out and 'riot'). *The Cosby Show* represented racial harmony as well as upward mobility as appropriate and available. Its message of meritocracy was nearly the inverse of the story unfolding amidst what some called a rebellion against both social and economic inequality.

20 The elevation of Cosby 'above race' in his public life, met with the utmost tragedy in his private experience, with the racially motivated murder of his beloved son, Ennis in 1997.

21 Another key figure in contemporary culture is Dave Chappelle. See Bambi Haggins (2009).

22 OWN, The Oprah Winfrey Network.

23 There are stories about Oprah Winfrey's extreme wealth (and generosity), along with stories about how she is turned away when unrecognized and shopping in Europe because she is Black.

24 Tyler Perry created, writes, directs, and executive produces, *The Haves and the Have Nots*, for OWN. It is set in Savannah, Georgia and follows the (African American) Cryer Family living an opulent lifestyle amidst their household staff. The second season finale in March 2014, broke network records, ranking no. 1 with 3.6 million viewers; furthermore, it ranked no. 1 among cable television networks and no. 4 for all of television for its timeslot that evening. The season finale saw an increase of 38% in total viewers, and of 24% in the target demographic (women, 25-54), according to THE WRAP, "Tyler Perry's 'Have and the Have Nots' Finale is OWN's Most-Watched Program Ever," Jethro Nededog (March 12, 2014).

REFERENCES

Acham, Christine (2005). *Revolution Televised: Prime Time and the Struggle for Black Power*. Minneapolis: University of Minnesota Press.

Caldwell, John (1995). *Televisuality: Style, Crisis, and Authority in American Television*. New Jersey: Rutgers University Press.

Coleman, Beth (2009). Race as Technology. *Camera Obscura* 70, Volume 24, Number 1. Duke University Press.

Hunt, Darnell (2004). *Channeling Blackness: Studies on Television and Race in America*. New York: Oxford University Press.

Fiske, John (1996). *Media Matters: Race and Gender in U.S. Politics*. Minneapolis: University of Minnesota Press.

Gray, Herman (2004). *Watching Race: Television and the Struggle for Blackness*. Minneapolis: University of Minnesota Press.

Haggins, Bambi (2009). In the Wake of 'The Nigger Pixie': Dave Chappelle and the Politics of Crossover Comedy. *Satire TV: Politics and Comedy in the Post-Network Era*. Jonathan Gray, Jeffrey P. Jones, and Ethan Thompson, eds. New York: NYU Press.

Havens, Timothy (2004). 'The Biggest Show in the World' Race and the Global Popularity of *The Cosby Show. The Television Studies Reader*, Robert C. Allen and Annette Hill, eds. London: Routledge.

Jhally, Sut and Justin Lewis (1992). *Enlightened Racism: The Cosby Show, Audiences, and the Myth of the American Dream*. Westview Press.

The Kerner Commission Report (1968). *The 1968 Report of the National Advisory* Commission on Civil Disorders. New York: Pantheon.

Leong, Russell (ed.) (1992). *Moving the Image: Independent Asian Pacific American Media Arts*. Los Angeles: UCLA Asian American Studies Center Press.

Mellencamp, Patricia (1996). Situation Comedy, Feminism, and Freud: Discourses of Gracie and Lucy, in Tania Modleski ed., *Studies in Entertainment: Critical Approaches to Mass Media*. Bloomington: University of Indiana.

Palumbo-Liu, David (1994). Los Angeles, Asians, and Perverse Ventriloquisms: On the Functions of Asian America in the Recent American Imaginary. *Public Culture*, 6: 365–381.

Riggs, Marlon (1991). *Color Adjustment* Distributed by California Newsreel.

28

Classed Audiences in the Age of Neoliberal Capitalism

Mike Wayne

INTRODUCTION

This chapter is primarily concerned with the way class shapes the consumption of television programmes, that is to say the way the class position of audience members predisposes them to consume programmes of a particular kind and in particular ways. This question has been far less popular with academic researchers studying audiences, who have preferred to explore gender, ethnicity, age and nationality as determinants for the way audiences consume symbolic goods (such as television programmes). The relative lack of interest in class by academics and other social researchers exploring cultural consumption is indicative of the entanglement of class as a category of analysis, in real-world politics and real-world struggles. Those struggles and politics go to the heart of what kind of society capitalist society is. Class struggle is uncomfortable for many, and is unwelcome in academia where it is frequently dismissed as a crude and reductive mode of analysis, out of date and incompatible with the various forms of methodological individualism that flourish there (see Skeggs 2004: 46–54).

In America, class analysis and class terms in the popular vernacular are to a very great extent, taboo. The invisibility of class in the public culture and the power of capital always go together, because the former is highly convenient to the latter. Resnick and Wolff suggest that the relatively impoverished nature of class consciousness in America is due to the fact that very high rates of exploitation have been historically coupled with high rates of *individual private consumption*, which has partly compensated for the problems and displaced the pain of such exploitation (Resnick and Wolff 2003). High rates of private consumption are in turn oiled by a massive advertising industry (which in turn funds the majority of media outlets in America) committed to selling the capitalist utopia that commodities equal happiness (Ewen and Ewen 2000). This social and historical context, as I shall briefly indicate later, has implications for how class determines the readings of popular television programmes by audiences in America. One aspect of the power of capital and the invisibility of class as an acknowledged fact of life within the US is the over-representation of the affluent and the wealthy in primetime television shows (compared to their real numeracy in the US population) and the corresponding under-representation of working-class characters in fictional shows (Holtzman 2000: 123).

In Europe and Latin America, class ways of looking at the world have been more successfully embedded into the cultures of what Gramsci called the subaltern classes. The relative independence and plurality of workers and peasant groups from both state and commodity power in Latin America, especially with the rise of the social movements in the late 1980s and 1990s, provided the essential background against which Martín-Barbero formulated his important distinction between popular culture and mass culture (Martín-Barbero 1993). For Martín-Barbero, popular culture (grounded in the everyday culture of the popular classes) and mass culture (the culture produced by the capitalist media organizations) have a complex relationship in which each side appropriates, re-contextualizes, borrows and adapts the cultural materials generated by the other. The popular is not a pure folk culture uncontaminated by mass culture and mass culture is not simply pure ideology manufactured only according to the logic of capital. Martín-Barbero's reformulation of the dynamic relationship between mass and popular culture has helped open up new approaches to the study of audiences, especially in the context of Latin America (Berry 2006). However, as we shall see later, Martín-Barbero's formulation of the mass culture/popular culture dynamic has a more general relevance.

Within Europe class achieved official recognition in the establishment of welfare states, which implicitly recognized the need to regulate the market and provide social goods for the working class that the market could not provide on a universal basis. However, within the UK, the dominant class, the class that represents the interests of capitalism, has been on the offensive for some thirty years. A new form of capitalism, neo-liberal capitalism, has been constructed, substantially dismantling the social-democratic structures of the past. In the process, working-class organizations, solidarities and consciousness have been broken up to a considerable extent, moving the UK closer to the United States in terms of the power of the dominant class and the unleashing of a culture of capitalism and the formation of neo-liberal subjectivities. We will find traces of this history in the way audiences relate to television news programmes within the UK.

In what follows I will ask five important questions that are pertinent to the issues of classed audiences and their relationships with television programmes. This will allow me to review some of the literature already existing on classed modes of consumption. In the second part of the chapter I will present and interpret focus group data from a research project I have been involved in that looked at the consumption of television news about politics in the UK. I situate the evidence of classed consumption within the context of the shift away from social democracy towards neo-liberal capitalism.

FIVE QUESTIONS ABOUT CLASS AND TELEVISION AUDIENCES

1 To What Extent Do Classed Audiences Consume Different Television Programmes?

The *locus classicus* of an approach that sees differential cultural consumption or appropriation as linked to class (and occupational position), is Pierre Bourdieu's book *Distinction* (1996). Here Bourdieu attacked the notion that taste in the field of culture was the product of individual sensibility and choice. Writing from a position somewhere between Weber and Marx, Bourdieu linked taste for cultural goods systematically to class belonging and class conditions. Class subjectivities are above all cultural formations and indicate the individual internalization of social conditions in the way people talk, their bodily movements and gestures, the clothes they wear, the ambitions they have, the friends they hang out with, the language they use, and many other ways of being in the world. Bourdieu has helped us understand the internalization of social conditions that then shapes an individual's ongoing and continual engagement with the world. He calls this mental construct or 'practice-unifying and practice-generating principle' the habitus (Bourdieu 1996: 101). The habitus is a product of internalizing social structures but it then provides a flexible guide for regulating the individual's encounters with a range of situations and institutions, including watching television programmes or engaging (or not) with political institutions and practices.

The habitus is structured around class differentiated dispositions and competences. Dispositions refer to distinct clusters of attitudes and actions that orientate the individual to seek out particular kinds of experiences. However, as we shall see, not all (cultural) experiences are exclusive, so the concept of dispositions must also refer to the distinct attitudinal positionings class subjects have in relation to cultural experiences that are shared or participated in by other classes. Competences refer to the socially conditioned resources individuals can mobilize to decode or interpret their experiences in various social encounters, whether that be a television programme or in relation to political institutions. Competences are acquired and passed on through the family, via interaction with peers, through school and formally accredited education certificates, and via cultural goods themselves. Competences are thus made up of 'assets' of various kinds – social (i.e. family and friends), educational, cultural and so forth – that help, Bourdieu argues, reproduce a class divided society because competences are unequally distributed.

Now, when we look at television we can certainly find programmes that have audience profiles that are skewed strongly towards one or other segment of the audience. A quiz show such as *University Challenge*, for example, is simply not going to be of much interest for most lower-middle-class and working-class audiences. Conversely, a quiz show like *Wheel of Fortune* is unlikely to attract many middle-class viewers, and, as with *University Challenge*, its assumed audience is reflected at every level of the programme (see Skovmand (1992: 91–92) who defines *Wheel of Fortune's* audience as lower middle class and female). Certainly multi-channel environments (cable and satellite) make it more possible for broadcasters to pitch their programmes towards the dispositions of particular classes and class fractions. One thinks of BSkyB's advertising push in the 1990s promoting their educational and cultural channels in a bid to expand their customer base into the middle class and rebrand satellite television away from being seen as 'council house television' (Hunt 1993).

However, because it is also a medium that is virtually universally accessible (in a way that theatre or opera or ballet are not) and still retains a sense of occupying a 'national' cultural space, television is also well placed to make programmes with more mixed audience profiles. A quiz show like *Who Wants To Be A Millionaire* for example, (in its UK production) deliberately set out to craft itself into the kind of product that would appeal to the dispositions of both middle- and working-class audiences and mobilize competences (in terms of the knowledge required to answer the questions) of both classes (see Wayne 2000). Of course class is not the only means by which broadcasters and advertisers try and construct a relationship between programmes and their audiences. Lifestyle categories, which have become popular from the 1980s onwards, but which were introduced back in the 1960s, try and construct audiences, around their leisure time activities, cultural tastes, attitudes and opinions (Plummer 1974). Nevertheless, such modes of 'market segmentation' often still map directly onto distinct class groupings. Irrespective of whether broadcasters relate to their audiences explicitly in class terms or what combination of classes are attracted to particular cultural experiences, class remains an active determinant in who consumes and what *sense* they make of the cultural goods or experiences on offer.

2 To What Extent Do Classed Audiences Consume the Same Programmes?

Perhaps though, a convergence between different classes on the same objects of consumption is not just a question of the specific quality of the medium of television, but a result of changing trends in class-defined tastes since the time of Bourdieu's research for *Distinction*? Chan and Goldthorpe argue that cultural consumption across a range of cultural goods does not display the kinds of sharp mutually exclusive patterns of taste correlating to class position that Bourdieu found. Instead they argue that the middle class consumers are characterized more as 'omnivores', i.e. grazing widely across the cultural landscape, while the working classes continue to consume within a narrower field of cultural goods (they are 'univores'). So that, for example, very few working-class people will go to the ballet, or even see a play, but they will go to the cinema. Conversely, middle-class consumers are more likely to do both (Chan and Goldthorpe 2005: 197–199). This indicates that perhaps what was historically true when Bourdieu conducted his work, has broken down, but only because the middle class have developed the dispositions (the socially internalized orientation towards certain kinds of cultural goods) to move into and appropriate popular cultural goods (something reflected in the rise of film studies within academia).

However, the class structure remains as restricted as it ever was for working-class cultural consumption (and the very terms 'omnivores' and 'univores' works to play down the class nature of the divisions under discussion). Moreover, once we move into qualitative analysis of what cultural goods mean to people (their signification), Bourdieu's categories still seem very relevant. While middle-class audiences may consume the same television programmes as working-class audiences, the evidence suggests that they do so in particular ways that indicate continuing struggles around 'distinction' and the mobilization of different forms of cultural assets. Thus Skeggs, Thumin and Wood find that middle-class women watching reality television shows such as *Wife Swap* and *Supernanny* do so with a degree of distance and superiority, filtering out or critiquing the most 'disreputable' aspects of the programmes (the fact that they are 'reality TV' shows) and citing their educational possibilities (2008: 11). By contrast, working-class women consume the same programmes explicitly for their entertainment value. Working-class women also, in the act of watching the programmes, have a more direct involvement with the characters and situations, mobilizing moral judgements around the particular situations on screen, while the middle-class women tend to distance themselves from the immediate narrative action and place moral judgments in the context of wider public debates (Skeggs, Thumin and Wood, 2008: 14–16). This capacity for a certain kind of 'abstraction', the ability to express in verbal language, structures, processes and complex interrelationships, is typical of the sort of competences the middle class can mobilize and is dependent on levels of education (both formal and informal) and mental training not routinely accessible to the working class (see Lawton 1970; Bernstein 1979) because school systems function to reproduce class divisions by preparing people for the existing divisions of labour.

3 Do Classed Audiences Consume the Same Programmes in the Same Way or in Different Ways?

We have already begun to suggest that the different competences and cultural assets possessed by differently classed audiences impacts on how audiences consume the same programmes in different ways. Bourdieu's category of cultural assets or 'capital' (a rather unhelpful term that has little relation to Marx's understanding of capital) tends to focus on the form of consumption (e.g. distant, superior, abstract for the middle class or direct, affective, immediately gratifying for the working class). But we also need to explore how class determines the production of specific meanings by audiences. Andrea L. Press's interview-based research into how American women from different classes and ages consume the same programmes in different ways is helpful here. Her research finds class differences in how the women respond to the 'realism' of fictional programmes. Press finds that the middle-class women filter their evident enjoyment of the shows through a 'high culture' optic of disdain, cynicism and rejection of their 'realist' credentials. Working-class women by contrast make a television show's supposed 'realism' an important part of their evaluation of whether they like the show or not. Their focus is less on the characters of the shows, in contrast with the middle-class women, and more on what Press calls 'the ambience' of the shows. This refers to plot, physical setting, moral issues and lessons learned. However, in an example of the domination of middle-class cultural values (what the Italian Marxist Antonio Gramsci called the struggle for hegemony or moral and intellectual leadership), Press finds many working-class women equating 'realism' with suburban middle-class life, as depicted for example in the 1960s comedy *Bewitched* or the 1980s *Cosby Show*. There is a strong aspirational component in the responses of many working-class women who want their families to live affluent and comfortable lives like the ones shown on television. In fact, the responses she gathers are poignant illustrations of the way the mass media images of success offer consolations and wish fulfilments that are unlikely to be translated into real-world changes for the audience members concerned (Press 1991: 108).

The emphasis of the working-class respondents on the *mise-en-scène* of middle-class life (the ambience of the show) indicates not a more critical reading, but an interest in the signifiers that constitute and construct this apparently alluring but unfamiliar lifestyle. The working-class women identify less strongly with television's construction of what it means to be a middle-class women, e.g. glamorous, ambitious and self-orientated. By contrast, middle-class women, who can take the 'ambience' of middle-class life for granted (because they live it), are much more focused on the middle-class female characters that dominate American television. In a programme such as *Dynasty*, the middle-class women tend to be drawn towards a character like Alexis who can be read within the terms of an individualistic pro-capitalist version of feminism. They were, by contrast, very disparaging towards Krystle, who was seen as lacking ambition. Conversely, the working-class women identified more with Krystle who could be integrated into their experience of working-class female life (caring for other people, husbands, children and family) while they thought Alexis was far too selfish to identify with.

4 To What Extent Do Classed Audiences Demonstrate an Awareness of Class-Consciousness in their Consumption of Television Programmes?

We have seen above that class makes a difference to both what audiences consume and how they consume television programmes. The *fact* of class, in itself, differentiates the way audiences appropriate cultural goods and integrate them into their lives, because their lives *are different*. Recalling Marx's distinction between a class in itself and a class for itself, we can make a distinction between attitudinal clusters in the consumption of cultural goods indicated above, and differences which reveal a critical consciousness of class difference in the consumption of cultural goods. Andrea Press found such class-consciousness amongst her working-class respondents significantly absent in her American study. She notes that this is in contrast to the audience research that has been carried out in the context of British Cultural Studies.

The seminal work here is David Morley's study of the 1970s UK television current affairs magazine programme *Nationwide*. This study was conducted as part of the Centre for Cultural Studies at Birmingham University, which played a key role in putting class (briefly) on the agenda of cultural analysis in the 1970s. Adapting the work of Frank Parkin on the relationship between meaning systems and class, and influenced by Gramsci's notion of the struggle for moral and intellectual leadership, the Centre for Cultural Studies developed an encoding-decoding model to think about the relationship between texts and their reception. A text encoded within the dominant institutions and meaning systems of a social order constructs a *preferred* meaning to the materials it organizes that is congruent with that social order. One option for viewers is to work primarily within the limits to meaning constructed in and by this preferred reading. A second option is to provide a 'negotiated' reading of the preferred reading, modifying aspects of its message and meaning. Morley explains the basis of a 'negotiated reading' thus:

> The thesis of these writers [e.g. Parkin and Mann] is that working-class consciousness is often characterized by an 'acceptance' of dominant ideological frameworks at an abstract level, combined with a tendency at a concrete, situated level to modify and re-interpret the abstractly dominant frameworks in line with localized meaning-systems erected on the basis of specific social experiences. (Morley 1992: 99)

In the absence of generating their *own* abstractions about the social order (an opportunity systematically denied by the education system), the working class utilize the abstractions of the dominant order to try and make sense of their lives. However, precisely because there is a rift between these abstractions and their own social being, a space is opened up, potentially, for reworking the dominant abstractions (e.g. law and order, wealth, justice, democracy, and so forth) according to other social experiences or 'localized meanings systems'. These social experiences might be direct lived experience, conversations with friends and other 'opinion-formers', other types of media consumption, and so forth. Morley found examples of such negotiated readings among the different audience groups he studied, such as trade union officials and university students. By contrast, Andrea Press found that American working-class women were rather more integrated into the preferred meaning and value system of the social order, even though on a *different* basis than the middle class women she interviewed. Press found that in the American context, the power of the mass media was graphically demonstrated when the clash between experience and images was resolved by viewers opting to believe the images over their own experience (1991: 111)! In other words, one potential corrective to aspects of the meanings produced by television, i.e. direct experience of empirically observable reality, was less compelling to the respondents than the wish-images (of increased private consumption) offered by mass mediated products. Although these studies are not comparing like with like (Press is concerned with fiction, Morley with news), the contrast is suggestive of the different national contexts and histories of the class struggle. It suggests that there is no natural recourse to (re)evaluating dominant meaning systems in the light of direct experience which always requires, as Morley notes, access to particular discourses whose production and circulation (availability) are historically and socially determined by political struggles (Morley 1992: 90).

Access to particular discourses is of course a product of where one is in the class structure, but also the historical context of the class struggle itself, where discourses are generated and disseminated. The third option for decoding is that of the *oppositional* reading.

This reading deconstructs the assumptions of the preferred reading, exposes its contradictions and strategic silences, and formulates its alternative understanding and response in the context of a coherently, if (often) implicitly, articulated value system that is in contradiction with the dominant social order. Thus Morley found trade union officials were more likely to offer negotiated readings of *Nationwide*, while working-class apprentices – perhaps because they had less experience of union activity – were more likely to recycle the preferred interpretations of events in line with the programme. From his sample, Morley found that Black predominantly female working class FE students and male and white trade union activists on the 'shop floor' were more likely to produce oppositional readings of *Nationwide*. Obviously then, oppositional readings are indicative of working-class class-consciousness, but due to class domination (hegemony) and the division of labour, with its strong fragmentation effect on consciousness and commodity fetishism (which conceals power relations), there can be no question of simply linking oppositional readings automatically to the working class en bloc or at all.

5 With What Effect Does Class Intersect with Other Social Identities, Such As Gender, Ethnicity and Age?

There is not space here to explore this question in any detail, but we can very briefly draw out some of the implications of this question in what has already been said. We have seen that there is some evidence to indicate, for example, that the intersection of class with national history and identity is important to the way audiences interact with television programmes. Within national-political states, even regional differences can be significant, cutting across the same class/occupational groups. Philo found middle-class respondents in the North of England and Scotland significantly more sceptical of the media reporting of the Miners' Strike (1984–85) than their counterparts in the South of England (Philo 1990: 55–69). Drawing on existing studies, we have also seen how women of different classes have consumed the same programmes in different ways. The case study I discuss below constructed its interpretation of audience responses along the lines of class and age, and some of the points of convergence between people of different classes is likely to be due to the shared determinant of what it means to be a young person in contemporary society. At the same time, significant differences emerge in audience responses to news that are the result of class differentiated dispositions and competences.

THE CONSUMPTION OF NEWS ABOUT POLITICS ON UK TELEVISION

The following case study of how young people from working- and middle-class backgrounds relate to news about politics on television, forms part of a broader study into the crisis of television news and politics in the UK (see Wayne, Petley, Murray and Henderson 2010). The broader context is that neo-liberal capitalism has hollowed out representative politics in the UK in much the same way as it has in the United States. The political class are so thoroughly integrated into a neo-liberal pro-big business agenda that beneath the surface of political life, there is considerable disconnection and withdrawal from mainstream party politics, across the classes. This is reflected in many ways, such as the declining participation in voting, the decline of parties with mass membership, and of course in the ageing profile of audiences watching television news, where news about domestic politics is the single biggest item. Young people are at the cutting edge of these trends of disconnection.

As part of the study, we conducted 12 focus groups (a total of 80 people) with young people aged 16–24 (but with more than half under 20). The gender mix was reasonably even (male = 46; female = 34) and the ethnic mix reflected the London and South-East locations for the interviews (White = 47; Black = 15; Asian = 13; Mixed race = 4; Hispanic = 1). We interviewed young people in small 'naturally occurring' groups, which is to say that each focus group was composed of people who knew each other, thus opening up the possibility of analysing the jokes, banter and anecdotes that people who are comfortable with each other can produce in an otherwise rather artificial situation. The focus groups were recorded and fully transcribed. Because the groups were naturally occurring and because of where the interviews were convened (in schools, youth groups, universities or, in one case, a prison) each focus group had a clear class profile: either working class or middle class.

The focus groups followed the same format in each case. After filling in a questionnaire, the young people broke into small groups and worked on an exercise using photographs taken from the television news. The groups were asked to put a selection of photographs together into a sequence that would form the visual track of the kind of television news story about young people that they thought was typical of the genre. The exercise provided evidence of both how literate the participants were in the language of television news and to what extent they internalized its norms and values. After discussing with each sub-group why they chose the pictures they did, the moderator asked a series of questions about how young people are represented on television news, what they think about politics and its portrayal on the news and how they themselves might make the news more interesting for young people if they were editors in charge of making the news.

HABITUS: HORIZONTAL AND VERTICAL RELATIONS

We have seen that Bourdieu's concepts of dispositions and competences are useful tools for thinking about how class subjects are formed with very different kinds of habitus. I want to argue that class-determined differences in dispositions and competences between the working-class and middle-class respondents will be demonstrated below both in relation to the world of politics and the world of television. We shall see that the classes have different attitudinal clusters (dispositions) in the way they engage with both the world of politics and the world of television as it mediates the political sphere. Similarly, class differences structure the competences that each class can mobilize in their 'decoding' of both television and the world of politics. We shall see that the middle-class respondents demonstrate, in general, considerably more skills in television literacy and in *empirical* knowledge of political processes than the working-class respondents. The latter, again, in general, demonstrate some relatively significant gaps in knowledge both about how television news works as a set of conventions (to represent politics) and about how politics works at Westminster. However, we shall also see that there are examples from the working-class respondents that show a degree of penetrative insight into the nature of *power* and *social interests* which are potentially more radical and critical than anything the middle-class respondents can generally muster.

We can differentiate the relations between the middle- and working-class respondents to the dominant institutions of both politics and television metaphorically in the following terms. Both the middle-class and working-class young people articulate a felt sense of distance from these institutions, a sense that the institutions of television news and politics are remote and unresponsive to them as young people and that they are dissatisfied with various aspects of their modes of representation (whether cultural in the case of television, or political in the case of Westminster). This distance has been exacerbated by the development of neo-liberalism, which neither wants nor needs an involved and participatory citizenship since its key category is the consumer. However, this distance manifests itself differently along class lines. For the middle-class respondents this distance takes place on what we might call a *horizontal* plane, while for the working-class respondents, this distance manifests itself along a *vertical* plane. This is to say that for the middle-class respondents, while there is dissatisfaction in the way these institutions relate to them and represent them, there is no sense that these institutions are actively hostile to their interests. By contrast, the working-class respondents articulate a sense of a huge gulf between them and the dominant institutions. This gulf means that for the working-class respondents there is a strong sense that their experiences, not as young people but as young *working-class people*, are unrecognized by the dominant institutions. The dominant institutions know little of their lives, and further, at least in some instances, there is a sense that the dominant institutions represent interests that are actively hostile and antithetical to their own social interests.

COMPETENCES, POLITICS AND TELEVISION

> **Moderator:** Think about the way ... when you see politics on TV news, do you think the way that TV news covers politics is interesting?
> **Girl:** No!
> **Girl:** No!
> **Girl:** It's boring.
> **Girl:** They all sit in a room and bark at each other.
> **Male 4:** They show you that room with the green chairs and they're all like...
> (State school, London, working-class respondents)

The phrase 'that room with the green chairs' is a little chilling. It demonstrates a very limited capacity (or competence) for recognition – the 'room' (the House of Commons) does not even have a name, let alone a role or function within a larger institutional complex. The most striking thing about the room is the colour of the chairs (and even this is technically incorrect, as the Commons has benches rather than individual chairs). The phrase speaks of a material position in relation to established political institutions that has very few resources available in order to effect a change in a relationship characterized by distance and incomprehension. It contrasts with a middle-class position that might, superficially appear to have the same perception of politics as 'remote'. Yet because of a different material position (for example due to social and educational assets), they might make of this remoteness something very different, something that does not stop them being political subjects of some sort. For example:

> **Female 3:** ... I feel quite strongly about the fact that politics shouldn't be pigeonholed into Westminster. I think it's about everything. I think it's about like just having the right to express what you think, what you feel. And I think that especially sort of in recent years there's been so many different arenas for political expression and I think part of the problem with politics today is the fact that people think of politics and they think of men in suits and Westminster ...
> (University students, London, middle-class respondents)

Note the very strong sense in which this respondent is able to articulate that what *she thinks* is important, that it has value. She firmly rejects the idea that politics belongs to a specialized institution over which a restricted constituency ('men in suits') have a monopoly. In doing so of course, she is implicitly criticizing television news representations, which assume precisely this conflation between politics and Westminster. Instead the respondent articulates a more liberal (non-hierarchical) vision of participation, of rights and diversity of opportunities or 'arenas' to express oneself politically. While this vision is certainly more attractive and progressive than the vision of politics that dominates television news, it does rather underplay the real hierarchies of power (including media power) that frustrate the possibilities of a more inclusive and diverse citizenry.

One area where differential competences in decoding both political processes and media representations of those processes became evident was in the exercise that focus group participants did at the beginning of the session. Asked to put together a typical news story using a selection of photographs, some of the working-class groups found it difficult to construct a rationale between images that connected with their immediate experience (images of 'the street') and the images which represented 'politics'. One way round this problem was to simply leave out the images of 'politics' (such as Tony Blair, Gordon Brown, the House of Commons and Downing Street).

> **Female 3:** ... It was quite difficult to choose the photos of the politics because ... I don't know. It's because we're young ourselves and we like ... because we watch a lot of the Six O'Clock News which is based more in London and there's been a lot of crimes and shootings and what not these days. Especially considering this year. You know a lot of black on black crime and knife and gun crime. So, as being young people, this attracts us more to talk about something that revolves around us basically instead of politics.
>
> (Community Centre, London, working-class respondents)

Here 'us' and the images associated with 'us' and politics (them) are seen as two different universes, and the group did not have the political or media literate skills to bring them together in some way. Note the way the speaker is confident that he can speak on behalf of a 'we'. There is no recourse here to individuation ('I think') which a middle-class speaker would more likely have used. Interestingly, there is a declared interest in the Six O'Clock News (BBC1). However, the respondent seems to be referring more to the local news ('based more in London'), which follows the national news at 6pm. This interest in local news in preference to national news, fits into the broader sociological work that has found working-class cognitive powers more 'restricted' to the immediacy of objects/events/things than middle-class people (Lawton 1970). The 'national' as a sphere of news events is a degree of abstraction that many *young* working-class respondents found difficult to grasp or relate to (and in general middle-class respondents expressed an interest in international news that contrasts strongly with the working-class respondents). There is a close, mutually reinforcing relationship then between competences and dispositions. Without the ability to relate to more 'abstract' issues, one is unlikely to be interested in them, which in turn maintains competences at a certain limited level. Note too the strong *vertical* dimension to the class subjectivity or habitus of this working-class respondent in relation to those areas of society such as the political sphere, dominated by the middle class.

Another group, confronted with this dilemma of linking lived experience and the apparently remote world of politics, created a fictional link, using a photograph of a women in a nightclub.

> **Moderator:** Okay. It's interesting that you made the young woman Tony Blair's daughter. What was behind that decision?
>
> **Male 4:** We had to have some kind of story to link all the pictures together.
>
> (State school, London, working-class respondents)

This invention of a fictional person (Tony Blair does not have a daughter) to create a link, demonstrates how the working-class respondents often struggled to organize and arrange the disparate images of politics and ordinary life into some sort of cause-effect relationship. In terms of constructing a news story, these gaps in the competences to understand both television news conventions and political processes are clearly not natural, but socially determined. This is powerfully illustrated by the apparently obvious, 'natural' and self-evident way a middle-class participant discusses how the images 'fit together'.

> **Female 4:** ... I think the photos just fit together, don't they, and kind of, they've obviously been chosen in order to sort of ... I don't know. They do just sort of make a story out of themselves really without thinking too much about it because, yeah like Female 1 said, this sort of comes up every year – every, every year – and nothing's ever done and then it's just, yeah, it's always the same, it's the same kind of images. I mean the story we've come up with is so, so standard, isn't it? That it's just, like, yeah, you know that they're gonna have, you know ... some report says this, we're gonna comment on it outside Downing St, we're gonna

> have someone from government [some] minister saying this, we're gonna have pictures, you know, of them debating it, we're gonna have someone from the Opposition saying this and then we're gonna have some youth worker saying this is what we need to do. But nothing ever gets done in the end. And they always put it down to stuff like, you know, let's put loads of money into sport. But they never do, so...
>
> (University student, London, middle-class respondent)

The middle-class participants had no difficulty in connecting the images of politics with the images of 'the street'. They begin at the 'top' of society, with a home office report that sparks a row. In contrast, the working-class participants tended to begin their reports by focusing on a particular concrete 'incident', such as in a nightclub, and often struggled to then link these images to those that designate the realm of mainstream party politics. In other words, there was again a problem of 'abstraction'. The middle-class participants by contrast proceed smoothly from the government report, to a debate within Parliament, and then move 'down' as it were towards the 'street' as they construct an interview with a youth worker. In order to perform this operation, they must have a working knowledge of how mainstream politics functions (the report, the debate in the House of Commons, etc.) as well as the conventions of television news in reporting mainstream politics (for example, the structure of the news item and the locations of the 'action' within it). It is also evident that this participant is extremely weary of the tedious formulaic nature of these conventions and this is coupled with a sense that the news is colluding with a cyclical problem that is never really addressed properly by political institutions.

Many of the middle-class respondents demonstrated frustration with the way television news was unable to sufficiently distance itself from powerful institutions in society in order to criticize them, such as Parliament or the military. They yearned for more 'negotiated' readings of the situation and in their absence produced them themselves, often detecting gaps in logic or discrepancies in rationales as the basis for doing so:

> **Female 1:** ... I hate the way that they always like report on like [in serious newsreader tone] 'a soldier died in Iraq' and it's like 50,000 civilians died, but that one soldier who decided to be in the Army and got paid, and knew what he was doing, died and boo-hoo. Not that it's not sad, but the way that it gets reported does annoy me.
>
> (University student, Brighton (Group B), middle-class respondent)

COMPETENCES AND HEGEMONY

One of the contradictions of television news is the enormous investment it has made in a middle-class association between formality and authority. Signifiers of formality are articulated in the language broadcasters use, the clothes they wear (for example, shirts and ties) and the felt rigidity of the conventions within which journalists operate. For working-class viewers in our sample, this rigidity was felt to be oppressive to the personality of the journalist, whereas for middle-class viewers, the rigidity was problematic at a more abstract level of inhibiting the expression of more complex or independent ideas or perspectives. Thus some working-class respondents expressed a preference for breakfast news programmes they had watched in the morning before going to school. The less formal and rigid format ('It's more relaxed isn't it?') appealed because it was felt to open up greater space for dialogue ('they sort of discuss things a bit more'). There was a stronger expression of opinion and personality ('he had sort of a slight sort of comic side to some of it but also serious when he needed to be') and a more informal style ('they have like guests come and they sit on the sofa and have like a drink of water and stuff') than the later, more prestigious evening news programmes. Thus while formality invests news programmes with authority and status it also excludes many younger working-class viewers. Conversely, middle-class audiences exhibited disdain for *Five News* precisely because of its lack of formality and perceived 'seriousness'.

> **Male:** I think Channel 5 News is really good (laughs). [Laughter] I don't really. [Laughter]
> **Moderator:** What's wrong with Five?
> **Male:** It's just really badly done and the presenters are just rubbish and ...
> **Female 3:** It's so casual. It is really painfully casual, it's just awful.
> **Female 4:** Isn't it only 5 minutes long?
> **Female 3:** It's terrible. It's really terrible.
> (University students, London, middle-class respondents)

Note the confident calculation by the first speaker that expressing a preference for *Five News* could constitute a 'joke' (this implies that news in general is important, too important in fact to be treated in a less than 'serious' manner). His calculation is not misplaced and the other respondents agree that the programme's brevity and casualness make it 'terrible'. There is a clear process of 'distinction' going on here in the sense meant by Bourdieu.

Nevertheless, working-class dispositions towards the middle-class paradigms that structure television news were decidedly ambivalent. The vertical relation with television and politics they have could be expressed in terms of rejection or acceptance. While on the one hand working-class respondents were often hostile to the implied exclusion of their own experiences and perspectives from television news ('They're all like posh and up themselves'), which in turn makes television news 'boring', they were also often in thrall to the moral and intellectual dominance of middle-class values:

> **Male 2:** I still think they should probably have an image about them. [Female 2 laughs] Not a yob one, but an actual middle-class one. I disagree with you on that.
> **Female 2:** No. I agree with you that they should have like an image that they should portray because that is ...
> **Male 1:** They should wear suits.
> **Female 2:** ... it reflects like maturity ... not maturity, but intelligence. Do you get what I mean?
> (State school, London, working-class respondents)

Interestingly, the only image that seems to be conceivable outside a middle-class one, is 'a yob one', which rather implies a judgement on the speaker's own self-image. Here the respondents equate intelligence with formal attire and identify both as the natural property of the middle class. Another group of working-class respondents, also under 18s, spoke of the value of formal dress and an assumed authority and knowledge associated with language.

> **Male 4:** The way they dress and the way they speak is like ...
> **Girl:** Sophisticated.
> **Male 4:** ... it's like proper English, it's like 'and now moving on to the scene where the ...'... If they were talking like he does about ... [puts on accent to impersonate Male 1] 'now we're like moving on, innit!' [Laughter]
> **Female 2:** Everyone would laugh at him!
> **Male 4:** [again impersonating]... 'And now we're moving on to our main man now'.
> **Male 1:** Yeah, but then young people would actually listen to what ...
> **Girl:** No!
> **Female 2:** I wouldn't!
> (State school, London, working-class respondents)

Here there is a very active sense of internalizing dominant value systems to the exclusion and denigration of the respondents' own plebeian voices. The presence of their voices as journalists would it seem cause a catastrophic breakdown in the authority of the news; it would become comic, a joke, in much the same way that *Five News* is a joke for the middle-class respondents quoted above. M1's attempt to validate these voices and suggest that their presence on television might increase the interest of young working-class people in the news is drowned out in a chorus of disagreement from his peers.

COMPETENCES, CONSPIRACY THEORY AND THE CHALLENGE TO HEGEMONY

However, the hegemony of television news and the dominant political institutions is very far from being complete and unchallenged – although the nature of the challenge, I suggest, does reflect the dominance of three decades of neo-liberal assault on working-class culture, organizations and material well-being. There were certain issues and/or political personalities who were particularly vulnerable to a counter-hegemonic critique. The war on Iraq in 2003 and its close association with the then Prime Minister Tony Blair, certainly provoked some of the strongest critical comments on both politics and its coverage by television news across the classes even three years later. The most 'radical' critique of television news, or the one which perceived news as a production complex most integrated into material interests antithetical to the audience, came from the working-class groups. The 'vertical' relation between this class subjectivity and dominant institutions is crucial. Consider this:

> **Female 3:** They [the politicians] are puppets. The real people we will never come to know. The really ... there's only probably about one or two people that's running ... that's at the top of the pyramid. And we're right – we're kind of down here. [Laughter] ... It's just a pyramid basically and everyone's on top of each other. And you know who's below you but you don't know who's above you. That's it. They're working for someone.
> **Female 12:** The production company.
> **Female 1:** Yeah. There's someone that tells Tony Blair what to do. There's obviously someone that tells him. But he's just the face of it. There's someone else ...
> **Female 3:** It's like a movie, innit? You watch it, you see the actors and they're playing it, but there's so much more put into it. Like, we don't see what is being put into it and how much hard work and how much millions of people have been put in it to make that one movie. We just see the puppets. We see the film; we see the story. We like the movie getting done. But how much has gone into it we don't see? And we'll probably never see it.
> (Community centre, London, working-class respondents)

There is an incredibly rich and intuitive insight here as these three speakers develop the metaphor of the 'production company', which they see as controlling and subverting the democratic process. Their sense of the 'distance' of the institutions of politics and television news is quite different from the middle-class respondents. The latter see 'distance' but there is an implicit sense that this is distance on the same plane as them. Nowhere in their discourse do we find the strong metaphors that conceive distance on a vertical plane, as here with notions of the pyramid, the chain of command and a strong sense of 'them', 'up there' and 'us', 'down here'. Furthermore, by placing themselves in a relation of inequality with power, these respondents are able to articulate a sense of power complexes very different from the theory of formal democratic power that dominates liberal accounts.

Looking at the development of this discussion, we see that with the metaphor of the 'production company' the state apparatus and the media apparatus appear to be merging almost into one. The metaphor is then developed further into a film or some such media spectacle whereby audiences are so caught up in the story that they do not see it as a construct; as an artefact designed for specific purposes and to serve specific interests. Audiences do not see the 'work' that goes into the spectacle and are thus controlled and manipulated by it. In short, this is, in a popular vernacular, a version of what Marx called commodity fetishism (Marx 1983: 76–87). It is extremely unlikely that these young women have in fact read Marx or heard of the concept of commodity fetishism. Clearly these respondents are drawing on a different source of knowledge than formally acquired education.

This base of knowledge is what we might call 'organic' to their social being, it is that acquisition of 'knowing' that people breathe in and absorb not through the conscious application of effort to master a given body of knowledge, but through their everyday life experience and conditions, their everyday social being, or their popular culture. Significantly perhaps, all the respondents from this community centre were Black or Asian, thus compounding exclusions and inequalities of class with ethnic discrimination. Although this is speculative, ethnic identities other than white working class may also have opened up access to discourses from outside the dominant national cultural formation that could be mobilized as sources of critique (see e.g. Gillespie 1995: 131–141). The middle-class respondents display in general far higher levels of articulacy in the language that they use and much greater density of empirical knowledge of both political processes and the conventions of television news than the working-class respondents generally do. They also have a greater ability to abstract particular issues or images into a broader context. However, despite their disadvantages, the working-class respondents draw on their organically derived knowledge of class inequality to critique the power relations dominant institutions are caught up in, much more powerfully and deeply than the middle-class respondents generally do. There is for example, little sense in the responses of the middle-class subjects that the dominant institutions are integrated into processes that are actively hostile to them. But because the working-class respondents lack access to formal education and the mental training for abstraction, their intuitive knowledge organic to their life experience and social being is limited and vague and lacks the conceptual instruments to really name power relations, processes and causal forces. Thus there is a tendency in the discourse of the powerless to veer towards conspiracy theories, which function as something akin to the poor person's class-consciousness in the absence of radical political parties that Frank Parkin saw as central to sustaining oppositional or radical value systems within the working class (Parkin 1981: 100–102).

Of course popular film culture is awash with conspiracy-orientated narratives and they have been theorized precisely as symptomatic of embryonic forms of class-consciousness (Wayne 2005). The above respondents do seem to be drawing on their knowledge of popular film culture to explain other institutions (television and politics), which they have only an intuitive understanding of. So here we are seeing something like that distinction and reciprocal interaction and borrowing between mass culture and popular culture that Martín-Barbero discussed in relation to Latin America. As with popular film narratives, the respondents' discourse greatly simplifies the processes involved, for example by individualizing power ('There's obviously someone that tells him'). This is precisely because the respondents do not have the competences to formulate the power relations they intuit in social scientific terms. They cannot articulate abstract institutional practices or the connections between what happens inside the political realm and powerful forces (such as economic power) outside it. Nevertheless, they *do* intuit such power relations and connections in a way that the middle-class respondents do not. The kind of abstractions or, to use Gramsci's term, philosophy, these working-class respondents *could* develop, under more propitious circumstances, would be one more congruent with their experience of a world structured around unequal power relations than the abstractions/philosophy associated with the middle class, who, as we have seen, have a different habitus less attuned to class inequality (Gramsci 1967: 58–59).

One variable which opens up respondents' access to frameworks with which they can critique media and social/political power is the internet, where conspiracy theory paradigms are particularly popular. Here the internet, as a tool of dissemination that easily crosses national borders, circumnavigating dominant corporate media systems and dominant *national-cultural* formations, opens up the possibility of various forms of audience participation in political critique. Henry Jenkins calls this 'convergence culture', which is less about convergence of technologies than it is about the complex intersection and reciprocal dynamics between old and new media, grassroots and corporate media, and media producers and consumers (2008: 2–3). In some ways these dynamics can be seen as a conceptual parallel to Martín-Barbero's distinction between mass and popular culture.

Although only a minority of our sample used the internet as an alternative source of news, it has provided this minority with alternative material that is in general excluded from the corporate media and is often explicitly critical of the corporate media.

> **Male 7:** Do you remember what I said about the illusion of power. That's what I'm saying – we're given the illusion of power, freedom of thought or whatever...
> **Male 4:** But then again, they only get told ... they only tell us what they get told to say. The news broadcasters – he or she has a script, which has been written up...
> **Male 5:** They've been told how to argue a case, innit? Because I watch a lot of stuff on YouTube about Fox's news broadcaster...
> **Male 2:** Fox News is terrible.
> **Male 5:** ... and I watch a lot of coverage and montages of different Fox reports and that. And that's the way I see Fox now – it's prejudiced, it's racist, it's all slanted and everything ... they've got a filter that holds information back that supports who they want to support and it makes other people look bad and that.
> (Community group, North London, working-class respondents)

Here media discourses about 'free speech' and expression become a metaphor for a broader set of power complexes that contradict such discourses ('the illusion of power ...'). Alternative sources of information cast established broadcasters in a new light, whereby systemic shaping of the material by various 'filters' (shades here of the Chomsky and Herman critique of media propaganda) pre-determines which people look 'good' and which people look 'bad'. A growing diversity of media sources and wider access to media production can help drive a critical media literacy beyond the middle class who dominate the formal study of the media in secondary and tertiary levels of education. Meanwhile, the old politics has been mobilizing the new media to speak directly with citizens through web videos, blogs, Facebook, Twitter, podcasts, and so forth. However, whether such new means of dissemination can overcome the deep gulf in class characteristic of capitalism, seems questionable:

> **Male 2:** It's the characters in politics as well. It's like, someone like David Cameron [then leader of the UK opposition], it's like he went to Eton. It's like 'what the fuck?'
> **Female 3:** Yeah, you want someone you can relate to.
> **Male 2:** What would he know about me and you?
> (Youth group, West London, working-class respondents)

Similarly the old established news media, both press and television, are using the internet to provide non-linear news content, and facilitate feedback and participation. Such innovations are likely to simply see new technologies integrated into and contained within the established socio-economic paradigms of the mainstream media providers. Technological fixes alone will not mend the disconnect between television news and its reportage of the political sphere on the one hand, and young class differentiated audiences on the other.

CONCLUSION

Class determined readings of television need to be situated within their historical and political contexts – contexts that may vary from the epochal to a very delimited time-span. If audience responses are context sensitive in terms of time and geography, these variables intersect with enduring structures and patterns of class power and difference. One way to catch these enduring structures and differentiations has been the encoding-decoding model (the preferred/dominant reading, negotiated reading, the oppositional reading), which explores the relationship between audiences and dominant interpretive paradigms and value systems. Variable factors here include not just historical contexts but also how class intersects with other identities, such as gender, ethnicity and age. This means that the content of a preferred reading or an oppositional reading at any given historical moment is itself likely to be differentiated and plural, with a number of possible positions being staked out within what might still fall under, for example, a 'preferred reading'. Working-class difference itself will differentiate and nuance what a preferred reading might look like when compared to a middle-class preferred reading of the same text. We saw that Andrea Press's working-class respondents were integrated into a dominant middle-class paradigm but on a differential basis than the middle-class respondents. We saw that some young working-class respondents accepted middle-class forms of authority on television news just as the middle-class respondents did. But the fact of class difference makes a difference in terms of the nature of that acceptance. In accepting such forms of authority, working-class respondents must deny validity to their own voices and images in much the same way that colonial subjects had to do in the past (Fanon 1986), thus producing a fissured sense of self. To be integrated into dominant power relations from a position of inequality is very different from being integrated into dominant power relations from a more materially and culturally advantageous position. I suggested that Bourdieu's concept of the habitus, with its dispositions and competences, can be given a useful metaphorical differentiation along class lines. I suggested that the working-class habitus and the middle-class habitus were differentiated from each other in their relations to dominant power complexes. The former has a 'vertical' relation, the latter a 'horizontal' relation. This impacts on both their embodied attitudes towards institutions such as television and politics and the competences they mobilize. The middle-class competences are orientated towards a more horizontal relation, stressing empirical detail, diversity, specialized roles and their own individual responses, while working-class linguistic competences stress subordination, hierarchy and restricted knowledge flows ('you know who's below you but you don't know who's above you') as well as a stronger sense of a 'we' or class differentiated 'us' identity. This verticality in dispositions and competences 'fixes' the working-class respondent in numerous ways (see Skeggs (2004: 50–52) for a discussion of 'fixing'). For example, while middle-class competences and dispositions have developed so as to 'move in' to popular culture in a significant way, working-class dispositions and competences remain as excluded from middle-class cultural experiences as ever. This lack of mobility plays itself out cognitively as well. Working-class respondents were less able to 'travel' imaginatively between television images that denoted their local surroundings and those drawn from the world of high politics. They lacked the televisual literacy and political knowledge to know how the news typically constructs a picture of politics and society. And this gap between their own immediate lives and politics with a capital 'P', tends to orient dispositions towards local news over national news and the national over the international. However, if in some respects the cognitive maps of working-class respondents were significantly poorer than middle-class respondents, in other respects their embodied vertical relation to power gave a sharper, more critical thrust to their discourse. Here they could not draw on assets acquired through formal education, as dominant education systems exist precisely to exclude them from potentially 'dangerous' critical knowledge. Instead they draw on the resources of popular culture, in film or on the internet, to makes sense of power relations, which they know by experience, simply *do not fit into the liberal paradigm of formal equality* assumed by television news. It has been important to hear their voices on these (academic) pages. Oppositional readings informed by a conspiracy framework are a long way from the kind of socialist informed critiques that David Morley found amongst working-class respondents in his study of *Nationwide* and this reflects the impact of neo-liberalism over three decades.

REFERENCES

Bernstein, B. (1979). 'Social Class, Language and Socialization', in Victor Lee (ed.), *Language and Development*. London: Open University/Croom Helm.

Berry, D. (2006). 'Popular Culture and Mass Media in Latin America: Some Reflections on the Works of Jésus Martín-Barbero and Nestor

García Canclini', in D. Berry & J. Theobald (eds), *Radical Mass Media Criticism: A Cultural Genealogy*. London: Black Rose Books.

Bourdieu, P. (1996). *Distinction: A Social Critique of Taste*. London: Routledge.

Chan, T.W. & Goldthorpe, J. (2005). 'The Social Stratification of Theatre, Dance and Cinema Attendance', *Cultural Trends* 14(3), pp. 193–212.

Ewen, S. & Ewen, E. (2000). *Channels of Desire, Mass Images and the Shaping of American Consciousness*. Minneapolis: University of Minnesota Press.

Fanon, F. (1986). *Black Skin/White Mask*. London: Pluto Press.

Gillespie, M. (1995). *Television, Ethnicity and Cultural Change*. London: Routledge.

Gramsci, A. (1967). *The Modern Prince and Other Writings*. London: Lawrence and Wishart.

Holtzman, L. (2000). *Media Message: What Film, Television, and Popular Music Teach Us About Race, Class, Gender and Sexual Orientation*. New York: M.E. Sharpe.

Hunt, R. (1993). 'Middle Classes, Try My Dish: David Elstein of BSkyB is Trying to Widen the Appeal of "Council-House Television"', *The Independent* www.independent.co.uk/news/media/media-middle-classes-try-my-dish-david-elstein-of-bskyb-is-trying-to-widen-the-appeal-of-councilhouse-television-he-talks-to-robin-hunt-1461738.html (accessed February 17 2010).

Jenkins, H. (2008). *Convergence Culture: Where Old and New Media Collide*. New York: New York University Press.

Lawton, D. (1970). *Social Class, Language and Education*. London: Routledge & Kegan Paul.

Martín-Barbero, J. (1993). *Communication, Culture and Hegemony: From the Media to Mediations*. London: Sage.

Marx, K. (1983). *Capital: A Critique of Political Economy*, Vol. 1. London: Lawrence and Wishart.

Morley, D. (1992). *Television, Audiences and Cultural Studies*. London: Routledge.

Parkin, F. (1981). *Class Inequality and Political Order*. London: Paladin

Philo, G. (1990). *Seeing is Believing*. London: Routledge.

Plummer, J.T. (1974). 'The Concept and Application of Life Style Segmentation', *American Marketing Association* 38(1), pp. 33–37.

Press, A.L. (1991). *Women Watching Television: Gender, Class and Generation in the American Television Experience*. Philadelphia: University of Pennsylvania Press.

Resnick, S. & Wolff, R. (2003). 'Exploitation, Consumption and the Uniqueness of US Capitalism', *Historical Materialism* 11(4), pp. 209–226.

Skeggs, B. (2004). *Class, Self, Culture*. London: Routledge.

Skeggs, B., Thumin, N. & Wood, H. (2008). 'Oh Goodness, I Am Watching Reality TV', *Cultural Studies* 11(1), pp. 5–24.

Skovmand, M. (1992). 'Barbarous TV International: Syndicated *Wheels of Fortune*', in Michael Skovmand and Kim Christian Schoder (eds), *Media Cultures:, Reappraising Transnational* Media. London: Routledge.

Wayne, M. (2000). *'Who Wants To Be A Millionaire?* Contextual Analysis and the Endgame of Public Service Television', in D. Fleming (ed.), *Formations: A 21st Century Media Studies Textbook*. Manchester: Manchester University Press.

Wayne, M. (2005). 'Jameson, Postmodernism and the Hermeneutics of Paranoia', in M. Wayne (ed.), *Understanding Film: Marxist Perspectives*. London: Pluto Press.

Wayne, M., Petley, J., Murray, C. & Henderson, L. (2010). *Generation Disconnected? Television News, Politics and Young People*. Basingstoke: Palgrave.

Index

Page references to Notes will include the letter 'n' following the page number